CW01648998

Nissan Cherry Owners Workshop Manual

A K Legg T Eng (CEI), AMIMI

Models covered
Nissan Cherry (N12 Series) Hatchback with 987 cc, 1269 cc and 1488 cc engines, including Turbo

Covers manual and automatic transmissions
Does not cover Cherry 'Europe' model

ABCDE
FGHIJ
KLMNO
PQRST

Haynes Publishing Group
Sparkford Nr Yeovil
Somerset BA22 7JJ England

Haynes Publications, Inc
861 Lawrence Drive
Newbury Park
California 91320 USA

Acknowledgements

Thanks are due to the Nissan Motor Company Limited of Japan for the provision of technical information and certain illustrations. The Champion Sparking Plug Company supplied the illustrations showing the various spark plug conditions. Sykes-Pickavant supplied some of the workshop tools. Thanks are also due to those people at Sparkford who helped in the production of this manual.

A book in the **Haynes Owners Workshop Manual Series**

Printed by J. H. Haynes & Co. Ltd, Sparkford, Nr Yeovil, Somerset BA22 7JJ, England

ISBN 1 85010 031 4

British Library Cataloguing in Publication Data
Legg, A. K.
Nissan/Datsun Cherry ('82 to '85) owners workshop manual.–
(Owners Workshop Manual/Haynes)
1. Cherry automobile
I. Title II. Series
629.28'722 TL215.C5/
ISBN 1–85010–031–4

Contents

Nissan Pulsar NX Coupe

Nissan Cherry 1.3 DX five-door

About this manual

Its aim

The aim of this manual is to help you get the best value from your vehicle. It can do so in several ways. It can help you decide what work must be done (even should you choose to get it done by a garage), provide information on routine maintenance and servicing, and give a logical course of action and diagnosis when random faults occur. However, it is hoped that you will use the manual by tackling the work yourself. On simpler jobs it may even be quicker than booking the car into a garage and going there twice, to leave and collect it. Perhaps most important, a lot of money can be saved by avoiding the costs a garage must charge to cover its labour and overheads.

The manual has drawings and descriptions to show the function of the various components so that their layout can be understood. Then the tasks are described and photographed in a step-by-step sequence so that even a novice can do the work.

Its arrangement

The manual is divided into twelve Chapters, each covering a logical sub-division of the vehicle. The Chapters are each divided into Sections, numbered with single figures, eg 5; and the Sections into paragraphs (or sub-sections), with decimal numbers following on from the Section they are in, eg 5.1, 5.2, 5.3 etc.

It is freely illustrated, especially in those parts where there is a detailed sequence of operations to be carried out. There are two forms of illustration: figures and photographs. The figures are numbered in sequence with decimal numbers, according to their position in the Chapter – eg Fig. 6.4 is the fourth drawing/illustration in Chapter 6. Photographs carry the same number (either individually or in related groups) as the Section or sub-section to which they relate.

There is an alphabetical index at the back of the manual as well as a contents list at the front. Each Chapter is also preceded by its own individual contents list.

References to the 'left' or 'right' of the vehicle are in the sense of a person in the driver's seat facing forwards.

Unless otherwise stated, nuts and bolts are removed by turning anti-clockwise, and tightened by turning clockwise.

Vehicle manufacturers continually make changes to specifications and recommendations, and these, when notified, are incorporated into our manuals at the earliest opportunity.

Introduction to the New Cherry and Pulsar

The New Cherry was introduced in the UK in September 1982 and the North American versions followed in 1983. All models covered in this manual are fitted with the E series overhead camshaft engine. E10, E13, E15 and E16 engines employ a carburettor fuel system, but the E15ET engine employs a fuel injection system and turbocharger.

The engine is mounted transversely at the front of the car with the transmission on the left-hand side. Drive is transmitted to the front wheels through driveshafts.

The car is quite conventional in design and the DIY home mechanic should find most work straight forward.

General dimensions, weights and capacities

Dimensions mm (in)

Overall length:	
Saloon (UK)	3960 (155.9)
Saloon (North America)	4115 (162.0)
Coupe (North America)	4125 (162.4)
Overall width:	
All models	1620 (63.8)
Overall height:	
Saloon (non-Turbo UK)	1390 (54.7)
Saloon (Turbo UK)	1385 (54.5)
Saloon (North America)	1390 (54.7)
Coupe (North America)	1355 (53.3)
Wheelbase:	
All models	2415 (95.1)
Ground clearance:	
All models	165 (6.5)
Turning circle (between walls):	
Manual transmission models	10.2 m (33.5 ft)
Automatic transmission models	10.6 m (34.8 ft)

Kerb weights kg (lb)

UK models	**Three-door**	**Five-door**
E10 engine models	780 (1720)	790 (1740)
E13 engine models	795 (1755)	805 (1775)
E15 engine models	825 (1280)	835 (1840)
E15ET engine models	885 (1950)	–

North American models

Refer to vehicle FMVSS certification label

Capacities

Fuel tank	11.0 Imp gal; 13.3 US gal; 50.0 litre
Cooling system (with heater):	
UK models:	
Manual transmission – except Turbo	9.8 Imp pt; 5.5 litre
Manual transmission – Turbo	10.8 Imp pt; 6.1 litre
Automatic transmission	10.6 Imp pt; 6.0 litre
North American models:	
Manual transmission – except Turbo	5.0 US qt; 4.7 litre
Manual transmission – Turbo	6.5 US qt; 6.1 litre
Automatic transmission	5.6 US qt; 5.3 litre
Engine oil (with oil filter):	
E15ET engine models	6.5 Imp pt; 3.9 US qt; 3.7 litre
Except E15ET engine models	6.9 Imp pt; 4.1 US qt; 3.9 litre
Manual transmission:	
Four-speed	4.0 Imp pt; 2.4 US qt; 2.3 litre
Five-speed	4.8 Imp pt; 2.8 US qt; 2.7 litre
Automatic transmission	10.6 Imp pt; 6.3 US qt; 6.0 litre
Power steering system	1.76 Imp pt; 1.06 US qt; 1.0 litre

Jacking, wheel changing and towing

Jacking

The jack supplied with the vehicle should only be used for emergency roadside wheel changing.

Chock the roadwheels on the side opposite to that from which the wheel is being removed. Engage the jack in one of the two cut-outs at the base of the sill, either front or rear according to which wheel is being removed (photo).

When carrying out overhaul or repair work use a trolley jack or a hydraulic bottle or screw jack. Locate the jack under the vehicle only at the positions indicated and **always** supplement the jack with axle stands placed under the side-members.

To avoid repetition, the procedure for raising the vehicle in order to carry out work under it is not included before each operation described in this Manual. It is to be preferred, and is certainly recommended, that the vehicle is positioned over an inspection pit or raised on a lift. Where these facilities are not available, use ramps or jack up the vehicle and supplement with axle stands, as described earlier.

Wheel changing

The removal and refitting of a roadwheel should be carried out in the following way.

Prise off the centre trim (where fitted) from the roadwheel. Unscrew the roadwheel nuts just enough to release them. On

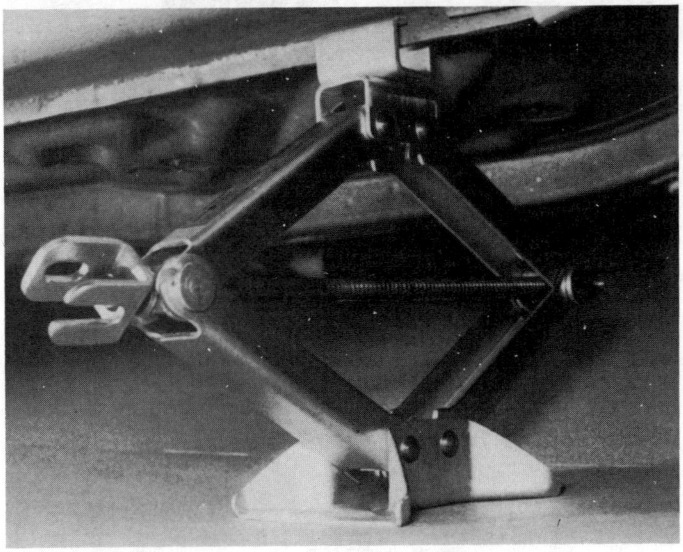

Vehicle tool kit jack

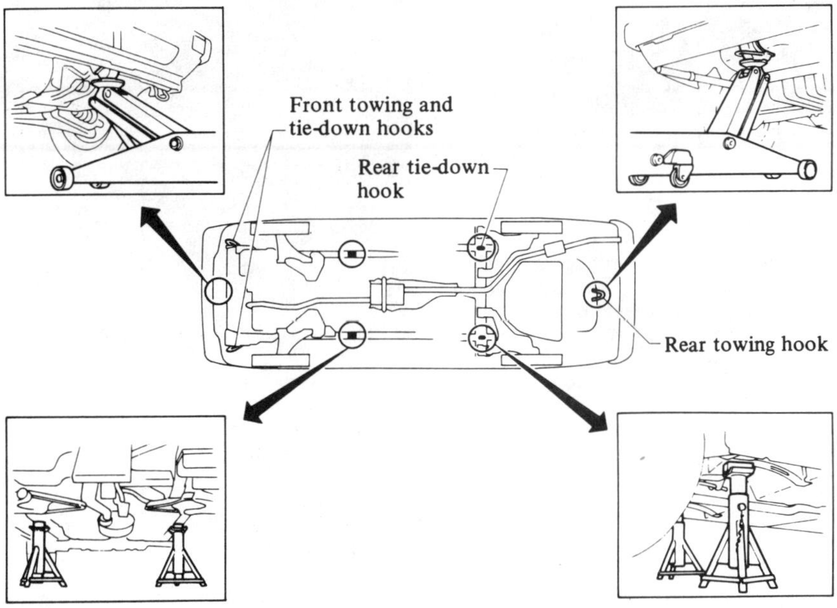

Jacking points for workshop jack

Removing roadwheel centre cap

Spare wheel and tool stowage

roadwheels having a small plastic centre cap, twist it through a quarter turn and remove it (photo). Raise the roadwheel from the ground and then remove the nuts completely and lift the roadwheel from the studs.

Refit the roadwheel, tighten the nuts as tightly as possible while the wheel is held against rotation with the foot. Lower the vehicle and tighten the nuts fully (95 Nm, 70 lbf ft). Fit the cap/trim to the centre of the roadwheel.

Towing

The front and rear towing hooks may be used in an emergency (photos). On vehicles with manual transmission, restrict the towing speed to below 80 kph (50 mph) and the distance towed to 80 km (50 miles). On vehicles with automatic transmission, restrict the speed to 30 kph (20 mph) and the distance towed to 30 km (20 miles).

If the transmission has a fault, then the front wheels of the vehicle must be raised and placed on a dolly.

Never tow a vehicle with automatic transmission by raising the rear wheels and leaving the front wheels in contact with the road.

Front towing hook

Rear towing hook

Buying spare parts and vehicle identification numbers

Buying spare parts

Spare parts are available from many sources, for example: Nissan garages, other garages and accessory shops, and motor factors. Our advice regarding spare part sources is as follows:

Official appointed Nissan garages – This is the best source of parts which are peculiar to your vehicle and are otherwise not generally available (eg, complete cylinder heads, internal gearbox components, badges, interior trim etc). It is also the only place at which you should buy parts if your vehicle is still under warranty – non-standard components may invalidate the warranty. To be sure of obtaining the correct parts it will always be necessary to give the storeman your vehicle's engine and chassis number, and if possible, to take the 'old' parts along for positive identification. Remember that some parts are available on a factory exchange scheme – any parts returned should always be clean! It obviously makes good sense to go straight to the specialists on your vehicle for this type of part for they are best equipped to supply you.

Other garages and accessory shops – These are often very good places to buy materials and components needed for the maintenance of your vehicle (eg, spark plugs, bulbs, drivebelts, oils and greases, touch-up paint, filler paste, etc). They also sell general accessories, usually have convenient opening hours, charge lower prices and can often be found not far from home.

Motor factors –Good factors will stock all of the more important components which wear out relatively quickly (eg clutch components, pistons, valves, exhaust systems, brake cylinders/pipes/hoses/seals shoes and pads etc). Motor factors will often provide new or reconditioned components on a part exchange basis – this can save a considerable amount of money.

Vehicle identification numbers

The chassis number is located on a plate in the rear corner of the engine compartment (photo). The number is repeated on the bulkhead upper panel in which it is stamped.

The engine number is stamped into a machined inclined surface on the crankcase just below the distributor mounting (photo).

On North American models, an emission control information label is located under the bonnet. An FMVSS certification label is located on the centre (B) pillar, and the vehicle identification number is located on the left-hand side facia.

The manual transmission serial number is located on the clutch release arm, and the automatic transmission serial number on the upper casing surface.

Engine number

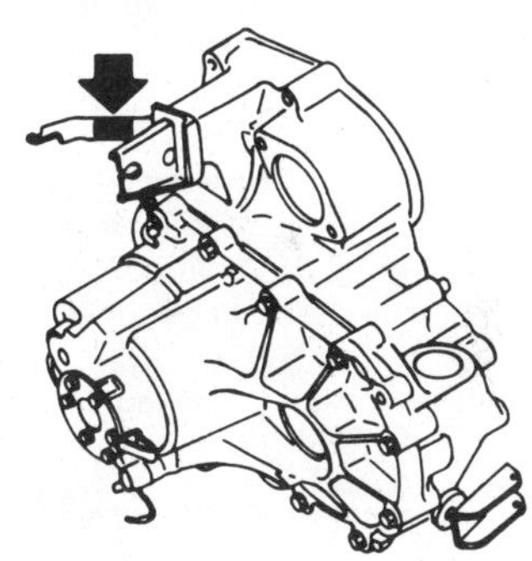

Manual transmission serial number location

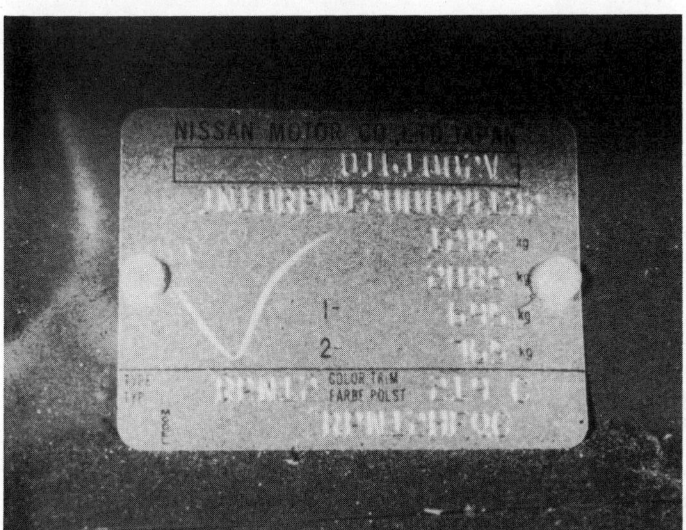

Vehicle identification plate

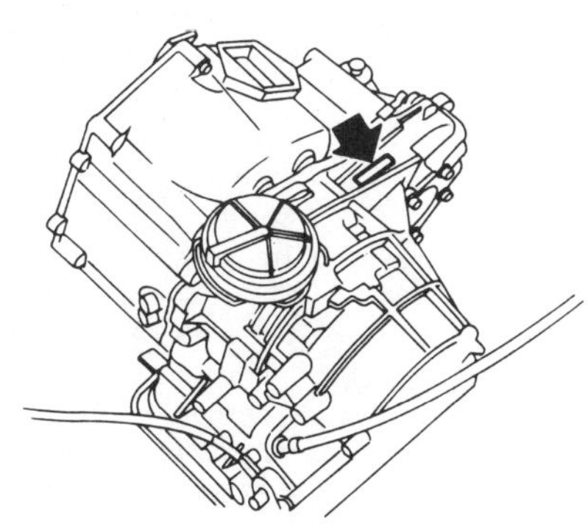

Automatic transmission serial number location

Use of English

As this book has been written in England, it uses the appropriate English component names, phrases, and spelling. Some of these differ from those used in America. Normally, these cause no difficulty, but to make sure, a glossary is printed below. In ordering spare parts remember the parts list may use some of these words:

English	American	English	American
Accelerator	Gas pedal	Leading shoe (of brake)	Primary shoe
Aerial	Antenna	Locks	Latches
Anti-roll bar	Stabiliser or sway bar	Methylated spirit	Denatured alcohol
Big-end bearing	Rod bearing	Motorway	Freeway, turnpike etc
Bonnet (engine cover)	Hood	Number plate	License plate
Boot (luggage compartment)	Trunk	Paraffin	Kerosene
Bulkhead	Firewall	Petrol	Gasoline (gas)
Bush	Bushing	Petrol tank	Gas tank
Cam follower or tappet	Valve lifter or tappet	'Pinking'	'Pinging'
Carburettor	Carburetor	Prise (force apart)	Pry
Catch	Latch	Propeller shaft	Driveshaft
Choke/venturi	Barrel	Quarterlight	Quarter window
Circlip	Snap-ring	Retread	Recap
Clearance	Lash	Reverse	Back-up
Crownwheel	Ring gear (of differential)	Rocker cover	Valve cover
Damper	Shock absorber, shock	Saloon	Sedan
Disc (brake)	Rotor/disk	Seized	Frozen
Distance piece	Spacer	Sidelight	Parking light
Drop arm	Pitman arm	Silencer	Muffler
Drop head coupe	Convertible	Sill panel (beneath doors)	Rocker panel
Dynamo	Generator (DC)	Small end, little end	Piston pin or wrist pin
Earth (electrical)	Ground	Spanner	Wrench
Engineer's blue	Prussian blue	Split cotter (for valve spring cap)	Lock (for valve spring retainer)
Estate car	Station wagon	Split pin	Cotter pin
Exhaust manifold	Header	Steering arm	Spindle arm
Fault finding/diagnosis	Troubleshooting	Sump	Oil pan
Float chamber	Float bowl	Swarf	Metal chips or debris
Free-play	Lash	Tab washer	Tang or lock
Freewheel	Coast	Tappet	Valve lifter
Gearbox	Transmission	Thrust bearing	Throw-out bearing
Gearchange	Shift	Top gear	High
Grub screw	Setscrew, Allen screw	Trackrod (of steering)	Tie-rod (or connecting rod)
Gudgeon pin	Piston pin or wrist pin	Trailing shoe (of brake)	Secondary shoe
Halfshaft	Axleshaft	Transmission	Whole drive line
Handbrake	Parking brake	Tyre	Tire
Hood	Soft top	Van	Panel wagon/van
Hot spot	Heat riser	Vice	Vise
Indicator	Turn signal	Wheel nut	Lug nut
Interior light	Dome lamp	Windscreen	Windshield
Layshaft (of gearbox)	Countershaft	Wing/mudguard	Fender

General repair procedures

Whenever servicing, repair or overhaul work is carried out on the car or its components, it is necessary to observe the following procedures and instructions. This will assist in carrying out the operation efficiently and to a professional standard of workmanship.

Joint mating faces and gaskets

Where a gasket is used between the mating faces of two components, ensure that it is renewed on reassembly, and fit it dry unless otherwise stated in the repair procedure. Make sure that the mating faces are clean and dry with all traces of old gasket removed. When cleaning a joint face, use a tool which is not likely to score or damage the face, and remove any burrs or nicks with an oilstone or fine file.

Make sure that tapped holes are cleaned, and keep them free of jointing compound if this is being used unless specifically instructed otherwise.

Ensure that all orifices, channels or pipes are clear and blow through them, preferably using compressed air.

Oil seals

Whenever an oil seal is removed from its working location, either individually or as part of an assembly, it should be renewed.

The very fine sealing lip of the seal is easily damaged and will not seal if the surface it contacts is not completely clean and free from scratches, nicks or grooves. If the original sealing surface of the component cannot be restored, the component should be renewed.

Protect the lips of the seal from any surface which may damage them in the course of fitting. Use tape or a conical sleeve where possible. Lubricate the seal lips with oil before fitting and, on dual lipped seals, fill the space between the lips with grease.

Unless otherwise stated, oil seals must be fitted with their sealing lips toward the lubricant to be sealed.

Use a tubular drift or block of wood of the appropriate size to install the seal and, if the seal housing is shouldered, drive the seal down to the shoulder. If the seal housing is unshouldered, the seal should be fitted with its face flush with the housing top face.

Screw threads and fastenings

Always ensure that a blind tapped hole is completely free from oil, grease, water or other fluid before installing the bolt or stud. Failure to do this could cause the housing to crack due to the hydraulic action of the bolt or stud as it is screwed in.

When tightening a castellated nut to accept a split pin, tighten the nut to the specified torque, where applicable, and then tighten further to the next split pin hole. Never slacken the nut to align a split pin hole unless stated in the repair procedure.

When checking or retightening a nut or bolt to a specified torque setting, slacken the nut or bolt by a quarter of a turn, and then retighten to the specified setting.

Locknuts, locktabs and washers

Any fastening which will rotate against a component or housing in the course of tightening should always have a washer between it and the relevant component or housing.

Spring or split washers should always be renewed when they are used to lock a critical component such as a big-end bearing retaining nut or bolt.

Locktabs which are folded over to retain a nut or bolt should always be renewed.

Self-locking nuts can be reused in non-critical areas, providing resistance can be felt when the locking portion passes over the bolt or stud thread.

Split pins must always be replaced with new ones of the correct size for the hole.

Special tools

Some repair procedures in this manual entail the use of special tools such as a press, two or three-legged pullers, spring compressors etc. Wherever possible, suitable readily available alternatives to the manufacturer's special tools are described, and are shown in use. In some instances, where no alternative is possible, it has been necessary to resort to the use of a manufacturer's tool and this has been done for reasons of safety as well as the efficient completion of the repair operation. Unless you are highly skilled and have a thorough understanding of the procedure described, never attempt to bypass the use of any special tool when the procedure described specifies its use. Not only is there a very great risk of personal injury, but expensive damage could be caused to the components involved.

Tools and working facilities

Introduction

A selection of good tools is a fundamental requirement for anyone contemplating the maintenance and repair of a motor vehicle. For the owner who does not possess any, their purchase will prove a considerable expense, offsetting some of the savings made by doing-it-yourself. However, provided that the tools purchased are of good quality, they will last for many years and prove an extremely worthwhile investment.

To help the average owner to decide which tools are needed to carry out the various tasks detailed in this manual, we have compiled three lists of tools under the following headings: *Maintenance and minor repair, Repair and overhaul,* and *Special.* The newcomer to practical mechanics should start off with the *Maintenance and minor repair* tool kit and confine himself to the simpler jobs around the vehicle. Then, as his confidence and experience grow, he can undertake more difficult tasks, buying extra tools as, and when, they are needed. In this way, a *Maintenance and minor repair* tool kit can be built-up into a *Repair and overhaul* tool kit over a considerable period of time without any major cash outlays. The experienced do-it-yourselfer will have a tool kit good enough for most repair and overhaul procedures and will add tools from the *Special* category when he feels the expense is justified by the amount of use to which these tools will be put.

It is obviously not possible to cover the subject of tools fully here. For those who wish to learn more about tools and their use there is a book entitled *How to Choose and Use Car Tools* available from the publishers of this manual.

Maintenance and minor repair tool kit

The tools given in this list should be considered as a minimum requirement if routine maintenance, servicing and minor repair operations are to be undertaken. We recommend the purchase of combination spanners (ring one end, open-ended the other); although more expensive than open-ended ones, they do give the advantages of both types of spanner.

Combination spanners - 10, 11, 12, 13, 14 & 17 mm
Adjustable spanner - 9 inch
Gearbox drain plug key
Spark plug spanner (with rubber insert)
Spark plug gap adjustment tool
Set of feeler gauges
Brake bleed nipple spanner
Screwdriver - 4 in long x $\frac{1}{4}$ in dia (flat blade)
Screwdriver - 4 in long x $\frac{1}{4}$ in dia (cross blade)
Combination pliers - 6 inch
Hacksaw (junior)
Tyre pump
Tyre pressure gauge
Grease gun
Oil can
Fine emery cloth (1 sheet)
Wire brush (small)
Funnel (medium size)

Repair and overhaul tool kit

These tools are virtually essential for anyone undertaking any major repairs to a motor vehicle, and are additional to those given in the *Maintenance and minor repair* list. Included in this list is a comprehensive set of sockets. Although these are expensive they will be found invaluable as they are so versatile - particularly if various drives are included in the set. We recommend the $\frac{1}{2}$ in square-drive type, as this can be used with most proprietary torque wrenches. If you cannot afford a socket set, even bought piecemeal, then inexpensive tubular box spanners are a useful alternative.

The tools in this list will occasionally need to be supplemented by tools from the *Special* list.

Sockets (or box spanners) to cover range in previous list
Reversible ratchet drive (for use with sockets)
Extension piece, 10 inch (for use with sockets)
Universal joint (for use with sockets)
Torque wrench (for use with sockets)
'Mole' wrench - 8 inch
Ball pein hammer
Soft-faced hammer, plastic or rubber
Screwdriver - 6 in long x $\frac{5}{16}$ in dia (flat blade)
Screwdriver - 2 in long x $\frac{5}{16}$ in square (flat blade)
Screwdriver - $1\frac{1}{2}$ in long x $\frac{1}{4}$ in dia (cross blade)
Screwdriver - 3 in long x $\frac{1}{8}$ in dia (electricians)
Pliers - electricians side cutters
Pliers - needle nosed
Pliers - circlip (internal and external)
Cold chisel - $\frac{1}{2}$ inch
Scriber
Scraper
Centre punch
Pin punch
Hacksaw
Valve grinding tool
Steel rule/straight-edge
Allen keys
Selection of files
Wire brush (large)
Axle-stands
Jack (strong trolley or hydraulic type)

Special tools

The tools in this list are those which are not used regularly, are expensive to buy, or which need to be used in accordance with their manufacturers' instructions. Unless relatively difficult mechanical jobs are undertaken frequently, it will not be economic to buy many of these tools. Where this is the case, you could consider clubbing together with friends (or joining a motorists' club) to make a joint purchase, or borrowing the tools against a deposit from a local garage or tool hire specialist.

The following list contains only those tools and instruments freely available to the public, and not those special tools produced by the vehicle manufacturer specifically for its dealer network. You will find occasional references to these manufacturers' special tools in the text of this manual. Generally, an alternative method of doing the job without the vehicle manufacturers' special tool is given. However, sometimes, there is no alternative to using them. Where this is the case and the relevant tool cannot be bought or borrowed, you will have to entrust the work to a franchised garage.

Valve spring compressor
Piston ring compressor
Balljoint separator
Universal hub/bearing puller
Impact screwdriver
Micrometer and/or vernier gauge
Dial gauge
Stroboscopic timing light
Dwell angle meter/tachometer
Universal electrical multi-meter
Cylinder compression gauge
Lifting tackle
Trolley jack
Light with extension lead

Buying tools

For practically all tools, a tool factor is the best source since he will have a very comprehensive range compared with the average garage or accessory shop. Having said that, accessory shops often offer excellent quality tools at discount prices, so it pays to shop around.

Remember, you don't have to buy the most expensive items on the shelf, but it is always advisable to steer clear of the very cheap tools. There are plenty of good tools around at reasonable prices, so ask the proprietor or manager of the shop for advice before making a purchase.

Care and maintenance of tools

Having purchased a reasonable tool kit, it is necessary to keep the tools in a clean serviceable condition. After use, always wipe off any dirt, grease and metal particles using a clean, dry cloth, before putting the tools away. Never leave them lying around after they have been used. A simple tool rack on the garage or workshop wall, for items such as screwdrivers and pliers is a good idea. Store all normal wrenches and sockets in a metal box. Any measuring instruments, gauges, meters, etc, must be carefully stored where they cannot be damaged or become rusty.

Take a little care when tools are used. Hammer heads inevitably become marked and screwdrivers lose the keen edge on their blades from time to time. A little timely attention with emery cloth, a file or grindstone will soon restore items like this to a good serviceable finish.

Working facilities

Not to be forgotten when discussing tools, is the workshop itself. If anything more than routine maintenance is to be carried out, some form of suitable working area becomes essential.

It is appreciated that many an owner mechanic is forced by circumstances to remove an engine or similar item, without the benefit of a garage or workshop. Having done this, any repairs should always be done under the cover of a roof.

Wherever possible, any dismantling should be done on a clean, flat workbench or table at a suitable working height.

Any workbench needs a vice: one with a jaw opening of 4 in (100 mm) is suitable for most jobs. As mentioned previously, some clean dry storage space is also required for tools, as well as for lubricants, cleaning fluids, touch-up paints and so on, which become necessary.

Another item which may be required, and which has a much more general usage, is an electric drill with a chuck capacity of at least $\frac{5}{16}$ in (8 mm). This, together with a good range of twist drills, is virtually essential for fitting accessories such as mirrors and reversing lights.

Last, but not least, always keep a supply of old newspapers and clean, lint-free rags available, and try to keep any working area as clean as possible.

Spanner jaw gap comparison table

Jaw gap (in)	Spanner size
0.250	$\frac{1}{4}$ in AF
0.276	7 mm
0.313	$\frac{5}{16}$ in AF
0.315	8 mm
0.344	$\frac{11}{32}$ in AF; $\frac{1}{8}$ in Whitworth
0.354	9 mm
0.375	$\frac{3}{8}$ in AF
0.394	10 mm
0.433	11 mm
0.438	$\frac{7}{16}$ in AF
0.445	$\frac{3}{16}$ in Whitworth; $\frac{1}{4}$ in BSF
0.472	12 mm
0.500	$\frac{1}{2}$ in AF
0.512	13 mm
0.525	$\frac{1}{4}$ in Whitworth; $\frac{5}{16}$ in BSF
0.551	14 mm
0.563	$\frac{9}{16}$ in AF
0.591	15 mm
0.600	$\frac{5}{16}$ in Whitworth; $\frac{3}{8}$ in BSF
0.625	$\frac{5}{8}$ in AF
0.630	16 mm
0.669	17 mm
0.686	$\frac{11}{16}$ in AF
0.709	18 mm
0.710	$\frac{3}{8}$ in Whitworth; $\frac{7}{16}$ in BSF
0.748	19 mm
0.750	$\frac{3}{4}$ in AF
0.813	$\frac{13}{16}$ in AF
0.820	$\frac{7}{16}$ in Whitworth; $\frac{1}{2}$ in BSF
0.866	22 mm
0.875	$\frac{7}{8}$ in AF
0.920	$\frac{1}{2}$ in Whitworth; $\frac{9}{16}$ in BSF
0.938	$\frac{15}{16}$ in AF
0.945	24 mm
1.000	1 in AF
1.010	$\frac{9}{16}$ in Whitworth; $\frac{5}{8}$ in BSF
1.024	26 mm
1.063	$1\frac{1}{16}$ in AF; 27 mm
1.100	$\frac{5}{8}$ in Whitworth; $\frac{11}{16}$ in BSF
1.125	$1\frac{1}{8}$ in AF
1.181	30 mm
1.200	$\frac{11}{16}$ in Whitworth; $\frac{3}{4}$ in BSF
1.250	$1\frac{1}{4}$ in AF
1.260	32 mm
1.300	$\frac{3}{4}$ in Whitworth; $\frac{7}{8}$ in BSF
1.313	$1\frac{5}{16}$ in AF
1.390	$\frac{13}{16}$ in Whitworth; $\frac{15}{16}$ in BSF
1.417	36 mm
1.438	$1\frac{7}{16}$ in AF
1.480	$\frac{7}{8}$ in Whitworth; 1 in BSF
1.500	$1\frac{1}{2}$ in AF
1.575	40 mm; $\frac{15}{16}$ in Whitworth
1.614	41 mm
1.625	$1\frac{5}{8}$ in AF
1.670	1 in Whitworth; $1\frac{1}{8}$ in BSF
1.688	$1\frac{11}{16}$ in AF
1.811	46 mm
1.813	$1\frac{13}{16}$ in AF
1.860	$1\frac{1}{8}$ in Whitworth; $1\frac{1}{4}$ in BSF
1.875	$1\frac{7}{8}$ in AF
1.969	50 mm
2.000	2 in AF
2.050	$1\frac{1}{4}$ in Whitworth; $1\frac{3}{8}$ in BSF
2.165	55 mm
2.362	60 mm

Safety first!

Professional motor mechanics are trained in safe working procedures. However enthusiastic you may be about getting on with the job in hand, do take the time to ensure that your safety is not put at risk. A moment's lack of attention can result in an accident, as can failure to observe certain elementary precautions.

There will always be new ways of having accidents, and the following points do not pretend to be a comprehensive list of all dangers; they are intended rather to make you aware of the risks and to encourage a safety-conscious approach to all work you carry out on your vehicle.

Essential DOs and DON'Ts

DON'T rely on a single jack when working underneath the vehicle. Always use reliable additional means of support, such as axle stands, securely placed under a part of the vehicle that you know will not give way.

DON'T attempt to loosen or tighten high-torque nuts (e.g. wheel hub nuts) while the vehicle is on a jack; it may be pulled off.

DON'T start the engine without first ascertaining that the transmission is in neutral (or 'Park' where applicable) and the parking brake applied.

DON'T suddenly remove the filler cap from a hot cooling system – cover it with a cloth and release the pressure gradually first, or you may get scalded by escaping coolant.

DON'T attempt to drain oil until you are sure it has cooled sufficiently to avoid scalding you.

DON'T grasp any part of the engine, exhaust or catalytic converter without first ascertaining that it is sufficiently cool to avoid burning you.

DON'T syphon toxic liquids such as fuel, brake fluid or antifreeze by mouth, or allow them to remain on your skin.

DON'T inhale brake lining dust – it is injurious to health.

DON'T allow any spilt oil or grease to remain on the floor – wipe it up straight away, before someone slips on it.

DON'T use ill-fitting spanners or other tools which may slip and cause injury.

DON'T attempt to lift a heavy component which may be beyond your capability – get assistance.

DON'T rush to finish a job, or take unverified short cuts.

DON'T allow children or animals in or around an unattended vehicle.

DO wear eye protection when using power tools such as drill, sander, bench grinder etc, and when working under the vehicle.

DO use a barrier cream on your hands prior to undertaking dirty jobs – it will protect your skin from infection as well as making the dirt easier to remove afterwards; but make sure your hands aren't left slippery.

DO keep loose clothing (cuffs, tie etc) and long hair well out of the way of moving mechanical parts.

DO remove rings, wristwatch etc, before working on the vehicle – especially the electrical system.

DO ensure that any lifting tackle used has a safe working load rating adequate for the job.

DO keep your work area tidy – it is only too easy to fall over articles left lying around.

DO get someone to check periodically that all is well, when working alone on the vehicle.

DO carry out work in a logical sequence and check that everything is correctly assembled and tightened afterwards.

DO remember that your vehicle's safety affects that of yourself and others. If in doubt on any point, get specialist advice.

IF, in spite of following these precautions, you are unfortunate enough to injure yourself, seek medical attention as soon as possible.

Fire

Remember at all times that petrol (gasoline) is highly flammable. Never smoke, or have any kind of naked flame around, when working on the vehicle. But the risk does not end there – a spark caused by an electrical short-circuit, by two metal surfaces contacting each other, or even by static electricity built up in your body under certain conditions, can ignite petrol vapour, which in a confined space is highly explosive.

Always disconnect the battery earth (ground) terminal before working on any part of the fuel system, and never risk spilling fuel on to a hot engine or exhaust.

It is recommended that a fire extinguisher of a type suitable for fuel and electrical fires is kept handy in the garage or workplace at all times. Never try to extinguish a fuel or electrical fire with water.

Fumes

Certain fumes are highly toxic and can quickly cause unconsciousness and even death if inhaled to any extent. Petrol (gasoline) vapour comes into this category, as do the vapours from certain solvents such as trichloroethylene. Any draining or pouring of such volatile fluids should be done in a well ventilated area.

When using cleaning fluids and solvents, read the instructions carefully. Never use materials from unmarked containers – they may give off poisonous vapours.

Never run the engine of a motor vehicle in an enclosed space such as a garage. Exhaust fumes contain carbon monoxide which is extremely poisonous; if you need to run the engine, always do so in the open air or at least have the rear of the vehicle outside the workplace.

If you are fortunate enough to have the use of an inspection pit, never drain or pour petrol, and never run the engine, while the vehicle is standing over it; the fumes, being heavier than air, will concentrate in the pit with possibly lethal results.

The battery

Never cause a spark, or allow a naked light, near the vehicle's battery. It will normally be giving off a certain amount of hydrogen gas, which is highly explosive.

Always disconnect the battery earth (ground) terminal before working on the fuel or electrical systems.

If possible, loosen the filler plugs or cover when charging the battery from an external source. Do not charge at an excessive rate or the battery may burst.

Take care when topping up and when carrying the battery. The acid electrolyte, even when diluted, is very corrosive and should not be allowed to contact the eyes or skin.

If you ever need to prepare electrolyte yourself, always add the acid slowly to the water, and never the other way round. Protect against splashes by wearing rubber gloves and goggles.

When jump starting a car using a booster battery, for negative earth (ground) vehicles, connect the jump leads in the following sequence: First connect one jump lead between the positive (+) terminals of the two batteries. Then connect the other jump lead first to the negative (–) terminal of the booster battery, and then to a good earthing (ground) point on the vehicle to be started, at least 18 in (45 cm) from the battery if possible. Ensure that hands and jump leads are clear of any moving parts, and that the two vehicles do not touch. Disconnect the leads in the reverse order.

Mains electricity

When using an electric power tool, inspection light etc, which works from the mains, always ensure that the appliance is correctly connected to its plug and that, where necessary, it is properly earthed (grounded). Do not use such appliances in damp conditions and, again, beware of creating a spark or applying excessive heat in the vicinity of fuel or fuel vapour.

Ignition HT voltage

A severe electric shock can result from touching certain parts of the ignition system, such as the HT leads, when the engine is running or being cranked, particularly if components are damp or the insulation is defective. Where an electronic ignition system is fitted, the HT voltage is much higher and could prove fatal.

Routine maintenance

The routine maintenance instructions listed are basically those recommended by the vehicle manufacturer. They are sometimes supplemented by additional maintenance tasks proven to be necessary.

The maintenance intervals recommended are those specified by the manufacturer. They are necessarily something of a compromise, since no two vehicles operate under identical conditions. The DIY mechanic, who does not have labour costs to consider, may wish to shorten the service intervals. Experience will show whether this is necessary.

Where the vehicle is used under severe operating conditions (extremes of heat or cold, dusty conditions, or mainly stop-start driving), more frequent oil changes may be desirable. If in doubt consult your dealer.

Engine compartment (air cleaner removed) – carburettor model

1 Front suspension top mounting
2 Brake fluid reservoir and master cylinder
3 Alternator
4 Inlet manifold
5 Carburettor
6 Brake pressure regulating valve
7 Oil filter
8 Fuel pump
9 Automatic choke
10 Speedometer cable
11 Heater hose
12 Steering gear
13 Fuel filter
14 Starter motor
15 Windscreen wiper intermittent relay
16 Windscreen wiper motor
17 Ignition coil
18 Windscreen/tailgate washer reservoir
19 Headlamp unit
20 Battery
21 Radiator top hose
22 Clutch release arm
23 Radiator
24 Distributor
25 Radiator filler/pressure cap
26 Radiator bottom hose
27 Exhaust manifold and hot air shroud
28 Engine oil filler cap
29 Headlamp washer reservoir
30 Engine mounting

View of rear underside of car

1 Exhaust pipe	*6 Fuel tank*	*10 Rear hub and brake assembly*
2 Rear suspension trailing arm	*7 Fuel drain plug*	*11 Fuel feed and return hoses*
3 Handbrake cable	*8 Fuel tank filler hose*	*12 Brake pipe*
4 Coil spring seat and buffer stop	*9 Shock absorber lower mounting*	*13 Handbrake equaliser*
5 Rear silencer		

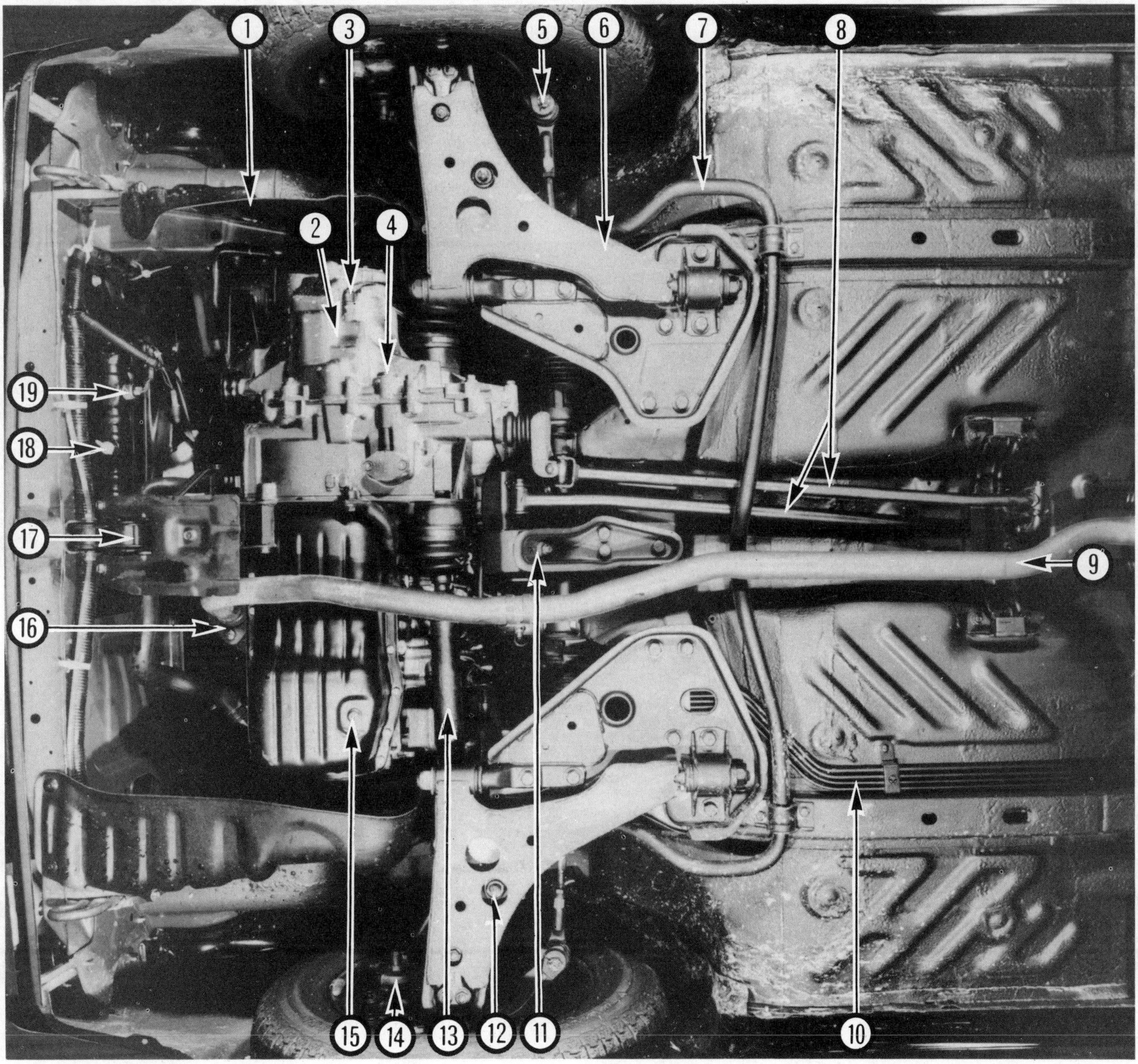

View of front underside of car

1 *Engine shield*
2 *Transmission*
3 *Reversing lamp switch*
4 *Gearbox drain plug (manual transmission)*
5 *Track rod end*
6 *Front suspension lower arm*
7 *Anti-roll bar*
8 *Gearchange rods*
9 *Front exhaust pipe*
10 *Brake and fuel lines*
11 *Rear engine mounting*
12 *Anti-roll bar mounting*
13 *Driveshaft*
14 *Brake caliper*
15 *Engine oil drain plug*
16 *Exhaust manifold flange*
17 *Front engine mounting*
18 *Radiator drain plug*
19 *Electrical cooling fan thermal switch*

Weekly or before a long journey

Check engine oil level (photos)
Check operation of all lights, direction indicators, horn, wipers and washers
Check engine coolant level (photo)
Check colour of battery test indicator (where applicable) (Chapter 10)
Check washer fluid level (photos)
Check tyre pressures (cold) including the spare (photo)
Check brake fluid level (photo)
Check automatic transmission fluid level (Chapter 7)

UK models

Every 3000 miles (5000 km) or 6 months whichever comes first

Turbo models only:
Change the engine oil

Every 6000 miles (10 000 km) or 6 months whichever comes first

Change engine oil and filter (photo) (Chapter 1)
Adjust idle speed and mixture (Chapter 3)

Checking engine oil level

Topping-up engine oil level

Topping-up coolant level

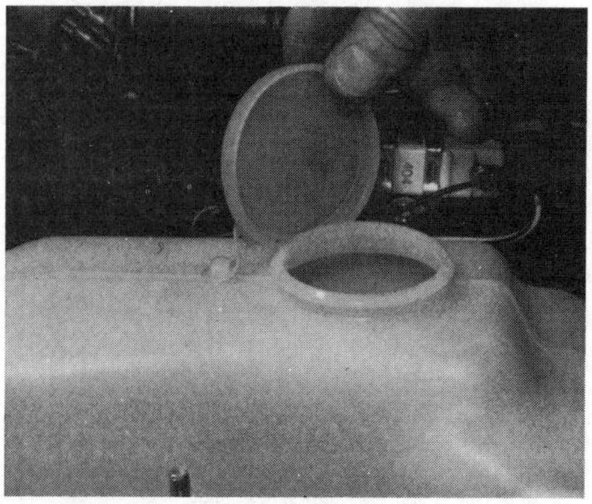

Checking windscreen/tailgate washer fluid level

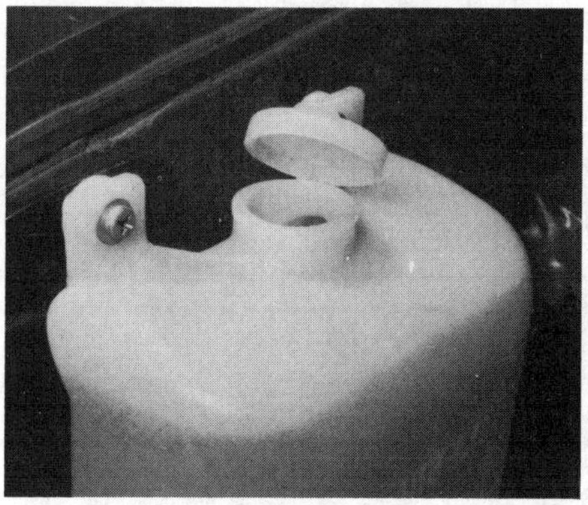

Checking headlamp washer fluid level

Checking tyre pressure

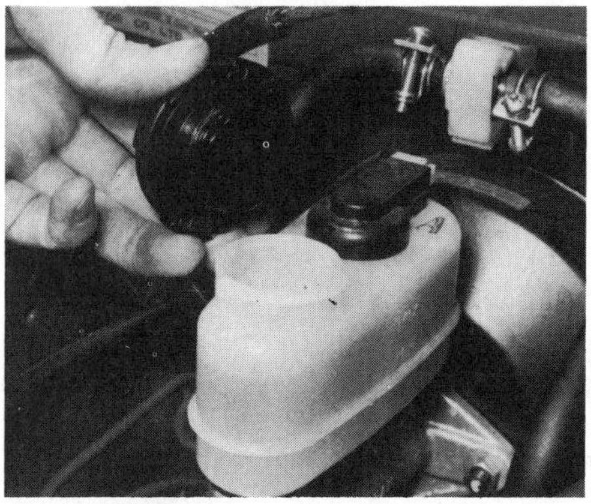

Checking brake fluid level

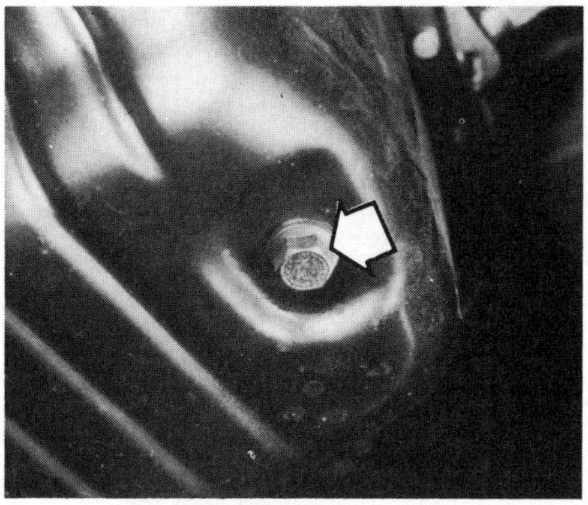

Engine oil drain plug

Checking tyre tread depth

Check contact points condition (Chapter 4)
Adjust ignition timing (Chapter 4)
Check spark plug condition (Chapter 4)
Check brake and automatic transmission fluid levels (Chapters 9 and 7)
Check power steering fluid level and fluid lines (Chapter 11)
Check brake, fuel, and exhaust systems condition
Check manual transmission oil level (photo) (Chapter 6)
Check front brakes and disc pads (Chapter 9)
Lubricate body locks and hinges
Check brake and clutch operation, and tyre condition

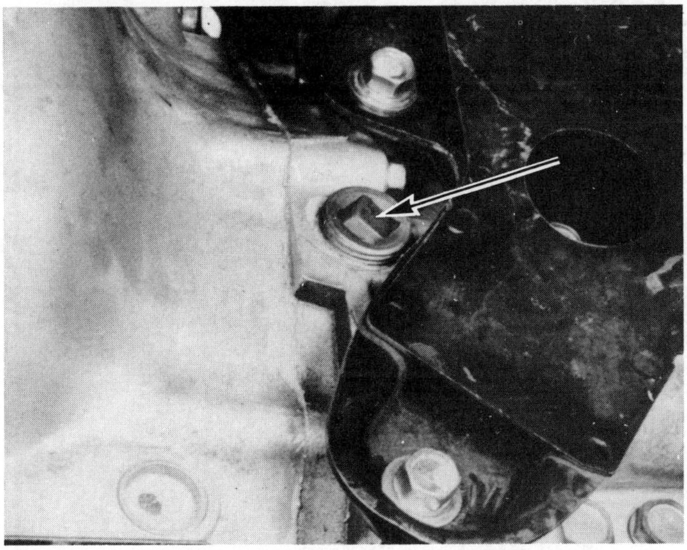

Manual transmission filler/level plug

Every 12 000 miles (20 000 km) or 12 months whichever comes first

Adjust valve clearances (Chapter 1)
Check and adjust drivebelts (Chapter 2)
Check cooling system hoses
Renew contact points (Chapter 4)
Renew spark plugs (Chapter 4)
Check crankcase ventilation system
Change brake fluid (Chapter 9)
Check steering, suspension and driveshafts for condition
Check wheel alignment and balance wheels (Chapter 11)
Check rear brakes and linings (Chapter 9)
Check seat belt condition

Every 24 000 miles (40 000 km) or 24 months whichever comes first

Change antifreeze (Chapter 2)
Renew fuel filter (Chapter 3)
Check fuel lines
Renew air cleaner (Chapter 3)
Check ignition wiring
Check brake servo unit (Chapter 9)
Grease front wheel bearings (Chapter 11)

North American models

Every 3750 miles (6000 km)

Turbo models only:
Change the engine oil

Every 7500 miles (12 000 km) or 6 months whichever comes first

Change the engine oil and filter

Every 15 000 miles (24 000 km) or 12 months whichever comes first

Adjust the valve clearances (Chapter 1)
Check idle speed (Chapter 3)
Renew spark plugs (Canada non-Turbo) (Chapter 4)
Adjust ignition timing (Canada non-Turbo) (Chapter 4)
Check brake lines and hoses
Check brake disc pads and rear shoe linings (Chapter 9)
Check manual transmission oil level (Chapter 6)
Check power steering lines and hoses
Check steering and suspension condition (Chapter 11)
Lubricate body locks and hinges
Check exhaust system
Check seat belt condition

Every 30 000 miles (48 000 km) or 24 months whichever comes first

Check drivebelts for condition and tension
Renew air cleaner (Chapter 3)
Renew air induction valve (Chapter 3)
Check choke operation
Check fuel lines
Renew coolant (Chapter 2)
Renew spark plugs (Chapter 4)
Check ignition wiring
Check air cleaner temperature control system (Chapter 3)
Check exhaust gas sensor (Chapter 3)
Check fuel tank relief valve
Renew brake fluid
Grease front wheel bearings (Chapter 11)
Renew vacuum control modulator filter (Chapter 3)
Renew fuel filter (Chapter 3)

Every 60 000 miles (96 000 km) or 48 months whichever comes first

Check steering and suspension balljoints for wear (Chapter 11)

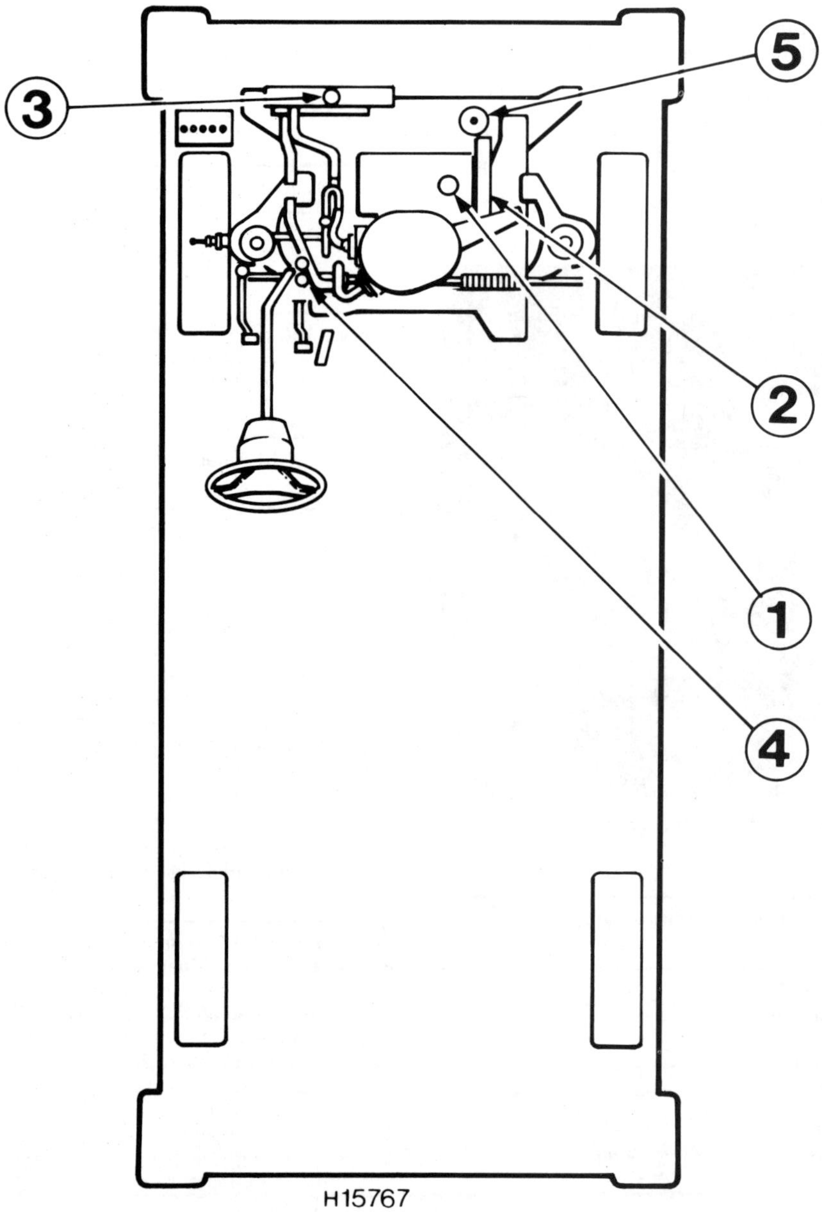

Recommended lubricants and fluids

Component or system	Lubricant type or specification
Engine 1	Engine oil to specification API SE or SF – SAE 10W/30, 10W/40 or 10W/50
Manual transmission 2	Gear oil to specification API GL-4
Automatic transmission 2	Automatic transmission fluid to Dexron II specification
Antifreeze 3	Ethylene glycol based
Brake fluid reservoir 4	Brake fluid to DOT 3 specification
Power-assisted steering 5	Automatic transmission fluid to Dexron II specification

Conversion factors

Length (distance)					
Inches (in)	X 25.4	= Millimetres (mm)	X 0.0394	= Inches (in)
Feet (ft)	X 0.305	= Metres (m)	X 3.281	= Feet (ft)
Miles	X 1.609	= Kilometres (km)	X 0.621	= Miles
Volume (capacity)				
Cubic inches (cu in; in^3)	X 16.387	= Cubic centimetres (cc; cm^3)	X 0.061	= Cubic inches (cu in; in^3)
Imperial pints (Imp pt)	X 0.568	= Litres (l)	X 1.76	= Imperial pints (Imp pt)
Imperial quarts (Imp qt)	X 1.137	= Litres (l)	X 0.88	= Imperial quarts (Imp qt)
Imperial quarts (Imp qt)	X 1.201	= US quarts (US qt)	X 0.833	= Imperial quarts (Imp qt)
US quarts (US qt)	X 0.946	= Litres (l)	X 1.057	= US quarts (US qt)
Imperial gallons (Imp gal)	X 4.546	= Litres (l)	X 0.22	= Imperial gallons (Imp gal)
Imperial gallons (Imp gal)	X 1.201	= US gallons (US gal)	X 0.833	= Imperial gallons (Imp gal)
US gallons (US gal)	X 3.785	= Litres (l)	X 0.264	= US gallons (US gal)
Mass (weight)				
Ounces (oz)	X 28.35	= Grams (g)	X 0.035	= Ounces (oz)
Pounds (lb)	X 0.454	= Kilograms (kg)	X 2.205	= Pounds (lb)
Force				
Ounces-force (ozf; oz)	X 0.278	= Newtons (N)	X 3.6	= Ounces-force (ozf; oz)
Pounds-force (lbf; lb)	X 4.448	= Newtons (N)	X 0.225	= Pounds-force (lbf; lb)
Newtons (N)	X 0.1	= Kilograms-force (kgf; kg)	X 9.81	= Newtons (N)
Pressure				
Pounds-force per square inch (psi; lbf/in^2; lb/in^2)	X 0.070	= Kilograms-force per square centimetre (kgf/cm^2; kg/cm^2)	X 14.223	= Pounds-force per square inch (psi; lbf/in^2; lb/in^2)
Pounds-force per square inch (psi; lbf/in^2; lb/in^2)	X 0.068	= Atmospheres (atm)	X 14.696	= Pounds-force per square inch (psi; lbf/in^2; lb/in^2)
Pounds-force per square inch (psi; lbf/in^2; lb/in^2)	X 0.069	= Bars	X 14.5	= Pounds-force per square inch (psi; lbf/in^2; lb/in^2)
Pounds-force per square inch (psi; lbf/in^2; lb/in^2)	X 6.895	= Kilopascals (kPa)	X 0.145	= Pounds-force per square inch (psi; lbf/in^2; lb/in^2)
Kilopascals (kPa)	X 0.01	= Kilograms-force per square centimetre (kgf/cm^2; kg/cm^2)	X 98.1	= Kilopascals (kPa)
Torque (moment of force)				
Pounds-force inches (lbf in; lb in)	X 1.152	= Kilograms-force centimetre (kgf cm; kg cm)	X 0.868	= Pounds-force inches (lbf in; lb in)
Pounds-force inches (lbf in; lb in)	X 0.113	= Newton metres (Nm)	X 8.85	= Pounds-force inches (lbf in; lb in)
Pounds-force inches (lbf in; lb in)	X 0.083	= Pounds-force feet (lbf ft; lb ft)	X 12	= Pounds-force inches (lbf in; lb in)
Pounds-force feet (lbf ft; lb ft)	X 0.138	= Kilograms-force metres (kgf m; kg m)	X 7.233	= Pounds-force feet (lbf ft; lb ft)
Pounds-force feet (lbf ft; lb ft)	X 1.356	= Newton metres (Nm)	X 0.738	= Pounds-force feet (lbf ft; lb ft)
Newton metres (Nm)	X 0.102	= Kilograms-force metres (kgf m; kg m)	X 9.804	= Newton metres (Nm)
Power				
Horsepower (hp)	X 745.7	= Watts (W)	X 0.0013	= Horsepower (hp)
Velocity (speed)				
Miles per hour (miles/hr; mph)	X 1.609	= Kilometres per hour (km/hr; kph)	X 0.621	= Miles per hour (miles/hr; mph)
*Fuel consumption**				
Miles per gallon, Imperial (mpg)	X 0.354	= Kilometres per litre (km/l)	X 2.825	= Miles per gallon, Imperial (mpg)
Miles per gallon, US (mpg)	X 0.425	= Kilometres per litre (km/l)	X 2.352	= Miles per gallon, US (mpg)

Temperature

Degrees Fahrenheit = (°C x 1.8) + 32

Degrees Celsius (Degrees Centigrade; °C) = (°F - 32) x 0.56

**It is common practice to convert from miles per gallon (mpg) to litres/100 kilometres (l/100km), where mpg (Imperial) x l/100 km = 282 and mpg (US) x l/100 km = 235*

Fault diagnosis

Introduction

The vehicle owner who does his or her own maintenance according to the recommended schedules should ot have to use this section of the manual very often. Modern component reliability is such that, provided those items subject to wear or deterioration are inspected or renewed at the specified intervals, sudden failure is comparatively rare. Faults do not usually just happen as a result of sudden failure, but develop over a period of time. Major mechanical failures in particular are usually preceded by characteristic symptoms over hundreds or even thousands of miles. Those components which do occasionally fail without warning are often small and easily carried in the vehicle.

With any fault finding, the first step is to decide where to begin investigations. Sometimes this is obvious, but on other occasions a little detective work will be necessary. The owner who makes half a dozen haphazard adjustments or replacements may be successful in curing a fault (or its symptoms), but he will be none the wiser if the fault recurs and he may well have spent more time and money than was necessary. A calm and logical approach will be found to be more satisfactory in the long run. Always take into account any warning signs or abnormalities that may have been noticed in the period preceding the fault – power loss, high or low gauge readings, unusual noises or smells, etc – and remember that failure of components such as fuses or spark plugs may only be pointers to some underlying fault.

The pages which follow here are intended to help in cases of failure to start or breakdown on the road. There is also a Fault Diagnosis Section at the end of each Chapter which should be consulted if the preliminary checks prove unfruitful. Whatever the fault, certain basic principles apply. These are as follows:

Verify the fault. This is simply a matter of being sure that you know what the symptoms are before starting work. This is particularly important if you are investigating a fault for someone else who may not have described it very accurately.

Don't overlook the obvious. For example, if the vehicle won't start, is there petrol in the tank? (Don't take anyone else's word on this particular point, and don't trust the fuel gauge either!) If an electrical fault is indicated, look for loose or broken wires before digging out the test gear.

Cure the disease, not the symptom. Substituting a flat battery with a fully charged one will get you off the hard shoulder, but if the underlying cause is not attended to, the new battery will go the same way. Similarly, changing oil-fouled spark plugs for a new set will get you moving again, but remember that the reason for the fouling (if it wasn't simply an incorrect grade of plug) will have to be established and corrected.

Don't take anything for granted. Particularly, don't forget that a 'new' component may itself be defective (especially if it's been rattling round in the boot for months), and don't leave components out of a fault diagnosis sequence just because they are new or recently fitted. When you do finally diagnose a difficult fault, you'll probably realise that all the evidence was there from the start.

Electrical faults

Electrical faults can be more puzzling than straightforward mechanical failures, but they are no less susceptible to logical analysis if the basic principles of operation are understood. Vehicle electrical wiring exists in extremely unfavourable conditions – heat, vibration and chemical attack – and the first things to look for are loose or corroded connections and broken or chafed wires, especially where the wires pass through holes in the bodywork or are subject to vibration.

All metal-bodied vehicles in current production have one pole of the battery 'earthed', ie connected to the vehicle bodywork, and in nearly all modern vehicles it is the negative (–) terminal. The various electrical components – motors, bulb holders etc – are also connected to earth, either by means of a lead or directly by their mountings. Electric current flows through the component and then back to the battery via the bodywork. If the component mounting is loose or corroded, or if a good path back to the battery is not available, the circuit will be incomplete and malfunction will result. The engine and/or gearbox are also earthed by means of flexible metal straps to the body or subframe; if these straps are loose or missing, starter motor, generator and ignition trouble may result.

Assuming the earth return to be satisfactory, electrical faults will be due either to component malfunction or to defects in the current supply. Individual components are dealt with in Chapter 10. If supply wires are broken or cracked internally this results in an open-circuit, and the easiest way to check for this is to bypass the suspect wire temporarily with a length of wire having a crocodile clip or suitable connector at each end. Alternatively, a 12V test lamp can be used to verify the presence of supply voltage at various points along the wire and the break can be thus isolated.

If a bare portion of a live wire touches the bodywork or other earthed metal part, the electricity will take the low-resistance path thus formed back to the battery: this is known as a short-circuit. Hopefully a short-circuit will blow a fuse, but otherwise it may cause burning of the insulation (and possibly further short-circuits) or even a fire. This is why it is inadvisable to bypass persistently blowing fuses with silver foil or wire.

Spares and tool kit

Most vehicles are supplied only with sufficient tools for wheel changing; the *Maintenance and minor repair* tool kit detailed in *Tools*

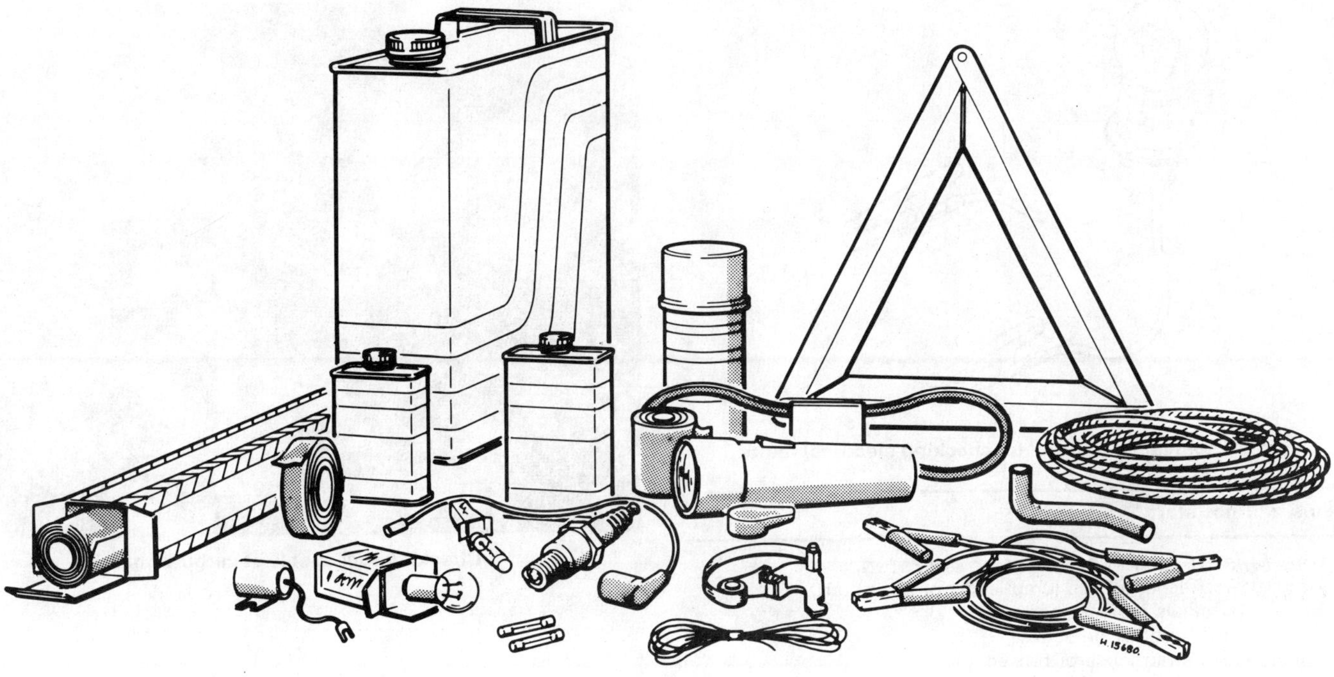

Carrying a few spares may save you a long walk!

and working facilities, with the addition of a hammer, is probably sufficient for those repairs that most motorists would consider attempting at the roadside. In addition a few items which can be fitted without too much trouble in the event of a breakdown should be carried. Experience and available space will modify the list below, but the following may save having to call on professional assistance:

Spark plugs, clean and correctly gapped
HT lead and plug cap – long enough to reach the plug furthest from the distributor
Distributor rotor, condenser and contact breaker points
Drivebelt(s) – emergency type may suffice
Spare fuses
Set of principal light bulbs
Tin of radiator sealer and hose bandage
Exhaust bandage
Roll of insulating tape
Length of soft iron wire
Length of electrical flex
Torch or inspection lamp (can double as test lamp)
Battery jump leads
Tow-rope
Ignition waterproofing aerosol
Litre of engine oil
Sealed can of hydraulic fluid
Emergency windscreen
'Jubilee' clips
Tube of filler paste

If spare fuel is carried, a can designed for the purpose should be used to minimise risks of leakage and collision damage. A first aid kit and a warning triangle, whilst not at present compulsory in the UK, are obviously sensible items to carry in addition to the above.

When touring abroad it may be advisable to carry additional spares which, even if you cannot fit them yourself, could save having to wait while parts are obtained. The items below may be worth considering:

Throttle cables
Cylinder head gasket
Alternator brushes
Fuel pump repair kit
Tyre valve core

One of the motoring organisations will be able to advise on availability of fuel etc in foreign countries.

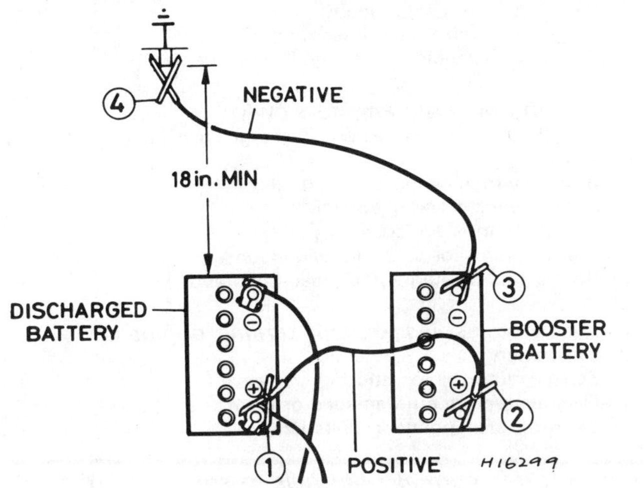

Jump start lead connections for negative earth – connect leads in order shown

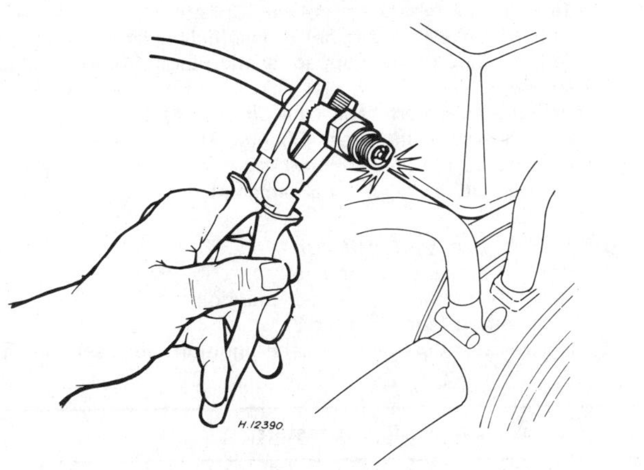

Crank engine and check for spark. Note use of insulated tool to hold plug lead

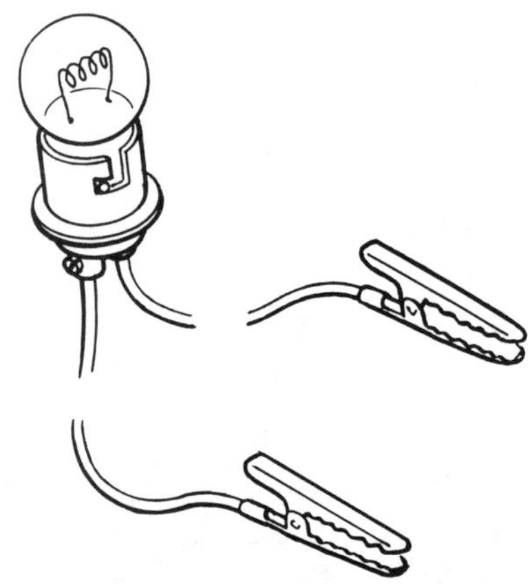

A simple test lamp is useful for checking electrical faults

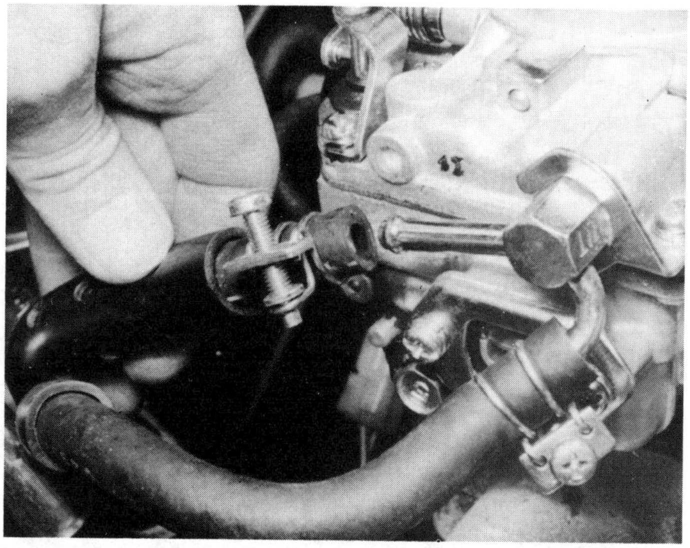

Checking fuel supply at carburettor

Engine will not start

Engine fails to turn when starter operated
Flat battery (recharge, use jump leads, or push start)
Battery terminals loose or corroded
Battery earth to body defective
Engine earth strap loose or broken
Starter motor (or solenoid) wiring loose or broken
Automatic transmission selector in wrong position, or inhibitor switch faulty
Ignition/starter switch faulty
Major mechanical failure (seizure)
Starter or solenoid internal fault (see Chapter 10)

Starter motor turns engine slowly
Partially discharged battery (recharge, use jump leads, or push start)
Battery terminals loose or corroded
Battery earth to body defective
Engine earth strap loose
Starter motor (or solenoid) wiring loose
Starter motor internal fault (see Chapter 10)

Starter motor spins without turning engine
Flat battery
Starter motor pinion sticking on sleeve
Flywheel gear teeth damaged or worn
Starter motor mounting bolts loose

Engine turns normally but fails to start
Damp or dirty HT leads and distributor cap (crank engine and check for spark)
Dirty or incorrectly gapped distributor points
No fuel in tank (check for delivery) (photo)
Excessive choke (hot engine) or insufficient choke (cold engine)
Fouled or incorrectly gapped spark plugs (remove, clean and regap)
Other ignition system fault (see Chapter 4)
Other fuel system fault (see Chapter 3)
Poor compression (see Chapter 1)
Major mechanical failure (eg camshaft drive)

Engine fires but will not run
Insufficient choke (cold engine)
Air leaks at carburettor or inlet manifold
Fuel starvation (see Chapter 3)
Ballast resistor defective, or other ignition fault (see Chapter 4)

Engine cuts out and will not restart

Engine cuts out suddenly – ignition fault
Loose or disconnected LT wires
Wet HT leads or distributor cap (after traversing water splash)
Coil or condenser failure (check for spark)
Other ignition fault (see Chapter 4)

Engine misfires before cutting out – fuel fault
Fuel tank empty
Fuel pump defective or filter blocked (check for delivery)
Fuel tank filler vent blocked (suction will be evident on releasing cap) – not N. American cars with sealed cap
Carburettor needle valve sticking
Carburettor jets blocked (fuel contaminated)
Other fuel system fault (see Chapter 3)

Engine cuts out – other causes
Serious overheating
Major mechanical failure (eg camshaft drive)

Engine overheats

Ignition (no-charge) warning light illuminated
Slack or broken drivebelt – retension or renew (Chapter 2)

Ignition warning light not illuminated
Coolant loss due to internal or external leakage (see Chapter 2)
Thermostat defective
Low oil level
Brakes binding
Radiator clogged externally or internally
Electric cooling fan not operating correctly
Engine waterways clogged
Ignition timing incorrect or automatic advance malfunctioning
Mixture too weak

Note: *Do not add cold water to an overheated engine or damage may result*

Low engine oil pressure

Gauge reads low or warning light illuminated with engine running
Oil level low or incorrect grade
Defective gauge or sender unit
Wire to sender unit earthed

Engine overheating
Oil filter clogged or bypass valve defective
Oil pressure relief valve defective
Oil pick-up strainer clogged
Oil pump worn or mountings loose
Worn main or big-end bearings

Note: *Low oil pressure in a high-mileage engine at tickover is not necessarily a cause for concern. Sudden pressure loss at speed is far more significant. In any event, check the gauge or warning light sender before condemning the engine.*

Engine noises

Pre-ignition (pinking) on acceleration

Incorrect grade of fuel
Ignition timing incorrect
Distributor faulty or worn
Worn or maladjusted carburettor
Excessive carbon build-up in engine

Whistling or wheezing noises

Leaking vacuum hose
Leaking carburettor or manifold gasket
Blowing head gasket

Tapping or rattling

Incorrect valve clearances
Worn valve gear
Worn timing belt
Broken piston ring (ticking noise)

Knocking or thumping

Unintentional mechanical contact (eg fan blades)
Worn drivebelt
Peripheral component fault (generator, water pump etc)
Worn big-end bearings (regular heavy knocking, perhaps less under load)
Worn main bearings (rumbling and knocking, perhaps worsening under load)
Piston slap (most noticeable when cold)

Chapter 1 Engine

Contents

Specifications

General

Type	Four-cylinder, in-line, overhead camshaft, mounted transversely	
Designation and capacity:		
E10	987 cc (60.23 cu in)	
E13	1269 cc (77.43 cu in)	
E15 and E15ET	1488 cc (90.80 cu in)	
E16	1597 cc (97.45 cu in)	
Bore:		
E10	73.0 mm (2.874 in)	
E13, E15, E15ET, E16	76.0 mm (2.992 in)	
Stroke:		
E10	59.0 mm (2.323 in)	
E13	70.0 mm (2.756 in)	
E15, E15ET	82.0 mm (3.228 in)	
E16	88.0 mm (3.465 in)	
Compression ratio:		
E10	9.0 : 1	
E13	9.0 : 1	
E15	9.0 : 1	
E15ET (UK)	7.4 : 1	
E15ET (Canada)	8.0 : 1	
E16	9.4 : 1	
Compression pressure – lbf/in^2 (bar)	**Normal**	**Minimum**
Except E15ET	181 (12.5)	142 (9.8)
E15ET	158 (10.9)	129 (8.9)
Firing order	1 – 3 – 4 – 2 (No 1 cylinder at timing belt end)	

Cylinder block

Material	Cast iron
Maximum bore out-of-round	0.015 mm (0.0006 in)
Maximum taper of bore	0.02 mm (0.0008 in)

Crankshaft

Number of main bearings	5
Main journal diameter	49.940 to 49.964 mm (1.9661 to 1.9671 in)
Crankpin diameter	39.954 to 39.974 mm (1.5730 to 1.5738 in)

Maximum journal and crankpin out-of-round	0.03 mm (0.0012 in)
Endfloat	0.05 to 0.18 mm (0.0020 to 0.0071 in)
Wear limit	0.30 mm (0.0118 in)
Main bearing running clearance:	
All models except E15ET:	
Nos 1 and 5	0.031 to 0.076 mm (0.0012 to 0.0030 in)
Nos 2, 3 and 4	0.031 to 0.092 mm (0.0012 to 0.0036 in)
Wear limit	0.10 mm (0.0039 in)
E15ET models:	
Nos 1 and 5	0.031 to 0.076 mm (0.0012 to 0.0030 in)
Nos 2 and 4	0.029 to 0.088 mm (0.0011 to 0.0035 in)
No 3	0.031 to 0.092 mm (0.0012 to 0.0036 in)
Wear limit	0.10 mm (0.0039 in)
Big-end bearing running clearance:	
E10, E13, E15 models	0.034 to 0.079 mm (0.0013 to 0.0031 in)
Wear limit	0.12 mm (0.0047 in)
E15ET models	0.016 to 0.058 mm (0.0006 to 0.0023 in)
Wear limit	0.10 mm (0.0039 in)
E16 models	0.030 to 0.060 mm (0.0012 to 0.0024 in)
Wear limit	0.10 mm (0.0039 in)
Bearing undersizes:	
Except E15ET:	
Main	0.25 mm (0.0098 in), 0.50 mm (0.0197 in) and 0.75 mm (0.0295 in)
Big-end	0.08 mm (0.0031 in), 0.25 mm (0.0098 in), 0.50 mm (0.0197 in) and 0.75 mm (0.295 in)
E15ET:	
Main	0.25 mm (0.0098 in)
Big-end	0.25 mm (0.0098 in)

Connecting rods

Side play at big-ends	0.1 to 0.37 mm (0.004 to 0.0146 in)
Wear limit	0.5 mm (0.020 in)

Gudgeon pin

Pin to piston clearance:	
E10 models	0.006 to 0.010 mm (0.0002 to 0.0004 in)
Except E10 models	0.008 to 0.012 mm (0.0003 to 0.0005 in)
Interference fit in small end	0.017 to 0.038 mm (0.0007 to 0.0015 in)

Piston rings

Oversizes	0.02 mm (0.0008 in) and 0.5 mm (0.020 in)
Clearance in block:	
E15ET	0.040 to 0.060 mm (0.0016 to 0.0024 in)
Except E15ET	0.023 to 0.043 mm (0.0009 to 0.0017 in)

Pistons

Type	Two compression and one oil control
Clearance in groove:	
Top compression	0.040 to 0.073 mm (0.0016 to 0.0029 in)
2nd compression	0.030 to 0.063 mm (0.0012 to 0.0025 in)
Oil control:	
Except E15ET	0.050 to 0.145 mm (0.0020 to 0.0057 in)
E15ET	0.050 to 0.125 mm (0.0020 to 0.0049 in)
End gap:	
E15ET models:	
Top compression	0.20 to 0.26 mm (0.0079 to 0.0102 in)
2nd compression	0.15 to 0.25 mm (0.0059 to 0.0098 in)
Oil control	0.20 to 0.60 mm (0.0079 to 0.0236 in)
Except E15ET models:	
Top compression	0.20 to 0.35 mm (0.0079 to 0.0138 in)
2nd compression	0.15 to 0.30 mm (0.0059 to 0.0118 in)
Oil control	0.30 to 0.90 mm (0.0118 to 0.0354 in)

Jack shaft

Maximum journal running clearance:	
E15ET, E16 models	0.15 mm (0.0059 in)
E10, E13, E15 models	0.20 mm (0.0079 in)
Journal diameter:	
Front	31.987 to 32.000 mm (1.2593 to 1.2598 in)
Rear	25.587 to 28.600 mm (1.1255 to 1.1260 in)
Shaft bearing inside diameter:	
Front	32.020 to 32.085 mm (1.2606 to 1.2632 in)
Rear	28.620 to 26.685 mm (1.1268 to 1.1293 in)
Shaft endfloat	0.045 to 0.105 mm (0.0018 to 0.0041 in)
Fuel pump cam height	27.8 to 27.9 mm (1.094 to 1.098 in)

Camshaft

Journal diameter:	
Nos 1, 3 and 5	41.949 to 41.965 mm (1.6515 to 1.6522 in)
Nos 2 and 4	41.906 to 41.922 mm (1.6498 to 1.6505 in)
Bearing inside diameter	42.000 to 42.025 mm (1.6535 to 1.6545 in)
Camshaft running clearance:	
Nos 1, 3 and 5	0.035 to 0.076 mm (0.0014 to 0.0030 in)
Wear limit	0.15 mm (0.0059 in)
Nos 2 and 4	0.078 to 0.119 mm (0.0031 to 0.0047 in)
Wear limit	0.20 mm (0.0079 in)

Cylinder head

Material	Aluminium alloy
Surface out-of-true (limit)	0.1 mm (0.004 in)

Valves

Clearance (hot), intake and exhaust	0.28 mm (0.011 in)
Valve seat angle:	
E15ET models – intake	60° 30′
E15ET models – exhaust	45° 30′
Except E15ET models – intake and exhaust	45° 15′ to 45° 45′
Valve spring free length	46.70 mm (1.8386 in)

Valve guides

Outside diameter:	
Standard	12.033 to 12.044 mm (0.4737 to 0.4742 in)
Oversize	12.256 to 12.274 mm (0.4825 to 0.4832 in)
Inside diameter (reamed)	7.005 to 7.020 mm (0.2758 to 0.2764 in)
Guide hole in cylinder head:	
Diameter for standard guide	11.970 to 11.988 mm (0.4713 to 0.4720 in)
Diameter for oversize guide	12.200 to 12.211 mm (0.4803 to 0.4807 in)
Interference fit in cylinder head	0.045 to 0.074 mm (0.0018 to 0.0029 in)
Valve stem to guide clearance:	
Wear limit	0.1 mm (0.004 in)

Lubrication

Oil capacity (with oil filter):	
Turbo (E15ET)	6.5 Imp pt; 3.9 US qt; 3.7 litre
Except Turbo	6.9 Imp pt; 4.1 US qt; 3.9 litre
Oil pressure (hot)	28 lbf/in^2 (2.0 bar) at 1050 rpm

Torque wrench settings

Torque wrench settings	lbf ft	Nm
Alternator bracket bolt	9	12
Alternator link bolt	15	20
Engine mounting bracket to cylinder block	26	35
Manifold nuts	15	20
Oil pump to crankcase	9	12
Power-assisted steering pump bracket	22	30
Power-assisted steering pump fixing bolt	30	40
Spark plug	18	25
Coolant pump bolt	4	5
Coolant pump pulley bolt	4	5
Air conditioner compressor bracket bolt	22	30
Fuel pump fixing bolts	9	12
Crankshaft pulley bolt	104	140
Thermostat housing bolt	4	5
Camshaft pulley bolt	6	8
Connecting rod cap nut	26	35
Cylinder head bolts (cold):		
Stage 1	22	29
Stage 2	51	69
Stage 3	Completely loosen all bolts	
Stage 4	22	29
Stage 5	51 to 54	69 to 74
Cylinder head front cover	4	5
Flywheel bolts (manual transmission)	63	85
Driveplate bolts (automatic transmission)	74	100
Timing belt cover bolts	4	5
Jack shaft pulley bolt	6	8
Main bearing cap bolt	40	55
Sump pan bolts and nuts	4	5

Torque wrench settings (continued)

	lbf ft	Nm
Oil pick-up fixing bolts	6	8
Rocker shaft bolts	15	20
Belt tensioner locknut	15	20
Rocker cover nut	6	8
Rocker arm adjuster screw locknut	15	20
Bellhousing to engine	15	20
Oil pump pressure regulating valve cap	33	45
Oil pump cover bolts	4	5
North American models:		
EGR valve to pipe	37	50
Air injection pipe union	37	50
Turbo models		
Air pipe to manifold	10	14
Air regulator to manifold	1	2
Collector to manifold	4	6
Converter to manifold	22	29
Oil cooler cover	14	20
Cooler/filter bracket – bolt	25	34
Cooler/filter bracket – nut	17	24
Pressure regulator	10	14
Oil feed tube	22	29
Turbocharger	25	34

1 General description

The engine is of four cylinder, in-line overhead camshaft type.

The cylinder block is of cast iron construction while the cylinder head, which is of cross-flow design, is made of light alloy.

A five bearing crankshaft is used.

The camshaft runs in bearings which are machined directly into the cylinder head.

Valve clearances are adjusted in the conventional way by means of a screw and locknut on the rocker arms.

2 Routine maintenance

1 At weekly intervals, check the engine oil level. Do this by withdrawing the dipstick, wiping it clean, reinserting it and withdrawing it for the second time.
2 The oil level should be between the L and H marks. Top up if necessary.
3 Change the engine oil and filter at the following intervals:

UK non-Turbo models: Every 6000 miles (10 000 km) or 6 months whichever comes first
UK Turbo models – oil: Every 3000 miles (5000 km) or 6 months whichever comes first
UK Turbo models – filter: Every 6000 miles (10 000 km) or 6 months whichever comes first
North American non-Turbo models: Every 7500 miles (12 000 km) or 6 months whichever comes first
North American Turbo models – oil: Every 3750 miles (6000 km) or 6 months whichever comes first
North American Turbo models – filter: Every 7500 miles (12 000 km) or 6 months whichever comes first

4 The oil should be drained when hot by removing the oil filler cap and the sump drain plug. Use a large bowl to catch the oil.
5 Using a suitable filter removal tool, unscrew the cartridge type oil filter which is located on the cylinder block just to the rear of the alternator (photo). Be prepared for some spillage of oil.
6 Wipe the filter mating face on the cylinder block clean and smear the rubber sealing ring of the new filter with a little oil. Screw on the filter hand-tight only.
7 Refit the sump drain plug.
8 Refill with the correcly quantity and type of engine oil. Refit the oil filler cap.
9 Every 12 000 miles (20 000 km) or 12 months whichever comes first on UK models, or every 15 000 miles (24 000 km) or 12 months whichever comes first on North American models, adjust the valve clearances and check the crankcase ventilation system hoses for condition and security.

2.5 Oil filter cartridge

3 Major operations possible without removing engine

The following operations may be carried out with the engine in position in the vehicle.

Removal and refitting of the cylinder head
Adjustment of the valve clearances
Removal and refitting of the timing belt
Removal and refitting of the oil pump
Removal and refitting of the sump
Renewal of piston rings and big-end bearings

4 Cylinder head – removal and refitting

1 Disconnect the battery.
2 Disconnect the HT leads from the spark plugs and the ignition coil. Disconnect all LT leads from the coil.
3 Unbolt and remove the distributor (Chapter 4).
4 Remove the air cleaner (Chapter 3).

5 Drain the cooling system, retaining the coolant if suitable for further use.
6 Remove the rocker cover and spark plugs (photos).
7 Unbolt and remove the hot air collector from the exhaust manifold.
8 Disconnect the exhaust downpipe from the exhaust manifold.
9 Refer to Chapter 2 and remove the drivebelts from the alternator, power steering pump and air conditioning compressor (where fitted). Remove the alternator adjuster link.
10 Disconnect the emission control hoses, pipes and leads as necessary to clear the cylinder head (refer to Chapter 3) (photo). On Turbo models disconnect the fuel injection and turbocharger equipment as necessary.
11 Disconnect the throttle, choke, and kickdown cables as applicable.
12 Disconnect the fuel hoses.
13 On models with an automatic choke, disconnect the electrical lead from the terminal on the automatic choke housing.
14 Unbolt and remove the intake manifold. On North American models remove the EGR and air injection pipes where fitted.
15 The upper timing belt cover must now be removed. Access to the bolts is by removal of the plastic shield and right-hand roadwheel after supporting the car on axle stands. The water pump pulley must also be unbolted. Removal of the lower timing belt cover is only possible after removing the crankshaft pulley (photos).

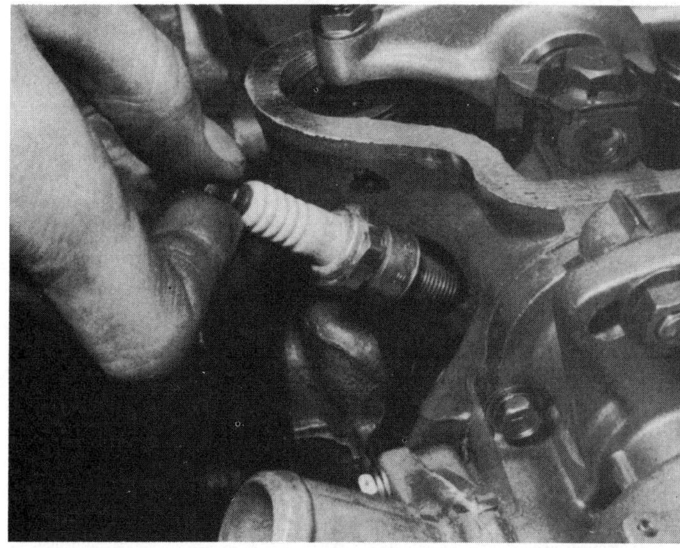
4.6C Removing a spark plug

4.6A Removing the rocker cover domed nut

4.10 Disconnecting the earth lead from the cylinder head

4.6B Lifting off the rocker cover

4.15A Right-hand engine shield side fixing

4.15B Right-hand engine shield front fixing

4.15C Right-hand engine shield bottom fixing

4.15D Right-hand engine shield top fixing

4.15E Removing the right-hand engine shield

4.15F Removing the water pump pulley

4.15G Timing belt cover top ...

4.15H ... middle ...

4.15I ... and side fixings

4.15J Removing the timing belt cover

16 Release the nut on the timing belt tensioner pulley (photo). Using a screwdriver in the slot on the eccentric hub, turn the hub clockwise to release it from the belt, then tighten the nut again.

17 Note the rotational direction of the belt. Original belts are marked with an arrow. Unmarked belts should be marked. Remove the belt (photo).

18 Loosen the cylinder head bolts progressively in reverse order to that shown in the tightening sequence diagram (Fig. 1.1).

4.16 Unscrewing the timing belt tensioner nut

4.17 Removing the timing belt from the camshaft pulley

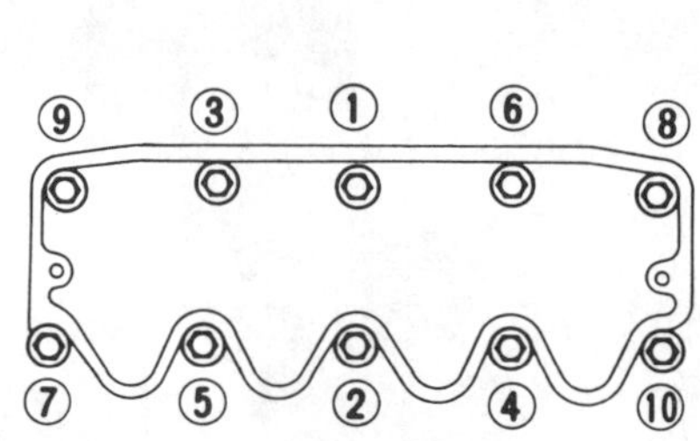

Fig. 1.1 Cylinder head bolt tightening sequence (Sec 4)

19 Remove the bolts and lift off the cylinder head (photos). If the cylinder head is to be dismantled or decarbonised, refer to Section 15.
20 Before refitting the cylinder head make quite sure that the mating surfaces of the cylinder head and block are perfectly clean and free from carbon or old pieces of gasket.
21 Mop out the oil from the cylinder head bolt holes in the block, clean the bolt threads and oil lightly.
22 Release the rocker arm adjuster screw locknuts and unscrew the screws a few turns.
23 Turn the crankshaft, by means of the pulley bolt until the timing mark on the crankshaft sprocket is at its lowest point, or the crankshaft pulley mark is aligned (photo).
24 Turn the camshaft sprocket until the timing marks are approximately in alignment according to engine type. This will eliminate the possibility of the valves digging into the piston crowns when the cylinder head is lowered onto the block.
25 Place a new gasket on top of the cylinder block (photo).
26 Lower the cylinder head carefully onto the block.
27 Screw the cylinder head bolts down finger tight. The shorter bolts go on the exhaust side. The longest bolt of all goes on the exhaust side, at the timing belt end.
28 Tighten the cylinder head bolts in the sequence shown in Fig. 1.1 in five stages to the specified torque setting (photo) (see Specifications).

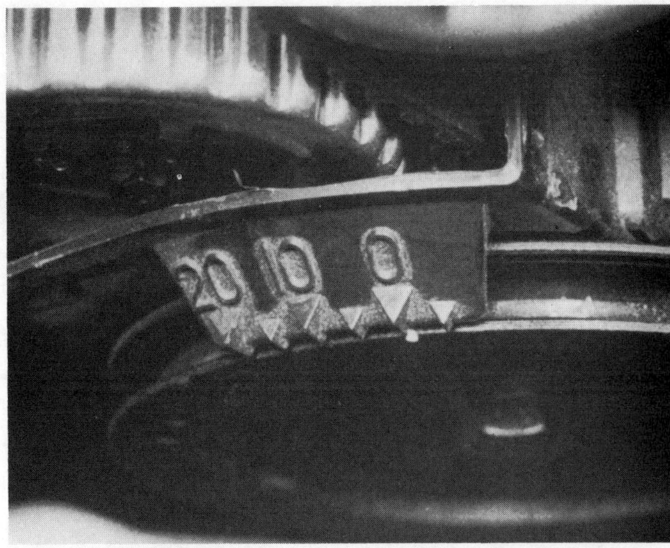

4.23 Crankshaft pulley TDC notch aligned with O (TDC) timing mark

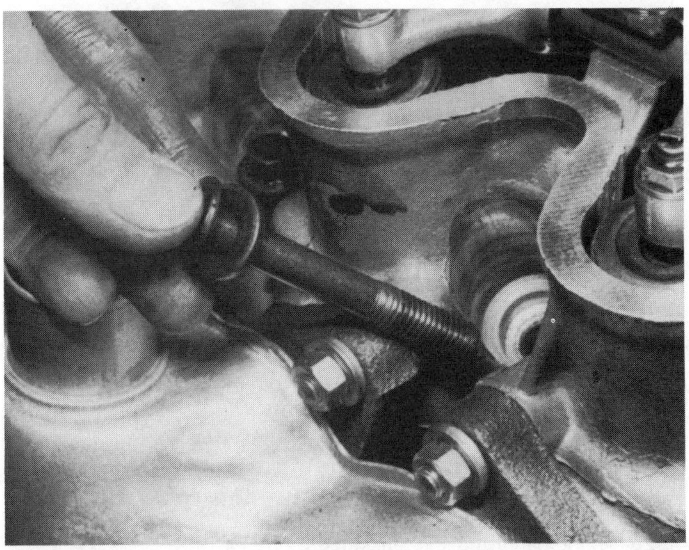

4.19A Removing a cylinder head bolt

4.25 New cylinder head gasket located on cylinder block

4.19B Removing the cylinder head from the cylinder block (manifolds still attached)

4.28 Tightening the cylinder head bolts

29 Using a new gasket, bolt on the exhaust manifold and reconnect the downpipe.
30 Using a new gasket, bolt on the intake manifold assembly. Reconnect the controls, leads and fuel hoses.
31 Check that the camshaft and crankshaft timing marks are correctly aligned and then engage the timing belt with them (photos).
32 Unscrew the timing belt tensioner nut so that the pulley pushes against the belt, then turn the camshaft pulley the equivalent of two cogs clockwise. Tighten the tensioner nut while using a screwdriver in the slot to prevent the hub moving. Check that the timing marks still align correctly (photos).
33 Refit the timing belt cover(s) (photo).
34 Refit any emission control components which were removed or disconnected, together with turbo equipment if applicable.
35 Refit the alternator adjuster link.
36 Refit the drivebelts, tensioning them as described in Chapter 2.
37 Refit the hot air collector to the exhaust manifold. This is held by two manifold bolts.
38 Refit the distributor (Chapter 4).
39 Reconnect the HT and LT ignition leads.
40 Adjust the valve clearances, as described in Section 5.
41 Using a new gasket, refit the rocker cover.
42 Refit the air cleaner and spark plugs.
43 Reconnect the battery.

4.32A Tightening the timing belt tensioner pulley nut

4.31A Camshaft pulley TDC alignment mark and notch

4.32B Timing belt fitted and tensioned

4.31B Crankshaft sprocket TDC alignment marks (pulley removed)

4.33 Showing timing belt top (A) and bottom (B) covers

44 Refill the cooling system (Chapter 2).
45 Check and top up the engine oil.
46 Check and adjust the valve clearances after the engine has reached normal operating temperature.
47 It is recommended that after the first 960 km (600 miles) the cylinder head bolt torque setting is checked. Do this with the engine cold. Unscrew the first bolt in the specified sequence through one quarter of a turn and then retighten to torque. Repeat one at a time on each of the remaining bolts.
48 Check the valve clearances (hot) on completion.

5 Valve clearances – adjustment

1 When this adjustment is being carried out at a routine servicing, the engine should be at normal operating temperature.
2 Remove the air cleaner and the rocker cover, also the spark plugs.
3 Using a socket wrench on the crankshaft pulley bolt, turn the crankshaft until No 1 piston (at timing belt end of the engine) is at TDC. This can be ascertained if a finger is placed over No 1 cylinder spark plug hole and the compression felt as it is being generated as the piston rises. The notch on the crankshaft pulley should be opposite the O mark on the timing scale.
4 Using a feeler blade of specified thickness check that it is a stiff sliding fit between the end of the valve stem and the ball end of the rocker arm adjuster screw on valves 1, 2, 3 and 6 (Fig. 1.2).

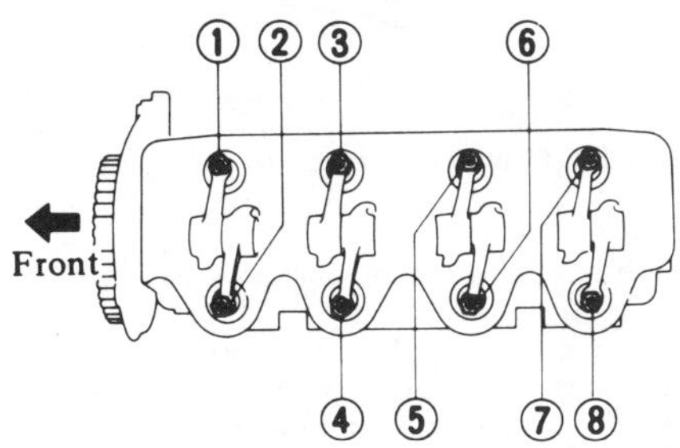

Fig. 1.2 Valve arrangement (Sec 5)

5 Where the clearance is not as specified, release the rocker arm adjuster screw locknut and turn the screw. Once the correct clearance is established, tighten the locknut without allowing the adjuster screw to turn (photo).
6 Now set No 4 piston at TDC on its compression stroke and repeat the operations on valves 4, 5, 7 and 8.
7 Check that the rocker cover gasket is in good order, refit the rocker cover, air cleaner and spark plugs.
8 The valve clearances are the same for both intake and exhaust valves.

6 Timing belt – removal and refitting

1 It is recommended that the timing belt is renewed after 50 000 miles (80 000 km) have been covered.
2 Removal, refitting and tensioning operations are as described in Section 4.

7 Oil pump – removal and refitting

1 Remove the alternator, as described in Chapter 10, and move it to one side of the engine compartment. On some later models the alternator need not be removed.
2 Pull the lead from the oil pressure switch.
3 Unbolt and remove the oil pump.
4 If the pump is to be checked for wear, refer to Examination and Renovation (Section 18).
5 Refitting is a reversal of removal, use a new gasket and clean the mating faces.

5.5 Adjusting the valve clearances

8 Sump pan – removal and refitting

1 Drain the engine oil.
2 Unbolt the reinforcement strut from the rear corner of the sump and transmission (photos).
3 Disconnect the exhaust downpipe from the manifold and release the exhaust pipe front mounting.
4 Unscrew and remove the sump pan bolts and nuts (photo). The nut nearest the transmission is very inaccessible and will require the use of a universally-jointed drive extension and socket to remove it.
5 Remove the sump pan by pulling the exhaust pipe downwards.
6 Remove and discard the old gaskets.
7 Before fitting the sump pan, stick new side gaskets and end sealing strips in position with gasket cement. Make sure that the ends of the sealing strips overlap the side gaskets. Cover the seam with a generous blob of cement.
8 Screw in the sump bolts and fit the nuts, tightening them evenly.
9 Reconnect the exhaust pipe.
10 Fill the engine with oil.

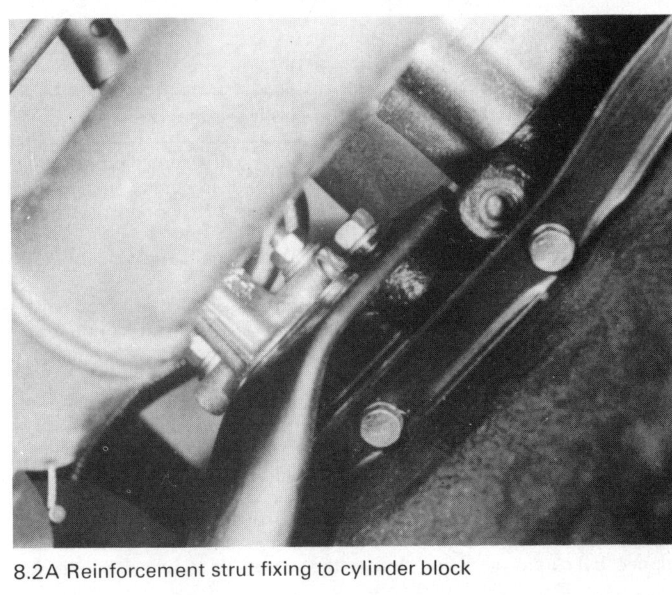
8.2A Reinforcement strut fixing to cylinder block

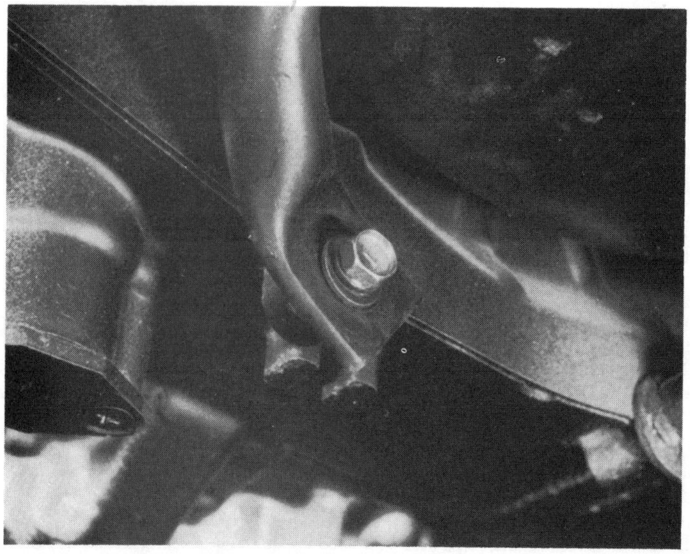

8.2B Reinforcement strut fixing to transmission

8.4 Showing the sump with exhaust downpipe removed

9 Piston rings and big-end bearing shells – renewal

1 The renewal of piston rings or substitution with special proprietary rings may be decided upon as a means of reducing heavy oil consumption without incurring the heavy cost of reboring and new oversize pistons.
2 Remove the cylinder head and the sump pan, as described in earlier Sections.
3 Note that the connecting rod big-end caps are numbered with matching numbers on adjacent machined surfaces on the connecting rod. Note to which side of the crankcase the numbers face, usually towards the jack shaft.
4 Unbolt and remove the oil pick-up pipe from within the crankcase.
5 Unscrew the big-end cap bolts and remove the caps. If the cap shell bearings are to be used again, tape them to their caps.
6 Using the wooden handle of a hammer, applied to the big-end of each connecting rod push the piston/rod assemblies out of the top of the cylinder block. If the cylinder bores are severely worn, and a wear ring can be felt at the top of the bores, the rings may have to be reduced by careful scraping before the piston rings will ride over it.

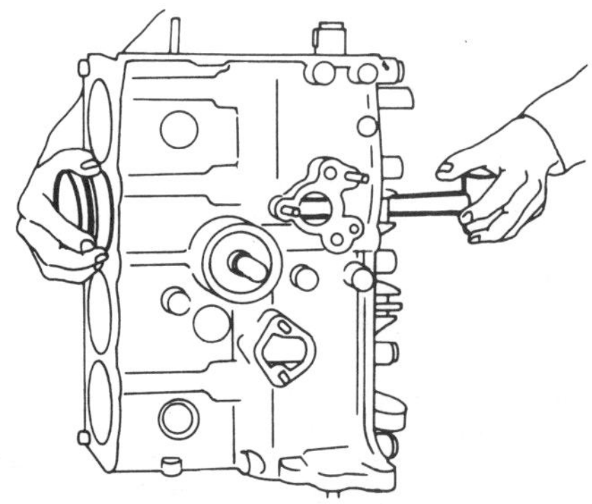

Fig. 1.3 Removing a piston/connecting rod (Sec 9)

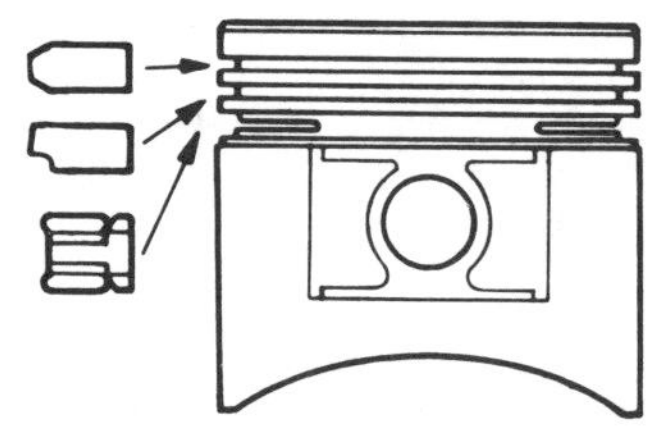

Fig. 1.4 Piston ring arrangement (Sec 9)

Retain the bearing shells with their respective rods if they are to be used again.
7 To remove the piston rings, slide three old feeler blades behind the top ring and position them at equidistant points. The ring can now be slid upwards off the piston using a twisting motion. Repeat on the remaining rings.
8 Clean the piston ring grooves completely free from carbon and other deposits. A piece of old piston ring makes an ideal tool for the purpose.
9 Check that the rings supplied have the correct groove clearance (photo) and end gap (see Specifications). Check the end gap by pushing the ring a little way down the cylinder bore, set it squarely and use a feeler blade to check the gap.
10 Fit the rings to the piston by reversing the removal operations. The

9.9 Checking piston ring groove clearance

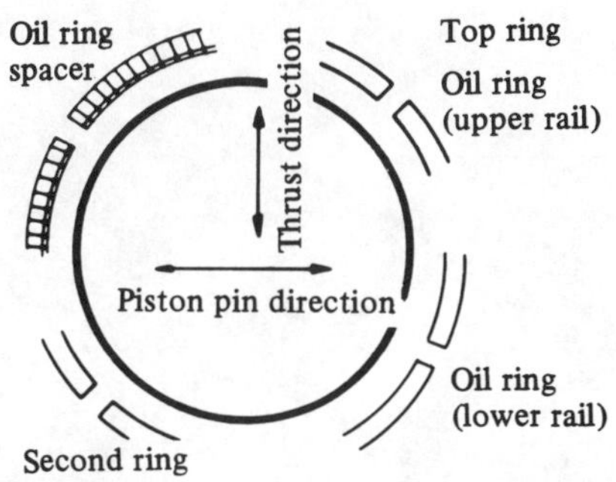

Fig. 1.5 Piston ring end gap setting diagram (Sec 9)

rings are marked on their top surfaces. The top compression ring has a chamfer on both edges.

11 If new rings are fitted then, to ensure rapid bedding in, the cylinder bores must have their hard glaze removed. This can be done using a rotary abrasive flap wheel. Alternatively, use fine glasspaper, rubbing it up and down at approximately 45° to the bore.

12 Stagger the piston ring gaps at equidistant points of a circle, oil the rings liberally and fit a piston ring compressor. These are available at most accessory stores. Oil the cylinder bores.

13 Insert the first piston/rod assembly into its original bore so that the bottom edge of the compressor rests on the top face of the cylinder block and the mark on the piston crown is facing the timing belt end of the engine (photos).

14 Again applying the wooden handle of a hammer, this time to the piston crown, push the piston rod assembly into the cylinder bore. The ring compressor will be released. Repeat on the remaining pistons.

15 Draw the connecting rod of the first piston down to connect with the crankshaft crankpin. Make sure that the crankpins have been liberally oiled and the bearing shells are returned to their original positions (photo). Unless the shells are in excellent condition, without any signs of the copper underlay showing through the white bearing material, they should be renewed with ones of identical size. The size is stamped on the back of the shell. Standard shells are unmarked or stamped STD or 0-00. If not standard it will be stamped with the undersize, for example 0.25 mm.

16 Fit the big-end cap with its shell so that the matching numbers are adjacent (photo) and towards the jack shaft.

17 Insert and tighten the cap bolts to the specified torque (photo).

18 Repeat on the remaining three piston/rod assemblies.

19 Refit the cylinder head and the sump pan, as described in earlier Sections.

20 Refill the engine with oil and coolant.

9.13B Piston crown front marking (arrowed)

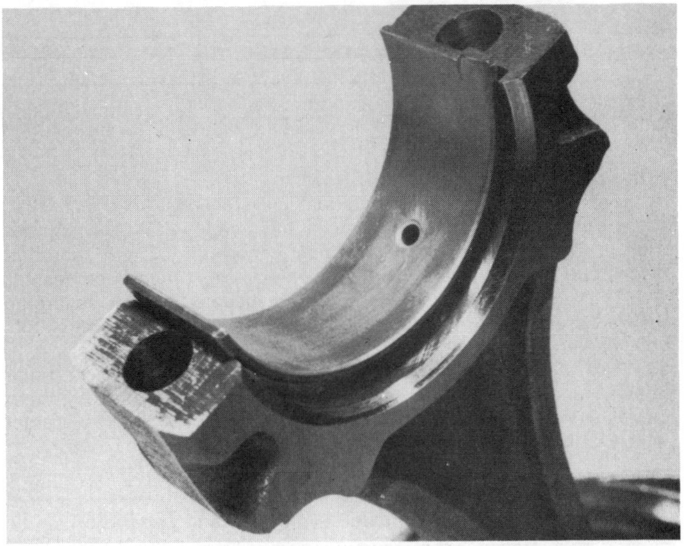

9.15 Connecting rod shell bearing

9.13A Piston ring clamp

9.16 Connecting rod and cap markings

9.17 Tightening big-end cap nut

11.8A Exhaust downpipe at manifold

10 Engine removal – method

The engine should be removed from the vehicle as a unit complete with transmission. The help of an assistant will definitely be required.

For vehicles with air conditioning

If components of the air conditioning system obstruct the overhaul of the engine and cannot be moved sufficiently within the limits of their flexible hoses to avoid such obstruction, the system should be discharged by your dealer or a competent refrigeration engineer.

As the system must be completely evacuated before recharging, the necessary vacuum equipment to do this is only likely to be held by your dealer.

The refrigerant fluid is Freon 12 and, although harmless under normal conditions, contact with eyes or skin must be avoided. If Freon comes into contact with a naked flame a poisonour gas is created which is injurious to health.

11 Engine/manual transmission – removal and separation

Refer to Chapter 6 where necessary

1 With the help of an assistant, unbolt and remove the bonnet (see Chapter 12).

2 Disconnect the battery and remove it together with its support bracket.

3 Remove the air cleaner.

4 Drain the cooling system, retaining the coolant if it is suitable for further use.

5 Remove the radiator complete with electric cooling fan (see Chapter 2).

6 If power steering is fitted, unbolt the pump and move it to one side of the engine compartment.

7 If air conditioning is fitted, unbolt the compressor and belt tensioner pulley and move them aside. *Do not disconnect the refrigerant circuit pipelines* (see Chapter 12).

8 Disconnect the exhaust downpipe from the manifold (photo) and the mounting bracket near the fuel pump (photo). Refer also to Chapter 3 and disconnect the turbocharger and fuel injection equipment as applicable.

9 Unbolt and disconnect the gearchange control rod and its stabiliser rod from the transmission.

10 Support the vehicle under its side-member and then disconnect the front suspension lower balljoints. Do this by unscrewing the three nuts which hold the balljoint to the suspension arm. It is recommended that new nuts are used at reassembly.

11 Remove the front roadwheels.

12 Drain the engine and transmission oils.

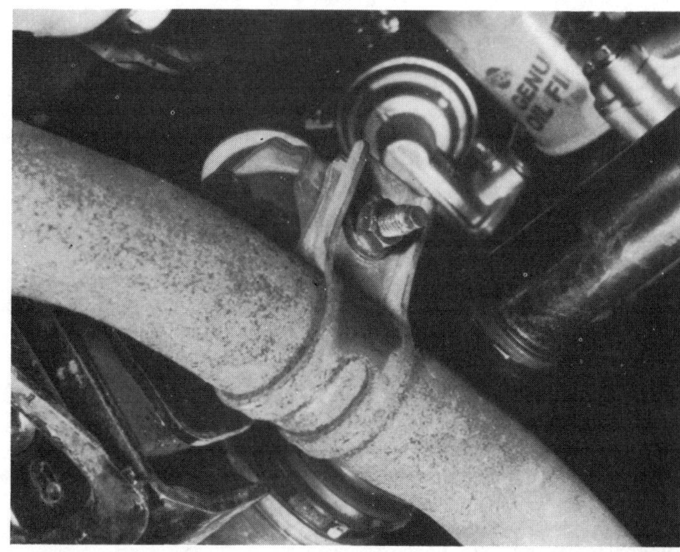

11.8B Exhaust front mounting

13 Unbolt the disc calipers and tie them up out of the way.

14 Unscrew, but do not remove, the nuts at the front suspension strut top mountings. This is to allow movement of the struts when the driveshafts are withdrawn from the transmission.

15 To disconnect a driveshaft, insert a large screwdriver or suitable lever behind the inboard joint flange and prise to overcome the resistance of the joint circlip. Take care not to damage the transmission oil seal and do not pull on the outer end of the driveshaft or the joints may come apart. Insert a rod into the side gears in the transmission casing to prevent them moving from the differential case.

16 Disconnect the clutch operating cable from the release lever.

17 Disconnect the speedometer drive cable from the transmission (photo).

18 Disconnect the throttle and choke cables as applicable.

19 Disconnect the fuel hoses from the fuel pump and plug them. Also disconnect all vacuum and air hoses from the engine.

20 Disconnect all electrical leads, including those from the coolant temperature switch, oil pressure switch, reverse lamp switch and alternator, also the HT and LT leads from the ignition coil (photo).

21 On models with an automatic choke, disconnect the electrical lead

11.17 Disconnecting the speedometer cable from the transmission

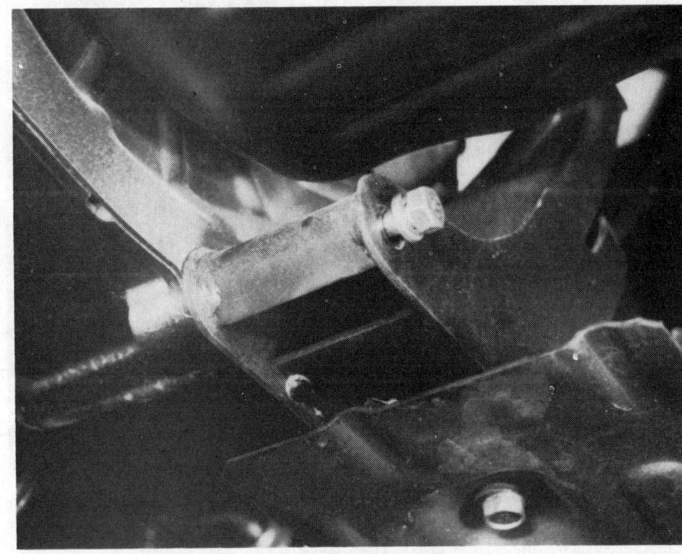

11.23A Lower view of front mounting

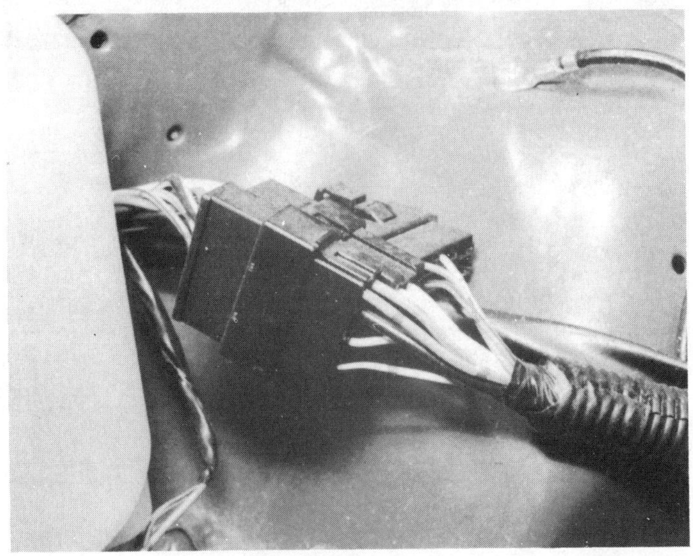

11.20 Engine wiring harness connector

11.23B Front mounting support bar

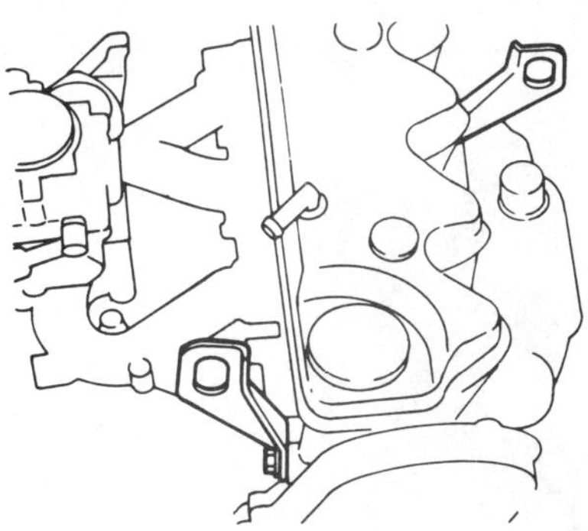

Fig. 1.6 Engine lifting eyes (Sec 11)

11.23C Front mounting fixing to cylinder block

from the choke terminal. Withdraw the complete wiring harness from under the intake manifold after releasing the clips.

22 Lifting eyes should be bolted to the engine. Attach suitable lifting gear and take the weight of the engine.

23 Disconnect the four engine/transmission flexible mountings by unscrewing either the bush pivot bolt or the nut according to location.

11.23D Right-hand side mounting

11.23E Left-hand side mounting

11.23F Upper view of rear mounting

11.23G Lower view of rear mounting

11.24 Removing the engine/transmission unit

The bolt at the bottom of the rear mounting bracket must be removed as well (photos).

24 Lift the engine/transmission up and out of the engine compartment (photo). Take care not to damage adjacent components or the wiring surface.

25 With the unit removed, clean away external dirt using a water soluble solvent or paraffin and a stiff brush.

26 To separate the engine from the transmission unbolt and remove the starter motor and then withdraw the bolts which connect the clutch bellhousing to the engine. Note that some of these bolts retain

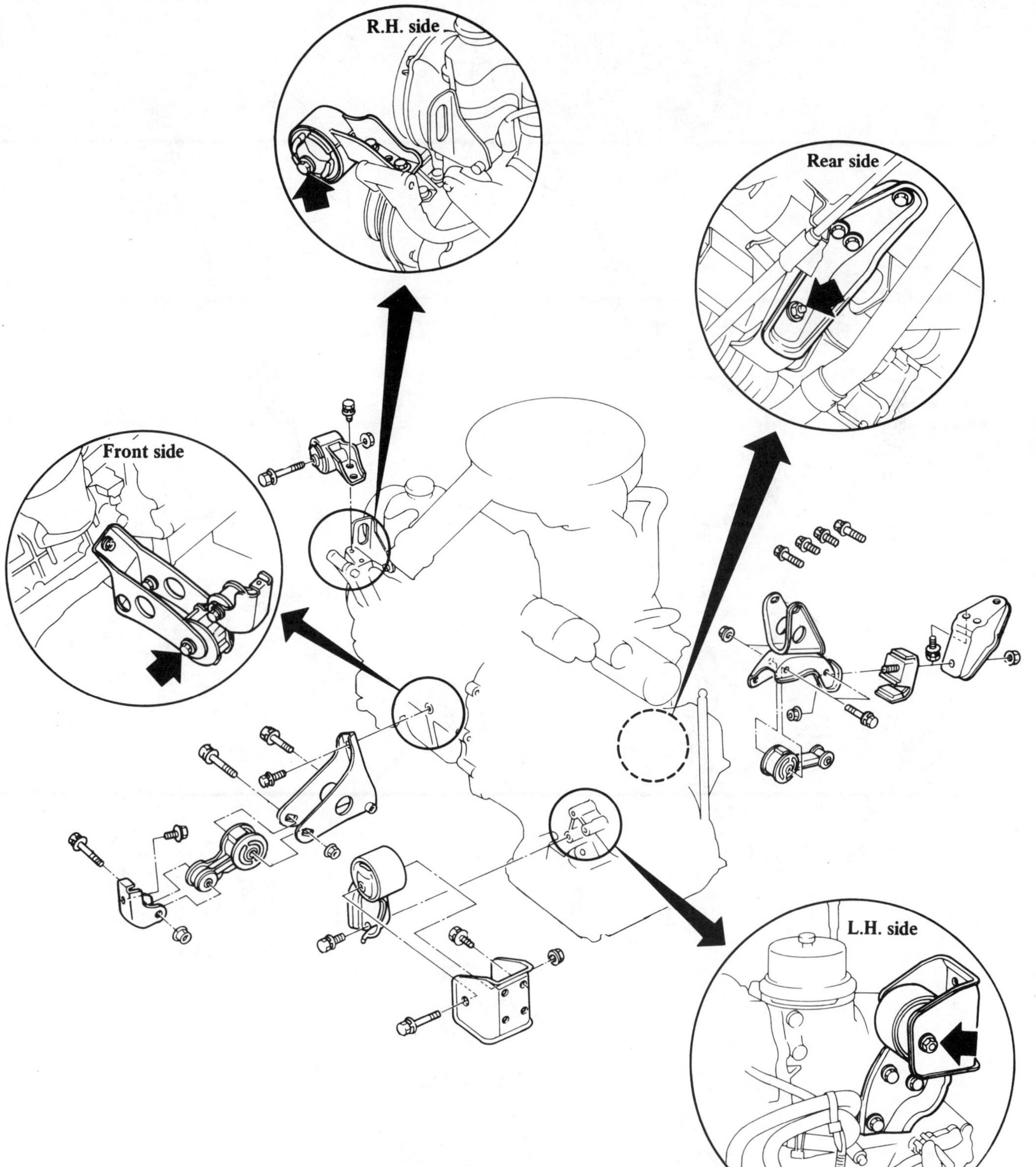

Fig. 1.7 Engine/transmission mountings on early models (Sec 11)

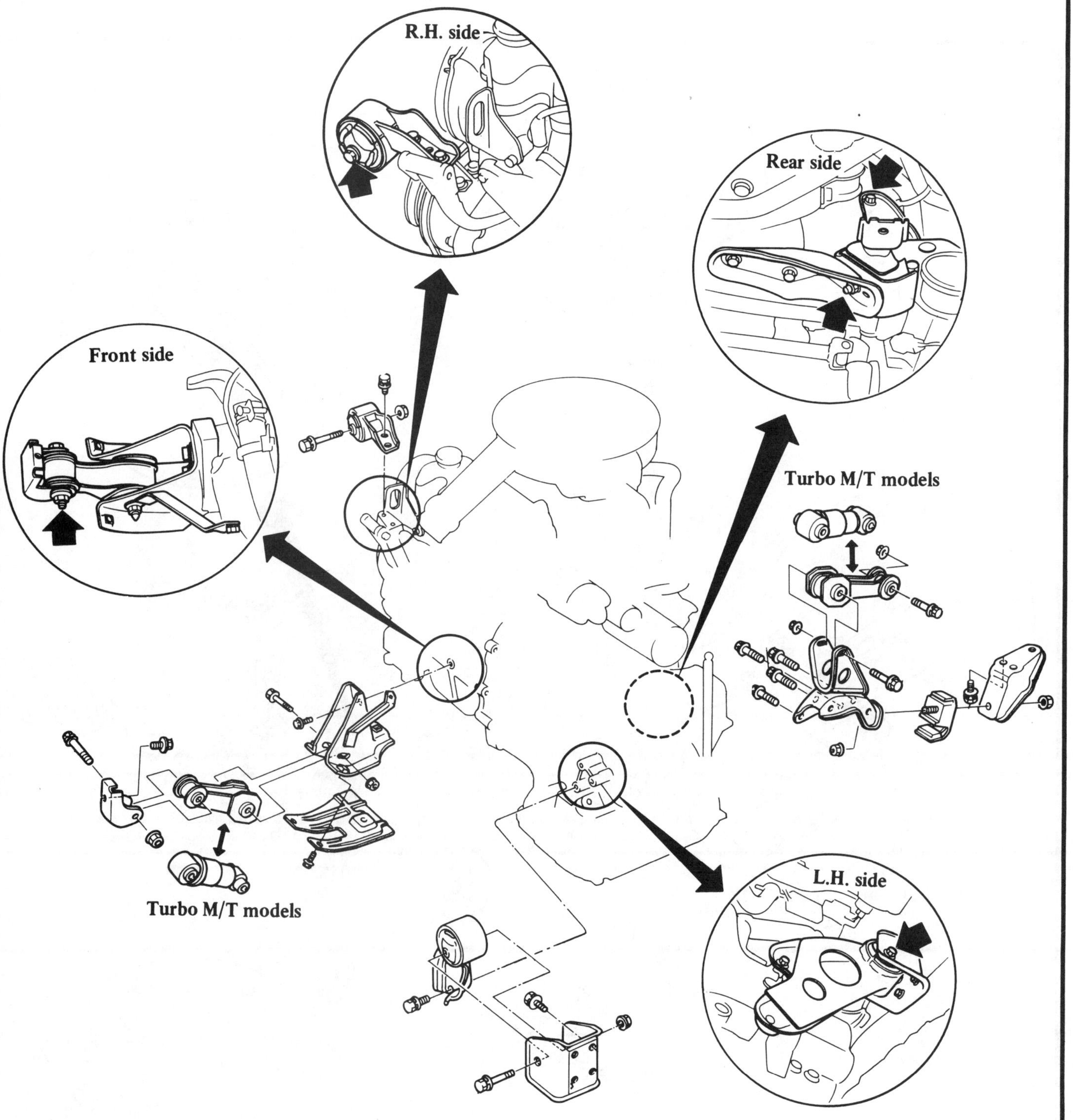

Fig. 1.8 Engine/transmission mountings on later models (Sec 11)

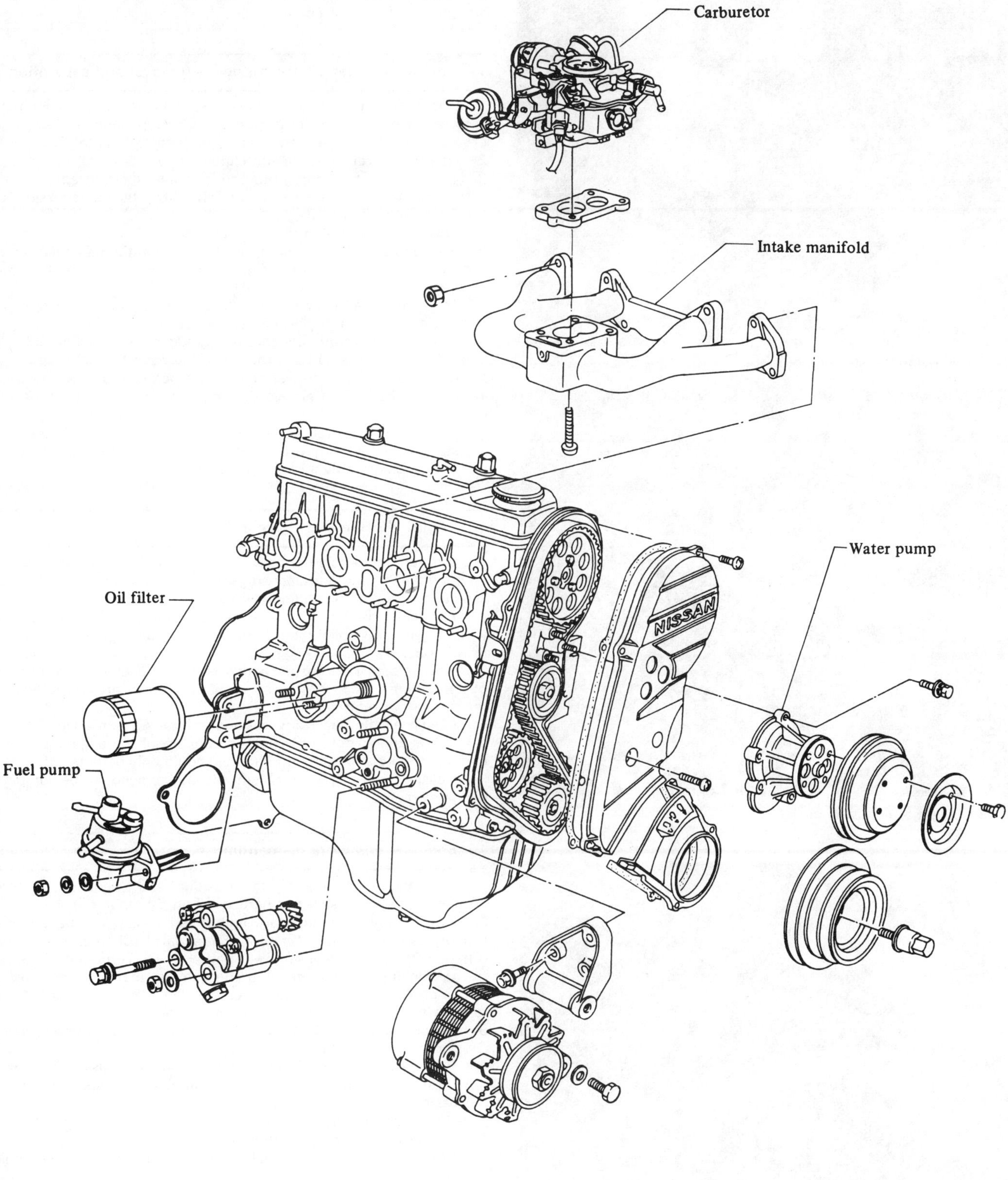

Fig. 1.9 Engine external components – UK non-Turbo models (Sec 14)

11.26A Removing the starter motor

11.26B Coolant tube at bellhousing

11.26C Removing the left-hand mounting bracket

the upper coolant tube, the sump pan-to-bellhousing reinforcement tube and the transmission mounting brackets. Mark their positions for ease of refitting (photos).

27 Support the weight of the transmission and then withdraw it in a straight line from the engine.

12 Engine/automatic transmission – removal and separation

1 The operations for removal are very similar to those described for vehicles with manual transmission in the preceding Section but observe the following differences. Refer also to Chapter 7.

2 Ignore any reference to the clutch cable.

3 Remove the front wing protective shield where fitted.

4 Disconnect the speed selector cable from the transmission, also the inhibitor switch leads.

5 Disconnect and plug the oil cooler hoses.

6 To separate the engine from the automatic transmission first unbolt and remove the starter motor. Also disconnect the kick-down cable.

7 Mark the relationship of the torque converter to the driveplate using a dab of quick-drying paint.

8 Unscrew the torque converter-to-driveplate connecting bolts. The crankshaft will have to be turned to bring each bolt into view in the cut-out in the torque converter housing before a spanner or socket wrench can be used. Remove the engine-to-transmission connecting bolts.

9 Withdraw the automatic transmission, at the same time have an assistant hold the torque converter in full engagement with the oil pump driveshaft to avoid loss of transmission fluid.

13 Engine dismantling – general

1 Before commencing a major engine overhaul, make sure that you have gathered together clean rags, brushes, freeing fluid and a good selection of tools – including a torque wrench.

2 A number of clean tins or other containers is useful to keep the various nuts and bolts safely. Mark the tins as a guide to where the fixings belong.

3 Have a pencil and paper handy to record sequences of assembly of small items, or to sketch an item which may present difficulty at reassembly or refitting.

4 Obtain all the necessary gaskets and oil seals in advance.

5 If it is known that only one component of the engine is worn or damaged the dismantling operations should only be pursued as far as is necessary to rectify the problem, the engine need not be completely dismantled.

14 Engine – complete dismantling

1 Place the engine in an upright position on the bench, or, if there is no alternative, on a sheet of hardboard on the floor.

2 First remove any ancillary components such as the alternator (Chapter 10), distributor (Chapter 4), fuel pump (Chapter 3), manifolds (Chapter 3) and clutch assembly (Chapter 5). On Turbo models remove the turbocharger complete with the exhaust manifold, the pressure regulator, and the oil cooler assembly. On North American models remove the EGR and air injection pipes where fitted.

3 Unscrew and remove the oil filter. Be prepared for some loss of oil. On Turbo models unbolt the oil filter bracket.

4 Remove the alternator mounting bracket and also those for the power steering pump and air conditioning compressor (where these are fitted).

5 Unbolt and remove the oil pump.

6 Unbolt and remove the thermostat housing.

7 Unbolt and remove the coolant pump pulley and then the coolant pump.

8 Carefully turn the engine onto its side and remove the sump pan (Section 8).

9 Using a piece of wood inserted between the crankcase and one of the crankshaft counterweights to prevent the crankshaft rotating, unscrew and remove the crankshaft pulley bolt and the pulley.

10 As the flywheel bolt holes are not offset, mark the relationship of the flywheel to the crankshaft hub. Unscrew the bolts and remove the

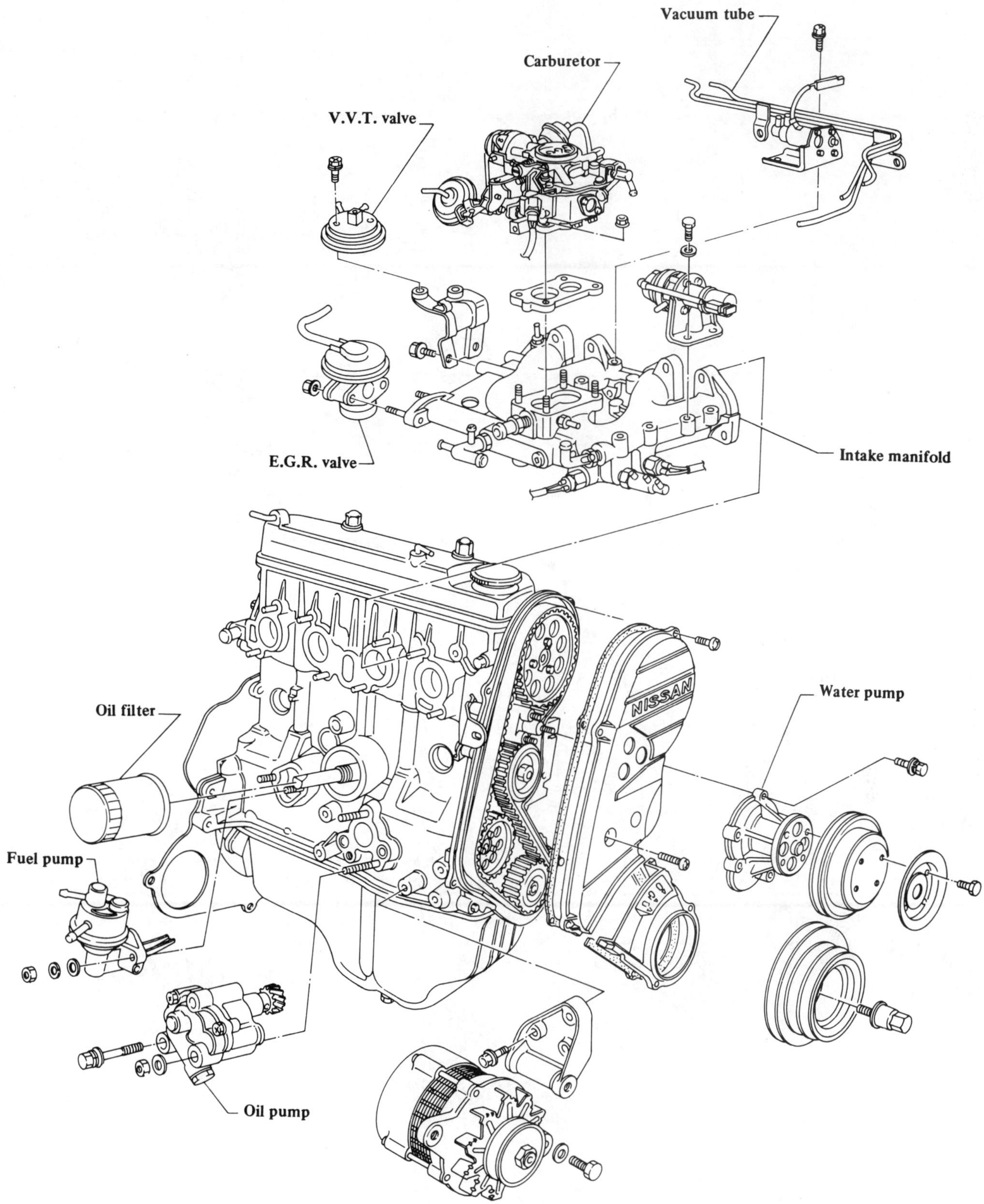

Fig. 1.10 Engine external components – North American non-Turbo models (Sec 14)

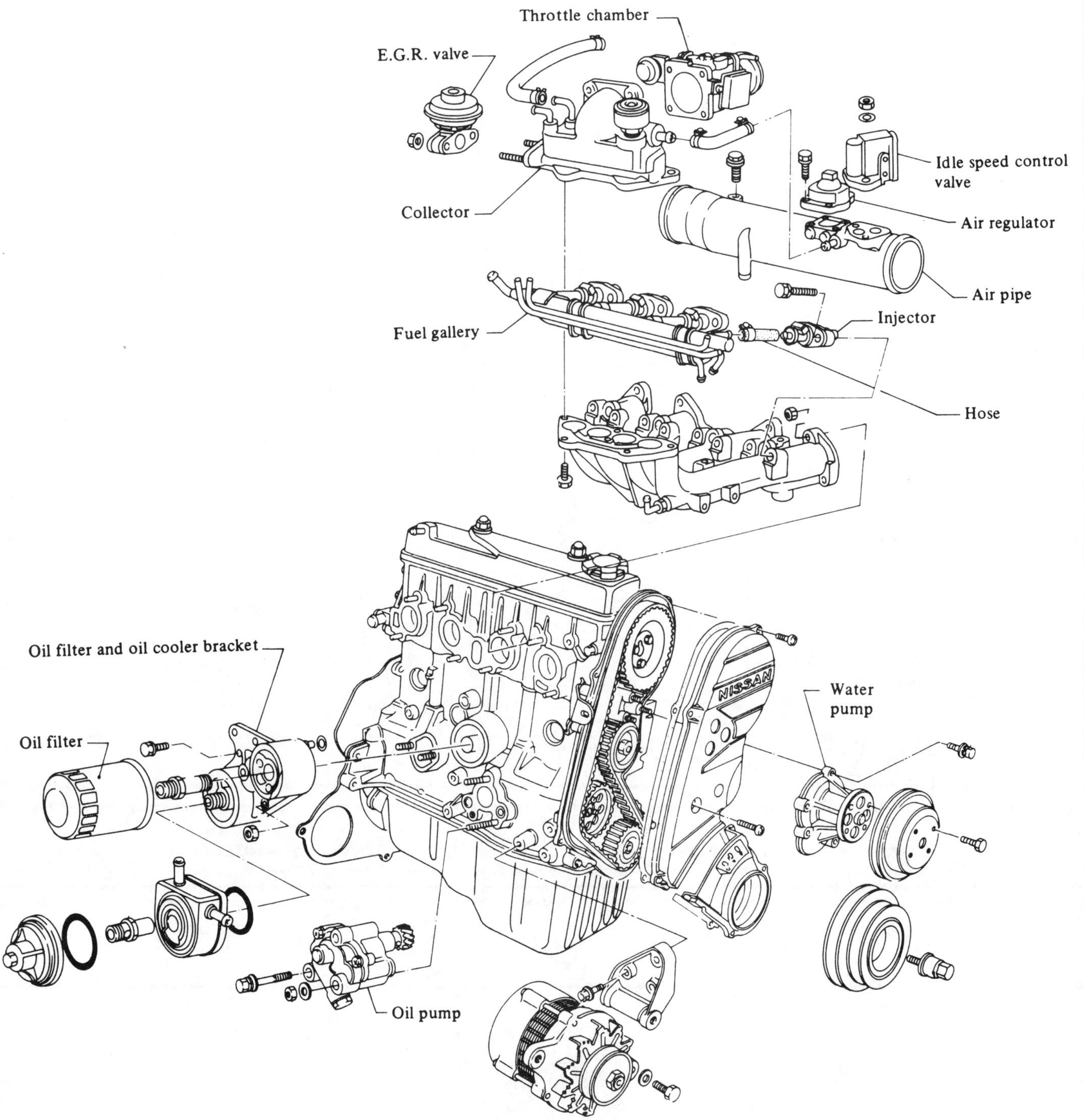

Fig. 1.11 Engine external components – Turbo models (Sec 14)

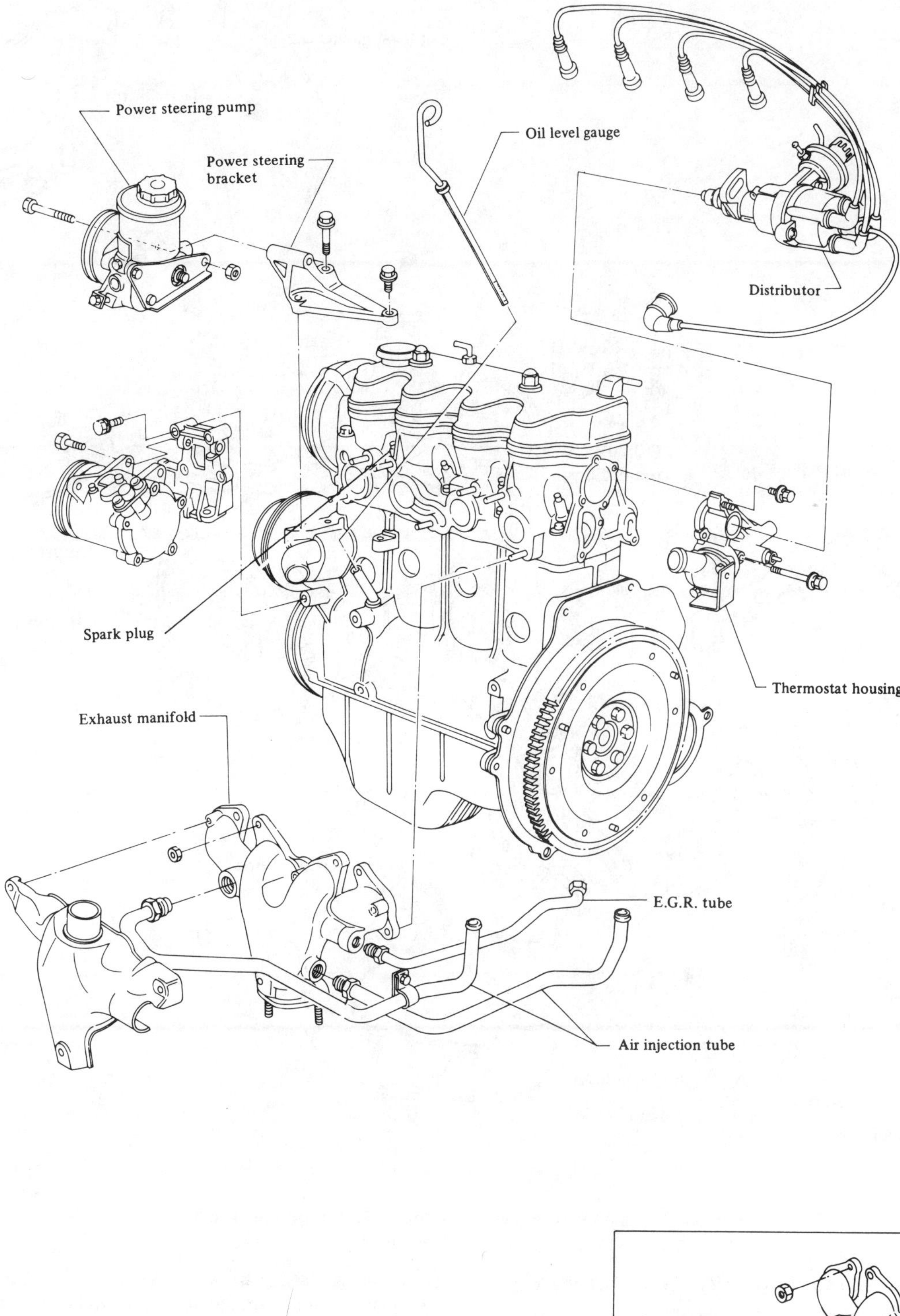

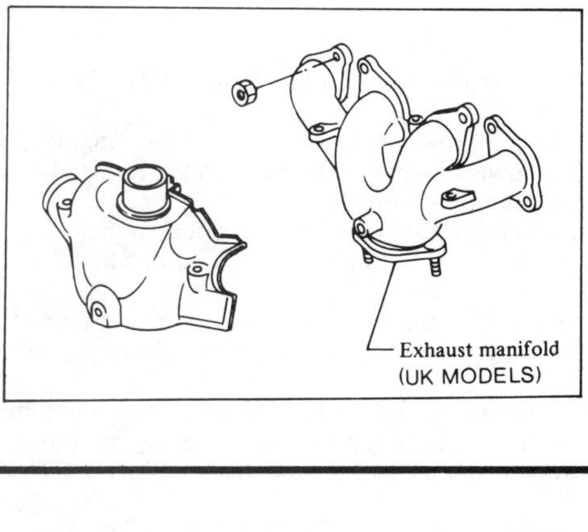

Fig. 1.12 Engine external components – non-Turbo models (Sec 14)

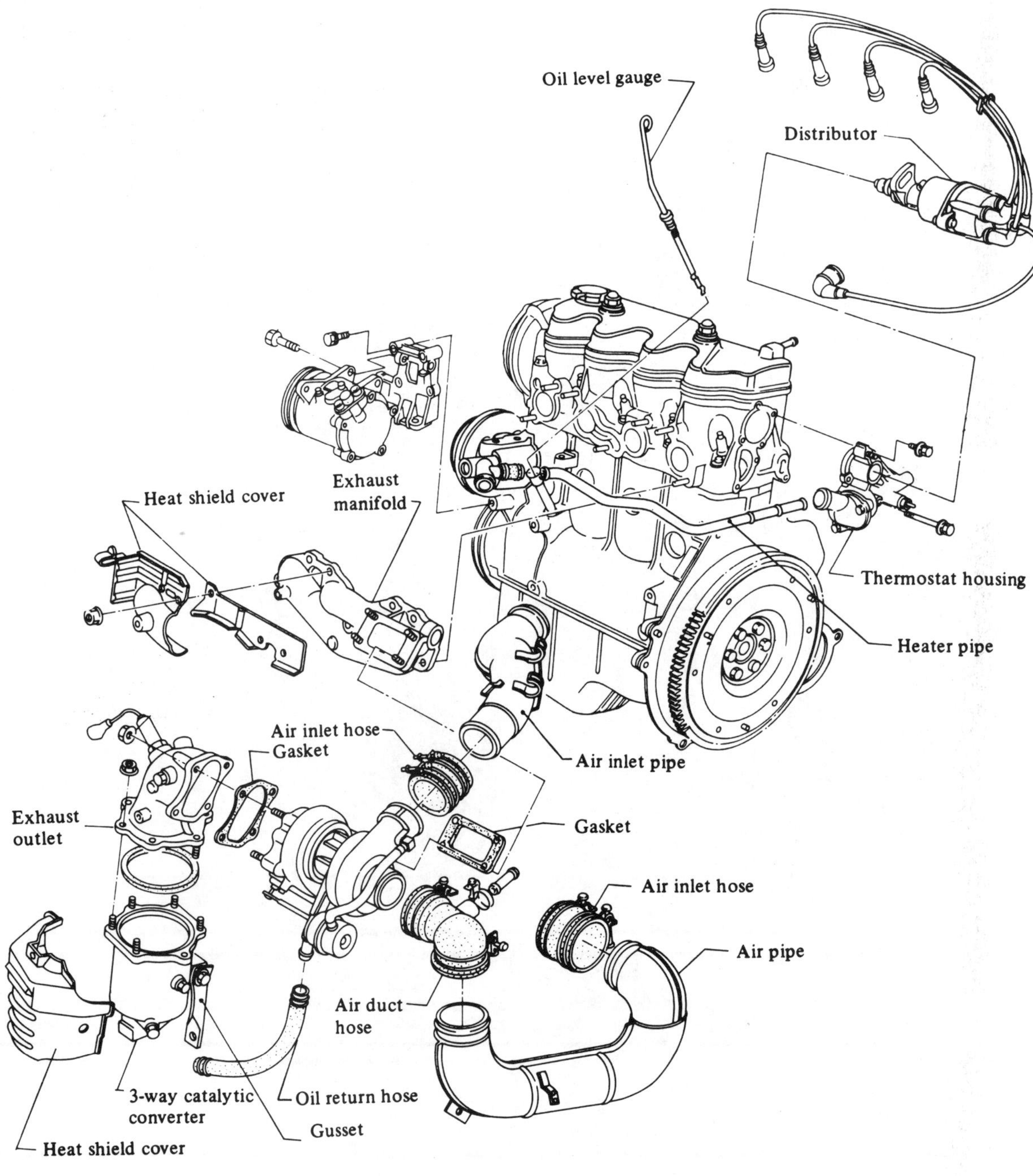

Fig. 1.13 Engine external components – Turbo models Sec 14)

flywheel. Take out the temporary wooden chock. Pull the engine endplate from the dowels.

11 Working at the timing belt end of the engine, remove the belt upper and lower covers. The cover screws are tight and may require the use of an impact driver.

12 Unbolt and remove the belt tensioner pulley.

13 Remove the timing belt, marking its rotational direction if it is to be used again.

14 Unbolt and remove the sprocket from the jack shaft. Unbolt the triangular mounting block used for attaching the alternator adjuster link.

15 Remove the crankshaft sprocket and the timing belt guide disc. If it is tight use a three-legged puller, taking care not to damage the teeth.

16 Remove the rocker cover.

17 Unscrew and remove the cylinder head bolts and remove the cylinder head and gasket, as described in Section 4. Unbolt the timing belt lower backplate. Take the oil slinger from the end of the crankshaft.

18 Extract the securing screw and remove the jack shaft retaining plate. Withdraw the jack shaft.

19 From within the crankcase unbolt and remove the oil pick-up pipe and strainer.

20 Remove the piston/connecting rod assemblies, as described in Section 9.

21 Unbolt and remove the crankshaft rear oil seal retainer.

22 Unscrew the main bearing cap bolts, working from the end ones towards the centre. The caps are numbered 1 to 5, number 1 being at

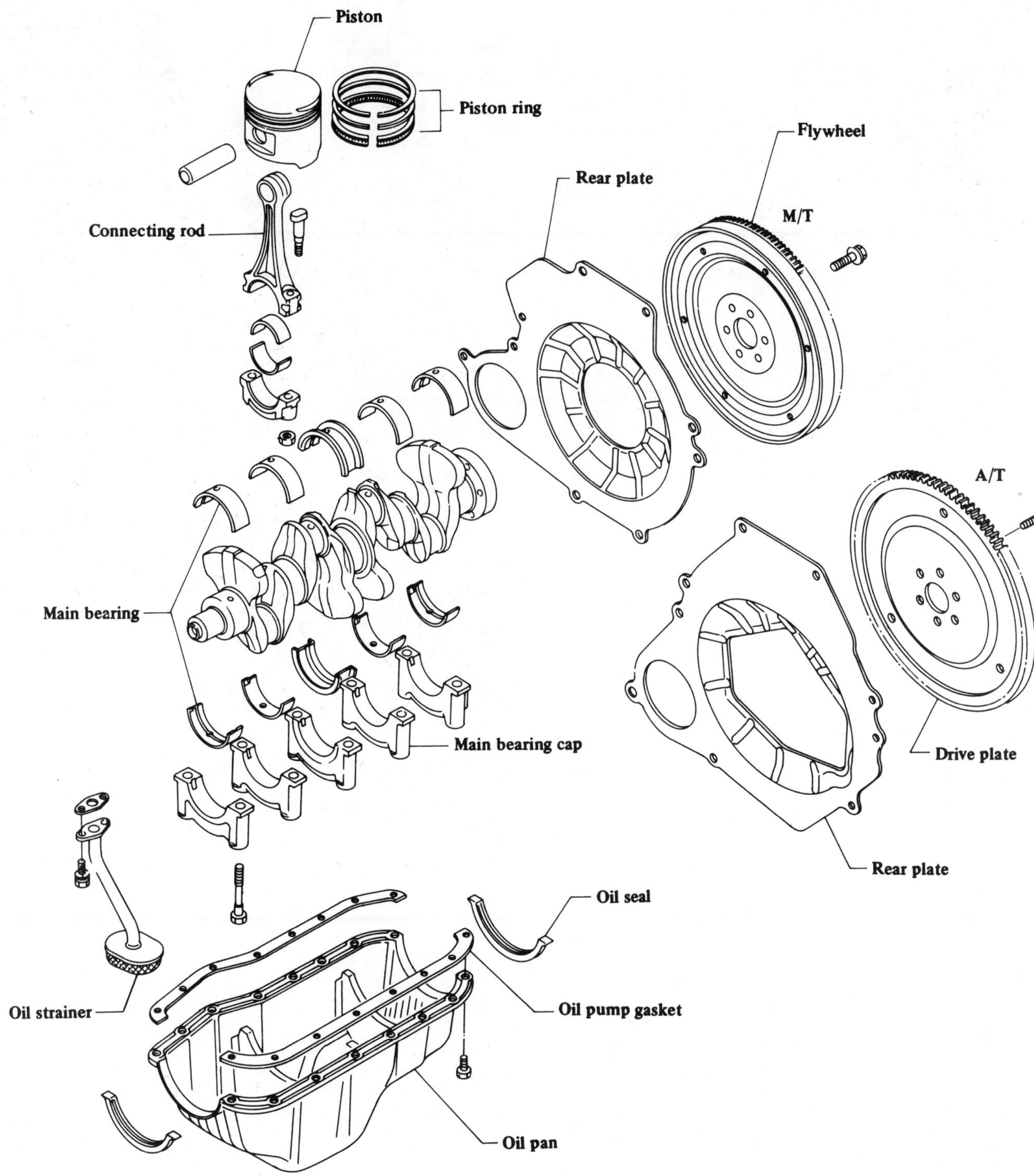

Fig. 1.14 Engine internal components (Sec 14)

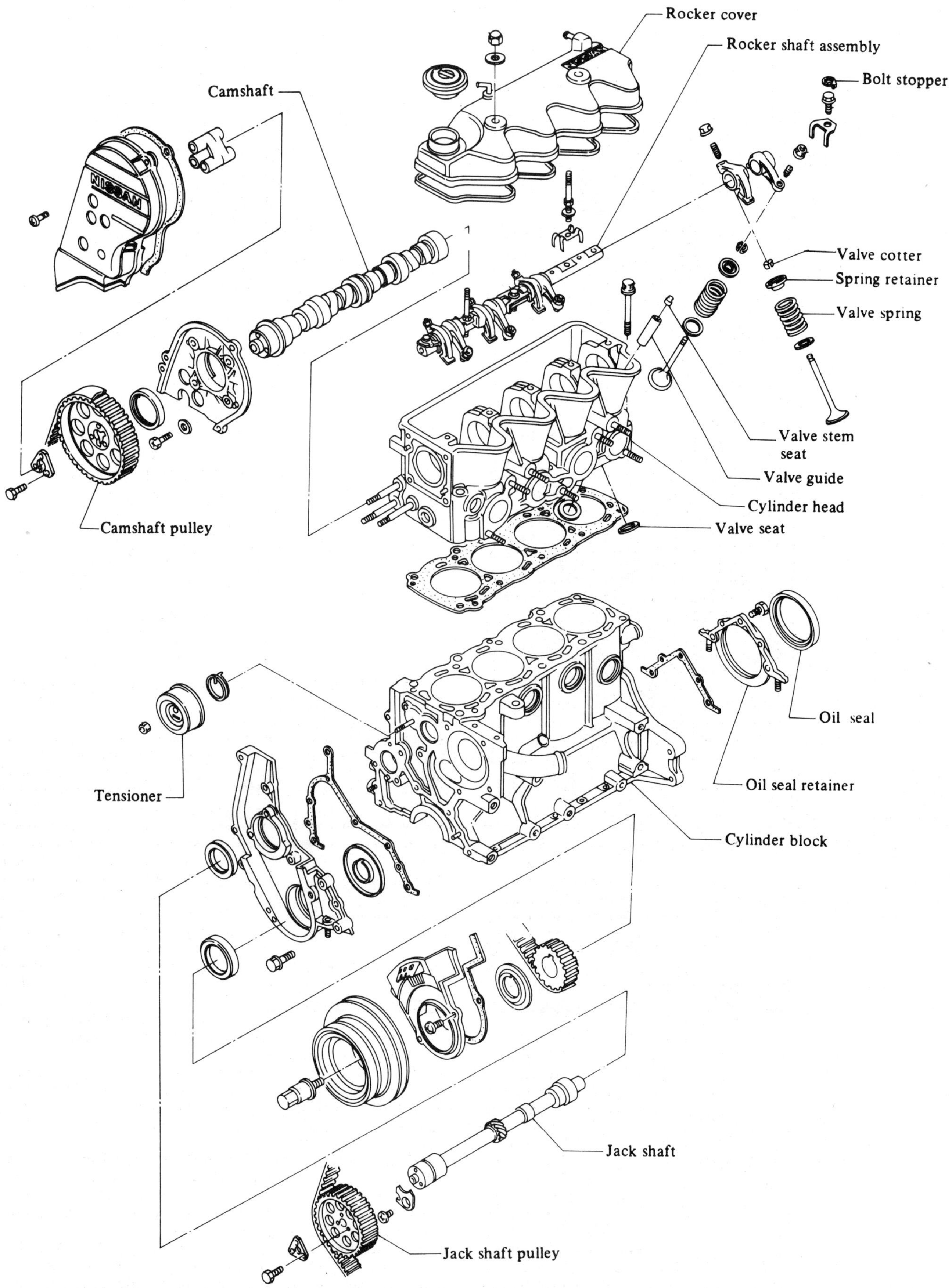

Fig. 1.15 Cylinder head and block components (Sec 14)

the timing belt end of the engine, but the numbers are legible from the flywheel end of the engine.
23 Lift the crankshaft from the crankcase. Refit the bearing shells with their respective caps. Note that the centre shell incorporates thrust flanges to control endfloat.
24 The engine is now completely dismantled and all parts should be cleaned and examined, as described in Section 18.

15 Cylinder head – dismantling and decarbonising

1 The manifolds and rocker cover will have been detached during removal of the cylinder head (see Section 4).
2 Unscrew the bolts and lift the rocker assembly from the cylinder head. Identify which way round the assembly is located. On late models note the lockplates and caps fitted to 1, 3 and 5 bolts. Discard the caps after prising them off 1 and 5 bolts.
3 Unbolt and remove the camshaft pulley.
4 Unbolt and remove the cylinder head end cover (timing belt backplate). This retains the camshaft. Unscrew and remove the spark plugs.
5 Carefully withdraw the camshaft, taking care not to damage the bearings as the lobes pass through.
6 The valves and their associated components should now be removed. Owing to the depth of the cylinder head a valve spring compressor having a long reach will be required. If this is not available, temporarily refit the rocker shaft and then make up a lever with a fork at one end to compress the valve spring by using the underside of the rocker shaft as a fulcrum.
7 Compress the first valve spring, extract the split cotters. If the valve spring refuses to compress, do not apply excessive force but remove the compressor and place a piece of tubing on the spring retainer and strike it a sharp blow to release the collets from the valve stem. Refit the compressor and resume operations.
8 Gently release the compressor, take off the spring retaining cap, the valve spring and the spring seat. Remove the valve. Keep the valve with its associated components together and in numbered sequence so that it can be refitted in its original position. A small box with divisions is useful for this purpose.
9 Remove the other valves in a similar way.
10 Bearing in mind that the cylinder head is of light alloy construction and is easily damaged use a blunt scraper or rotary wire brush to clean all traces of carbon deposits from the combustion spaces and the ports. The valve head stems and valve guides should also be freed from any carbon deposits. Wash the combustion spaces and ports down with a suitable solvent and scrape the cylinder head surface free of any foreign matter with the side of a steel rule, or a similar article.
11 If the engine is installed in the car, clean the pistons and the top of the cylinder bores. If the pistons are still in the block, then it is essential that great care is taken to ensure that no carbon gets into the cylinder bores as this could scratch the cylinder walls or cause damage to the piston and rings. To ensure this does not happen, first turn the crankshaft so that two of the pistons are at the top of their bores. Stuff rag into the other two bores or seal them off with paper and masking tape. The waterways should also be covered with small pieces of masking tape to prevent particles of carbon entering the cooling system and damaging the coolant pump.
12 Press a little grease into the gap between the cylinder walls and the two pistons which are to be worked on. With a blunt scraper carefully scrape away the carbon from the piston crown, taking great care not to scratch the aluminium. Also scrape away the carbon from the surrounding lip of the cylinder wall. When all carbon has been removed, scrape away the grease which will now be contaminated with carbon particles, taking care not to press any into the bores. To assist prevention of carbon build-up the piston crown can be polished with a metal polish. Remove the rags or masking tape from the other two cylinders and turn the crankshaft so that the two pistons which were at the bottom are now at the top. Place rag in the cylinders which have been decarbonised, and proceed as just described.
13 Examine the head of the valves for pitting and burning, especially the heads of the exhaust valves. The valve seatings should be examined at the same time. If the pitting on the valve and seat is very slight, the marks can be removed by grinding the seats and valves together with coarse, and then fine valve grinding paste.
14 Where bad pitting has occurred to the valve seats it will be necessary to recut them and fit new valves. This latter job should be entrusted to the local agent or engineering works. In practice it is very seldom that the seats are so badly worn. Normally it is the valve that is too badly worn for refitting, and the owner can easily purchase a new set of valves and match them to the seats by valve grinding.
15 Valve grinding is carried out as follows. Smear a trace of coarse carborundum paste on the seat face and apply a suction grinding tool to the valve head. With a semi-rotary motion, grind the valve head to its seat, lifting the valve occasionally to redistribute the grinding paste. When a dull matt even surface is produced on both the valve seat and the valve, wipe off the paste and repeat the process with fine carborundum paste, lifting and turning the valve to redistribute the paste as before. A light spring placed under the valve head will greatly ease this operation. When a smooth unbroken ring of light grey matt finish is produced on both valve and valve seat faces, the grinding operation is complete. Carefully clean away every trace of grinding compound, take great care to leave none in the ports or in the valve guides. Clean the valves and valve seats with a solvent-soaked rag, then with a clean rag, and finally, if an air line is available, blow the valves, valve guides and valve ports clean.
16 Check that all valve springs are intact. If any one is broken, all should be renewed. Check the free height of the springs against new ones. If some springs are not within specifications, replace them all. Springs suffer from fatigue and it is a good idea to renew them even if they look serviceable.
17 Check that the oil supply holes in the rocker arms are clear.
18 The cylinder head can be checked for warping either by placing it on a piece of plate glass or using a straight-edge and feeler blades. If there is any doubt or if its block face is corroded, have it re-faced by your dealer or motor engineering works.
19 Test the valves in their guides for side-to-side rock. If this is any more than almost imperceptible new guides must be fitted. This, as with valve seat renewal, is really a job for your dealer as the cylinder head must be warmed and the old guide driven out. New guides should be pressed in to protrude 15.0 mm (0.59 in) above the cylinder head and then reamed using a 7.005 to 7.020 mm (0.2758 to 0.2764 in) reamer.

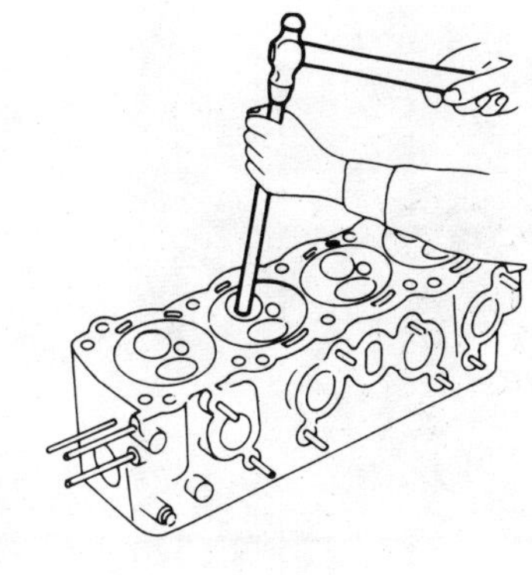

Fig. 1.16 Removing a valve guide (Sec 15)

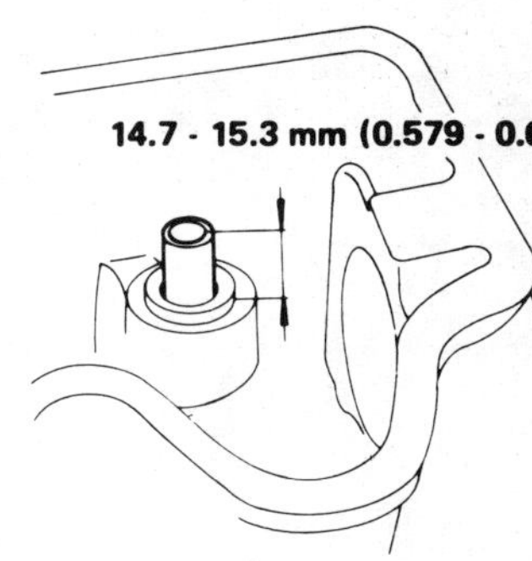

Fig. 1.17 Valve guide projection from cylinder head (Sec 15)

15.20 Valve oil seal on the guide

15.21 Fitting a valve into its guide

15.22A Valve spring seat

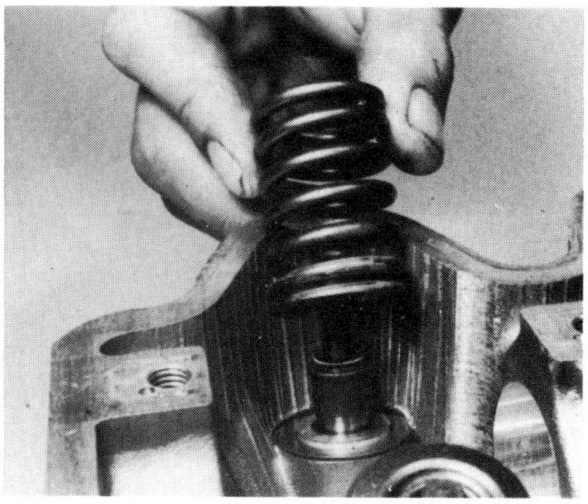

15.22B Valve spring

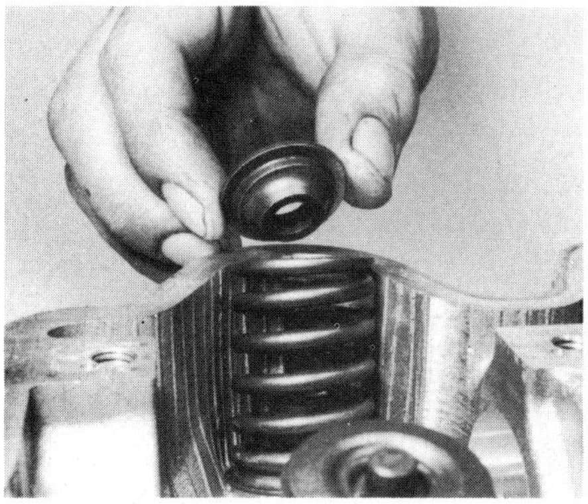

15.22C Valve spring cap

15.23 Compressing valve spring and locating collets

20 Renew the valve stem oil seals (photo).

21 Commence reassembly by oiling the stem of the first valve and pushing it into its guide (photo).

22 Fit the spring seat (photo), the valve spring (photo) and the spring cap (photo). The closer coils of the valve spring must be towards the cylinder head.

23 Compress the valve spring and locate the split collets in the valve stem cut-outs (photo).

24 Gently release the compressor, checking to see that the collets are not displaced.

25 Fit the remaining valves in the same way.

26 Tap the end of each valve stem with a plastic or copper-faced hammer to settle the components.

27 Lubricate the camshaft bearings and insert the camshaft into the cylinder head (photo).

28 Fit the cylinder head end cover to the timing belt end of the cylinder head, complete with new oil seal and new gasket (photos).

29 Bolt on the camshaft sprocket (photo). The bolt holes are offset so it will only go on one way.

30 Before refitting the rocker gear, check the shaft for wear and the rocker arms for general condition (photo). Renew any worn components, but make sure when reassembling that they are kept in their original order.

31 The hole in the rocker shaft must face downwards and the cut-out in the centre retainer must be towards the exhaust manifold side.

32 With the rocker arm adjuster screws fully released, bolt the rocker gear to the cylinder head.

33 The rocker shaft fixing bolts on later models incorporate lockplates

15.27 Inserting camshaft into cylinder head

15.28A Cylinder head end cover gasket

15.28B Cylinder head end cover

15.28C Tightening cylinder head end cover bolts

15.29 Camshaft sprocket

15.30 Rocker arms and shaft

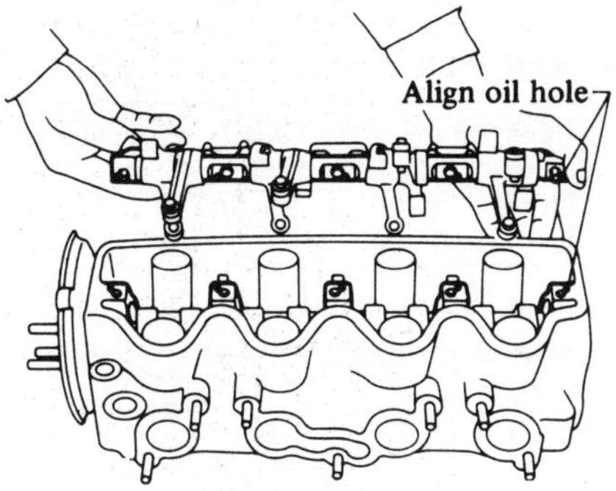

Fig. 1.18 Rocker shaft oil hole alignment (Sec 15)

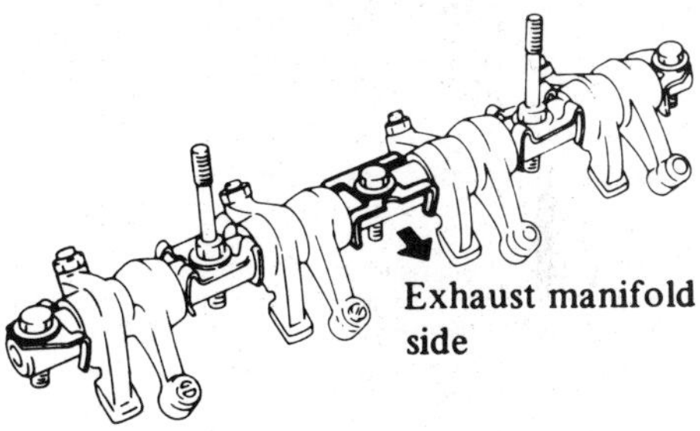

Fig. 1.19 Camshaft centre retainer cut-out location (Sec 15)

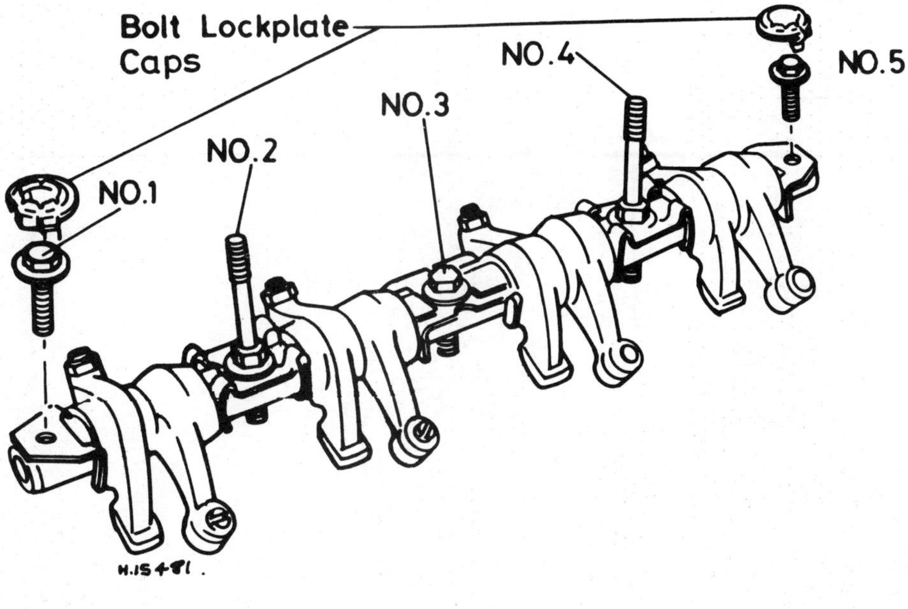

Fig. 1.20 Rocker shaft bolt lockplates and caps (Sec 15)

15.33 Rocker shaft bolt lockplate (arrowed)

to eliminate the possibility of their working loose (photo). Always fit new lock caps to the end bolts (1 and 5) by tapping on using a suitable socket.

16 Lubrication system – general

1 The engine lubrication system depends upon oil contained in the sump pan being drawn into an externally mounted oil pump, which is driven from a gear on the jack shaft, and then pressurised to supply all the engine working parts.
2 Oil from the pump passes through an externally mounted full-flow disposable type oil filter.
3 An oil pressure regulator valve is incorporated in the oil pump and an oil pressure switch is located close to the filter which actuates a warning lamp if pressure loss occurs.
4 Intermittent flickering of the oil pressure warning lamp may occur when the engine is idling after a long high speed run. This should be ignored as long as the lamp goes out immediately the engine speed is increased.
5 A pressure relief valve is located in the oil filter mounting base (photo). Should it be seen to be cracked or broken when the oil filter is removed, the valve may be prised out with a screwdriver and a new one tapped into position. The purpose of this valve is to open and bypass the filter should the filter become clogged.
6 On Turbo models an oil cooler is incorporated and located beneath the inlet manifold. The Turbocharger is lubricated from the engine lubrication system.

17 Crankcase ventilation system (PCV)

1 This is of positive, dual-line type which returns blow-by gas (which has passed the piston rings) from the crankcase to either the air cleaner or the intake manifold, according to manifold vacuum. According to engine load conditions a valve regulates the routing of the gas (photo).
2 Check the system connecting hoses regularly and clean them out.
3 To test the operation of the valve, have the engine idling and disconnect the hose from the side of the valve which is furthest from the intake manifold. Vacuum hiss should be heard coming from the open end of the valve and a strong suction felt if a finger is placed over it. If this is not so, renew the valve.

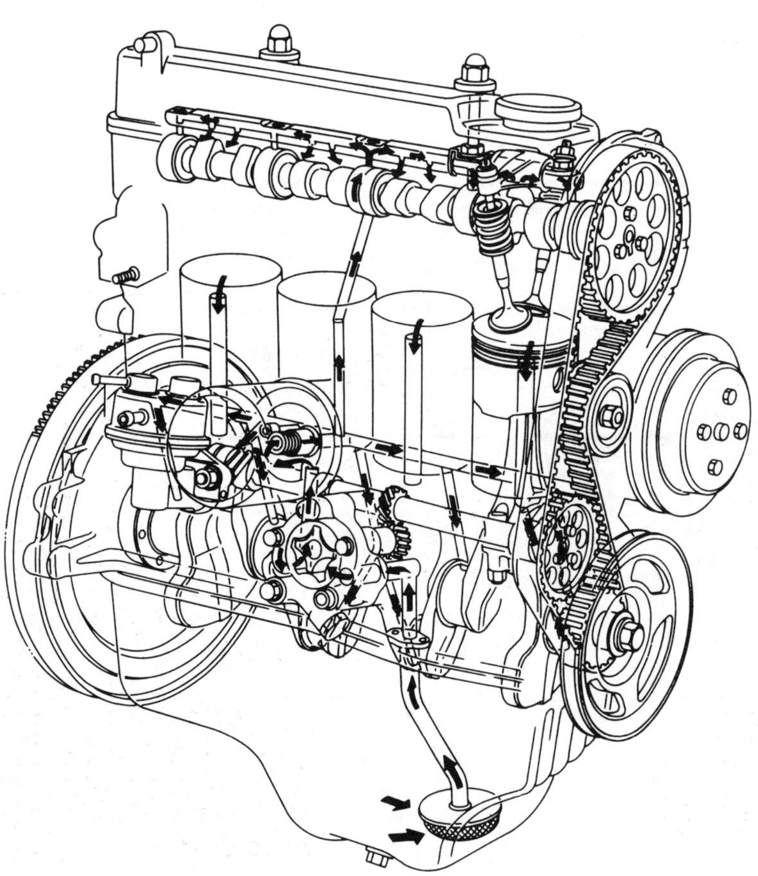

Fig. 1.21 Engine lubrication circuit – non-Turbo models (Sec 16)

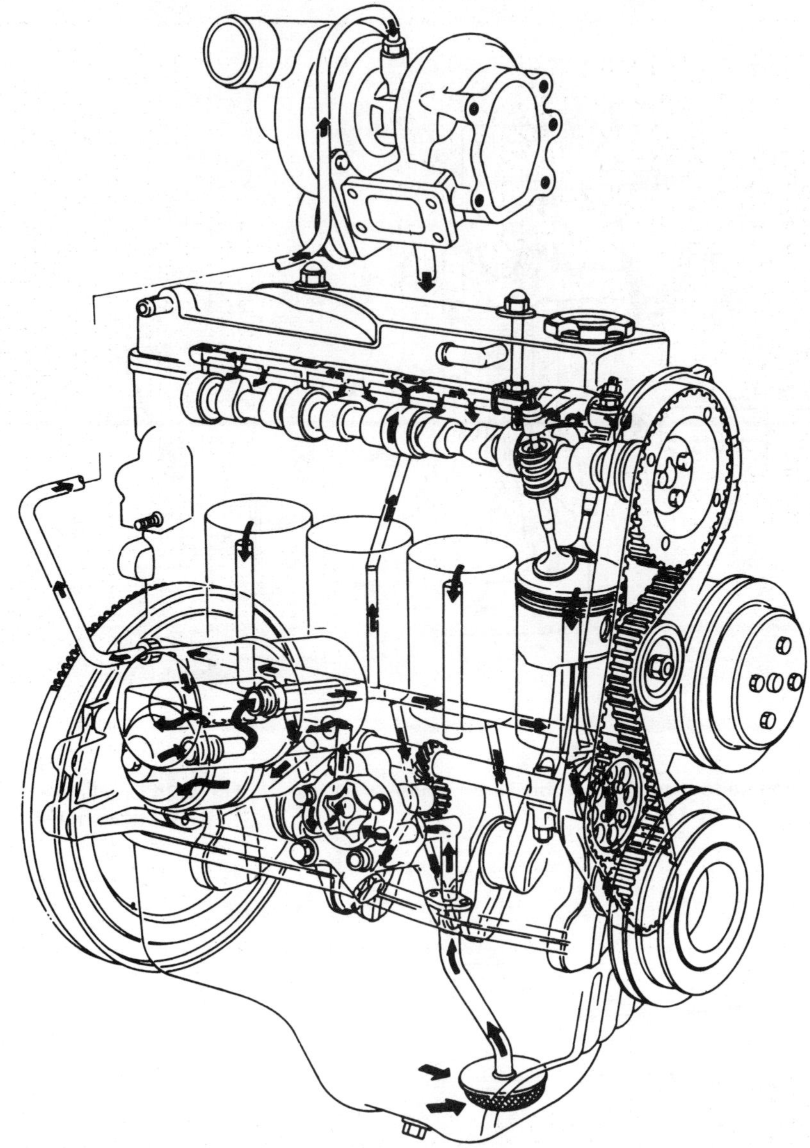

Fig. 1.22 Engine lubrication circuit – Turbo models (Sec 16)

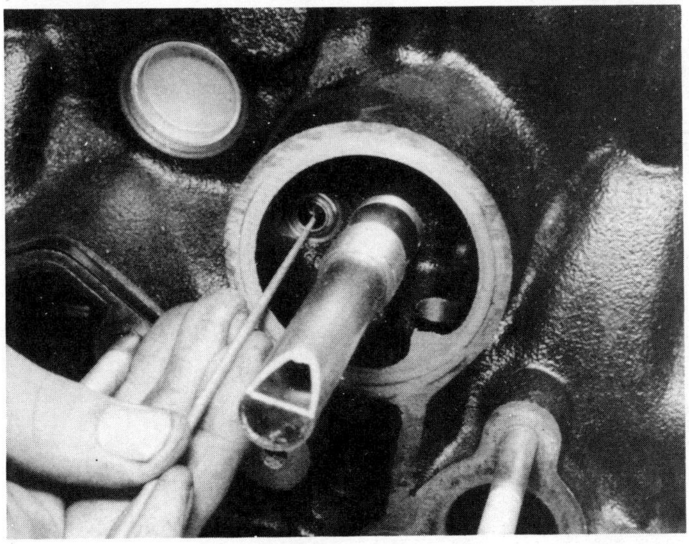

16.5 Pressure relief valve location

17.1 Crankcase vent hose and PCV valve

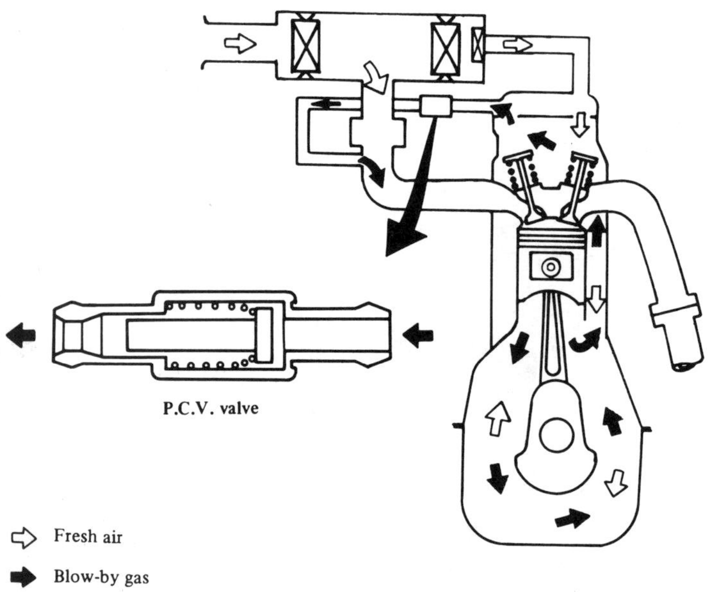

Fig. 1.23 Crankcase ventilation system – UK non-Turbo models (Sec 17)

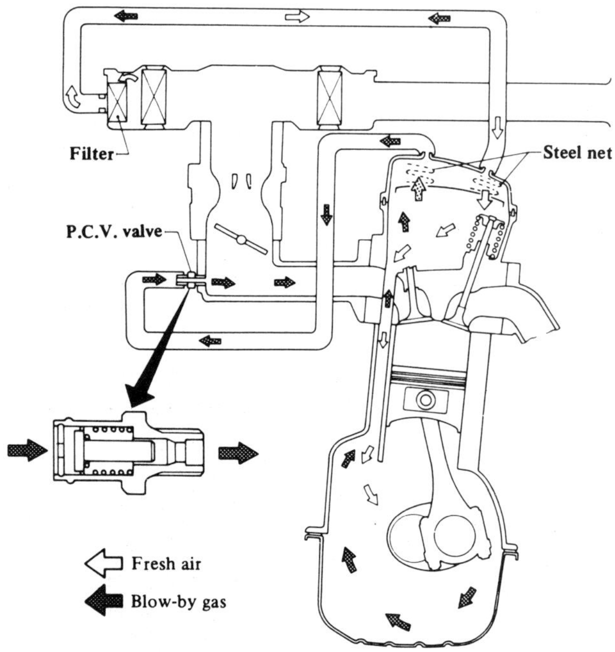

Fig. 1.24 Crankcase ventilation system – North American non-Turbo models (Sec 17)

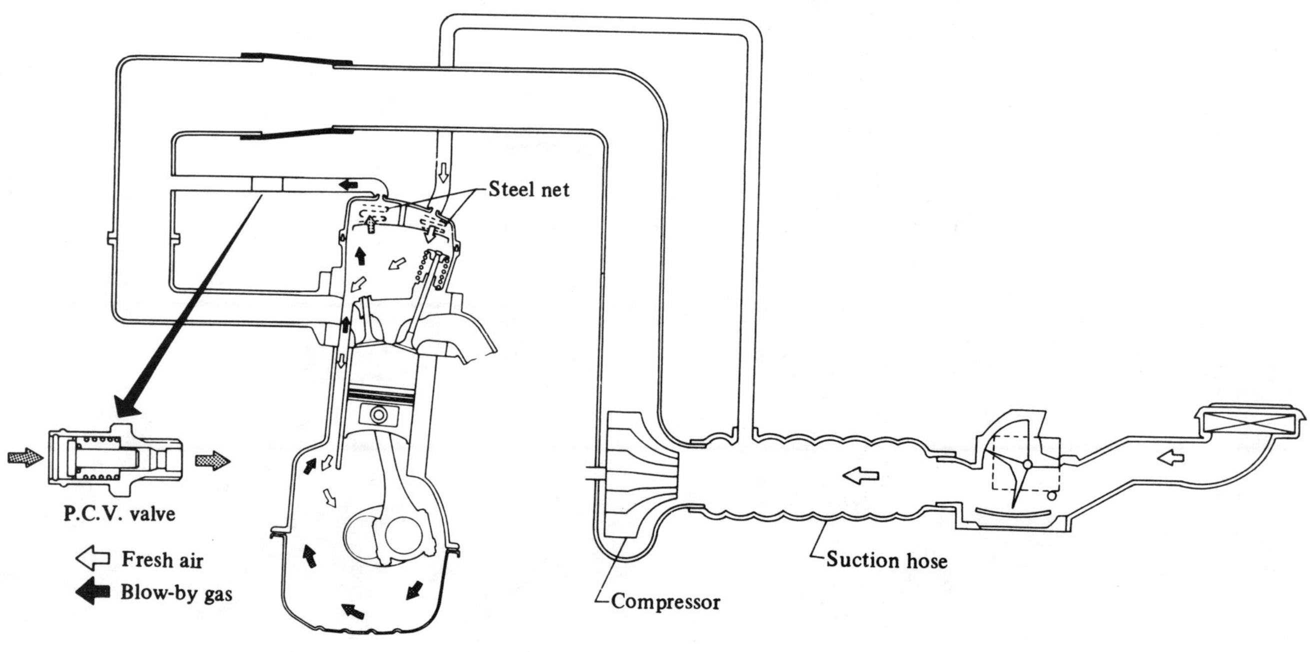

Fig. 1.25 Crankcase ventilation system – Turbo models (Sec 17)

18 Examination and renovation

Cylinder block and crankcase

1 Examine the castings carefully for cracks, especially around the bolt holes and between cylinders.
2 The cylinder bores must be checked for taper, ovality, scoring and scratching. Start by examining the top of the cylinder bores. If they are at all worn, a ridge will be felt on the thrust side. This ridge marks the top of piston travel. The owner will have a good indication of bore wear prior to dismantling by the quantity of oil consumed and the emission of blue smoke from the exhaust, especially when the engine is cold.
3 An internal micrometer or dial gauge can be used to check bore wear and taper against Specifications, but this is a pointless operation if the engine is obviously in need of reboring – indicated by excessive oil consumption.
4 Your engine reconditioner will be able to rebore the block for you and supply the correct oversize pistons to give the correct running clearance.
5 If the engine has reached the limit for reboring, cylinder liners can be fitted, but here again this is a job for your engine reconditioner.
6 To rectify minor bore wear it is possible to fit proprietary oil control rings, as described in Section 9. A good way to test the condition of the engine is to have it at normal operating temperature with the spark plugs removed. Screw a compression tester (available from most motor accessory stores) into the first plug hole. Hold the accelerator fully depressed and crank the engine on the starter motor for several revolutions. Record the reading. Zero the tester and check the remaining cylinders in the same way. All four compression figures should be approximately equal and within the tolerance given in Specifications. If they are all low, suspect piston ring or cylinder bore wear. If only one reading is down, suspect a valve not seating.

Crankshaft and bearings

7 Examine the surfaces of the crankpins and journals for signs of scoring or scratching, and check for ovality or taper. If a crankpin or journals are not within the dimensional tolerances given in the Specifications Section at the beginning of this Chapter the crankshaft will have to be reground.
8 Wear in a crankshaft can be detected while the engine is running. Big-end bearing and crankpin wear is indicated by distinct metallic, knocking, particularly noticeable when the engine is pulling from low engine speeds. Low oil pressure will also occur.
9 Main bearing and journal wear is indicated by engine rumble increasing in severity as the engine speed increases. Low oil pressure will again be an associated condition.
10 Crankshaft grinding should be carried out by specialist engine reconditioners who will supply the matching undersize bearing shells to give the required running clearance.
11 Inspect the connecting rod big-end and main bearing shells for signs of general wear, scoring, pitting and scratching. The bearings should be matt grey in colour. If a copper colour is evident, then the bearings are badly worn and the surface material has worn away to expose the underlay. Renew the bearings as a complete set.
12 At the time of major overhaul it is worthwhile renewing the bearing shells as a matter of routine even if they appear to be in reasonably good condition.
13 Bearing shells can be identified by the marking on the back of the shell. Standard sized shells are usually marked STD or 0.00. Undersized shells are marked with the undersize, such as 0.25 mm.

Connecting rods

14 Check the alignment of the connecting rods visually. If you suspect distortion, have them checked by your dealer or engine reconditioner on the special jig which he will have.
15 The gudgeon pin is an interference fit in the connecting rod small-end and removal or refitting and changing a piston is a job best left to your dealer or engine reconditioner due to the need for a press and jig.

Pistons and piston rings

16 If the engine is rebored then new oversize pistons with rings and gudgeon pins will be supplied. Have the supplier fit the pistons to the rods so that the oil hole in the connecting rod is located as shown with reference to the front facing mark on the piston crown (see Fig. 1.26).
17 Removal and refitting of piston rings is covered in Section 9.

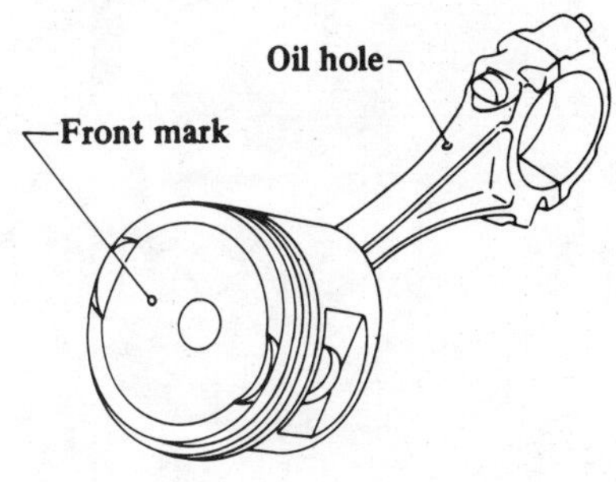

Fig. 1.26 Connecting rod/piston alignment (Sec 18)

Flywheel

18 Check the clutch mating surface of the flywheel. If it is deeply scored (due to failure to renew a worn driven plate) then it should be renewed. Slight roughness may be smoothed with fine emery cloth.
19 If lots of tiny cracks are visible on the surface of the flywheel this will be due to overheating caused by slipping the clutch or 'riding' the clutch pedal.
20 With a pre-engaged type of starter motor it is rare to find the teeth of the flywheel ring gear damaged or worn, but if they are the ring gear will have to be renewed.
21 To remove the ring gear, drill a hole between the roots of two teeth, taking care not to damage the flywheel, and then split the ring with a sharp cold chisel.
22 The new ring gear must be heated to between 180 and 220°C (356 and 428°F) which is very hot, so if you do not have facilities for obtaining these temperatures, leave the job to your dealer or engine reconditioner.

Driveplate (automatic transmission)

23 Should the starter ring gear on the driveplate require renewal, the driveplate should be renewed complete.

Camshaft

24 Examine the camshaft bearings for wear, scoring or pitting. If evident then the complete cylinder head will have to be renewed as the bearings are machined directly in it.
25 The camshaft itself should show no marks or scoring on the journal or cam lobe surfaces. Where marks are evident, renew the camshaft or have it reprofiled by a specialist reconditioner.
26 Check the teeth of the camshaft sprocket for wear. Renew the sprocket if necessary.

Timing belt and tensioner

27 Examine the belt for cracking or fraying and tooth wear. If any of these conditions is evident or if the belt has been in service for 80 000 km (50 000 miles) it is recommended that it is renewed.
28 The tensioner should not be noisy or shaky when turned, and have good spring action. Where these conditions are not met with, renew the tensioner complete (photo).

Jack shaft

29 The jack shaft journals should be smooth – without scoring or scratches. Wear can be checked using a micrometer.
30 The bearings are renewable, but make sure that the lubrication holes line up with those in the crankcase as the new bearings are pressed in.
31 When refitting the jack shaft end plug (removed to extract the bearings) coat the edges with a suitable sealant.

Oil pump

32 To dismantle the oil pump, remove the cover screws and take off the cover.
33 Lift out the outer rotor (photo). The inner rotor with drivegear cannot be separated from the pump casing.

18.28 Timing belt tensioner and spring

18.33 Oil pump rotors (note directional mark for correct refitting)

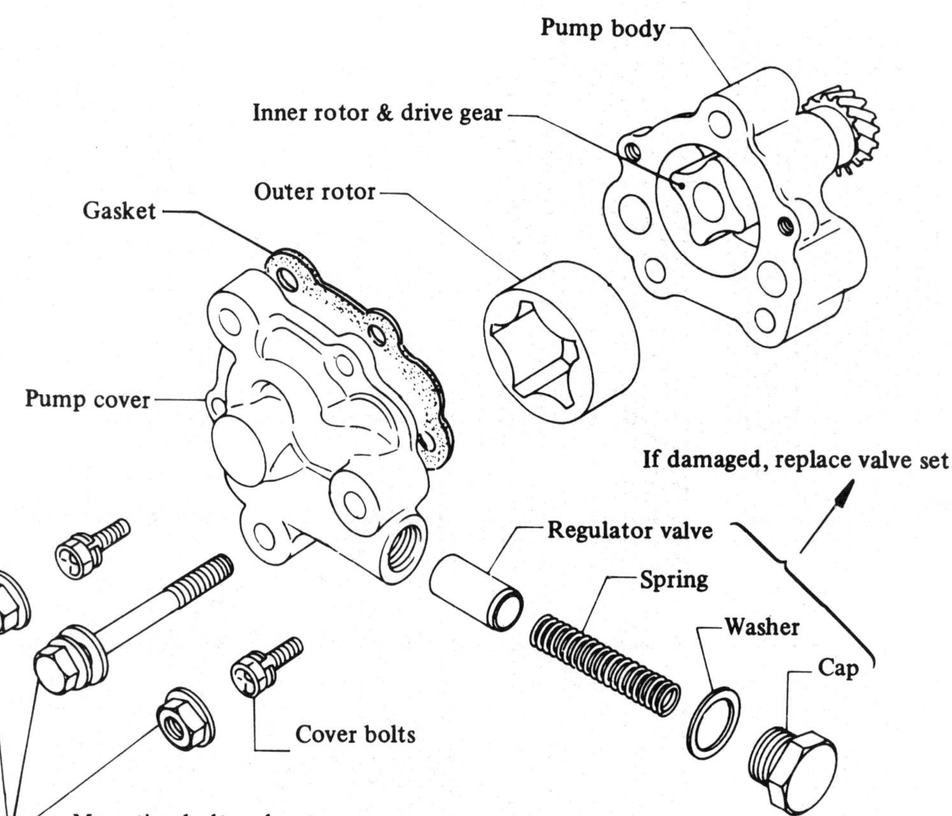

Fig. 1.27 Exploded view of oil pump (Sec 18)

18.34A Checking rotor tip clearance

18.34B Checking outer rotor clearance

34 Clean and dry the pump components and refit the outer rotor. Using a feeler blade check the following clearances (photos):

Rotor tip clearance	0.12 mm (0.0047 in) maximum
Outer rotor to pump body clearance	0.15 to 0.21 mm (0.0059 to 0.0083 in)

35 Using a straight-edge across the top of the outer rotor check the following clearances, again using a feeeler blade:

Oil pump body to straight edge	0.02 mm (0.0008 in) maximum
Inner rotor endfloat	0.05 mm (0.002 in) maximum

36 If, after these checks, the pump proves to be worn, renew it.
37 The pressure regulator components are seldom found to be faulty, but if they are to be inspected, unscrew the cap and extract the spring and valve plunger (photo). When refitting, remember to fit the sealing washer and tighten the cap to the specified torque.

Oil seals and gaskets

38 It is recommended that all gaskets and oil seals are renewed at major engine overhaul (photo). Sockets are useful for removing or

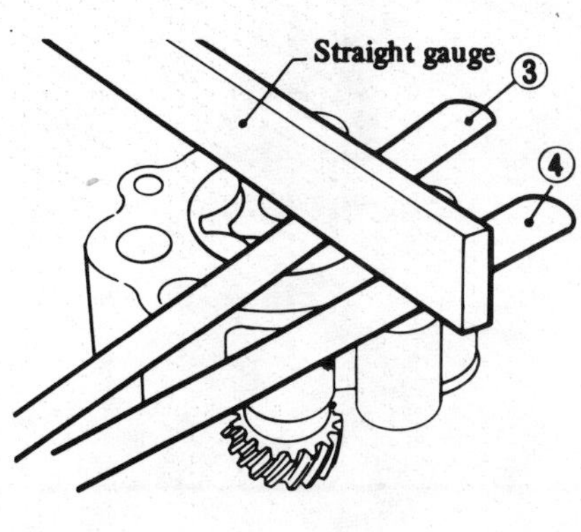

Fig. 1.28 Checking oil pump clearances (Sec 18)

3 Inner rotor endfloat *4 Oil pump body to straight-edge*

18.37 Oil pump pressure regulating valve

18.38A Prising out an oil seal

18.38B Using a socket to fit an oil seal

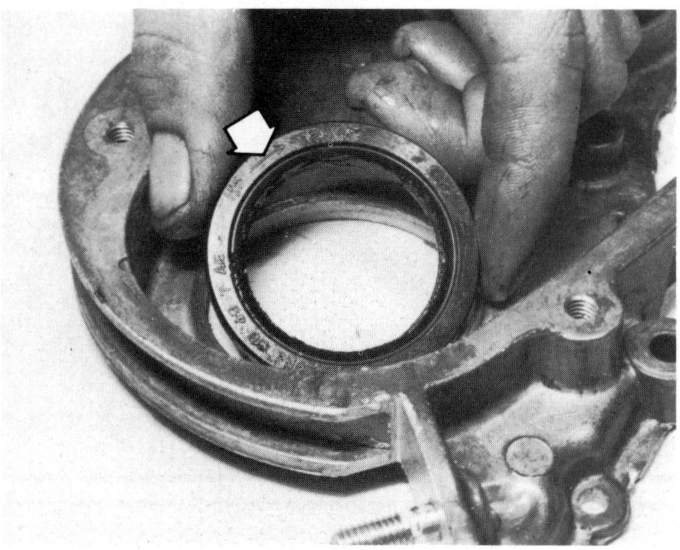

18.38C Showing running direction arrow on oil seal

refitting oil seals (photo). An arrow is moulded onto the seals to indicate the rotational direction of the component which it serves (photo). Make sure that the seal is fitted the correct way round to comply with the arrow.

Cylinder head

39 This is covered in Section 15, during dismantling and decarbonising.

19 Engine reassembly – general

1 To ensure maximum life with minimum trouble from a rebuilt engine, not only must everything be correctly assembled, but everything must be spotlessly clean, all the oilways must be clear, locking washers and spring washers must always be fitted where indicated and all bearing and other working surfaces must be thoroughly lubricated during assembly.

2 Before assembly begins renew any bolts or studs, the threads of which are in any way damaged, and whenever possible use new spring washers.

3 Apart from your normal tools, a supply of clean rag, an oil can filled with engine oil (an empty plastic detergent bottle thoroughly cleaned and washed out, will do just as well), a new supply of assorted spring washers, a set of new gaskets, and a torque wrench, should be collected together.

20 Engine – reassembly

1 Stand the cylinder block in an inverted position on the bench.

2 Clean the shell bearing recesses in the crankcase free from all grit and dirt and fit the bearing shells. The centre shell is of thrust flange type (photo). Nos 1 and 5 are of grooved type (photo) and Nos 2 and 4 are plain.

3 Oil the shells and lower the crankshaft into position (photos).

4 Clean the recesses in the main bearing caps free from grit and dirt and fit the bearing shells to match those already fitted to the crankcase. Oil the shells.

5 Fit the main bearing caps in their numbered sequence so that the numbers can be read from the flywheel end. Cap No 1 is at the timing belt end. The caps will only seat one way round (photo).

6 Insert and tighten the cap bolts to the specified torque – working from the centre ones towards each end (photo).

7 Now check the crankshaft endfloat using either a dial gauge or feeler blades inserted between the flange of the centre bearing and the machined shoulder of the crankshaft (photo). Make sure that the

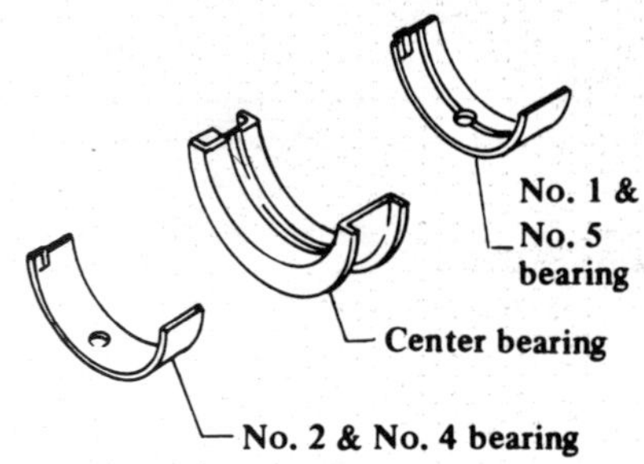

Fig. 1.29 Main bearing shell identification (Sec 20)

20.2A Main bearing centre shell bearing with thrust flanges

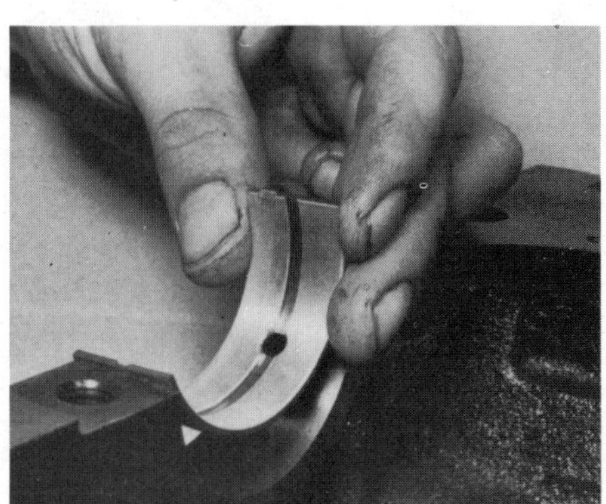

20.2B Main bearing end shell bearing

20.3A Oiling main bearing shells

20.3B Lowering crankshaft into position

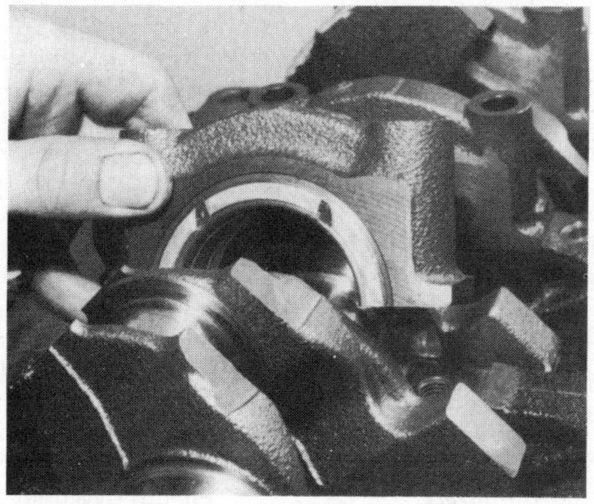

20.5 Fitting centre main bearing cap

20.6 Tightening main bearing cap bolts

crankshaft is pushed fully in one direction and then the other when measuring. If the endfloat is excessive (see Specifications) and new bearing shells have been fitted then it can only be due to an error at regrinding.

8 Bolt on the crankshaft oil seal retainer complete with a new oil seal and gasket (photos). Apply grease to the oil seal lips before pushing it over the flywheel mounting flange and check that the lip of the seal is not doubled under. Cut the ends of the gasket flush (photo).

9 Oil the cylinder bores and fit the piston/connecting rods (complete with bearing shells) into their respective cylinders, as described in

20.7 Checking crankshaft endfloat

20.8B Crankshaft oil seal and retainer

20.8A Crankshaft oil seal retainer gasket

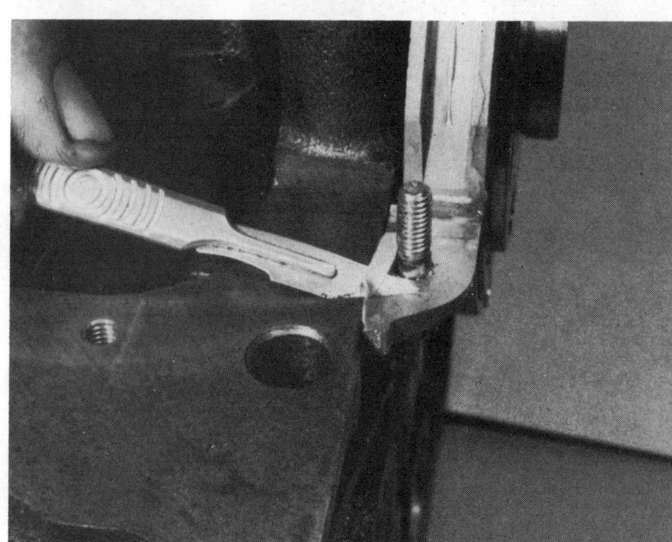

20.8C Cutting oil seal retainer gasket flush

Section 9. Make sure that the front mark on the piston crown is towards the timing belt of the engine.
10 Oil the crankpins.
11 Wipe out the recesses in the big-end caps and fit the bearing shells. Oil the shells.
12 Draw the connecting rods down onto the crankpins, fit the caps (numbers adjacent) and screw in and tighten the bolts to the specified torque (photo).
13 Oil the jack shaft journals, insert the jack shaft into its bearings and fit the retaining plate and screw, having applied locking fluid to the threads (photos).
14 Fit the cylinder head, as described in Section 4.
15 Fit the oil slinger to the front end of the crankshaft (photo).
16 Fit the timing belt lower backplate using a new gasket and oil seals (photos). Note that the two longer bolts go into the lower holes.
17 Refit the timing belt guide disc to mate with the concave side of the crankshaft sprocket which should now be fitted (photos).
18 Bolt the spocket to the jack shaft (photo). This sprocket does not go in any special position.
19 Fit the triangular mounting block for the alternator adjuster link (photo).
20 The camshaft sprocket will normally have been fitted during reassembly of the cylinder head. If not, fit it now.
21 Fit and tension the timing belt, as described in Section 4. Bolt on the oil pick-up pipe using a new gasket (photo).
22 Stick new gaskets and seals to the sump pan mating flange of the crankcase. Apply a bead of gasket sealant to the joints of the gasket strips and sealing strips before locating the sump pan (photos). Screw in the bolts progressively – do not overtighten (photo).
23 Fit the timing belt upper and lower covers using new sealing strips (photos). Fit the reinforcement plate (photo).
24 Fit the engine endplate to its dowels at the flywheel end of the engine (photo).
25 Fit the flywheel, aligning the marks made before removal (photo).

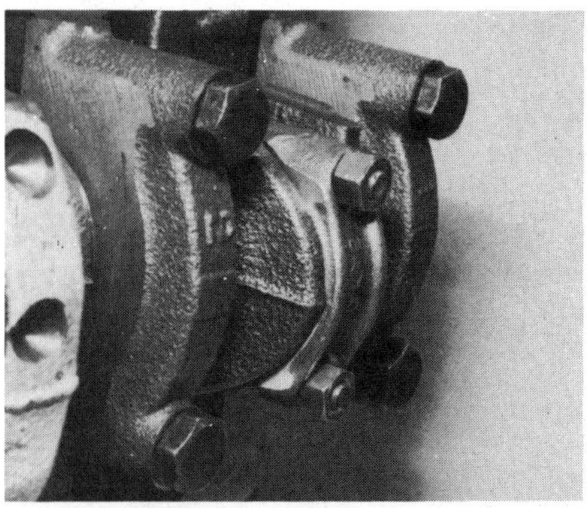
20.12 Main and big-end caps

20.13A Inserting jack shaft into crankcase

20.13B Jack shaft retaining plate

20.13C Jack shaft retaining plate screw

20.15 Crankshaft oil slinger

20.16A Timing belt lower backplate gasket

20.16B Timing belt lower backplate

20.17A Timing belt guide disc

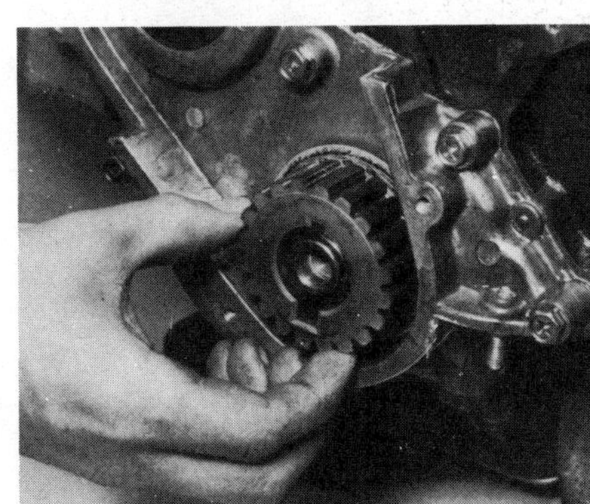
20.17B Crankshaft sprocket

20.18 Jack shaft sprocket

20.19 Alternator adjuster link mounting block

20.21 Oil pick-up pipe and gasket

20.22A Sump pan gasket and sealing strip overlap

20.22B Fitting the sump pan

20.22C Tightening sump pan bolt

20.23A Timing belt lower cover

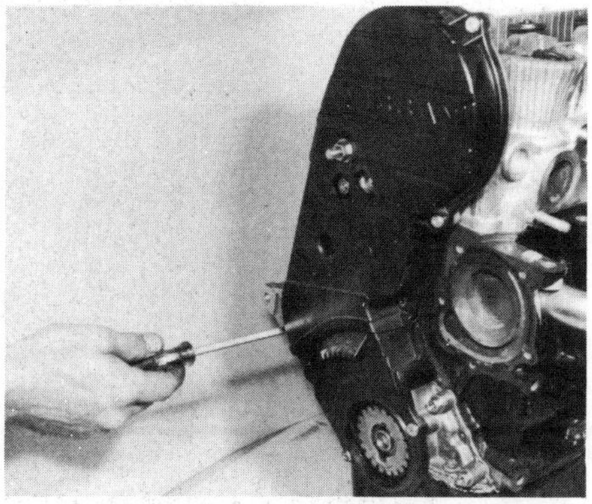

20.23B Timing belt upper cover

20.23C Cylinder head/crankcase reinforcement plate

20.24 Engine endplate

20.25 Fitting flywheel

20.26 Tightening flywheel bolts

26 Apply thread locking fluid to the bolt threads and tighten them to the specified torque (photo). Jam the flywheel starter ring teeth to prevent the flywheel from turning as the bolts are tightened.
27 Fit the crankshaft pulley and tighten to the specified torque, again jamming the flywheel teeth (photo).

20.27 Tightening crankshaft pulley bolt

20.29B Fitting thermostat housing – coolant temperature switch arrowed

28 Refit the coolant pump and pulley (Chapter 2).
29 Refit the thermostat housing using a new gasket (photos).
30 Using a new gasket bolt on the oil pump (photo).
31 Refit the alternator mounting bracket.
32 Where an air conditioner or power-assisted steering is fitted, bolt on the compressor and pump brackets.
33 Fit a new oil filter, as described in Section 2, but on Turbo models first fit the oil filter bracket.
34 Refit the ancillary items including the alternator, the distributor, the fuel pump, the manifolds and the clutch. Use new gaskets and refer to the appropriate Chapter for specific details.
35 Adjust the valve clearances (Section 5), fit a new gasket to the rocker cover and tighten the domed nuts (photo).
36 Screw in the spark plugs and connect the ignition leads.

20.30 Fitting the oil pump – oil pressure switch arrowed

20.29A Thermostat housing gasket

20.35 Fitting rocker cover

21 Engine/manual transmission – reconnection and refitting

Reconnection

Refer also to Chapter 6

1 Refer to Chapter 5 and make sure that the clutch driven plate has been centralised.

2 Apply a smear of molybdenum disulphide grease to the input shaft splines and then offer the transmission to the engine. As the input shaft splines pass into the hub of the driven plate, the transmission may require turning in either direction to align the splines.

3 Insert and tighten the flange connecting bolts, but remember to locate the coolant tube, the mounting brackets, and the sump-to-bellhousing reinforcement strut.

4 Bolt on the starter motor.

Refitting

5 Connect the hoist to the lifting lugs, then raise the engine/transmission and lower it carefully into the engine compartment. On RHD vehicles pass the alternator under the brake master cylinder as the engine is lowered.

6 Engage the flexible mountings and fit the connecting bolts. The bracket plate at the rear lower mounting incorporates a cut-out with a locating pin which may require the careful use of a jack under the sump pan to engage it.

7 Remove the hoist.

8 Reconnect the alternator leads and the lead to the automatic choke (where fitted).

9 Reconnect the ignition HT and LT leads, and the earth leads to the side of the cylinder head. Note that the wiring harness is located in a protective sleeve and is routed under the intake manifold. It is held in position by plastic straps on the underside of the manifold.

10 Reconnect the lead to the oil pressure switch.

11 Reconnect the leads to the coolant temperature switch and reverse lamp switch.

12 Reconnect the fuel hoses to the fuel pump. Also all vacuum and air hoses.

13 Reconnect the throttle cable and choke operating cables as applicable, also the rocker cover vent hose (photo).

21.13 Rocker cover vent hose

14 Reconnect the speedometer drive cable to the transmission.

15 Reconnect the clutch operating cable to the release lever. Adjust the clutch cable, as described in Chapter 5.

16 Remove the temporary side gear locating rods and then reconnect the inboard ends of the driveshafts with the final drive. Take care not to damage the oil seals. Use new circlips on the driveshafts and push the splined ends of the shafts into the transmission as far as they will go. Strike the weld bead of the joint cover to drive the shaft fully home. Test that the circlip has locked in its groove by pulling on the joint cover.

17 Using new nuts, reconnect the suspension bottom balljoints.

18 Tighten the suspension top strut mounting nuts to the specified torque.

19 Refit the disc calipers.

20 Refit the front roadwheels, lower the vehicle.

21 Reconnect the gearchange rod and its stabiliser rod.

22 Reconnect the exhaust downpipe to the manifold, also the turbocharger if applicable.

23 Refit the air conditioner compressor and the power steering pump (if fitted).

24 Refit the radiator and electric fan. Reconnect the fan and thermal switch leads.

25 Refit the air cleaner.

26 Refit and reconnect the battery.

27 Refit the bonnet.

28 Fill the cooling system, and the engine and transmission with the specified oil.

29 Refit the drivebelts and tension them, as described in Chapter 2.

22 Engine/automatic transmission – reconnection and refitting

Reconnection

Refer also to Chapter 7

1 The operations are very similar to those described in the preceding Section, but the following special points should be noted.

2 Before connecting the driveplate to the torque converter, check to see that the converter is pushed fully home by referring to Chapter 7.

3 Align the marks on the driveplate and torque converter (made before dismantling). Apply thread locking fluid to the clean threads of the connecting bolts and tighten them to the specified torque. Bolt on the starter motor, and reconnect the kick-down cable.

Refitting

4 Reconnect the speed selector control cable and adjust it, if necessary.

5 Reconnect the inhibitor switch leads.

6 Reconnect the oil cooler hoses.

7 Refit the front wing protective shield where fitted.

8 Top up the automatic transmission fluid.

23 Initial start-up after major overhaul

1 Set the idle speed screw to a higher setting than normal to offset the drag caused by new engine components.

2 Start the engine. This may take rather longer than usual as the fuel pump has to fill the carburettor with fuel.

3 Once the engine starts, allow it to warm without racing and then check for oil leaks.

4 There will be some odd smells caused by oil and grease burning off metal surfaces.

5 Treat the engine as a new unit for the first few hundred miles by restricting speed and load.

6 Once the engine is run-in after 1000 km (600 miles), check the idle speed, the valve clearances and the tightening torque of all engine nuts and bolts. Change the engine oil and filter at the same time.

24 Oil cooler (Turbo models) – removal and refitting

1 Drain the cooling system as described in Chapter 2.

2 Disconnect the water hoses at the oil cooler.

3 Unbolt the cover and remove the O-ring.

4 Unscrew the central stud.

5 Withdraw the oil cooler and remove the O-ring.

6 Refitting is a reversal of removal, but fit new O-rings. Fill the cooling system with reference to Chapter 2. Run the engine and check for oil leaks.

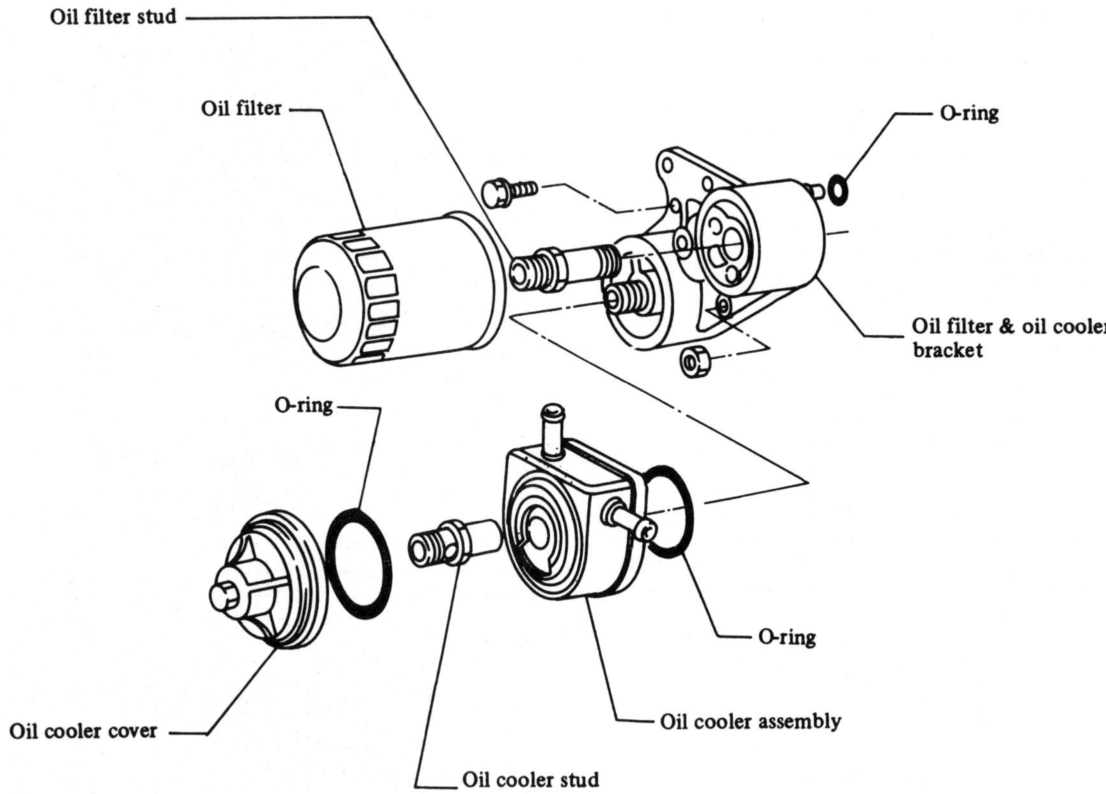

Fig. 1.30 Oil cooler components – Turbo models (Sec 24)

25 Fault diagnosis – engine

Symptom	Reason(s)
Engine fails to turn when starter control operated	
No current at starter motor	Flat or defective battery Loose battery leads Defective starter solenoid or switch or broken wiring Engine earth strap disconnect
Current at starter motor	Jammed starter motor drive pinion Defective starter motor
Engine turns but will not start	
No spark at spark plug	Ignition leads or distributor cap damp or wet Ignition leads to spark plugs loose Shorted or disconnected low tension leads Dirty, incorrectly set, or pitted contact breaker points* Faulty condenser* Defective ignition switch Ignition leads connected wrong way round Faulty coil Contact breaker point spring earthed or broken*
No fuel at engine	No petrol in petrol tank Vapour lock in fuel line (in hot conditions or at high altitude) Blocked float chamber needle valve Fuel pump filter blocked Choked or blocked carburettor jets (where applicable) Faulty fuel pump
Engine stalls and will not restart	
Excess of petrol in cylinder or carburettor flooding	Too much choke allowing too rich a mixture or wet plugs Float damaged or leaking or needle not seating Float lever incorrectly adjusted

Symptom	Reason(s)
No spark at spark plug	ıgnition failure – sudden Ignition failure – misfiring precedes total stoppage Ignition failure – in severe rain or after traversng water splash
No fuel at jets	No petrol in petrol tank Petrol tank breather choked Sudden obstruction in carburettor (where applicable) Water in fuel system
Engine misfires or idles unevenly	
Intermittent spark at spark plug	Ignition leads loose Battery leads loose on terminals Battery earth strap loose on body attachment point Engine earth lead loose Low tension leads on coil loose Low tension lead to distributor loose Dirty or incorrectly gapped plugs Dirty, incorrectly set, or pitted contact breaker points* Tracking across inside of distributor cover Ignition too retarded Faulty coil Slack timing belt
Fuel shortage at engine	Mixture too weak Air leak in carburettor Air leak at inlet manifold to cylinder head, or inlet manifold to carburettor
Lack of power and poor compression	
Mechanical wear	Burnt out valves Sticking or leaking valves Weak or broken valve springs Worn valve guides or stems Worn pistons and piston rings

** Applies only to ignition systems with a mechanical contact breaker*

Chapter 2 Cooling system

Contents

Specifications

System type	Pressurised with belt-driven pump, radiator, thermostat and electric cooling fan	
Radiator cap pressure	0.9 bar (13 lbf/in^2)	
Thermostat		
Opening temperature:		
Temperate climate	82°C (180°F)	
Cold climate	88°C (190°F)	
Tropical climate	76.5°C (170°F)	
Maximum valve lift:		
Temperate climate	8.0 mm (0.31 in) at 95°C (203°F)	
Cold climate	8.0 mm (0.31 in) at 100°C (212°F)	
Tropical climate	8.0 mm (0.31 in) at 90°C (194°F)	
Coolant temperature switch		
Operating temperature	90°C (194°F)	

Drivebelt tension

Deflection is measured with moderate thumb pressure applied at the centre of the belt's longest run. The tension of a new drivebelt should be checked after the vehicle's first operating period.

Component:	**Deflection (old belt)**	**Deflection (new belt)**
Alternator	13.0 to 17.0 mm (0.51 to 0.67 in)	10.0 to 14.0 mm (0.39 to 0.55 in)
Compressor	9.0 to 11.0 mm (0.35 to 0.43 in)	7.0 to 9.0 mm (0.28 to 0.35 in)
Power steering pump	7.0 to 9.0 mm (0.28 to 0.35 in)	6.5 to 8.5 mm (0.256 to 0.335 in)

System capacity*

	With heater	Without heater
UK models (Imp pt; Litre):		
Manual transmission:		
Except Turbo	9.8; 5.5	8.0; 4.6
Turbo	10.8; 6.1	9.8; 5.5
Automatic transmission	10.6; 6.0	9.0; 5.1
North American models (US qt; litre):		
Manual transmission:		
Except Turbo	5.0; 4.7	4.4; 4.1
Turbo	6.5; 6.1	5.9; 5.5
Automatic transmission	5.6; 5.3	5.0; 4.7

***Note**: *For models with a remote expansion tank add 1.2 Imp pt; 0.8 US qt; 0.7 litre*

Torque wrench settings

	lbf ft	Nm
Water pump	7 to 10	9 to 14
Thermostat cover	6 to 8	8 to 11
Temperature switch in radiator	2 to 4	3 to 5

1 General description

The cooling system consists of a front-mounted radiator, a belt-driven coolant pump and an electric radiator cooling fan.

Some models have a remotely-sited expansion tank which accepts displaced coolant when the engine is hot. When the engine cools down, excess coolant is drawn back into the radiator, so eliminating the need for regular topping-up.

Coolant from the engine cooling system is used for the vehicle interior heater and also to warm the intake manifold.

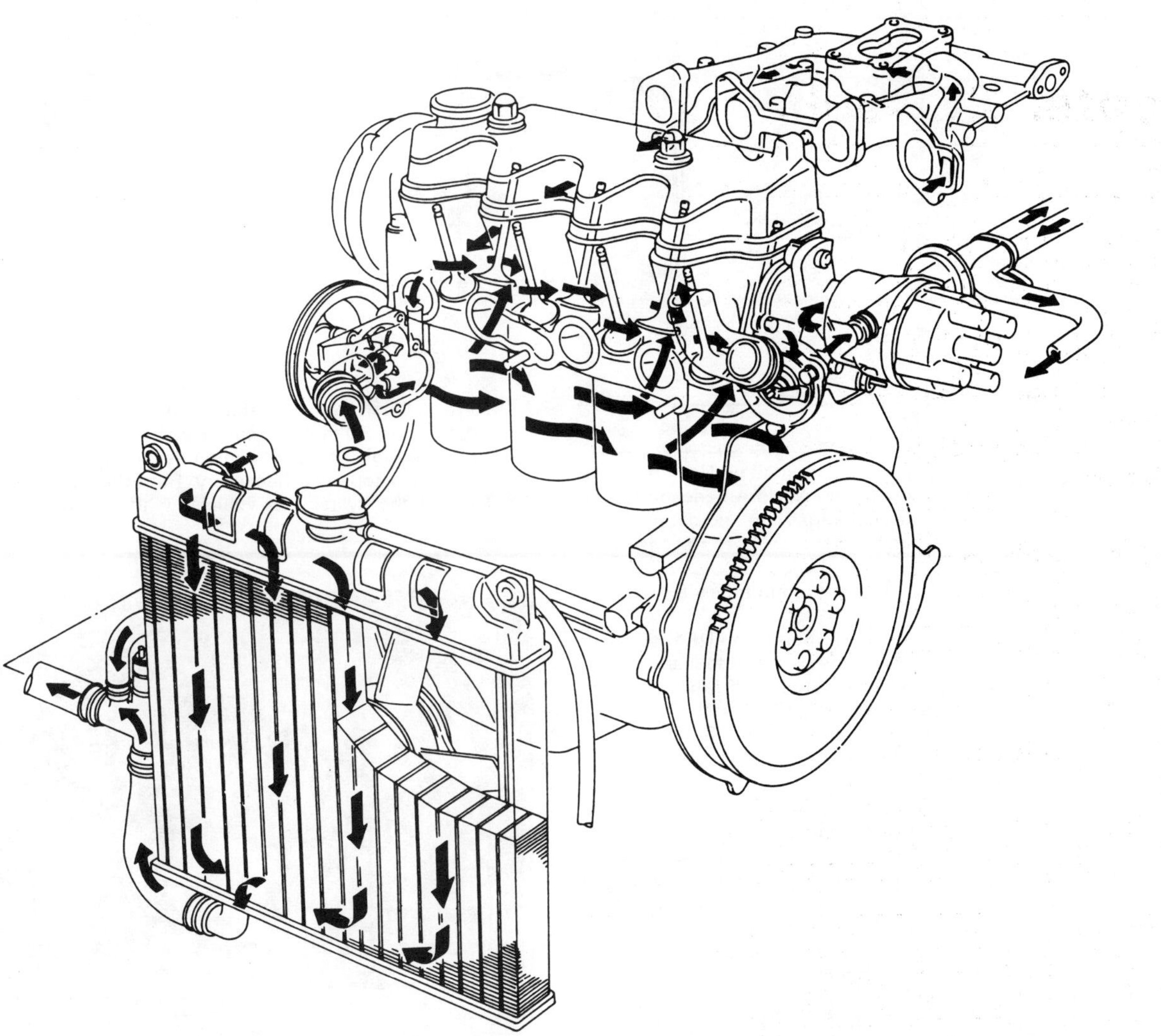

Fig. 2.1 Cooling system flow diagram for non-Turbo models (Sec 1)

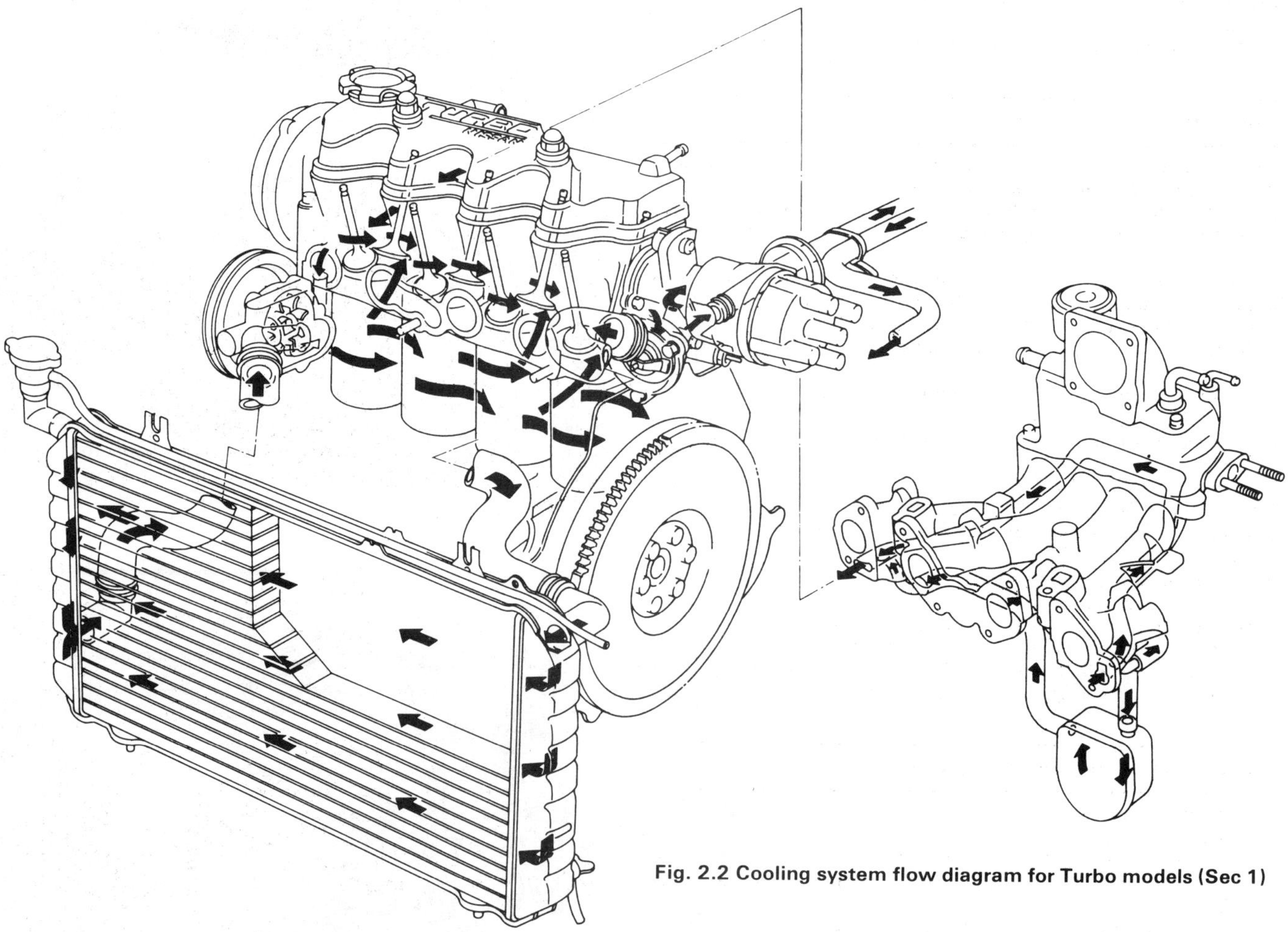

Fig. 2.2 Cooling system flow diagram for Turbo models (Sec 1)

2 Routine maintenance

1 Every 12 000 miles (20 000 km) or 12 months on UK models, or 30 000 miles (48 000 km) or 24 months on North American models, whichever comes first, check the drivebelts for wear and deterioration and renew them as necessary. Re-tension the drivebelts at the same time.
2 At the same interval check the cooling system hoses for damage and deterioration and security.
3 Every 24 000 miles (40 000 km) or 24 months on UK models, or 30 000 miles (48 000 km) or 24 months on North American models, whichever comes first, drain the cooling system and refill with new anti-freeze mixture. If for any reason plain water is used in the cooling system this must be changed every 6000 miles (10 000 km) or 6 months, whichever comes first.

3 Cooling system – draining, flushing and refilling

1 Preferably drain the cooling system when the engine is cold. Set the heater control to HOT, remove the radiator cap or expansion tank cap and then unscrew the drain tap at the base of the radiator (photo).
2 If the system has been well maintained the coolant should run clear to the last drop, in which case the radiator tap may be closed and the system refilled immediately with fresh antifreeze mixture.
3 If the system has been neglected, and the coolant is badly contaminated with rust and sediment, flush it through by inserting a cold water hose in the radiator filler neck.
4 In severe cases of neglect it may be necessary to remove the radiator, invert it and reverse flush it, or even use a cleaning and descaling fluid in accordance with the manufacturer's instructions.
5 To refill the system on vehicles without an expansion tank, fill the radiator to 20 to 30 mm (0.8 to 1.2 in) below the base of the filler neck then run the engine for a few minutes with the radiator cap removed. Switch off and top up if necessary. Refit the radiator cap.
6 To refill the system on vehicles with an expansion tank, fill the

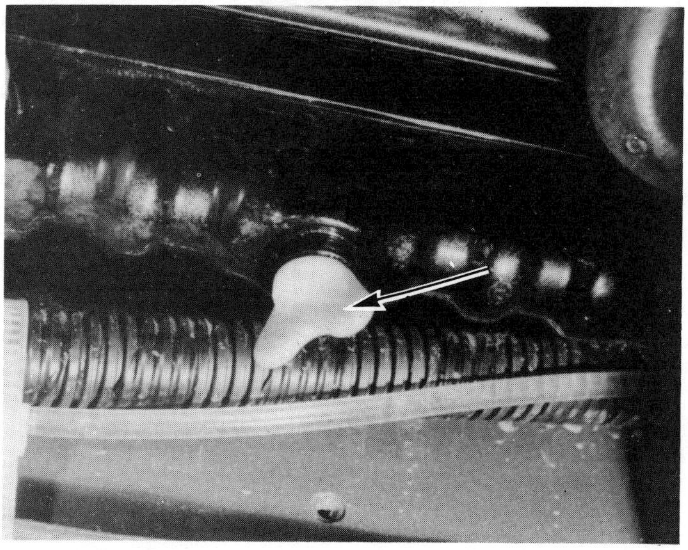

3.1 Radiator drain tap

radiator brim full and fit the cap. Fill the expansion tank to the MAX mark. Run the engine for a few minutes and top up the expansion tank if necessary.
7 A cylinder block drain plug is not fitted.

4 Coolant mixture – general

1 It is essential to keep the cooling system filled with antifreeze mixture instead of plain water. Apart from the obvious protection against damage caused by low temperatures, suitable antifreeze liquids contain an inhibitor against corrosion which, of course, works all year round.
2 Antifreeze should be mixed in accordance with the manufacturer's instructions on the container and the proportion of liquid to water should be chosen to meet the local weather conditions although a 50/50 mixture will ensure maximum protection against freezing and corrosion.
3 Before filling the system with antifreeze mixture check that all the hoses are in sound condition and that all hose clips are tight. Antifreeze has a searching action and will leak more readily than plain water.
4 Renew the coolant mixture every 2 years, as the inhibitor in the solution will by then be of little value. When it is necessary to top up the cooling system, use a mixture of the same strength as that in the system.
5 Always buy a top quality antifreeze of glycol base. Other cheaper antifreeze products usually contain chemicals which evaporate during service and soon provide little protection.
6 If the car is operating in climatic conditions which do not require antifreeze, always use a corrosion inhibitor to prevent corrosion of the system, particularly its light alloy content.

5 Thermostat – removal, testing and refitting

1 Drain sufficient coolant to bring the coolant level below the thermostat housing.
2 Disconnect the coolant hose from the thermostat housing (photo).
3 On North American models, release the exhaust air injection tube clamp bolts.
4 Unbolt and remove the thermostat housing cover, discard the gasket and withdraw the thermostat (photo). If the thermostat is stuck tight, do not lever it out by inserting a screwdriver under its bridge piece, but cut around the edge of its rim using a sharp knife.
5 To check the thermostat, suspend it in water at a temperature near that of the opening temperature of the unit. Observe that it opens fully as the temperature of the water is increased. Allow the thermostat to cool, when it should be fully closed.
6 If a thermostat is seized in the closed position leave it out until a replacement can be obtained.
7 Refit the thermostat into a clean seat with its air bleed hole or jiggle pin uppermost. Use a new gasket, fit the cover (photo) and tighten the bolts securely, but do not overtighten them.
8 Reconnect the hose and refill the cooling system.

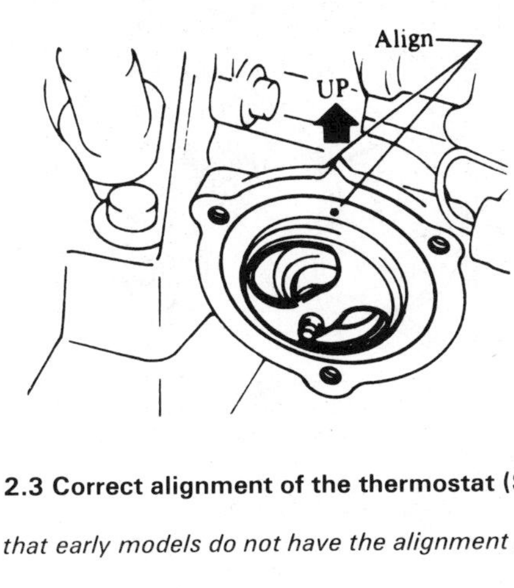

Fig. 2.3 Correct alignment of the thermostat (Sec 5)

Note that early models do not have the alignment pointer

5.2 Hose connection to thermostat housing

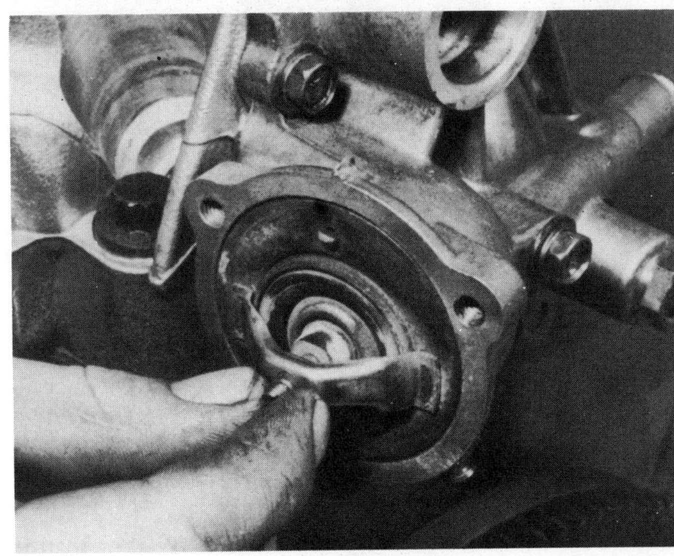

5.4 Removing the thermostat

5.7 Fitting the thermostat cover

6.3 View of radiator bottom hose connection from beneath the engine

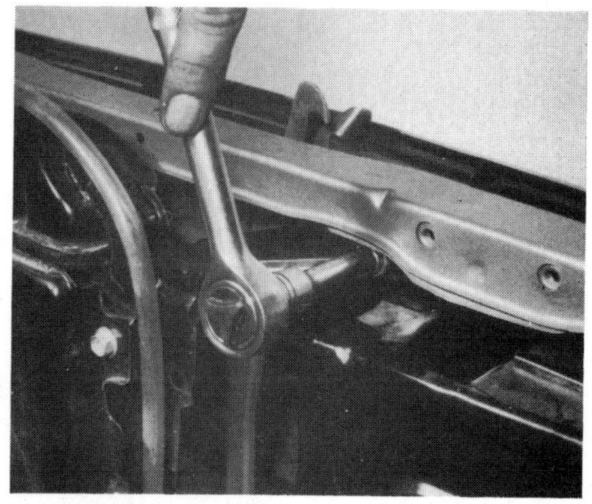

6.7A Removing a radiator mounting bolt

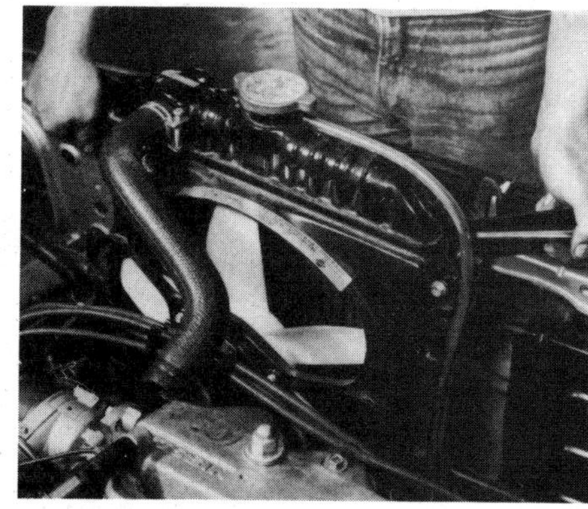

6.7B Removing the radiator and fan assembly

6 Radiator – removal, repair and refitting

1 Drain the cooling system, as described in Section 3. Retain the coolant if required for further use.
2 Where applicable unbolt the power steering pump and place it to one side without disconnecting the fluid lines.
3 Disconnect the radiator top and bottom hoses (photo), and where applicable unbolt the bottom hose adaptor.
4 On automatic transmission models disconnect the fluid cooler lines from the bottom of the radiator and plug them.
5 On Turbo models unbolt and remove the turbocharger cover.
6 Disconnect the wiring from the cooling fan motor and water temperature switch as applicable.
7 Unbolt the radiator top and side mountings (photo) and lift the radiator from the engine compartment, complete with cooling fan (photo).
8 If the radiator was removed because of a leak, it is best to leave its repair to the professional radiator repairer, although it may be possible to use fibreglass or a similar product to seal it. Loss of coolant may sometimes be caused by a defective radiator pressure cap. Your dealer can test this for you.
9 Refitting is a reversal of removal, but fill the system as described in Section 3.

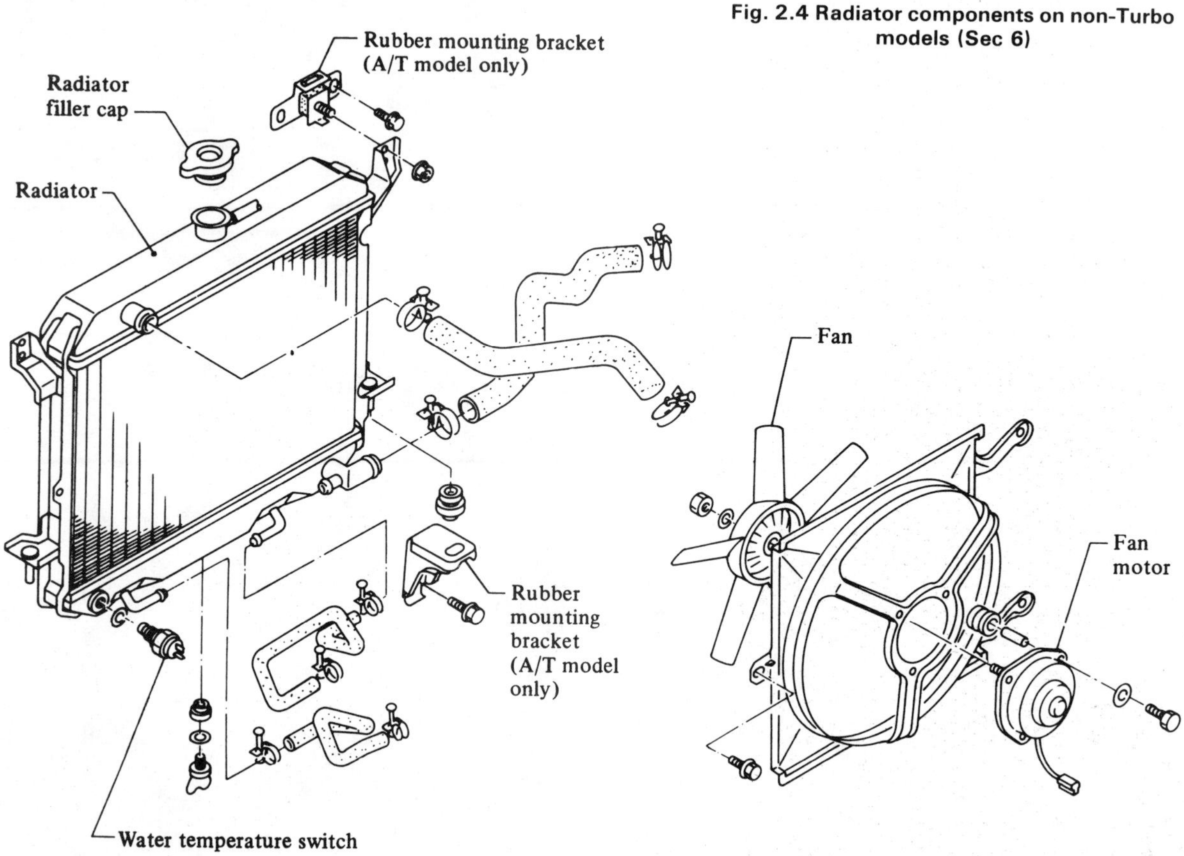

Fig. 2.4 Radiator components on non-Turbo models (Sec 6)

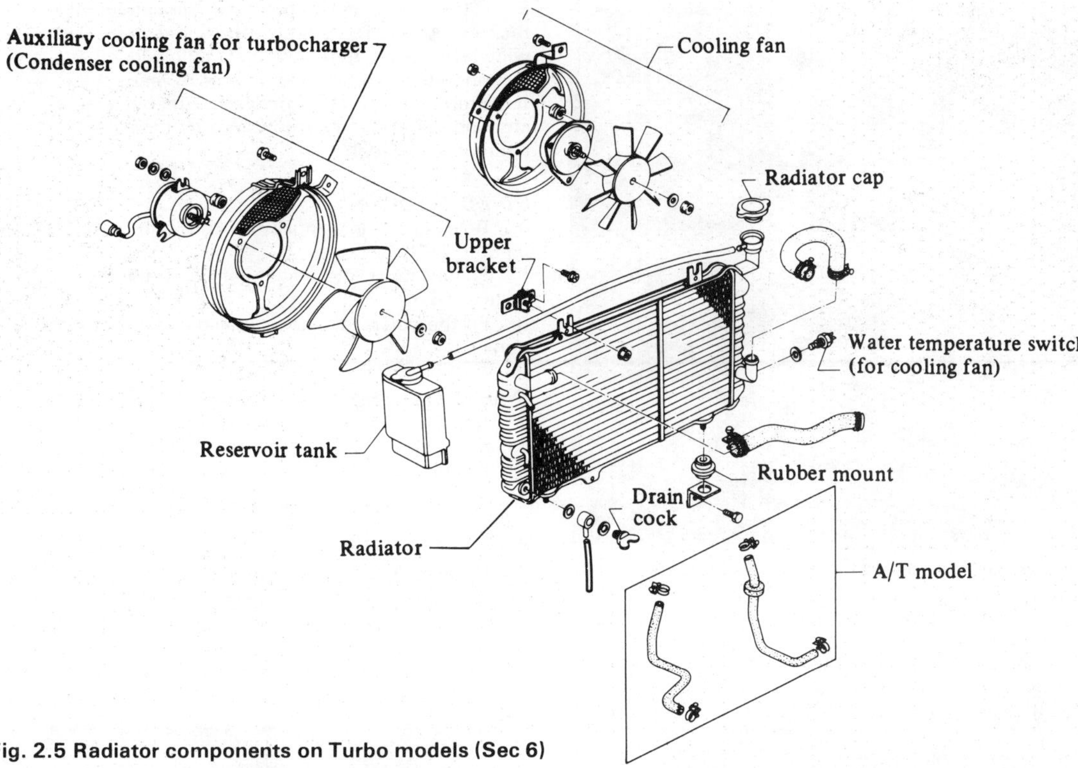

Fig. 2.5 Radiator components on Turbo models (Sec 6)

7 Radiator cooling fan and switch – removal and refitting

Fan

1 Open the bonnet and disconnect the battery negative lead.
2 On Turbo models remove the front grille.
3 Disconnect the leads from the fan motor.
4 Unbolt the fan mounting struts and lift the assembly from the engine compartment.
5 Refitting is a reversal of removal.

Switch

6 Open the bonnet and disconnect the battery negative lead.
7 Drain the cooling system as described in Section 3.
8 Disconnect the wiring from the switch located on the bottom radiator tank (photo – UK models except Turbo), RH side tank (Turbo models), or RH adaptor (North American models).

7.8 Radiator cooling fan switch location

9 Unscrew the switch and remove the washer.
10 The switch can be tested using a simple test bulb and battery if it is lowered into water and the temperature raised to the specified operational level.
11 Refitting is a reversal of removal but use a new sealing washer and refill the cooling system as described in Section 3.

8 Water pump – removal and refitting

1 Drain the cooling system as described in Section 3.
2 Where applicable remove the power steering drivebelt and unbolt the power steering pump. Place the pump on one side without disconnecting the hoses.
3 Loosen the alternator mounting bolts, pivot the alternator toward the engine then remove the drivebelt.
4 Unbolt and remove the pulley from the water pump drive flange (photo).

8.4 Water pump pulley

8.5A Showing water pump and retaining bolts

8.5B Water pump seating on the cylinder block

5 Unbolt the water pump from the cylinder block and remove the gasket (photos).
6 The coolant pump cannot be overhauled, as new seals and internal components are not available. A new pump will therefore have to be obtained if the original one is worn or leaking.
7 Refitting is a reversal of removal, but use a new gasket and tighten the bolts to the specified torque.
8 Tension the drivebelts (Section 9).
9 Refill the cooling system as described in Section 3.

9 Drivebelts – removal, refitting and adjustment

1 The number of drivebelts fitted and their configuration will depend upon the units with which the particular vehicle is equipped. These include:

Power steering
Air conditioning

2 An alternator is fitted to all models.
3 If the vehicle is equipped with air conditioning and power steering then except for 1985 on North American models the compressor belt will have to be removed first, followed by the one for the power steering pump, in order to be able to reach the inner belt which drives the alternator. For 1985 on North American models remove the power steering pump drivebelt followed by the air conditioning drivebelt and alternator drivebelt.

Compressor belt
4 Release the idler pulley locknut and the adjuster nuts on the eye bolt.
5 Fully slacken the adjustment until the belt can be slipped off the pulley.
6 Fit the belt and tension in accordance with the table in Specifications.

Power steering pump belt
7 Release the power steering pump bracket lockbolt. Turn the adjuster bolt to fully slacken the belt and then slip it off the pulley.
8 Fit the belt and tension in accordance with the table in Specifications.

Alternator belt
9 Release the alternator mounting and adjuster link bolts and push the alternator in towards the engine until the belt can be slipped off the pulleys.
10 Fit the belt and tension in accordance with the table in Specifications (photo).

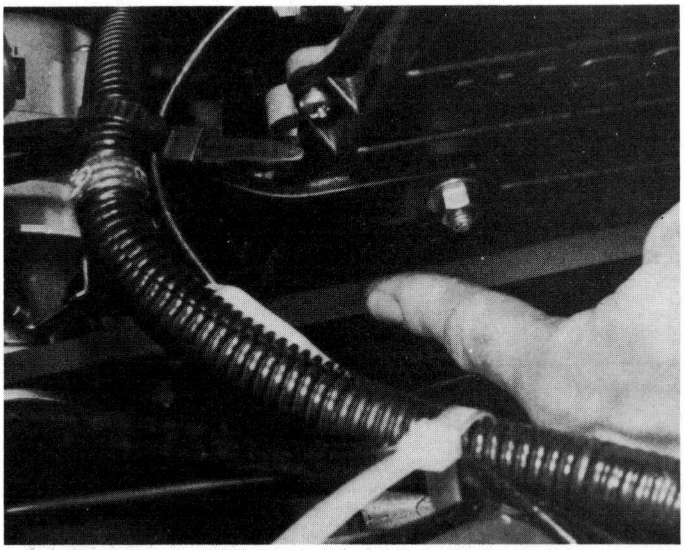

9.10 Checking the water pump/alternator drivebelt tension

General
11 Never overtension a drivebelt, or the coolant pump or alternator bearings may be damaged.

10 Coolant temperature switch and gauge – general

1 The coolant temperature switch can only be removed from its thermostat housing location after the cooling system has been partially drained (2.0 litres).
2 To test a coolant temperature switch requires the use of an ohmmeter, so this is a job best left to your dealer or auto-electrician.
3 Faulty indication of the temperature gauge may be caused by the switch-to-gauge lead earthing due to damaged insulation, check this first.
4 Immediate indication of maximum temperature when the ignition is switched on will be due to a fault in either the gauge or the switch.

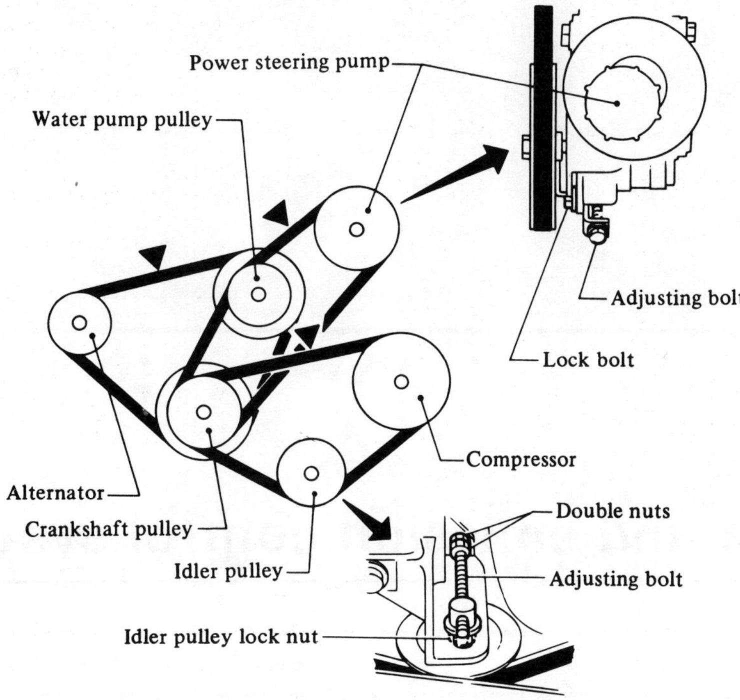

Fig. 2.6 Drivebelt adjustment (Sec 9)

11 Fault diagnosis – cooling system

Symptom	Reason(s)
Overheating	Coolant loss due to leakage Faulty electric cooling fan or switch Water pump/alternator drivebelt slack or broken Faulty thermostat Radiator matrix clogged internally or externally
Overcooling	Faulty thermostat Faulty electric cooling fan switch
Coolant loss	External leakage (hose joints etc) Overheating (see above) Internal leakage (head gasket)

Chapter 3
Fuel, exhaust and emission control systems

Contents

Specifications

System type

Non-Turbo models	Mechanical fuel pump, downdraught carburettor
Turbo models	Electric fuel pump, fuel injection system, turbocharger

Fuel tank capacity

11.0 Imp gal; 13.3 US gal; 50.0 litre

Fuel pump pressure

Early models	2.4 to 3.4 lbf/in^2 (0.167 to 0.235 bar)
Late models	2.8 to 3.8 lbf/in^2 (0.20 to 0.27 bar)

Carburettor

Type	Nikki dual barrel downdraught

General

UK non-Turbo models

Idle speed:	
Manual transmission	700 ± 50 rpm
Automatic transmission	750 ± 50 rpm
CO content at idle	1.5 ± 0.5%
Fast idle (automatic choke):	
Manual transmission:	
E10 engine	1700 to 1900 rpm
E13 engine	1900 to 2300 rpm
E15 engine	2000 to 2400 rpm
Automatic transmission:	
E15 engine	2200 to 2600 rpm

Dashpot touch-speed:	
E13 engine	1700 to 2100 rpm
E15 engine	1800 to 2200 rpm

UK and Turbo models

Pressure regulator control pressure	36.3 lbf/in² (2.501 bar)
Fuel pump pressure	61 to 71 lbf/in² (4.22 to 4.90 bar)
Throttle valve switch control speed	approx. 1100 rpm

North American non-Turbo models

Idle speed:	
Manual transmission (USA except California)	800 ± 100 rpm
Manual transmission (California and Canada)	750 ± 50 rpm
Automatic transmission	650 ± 50 rpm
CO content at idle	2.0 ± 1%
Fast idle (automatic choke):	
Manual transmission (USA except California)	2400 to 3200 rpm
Manual transmission (California)	2600 to 3400 rpm
Manual transmission (Canada)	1900 to 2700 rpm
Automatic transmission (USA except California)	2700 to 3500 rpm
Automatic transmission (California)	2900 to 3700 rpm
Automatic transmission (Canada)	2400 to 3200 rpm
Dashpot touch-speed:	
Automatic transmission	1900 to 2100 rpm
Manual transmission (Canada)	2250 to 2450 rpm

Torque wrench settings

	lbf ft	Nm
Solenoid fuel cut-off valve	16	22
Fuel pump nuts	9	12
Manifold nuts	15	20
Thermal vacuum valve	16	22
Neutral switch	16	22
Catalytic converter end flange bolts	30	40
Converter shield bolts	6	8
Carburettor mounting nuts	15	20
Fuel tank mounting bolts	30	40
Exhaust downpipe to manifold	18	24
Turbo models		
Throttle chamber	14	20
Exhaust gas sensor	36	49
EGR tube nut	33	44
Thermal vacuum valve	Less than 16	Less than 22

1 General description

The fuel system is either of downdraught carburettor type fitted to non-Turbo models, or of turbocharged fuel injection type fitted to Turbo models.

A temperature controlled air cleaner is fitted to most models. The fuel pump is either mechanical (non-Turbo models) actuated by a cam on the engine jack shaft, or electrical (Turbo models).

The fuel tank is mounted at the rear of the vehicle.

The carburettor may be of manual or automatic choke type, depending upon vehicle model and operating territory.

All models have emission control systems, but those destined for operation in N America have a very comprehensive arrangement with modified carburettors and other components.

2 Routine maintenance

1 Every 6000 miles (10 000 km) or 6 months whichever comes first on UK models, or every 15 000 miles (24 000 km) or 12 months whichever comes first on North American models check and adjust the engine idling and mixture settings. Also check the condition and security of the fuel lines and exhaust system.

2 Renew the air cleaner every 24 000 miles (40 000 km) or 24 months whichever comes first on UK models, or every 30 000 miles (48 000 km) or 24 months whichever comes first on North American models. Also renew the fuel filter.

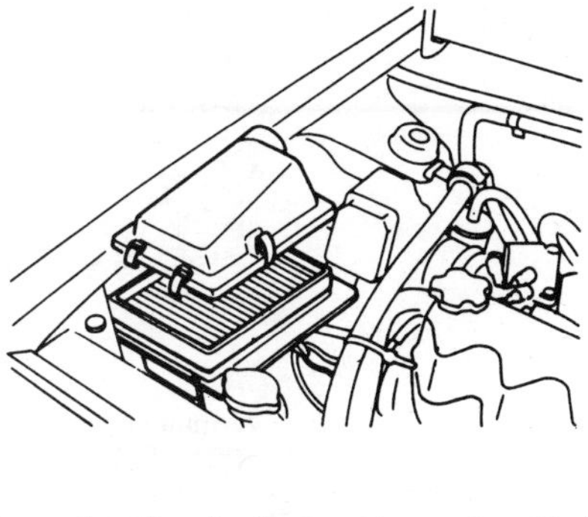

Fig. 3.1 Removing the air cleaner element on Turbo models (Sec 3)

3 Air cleaner – removal and refitting

1 On Turbo models loosen the air flow meter mounting screw.

2 Unscrew the wing nut (non-Turbo models) or prise open the spring clips (Turbo models). Lift the cover from the air cleaner, then take out the element and discard it (photos).

3 Wipe out the interior of the air cleaner casing, fit a new element and where applicable replace the cover so that the arrows are in alignment (photo). Where applicable check that the cover sealing ring is correctly located.

3.2A Air cleaner cover wing nut

3.2B Removing the air cleaner element

3.3 Air cleaner alignment arrows

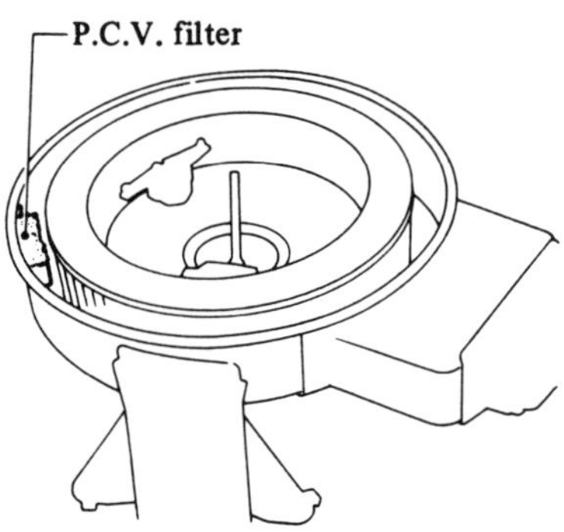

Fig. 3.2 Crankcase ventilation filter (Sec 3)

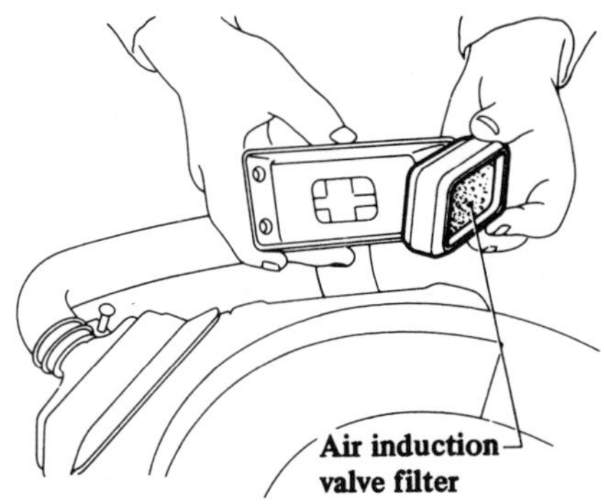

Fig. 3.3 Air induction valve filter (Sec 3)

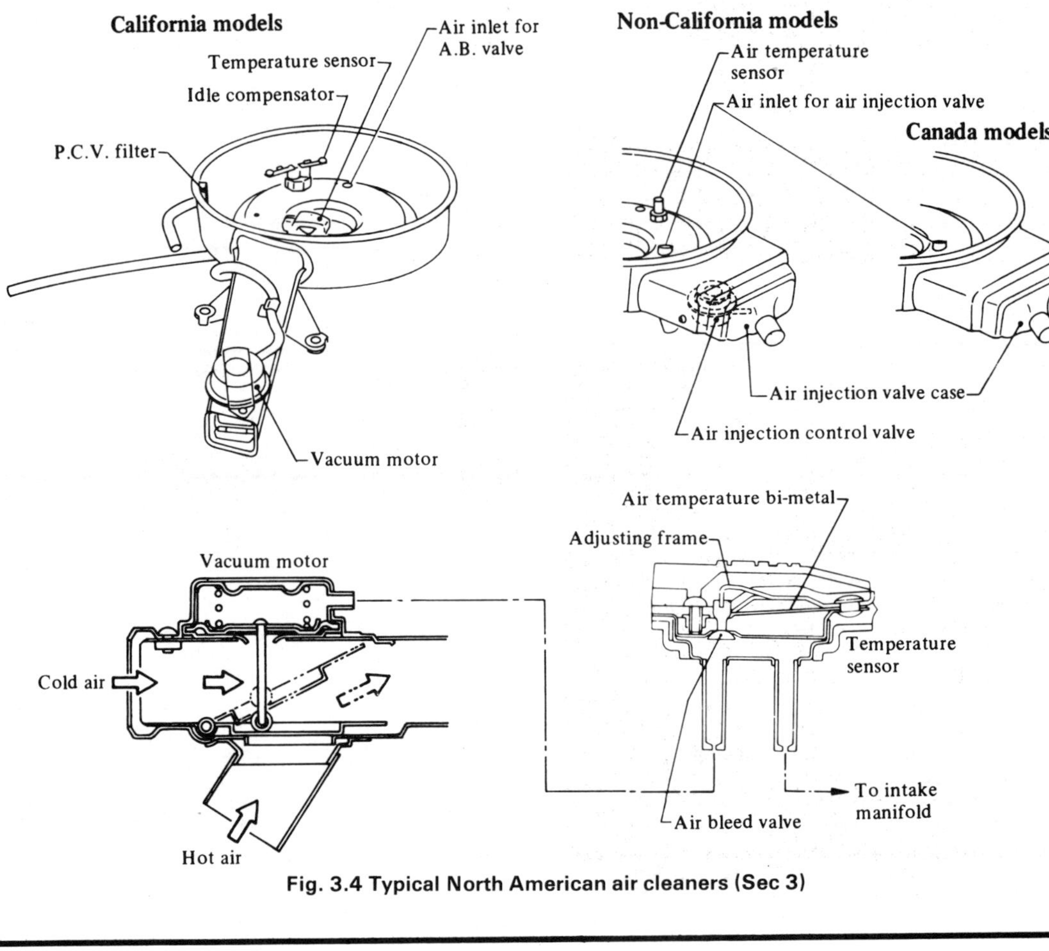

Fig. 3.4 Typical North American air cleaners (Sec 3)

4 On North American models, two additional filter elements are located within the air cleaner casing, these include one for the crankcase PCV intake and one for the air induction valve (refer to Section 21).

5 Depending upon the model, the air cleaner may be fitted with one of the following devices:

Idle compensator (photo). This is basically a thermostatic valve which at high under-bonnet temperatures when the engine is idling admits extra air to weaken the excessively rich mixture. The idle compensator can be assumed to be functioning correctly if a distinct hiss can be heard when it opens.

Temperature sensor. This is used to monitor the temperature of the intake air.

Air control valve. A flap valve to vary the source of intake air, either hot from around the exhaust manifold or cold from the front of the engine compartment. Its operation may be checked by using a mirror to see if the flap is closed to cold air entry when the engine is cold, or open to cold air when the engine is at normal operating temperature.

Vacuum capsule. This is used also to vary the setting of the air intake control valve, but its operation is controlled by engine vacuum which varies with engine load.

6 To remove the air cleaner casing, on non-Turbo models disconnect the hot air duct (photo) and identify and disconnect the other flexible hoses. Remove the cover and the filter element.

3.5 Air cleaner idle compensator

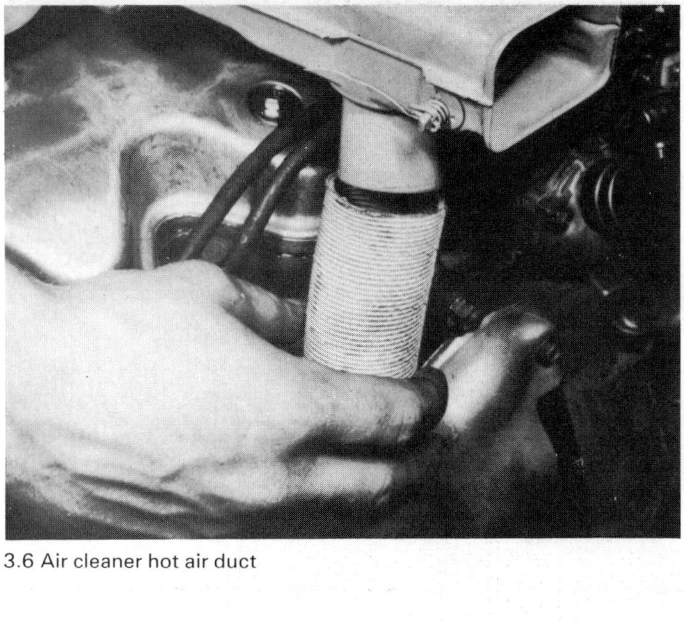

3.6 Air cleaner hot air duct

7 Extract the casing strut screws (photo) and lift the air cleaner from the carburettor. Refitting is a reversal of removal.

8 On models which have an air cleaner without automatic temperature control, the intake air deflector lever should be moved to the Winter or Summer position according to season (photo). In the Winter setting, warmed air is drawn in from the space between the exhaust manifold and the hot air collector plate.

3.7 Air cleaner mounting screws

3.8 Air cleaner intake air control lever

4 Fuel filter – renewal

1 An in-line type of fuel filter is used (photo).

2 On Turbo models carry out the following preliminary procedure in order to reduce the fuel pressure to zero. With the engine idling disconnect the fuel pump relay wiring plug (Figs. 3.5 and 3.6). After the engine has stopped turn it on the starter briefly then switch the ignition off. Reconnect the wiring plug.

3 Disconnect the hoses from the filter, and remove the unit from the clip.

4 Fit the new filter then start the engine and check for leaks.

4.1 In-line fuel filter

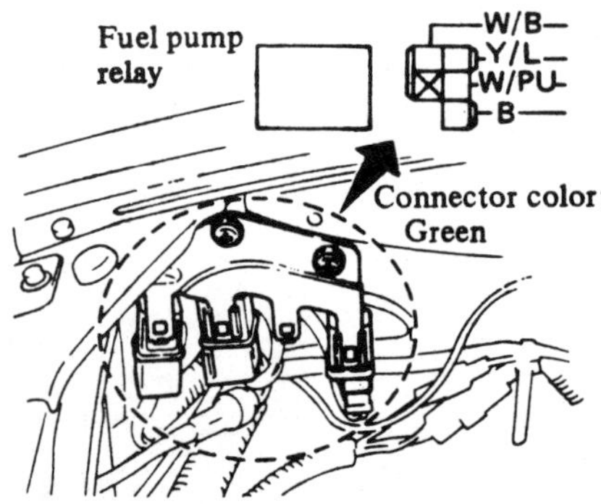

Fig. 3.5 Fuel pump relay location – right-hand drive models (Sec 4)

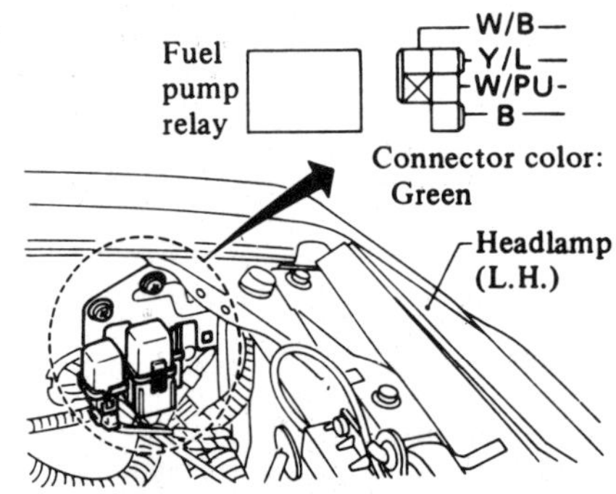

Fig. 3.6 Fuel pump relay location – left-hand drive models (Sec 4)

5 Fuel tank – removal, repair and refitting

1 Jack up the rear of the car and support on axle stands, then drain the fuel by unscrewing the drain plug from the base of the tank (photo).

2 Working within the luggage compartment, remove the cover plate from the tank transmitter unit and then disconnect the wiring harness plug (photo).

3 On Turbo models release any fuel pressure in the system by following the preliminary procedure given in Section 4.

4 Disconnect the battery.

5 Working under the rear wing, disconnect the fuel filler and ventilation hoses (photo). Stuff a piece of rag into the openings to prevent the entry of dirt. If required, the filler pipe can be removed by removing the upper and lower mounting plates (photo).

6 Working at the front of the fuel tank, disconnect the fuel flow and return hoses (photo), and the evaporation control system hoses.

7 Support the tank and then unscrew and remove the six mounting bolts from the tank flange (photo). Lower and remove the tank from the vehicle.

8 A leak in a fuel tank can be sealed using one of the several products available at motor accessory stores. For a permanent repair the tank will have to be soldered or brazed, but *on no account attempt to do this work yourself* due to the risk of explosion unless the tank has been steamed out thoroughly. Radiator repairers can usually undertake fuel tank repair work.

9 Removal of sediment, water or sludge can be carried out after first having removed the tank transmitter unit. Pour in some paraffin, or petrol, and shake the tank vigorously. Empty the tank and repeat the operations as many times as is necessary to clean it and then give a final rinse with clean fuel.

10 Refitting is a reversal of removal.

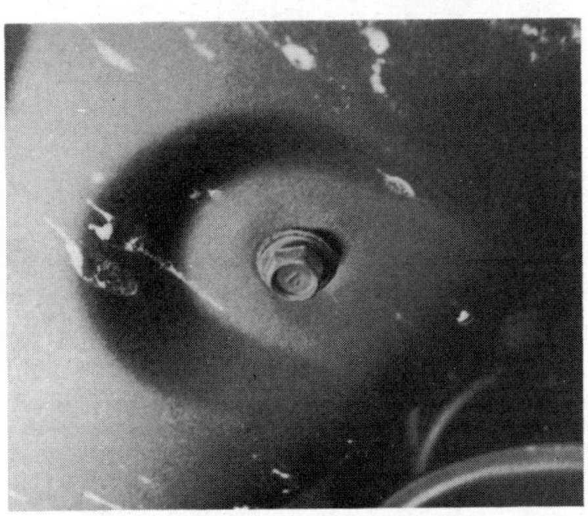

5.1 Fuel tank drain plug

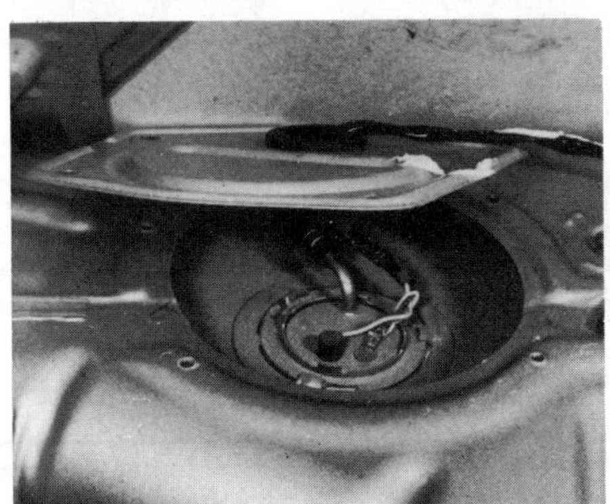

5.2 Fuel level transmitter

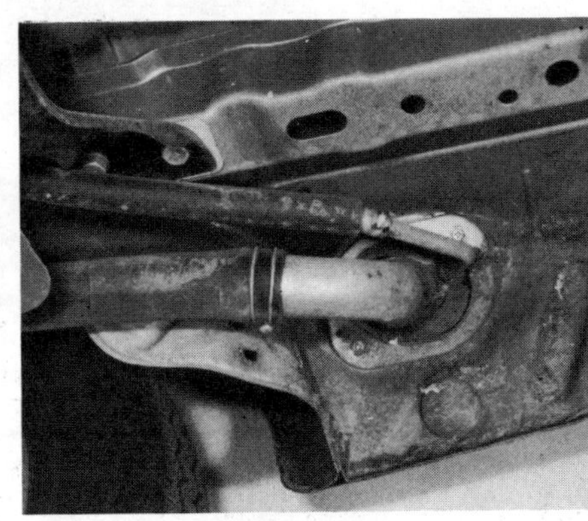

5.5A Fuel tank filler and vent hoses

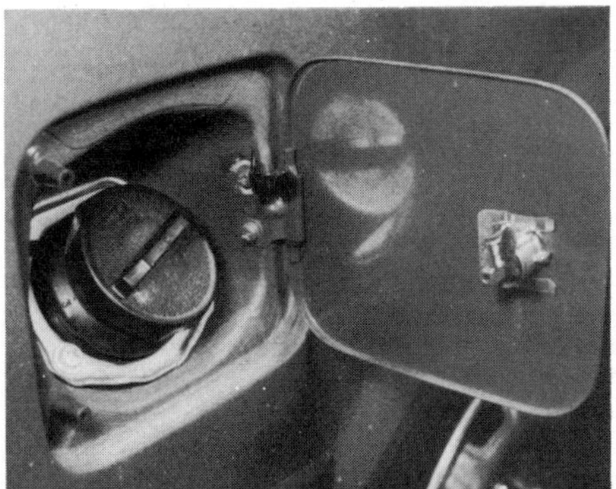

5.5B Showing filler pipe upper mounting plate

5.6 Fuel tank hoses

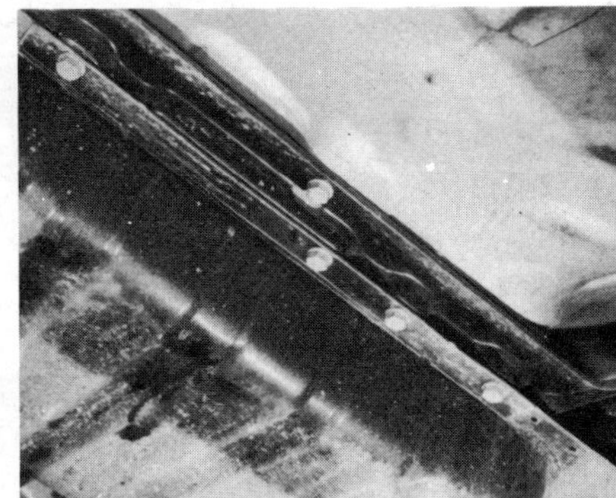

5.7 Fuel tank mounting bolts

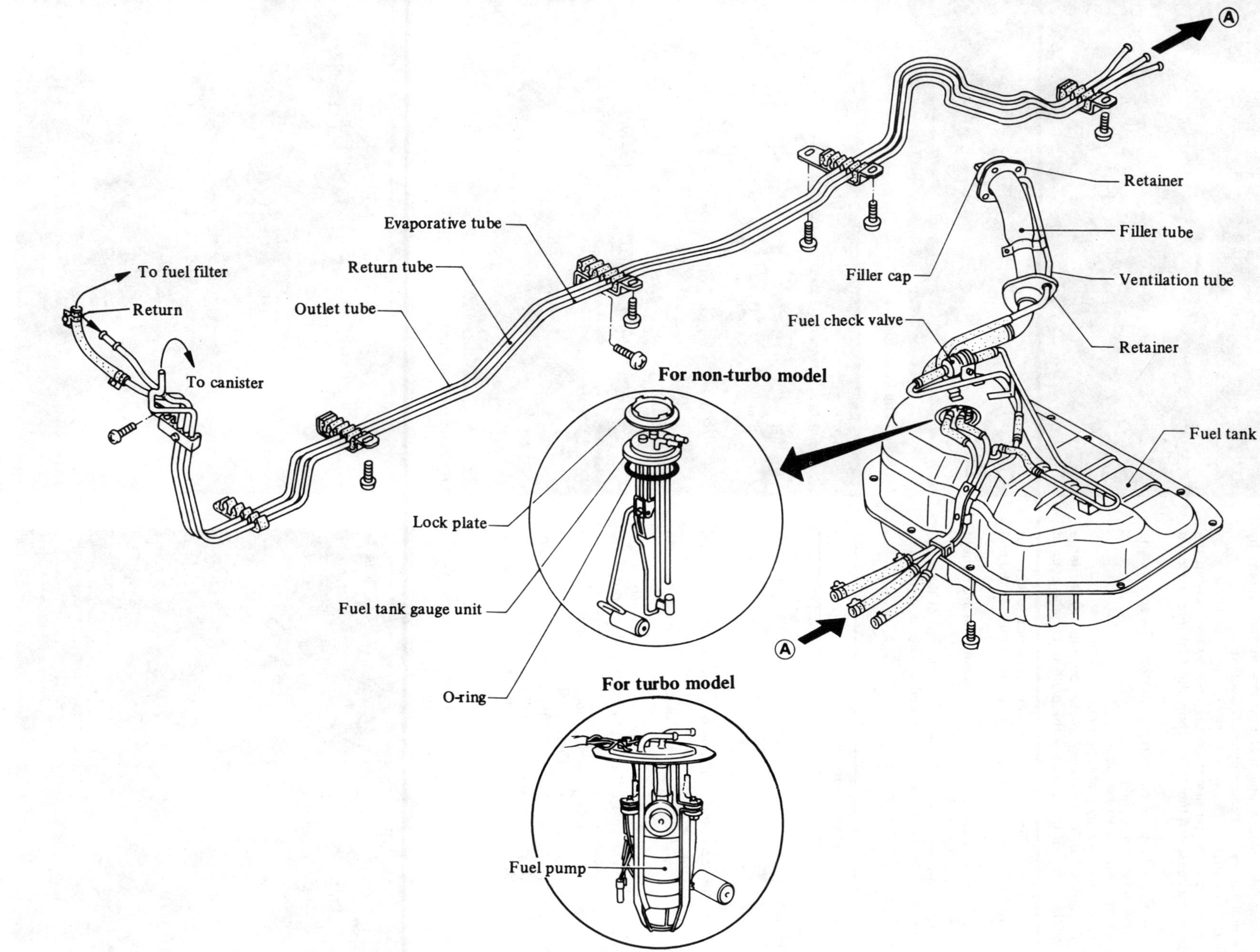

Fuel tank gauge unit with fuel pump

Fig. 3.7 Fuel tank components (Sec 5)

6 Fuel level transmitter and gauge – removal and refitting

1 On Turbo models release any fuel pressure in the system by following the preliminary procedure given in Section 4.
2 Disconnect the battery.
3 To remove the transmitter first drain the tank by unscrewing the drain plug from the base of the tank. Drain the fuel into a closed container.
4 Working inside the luggage compartment, extract the screws and remove the transmitter cover plate.
5 Disconnect the wiring harness plug.
6 Disconnect the fuel supply and return hoses.
7 Twist the sender unit mounting plate by placing a large screwdriver or flat piece of metal across the plate to engage against two tags.
8 Withdraw the transmitter carefully, taking care not to bend or damage the float. On Turbo models the assembly incorporates the fuel pump.
9 The fuel gauge can be removed after withdrawing the instrument panel, as described in Chapter 10.
10 Refitting is a reversal of removal, but make sure that the transmitter O-ring is in good order, and make sure that the tab engages the notch in the tank aperture.

7 Fuel pump – removal and refitting

Turbo models

1 The procedure is as described for the fuel level transmitter in Section 6.

Non-Turbo models

2 The mechanically-operated fuel pump is of rocker arm type, actuated by an eccentric cam on the jack shaft (photo).
3 Disconnect the fuel hoses from the pump, and plug them to prevent loss of fuel.
4 Unscrew and remove the pump mounting nuts and washers and lift the pump from the crankcase. Discard the joint gaskets, but retain the insulator, if fitted (photo).
5 The fuel pump is of sealed type and, if clogged or faulty, will have to be renewed, cleaning and repair is not possible.
6 Clean the pump and crankcase mating faces, use new joint gaskets and bolt the pump into position (photo).
7 Reconnect the fuel hoses (photo), run the engine and test for leaks.

7.4 Fuel pump insulator

7.6 Fitting the fuel pump

7.2 Fuel pump (non-Turbo models)

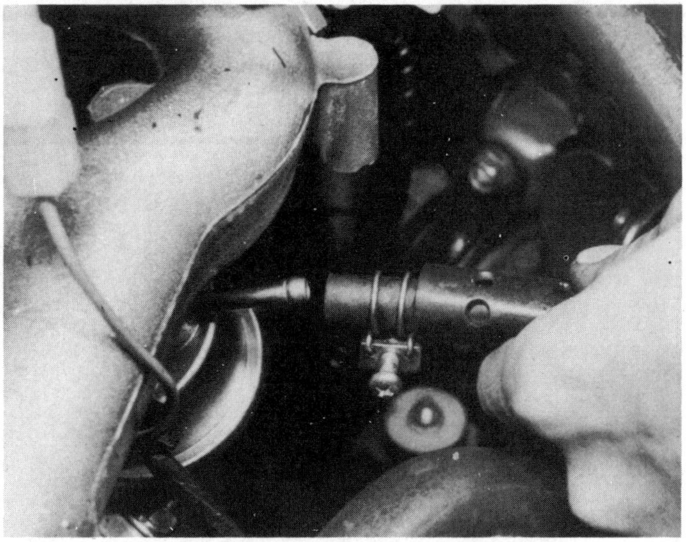

7.7 Connecting fuel hose to pump

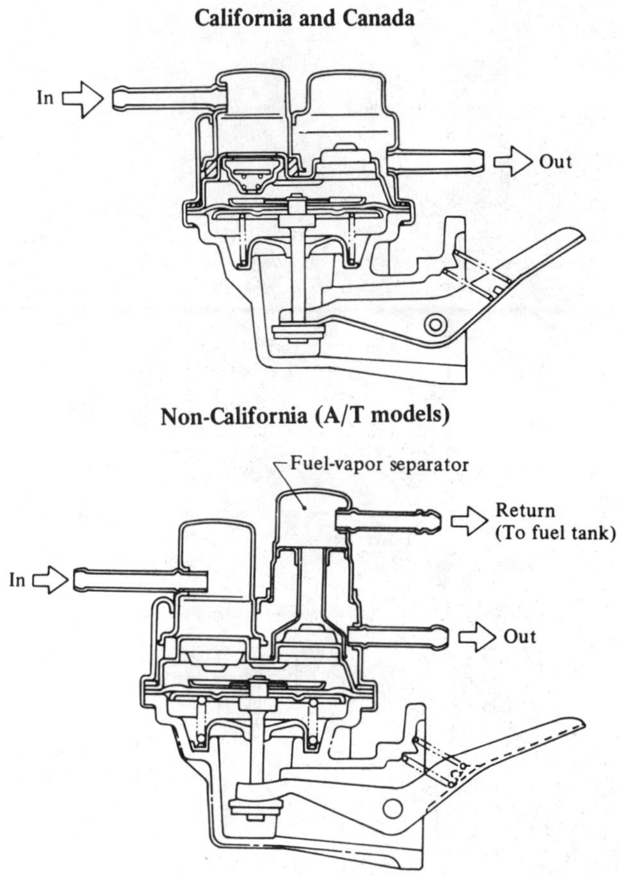

Fig. 3.8 Sectional view of the two types of mechanical fuel pump (Sec 7)

8 Carburettor – description

1 The carburettor on all models is of dual barrel downdraught type, but the calibration, ancillary devices and whether it is of manual choke or automatic choke type depends upon the particular vehicle model and operating territory.

2 **Automatic choke.** This is of bi-metal spring type, the heating being carried out electrically (photo). A relay is located within the engine compartment for operating the automatic choke.

3 **Fuel cut-off solenoid valve.** The purpose of this valve (photo) is to prevent the engine running on when the ignition is switched off. This is done by cutting off the fuel supply to the idle circuit.

4 **Secondary throttle vacuum diaphragm.** This is actuated by vacuum conditions in the carburettor venturi. The diaphragm is linked to the secondary throttle in order to open it after the primary throttle valve plate has opened through an angle of 48°.

5 **Dashpot.** On certain models equipped with automatic transmission, a dashpot is fitted to the carburettor to prevent the engine stalling during sudden braking or quick release of the accelerator pedal.

6 **Choke unloader.** This device opens the choke valve plate slightly when increasing the engine speed during the warm-up period to provide a suitable fuel/air mixture which would otherwise be too rich (photo).

7 **Vacuum break diaphragm.** This is a double-acting type diaphragm which opens the choke valve plate immediately after cold starting to create a suitable fuel/air mixture in accordance with the prevailing engine vacuum conditions.

8 **Altitude compensator.** The purpose of this device is to weaken the fuel/air mixture at high altitudes when it would otherwise be too rich. The compensator is a barometric pressure-sensitive bellows which opens and closes a needle type air bleed valve.

8.2 Automatic choke location

8.3 Fuel cut-off valve location

8.6 Choke unloader device

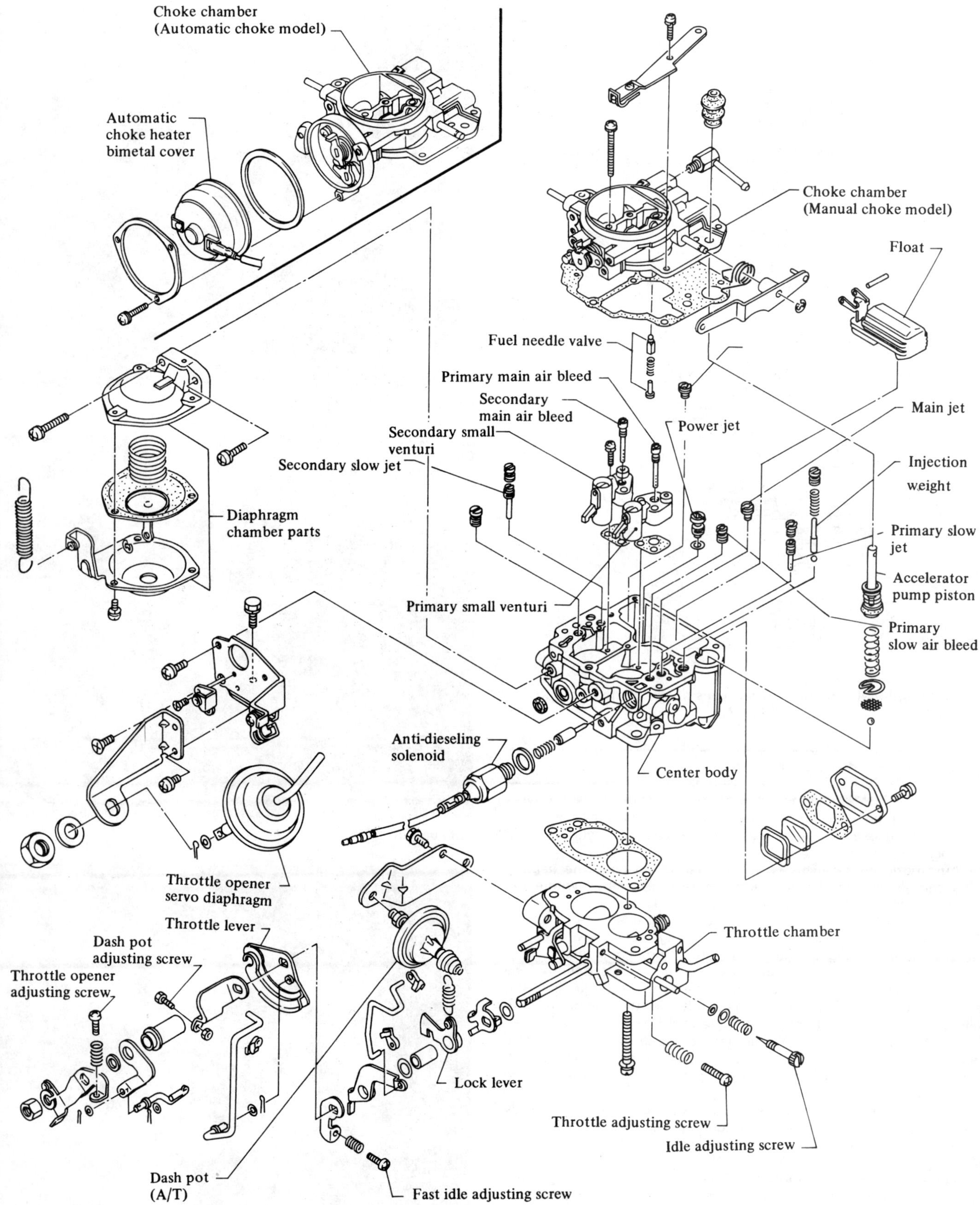

Fig. 3.9 Exploded view of the E10 and E13 engine carburettor (Sec 8)

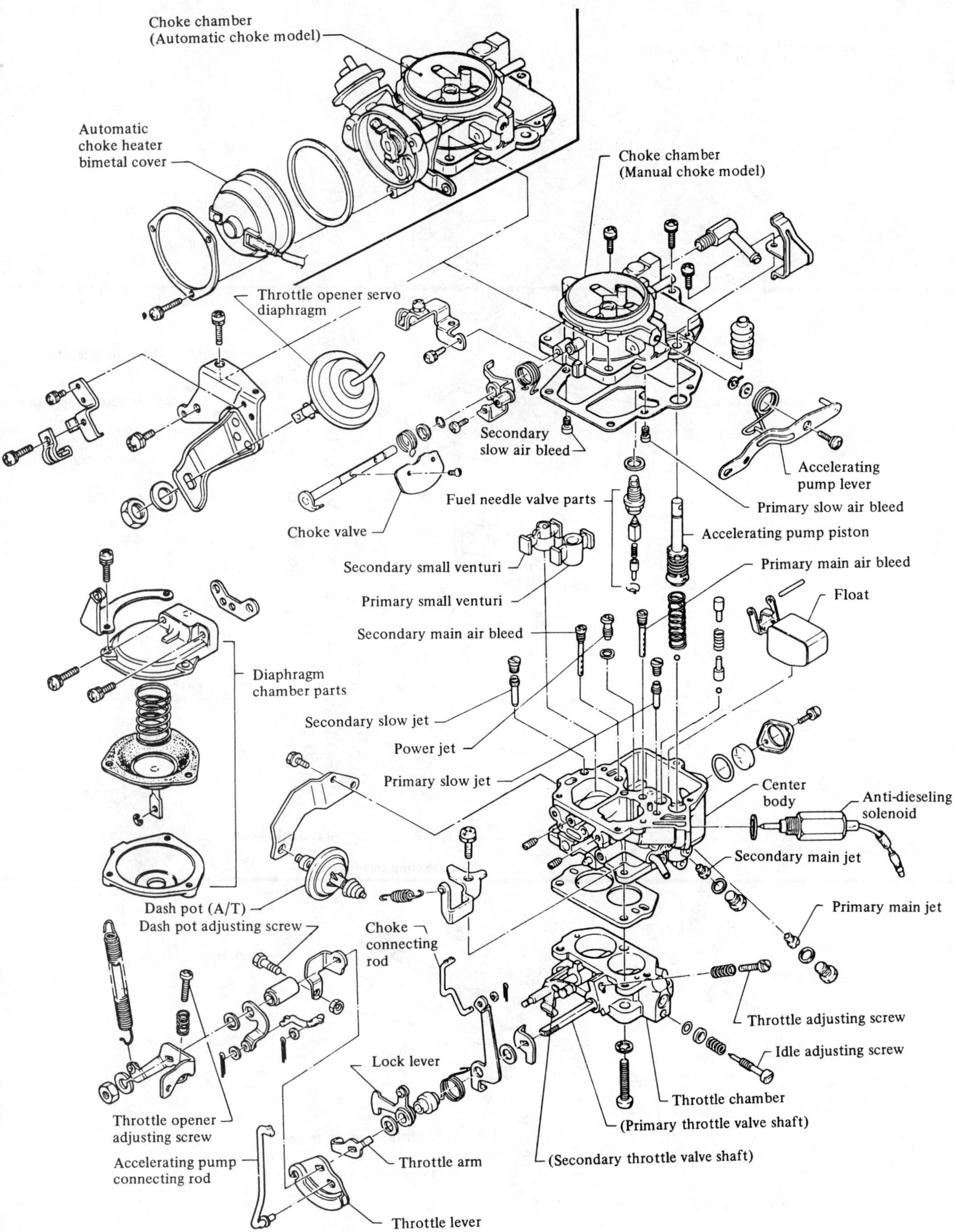

Fig. 3.10 Exploded view of the E15 engine carburettor (Sec 8)

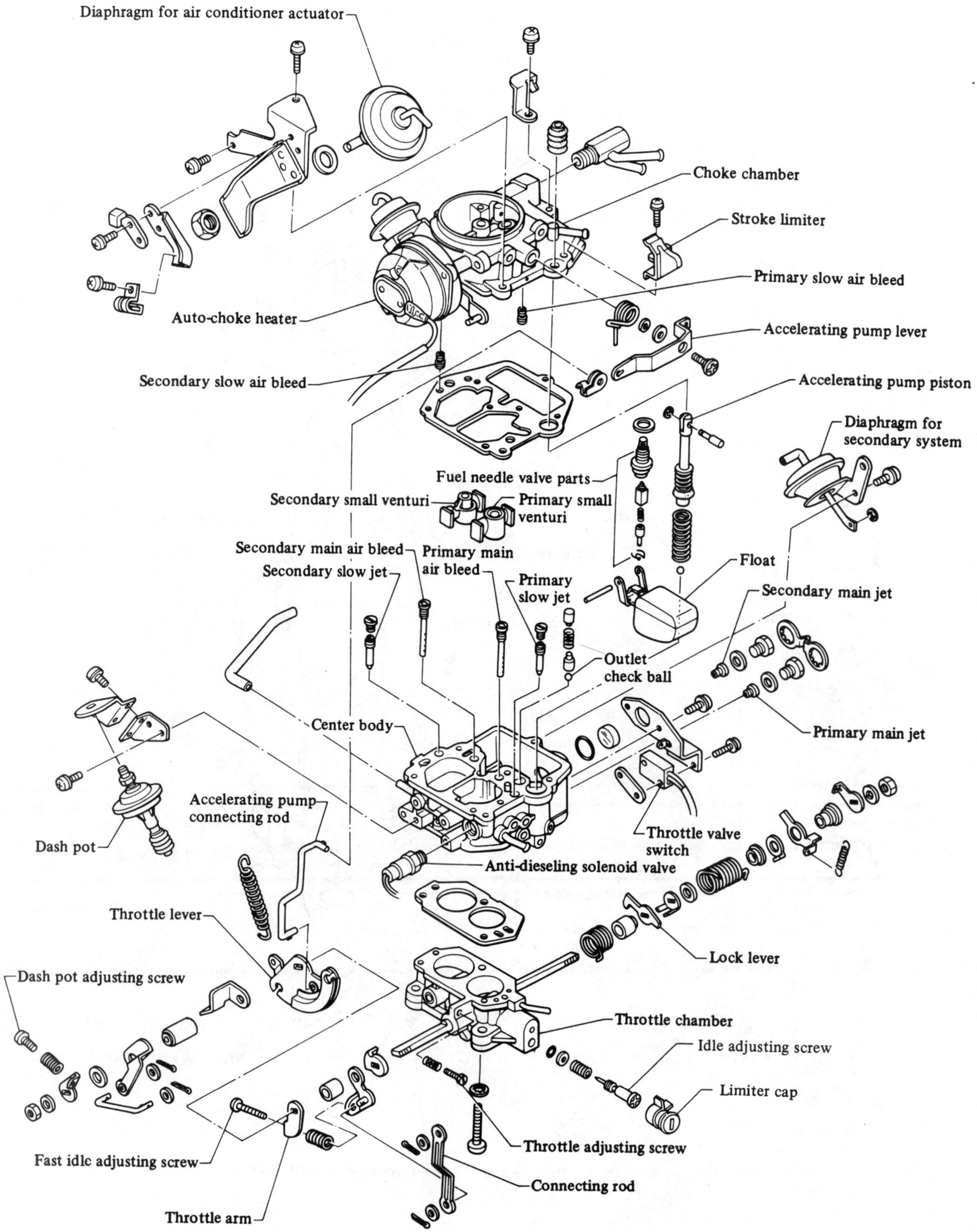

Fig. 3.11 Exploded view of the E16 engine carburettor for Canada (Sec 8)

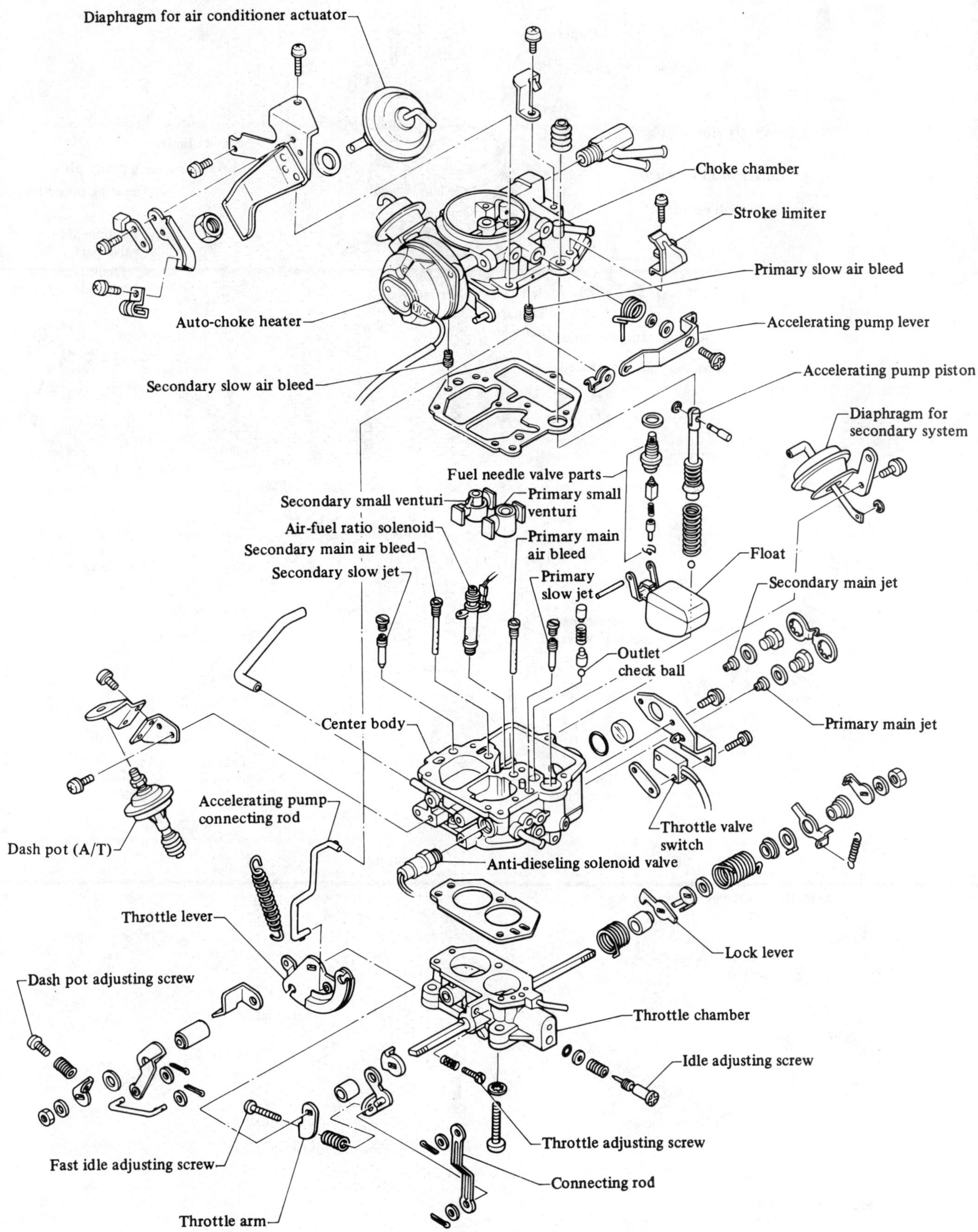

Fig. 3.12 Exploded view of the E16 engine carburettor for California (Sec 8)

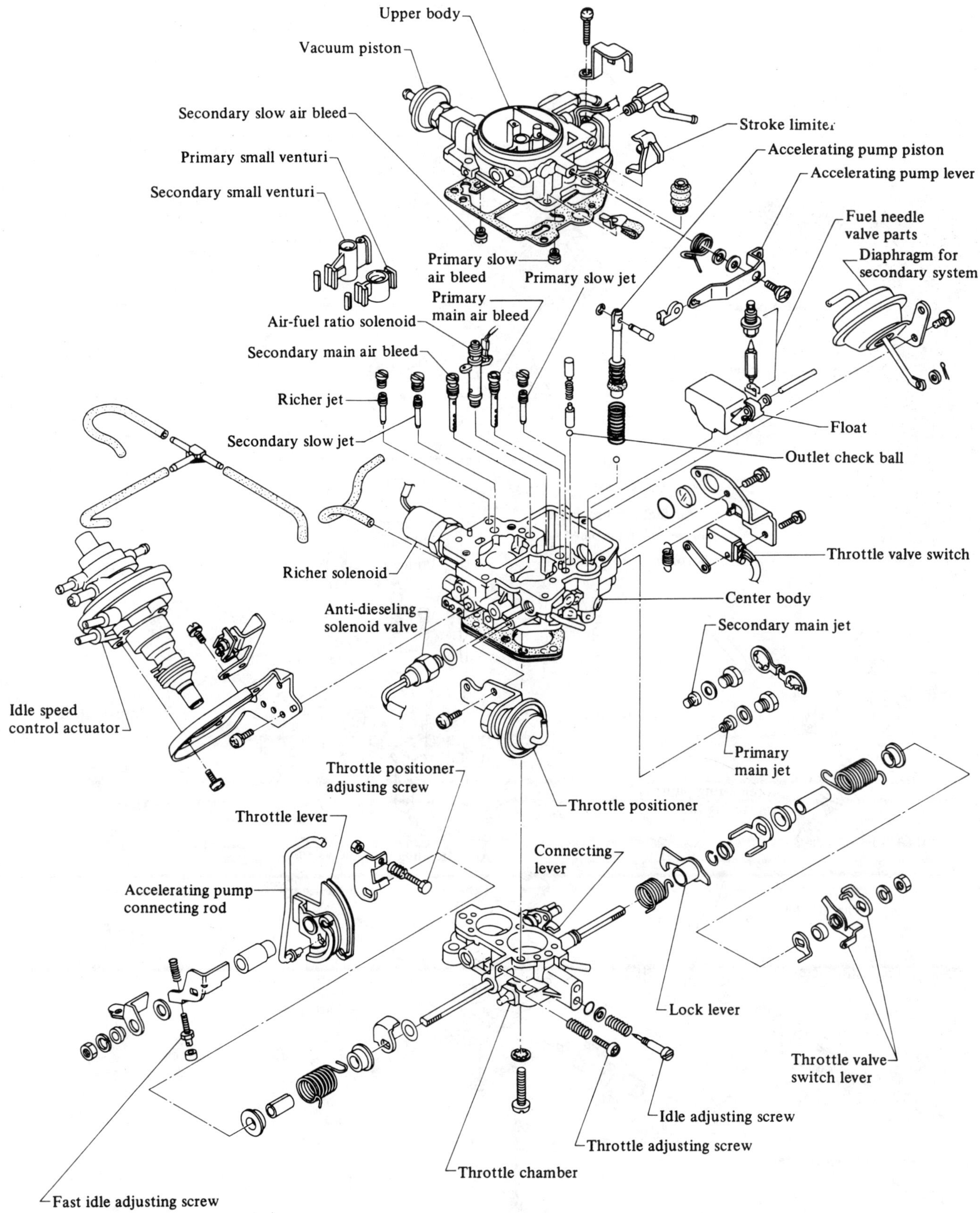

Fig. 3.13 Exploded view of the E16 engine carburettor for USA (except California) (Sec 8)

9 Carburettor – idle speed and mixture adjustment (UK models)

1 The mixture is preset during production and should not normally require altering. However, adjustment may be necessary if the carburettor has been overhauled or after a high mileage when the engine characteristics may have changed slightly due to the build-up of carbon or wear in the engine components.
2 On some models, the idle mixture screw is fitted with a limiter cap (photo). The screw can be turned if a screwdriver blade is ground to a shape similar to that shown in Fig. 3.14.

9.2 Idle mixture screw (A) and throttle speed screw (B)

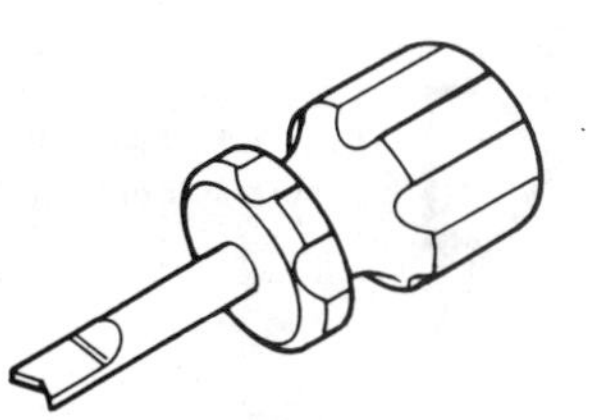

Fig. 3.14 Special screwdriver used to turn idle mixture screw (Sec 9)

3 Have the engine at normal operating temperature with the ignition timing and valve clearances correctly set.
4 Connect a tachometer to the engine in accordance with the manufacturer's instructions.
5 If an air conditioner is fitted, make sure that it is switched off.
6 With the engine idling, turn the throttle speed screw (photo 9.2) as necessary to bring the speed within the range given in Specifications.
7 For accuracy, the idle mixture should be adjusted using a CO meter (exhaust gas analyser). Rev up the engine two or three times to clear it and then let it idle. Turn the idle mixture screw until the meter indicates a CO content within the specified tolerance. This adjustment should be carried out quickly. If it extends over more than two minutes, rev the engine again before resuming adjustment.
8 If an exhaust gas analyser is not available, carry out the following alternative method of adjusting the idle mixture. Turn the idle mixture screw until the engine speed is at its highest level and does not increase any further. Make sure that the engine is idling smoothly and then readjust the throttle speed screw to bring the idle speed within the specified range.
9 If the territory in which the vehicle is being used is subject to strict emission regulations the idle mixture CO content should **always** be checked with an exhaust gas analyser.
10 If the idle mixture screw has been removed during carburettor overhaul, a starting point for mixture adjustment can be established if the screw is turned in very gently until it just seats and then unscrewed two full turns.
11 Note that on power steering models the front wheels should be in the straight ahead position to ensure the pump does not affect the idle speed.

10 Carburettor – idle speed and mixture adjustment (North American models)

1 The operations are similar to those described in the preceding Section, but observe the following special points.
2 To remove the idle mixture screw plug, drill a hole in it and prise it out using a thin rod or screwdriver. It is recommended that the carburettor is removed from the engine to do this.
3 Make sure that all electrical accessories are switched off. If the radiator cooling fan comes on during adjustment, wait until it switches off before continuing.
4 Where fitted the air injection hoses should be disconnected at the air cleaner end of the pipes, and the pipes plugged.
5 On automatic transmission models check that the idle speed is correct with D engaged, and if necessary make final adjustments to compensate.
6 On California models disconnect the air/fuel ratio solenoid harness connector (Fig. 3.16) only while adjusting the idle mixture setting.
7 Where fitted on non-Californian models disconnect the vacuum control modulator harness connector (Fig. 3.17) only while adjusting the idle speed setting.
8 Fit a new sealing plug to the mixture screw hole on completion.

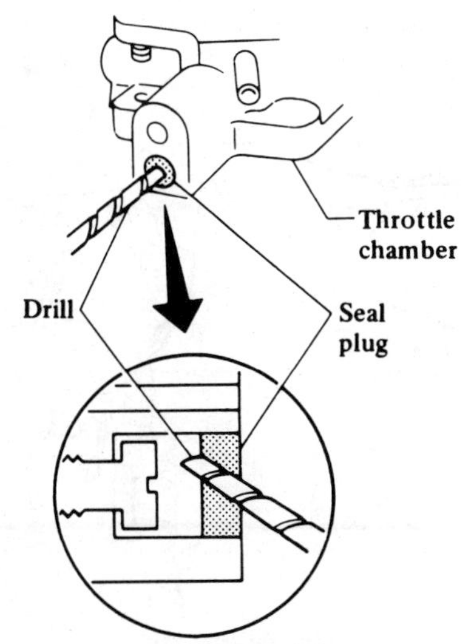

Fig. 3.15 Idle mixture screw plug removal (Sec 10)

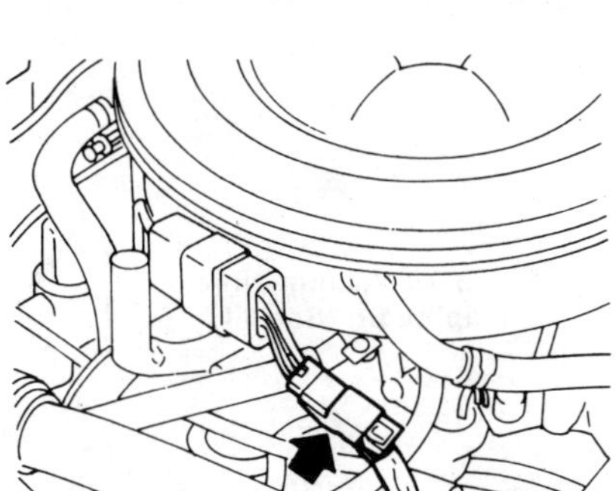

Fig. 3.16 Air/fuel solenoid connector on California models (Sec 10)

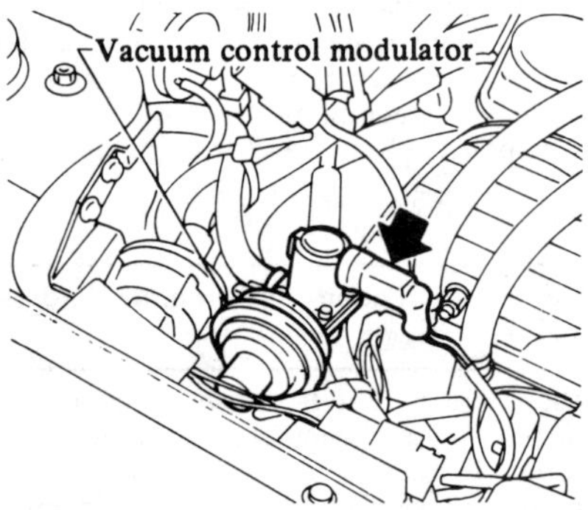

Fig. 3.17 Vacuum control modulator and connector on non-California models (Sec 10)

11 Carburettor – adjustments (UK models)

Fast idle (automatic choke)

1 Bring the engine to normal operating temperature. Remove the choke housing cover.

2 Set the fast idle lever on the second step of the fast idle cam.

3 Check that the fast idle engine speed is as given in the Specifications. If it is not, turn the fast idle screw as necessary.

Vacuum break

4 With the engine cold, remove the air cleaner and close the choke valve plate.

5 Depress the diaphragm rod fully and then check the clearance R is as specified (see Fig. 3.20). The clearance is measured between the edge of the valve plate and the wall of the carburettor. Where necessary, bend the choke lever.

Choke unloader (automatic choke)

6 With the engine cold remove the air cleaner and automatic choke housing cover, and close the choke valve plate.

7 Turn the throttle lever until the primary throttle valve is fully open. If it is not, bend the tongue of the unloader. The clearance C should be as specified (see Fig. 3.21).

Primary and secondary valve plate interlock

8 Turn the throttle lever until the throttle arm contacts the lock lever at point A (see Fig. 3.22).

9 Check that the clearance G is as specified. If it is not, bend the tongue on the throttle arm as necessary.

Dashpot (automatic transmission)

10 Have the engine at normal operating temperature and idling.

11 Turn the throttle lever on the carburettor by hand and have an

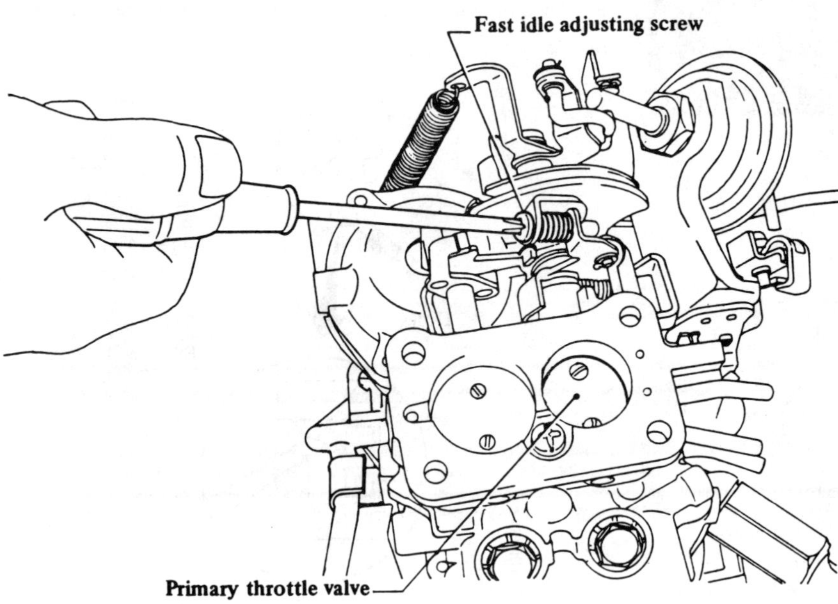

Fig. 3.18 Fast idle adjusting screw location (Sec 11)

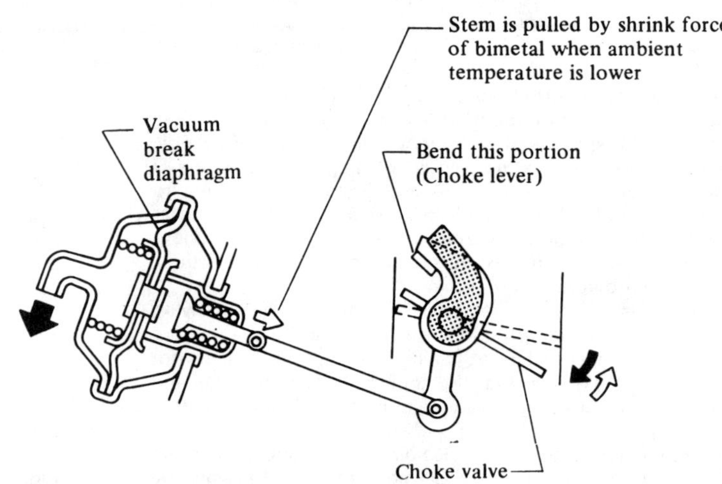

Fig. 3.19 Vacuum break diaphragm (Sec 11)

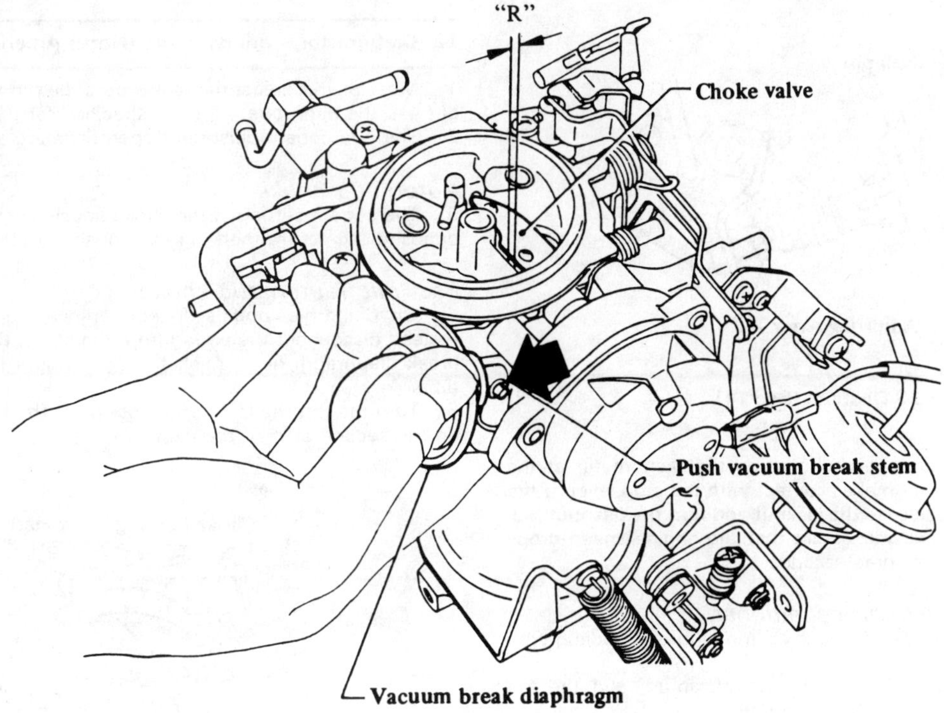

Fig. 3.20 Checking vacuum break adjustment (Sec 11)

R = 1.27 to 1.45 mm (0.05 to 0.06 in) for E10 and E13 engine
R = 1.18 to 1.36 mm (0.046 to 0.054 in) for E15 engine

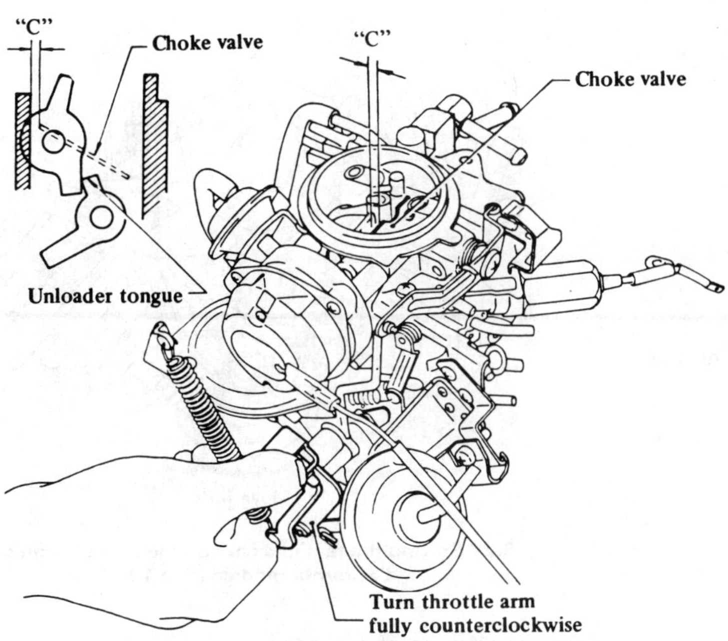

Fig. 3.21 Checking setting of choke unloader (Sec 11)

C = 1.74 mm (0.685 in) for E10 and E13 engine
C = 2.01 mm (0.0791 in) for E 15 engine

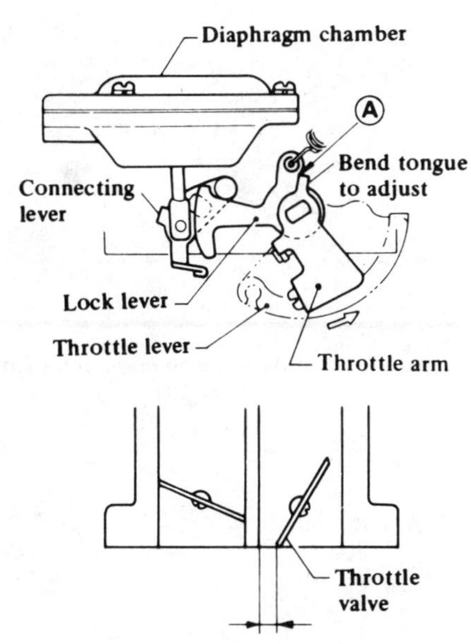

Fig. 3.22 Valve plate interlock setting diagram (Sec 11)

G = 5.84 to 5.90 mm (0.230 to 0.232 in) for E10 and E13 engine
G = 5.80 to 5.86 mm (0.228 to 0.231 in) for E15 engine

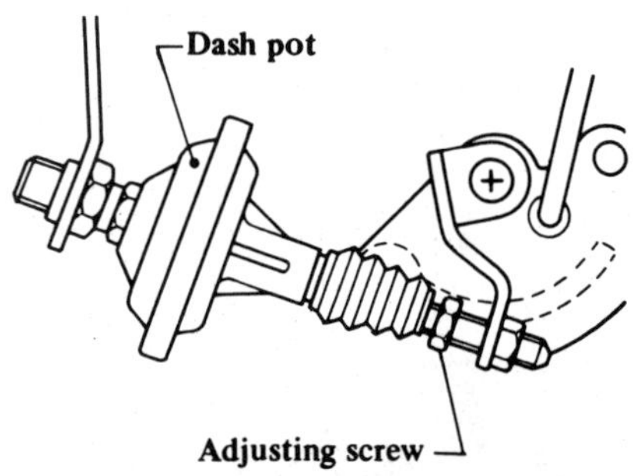

Fig. 3.23 Dashpot (Sec 11)

assistant record the engine speed shown on the tachometer at the point where the dashpot just makes contact with the stop lever. If the speed is not as specified, release the locknut and turn the dashpot rod.
12 Tighten the locknut and make sure that the engine speed drops from 2000 to 1000 rpm in three seconds.

Fast idle control device (air conditioner)

13 Have the engine at normal operating temperature and idling with the air conditioner off.
14 Set the air conditioner intake lever to maximum and the fan control to position 3, then switch on the unit.
15 Check that the engine idles at 800 ± 50 rpm for manual transmission models, and 620 ± 40 rpm for automatic transmission models (D engaged). If not, adjust the FICD unit as necessary (Figs. 3.24 and 3.25).

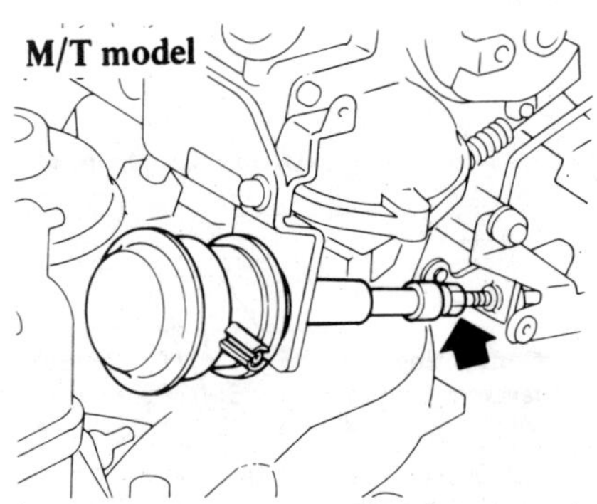

Fig. 3.24 Fast idle control device on manual transmission models (Sec 11)

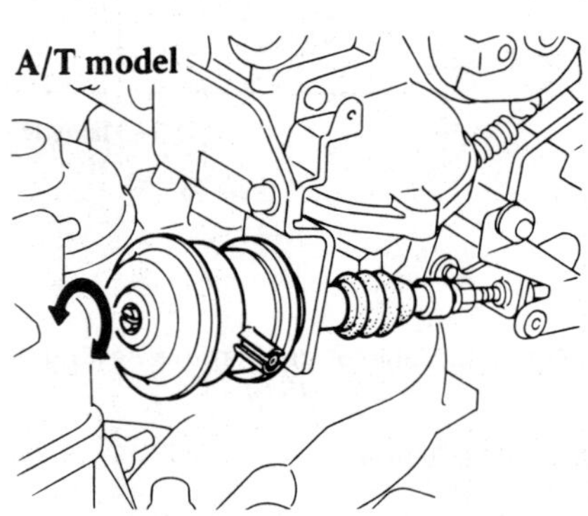

Fig. 3.25 Fast idle control device on automatic transmission models (Sec 11)

12 Carburettor – adjustments (North American models)

1 Most of the adjustments are described in the preceding Section, but use the appropriate figures specified for these models.
2 The additional adjustment operations are as follows.

Automatic choke

3 The correct setting of the choke housing cover is for its mark to be opposite the centre mark on the housing scale.

Fast idle (automatic choke)

4 On California models do not remove the choke housing cover. Instead disconnect the wiring from the cover. Disconnect the vacuum break diaphragm hose, push the unit control rod, and plug the inlet tube.
5 Turn the throttle lever anti-clockwise. The fast idle cam is now set on the second step of the cam.

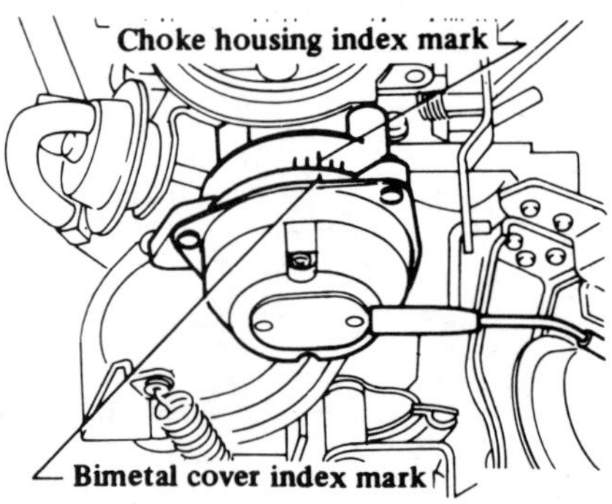

Fig. 3.26 Automatic choke housing and cover markings (Sec 12)

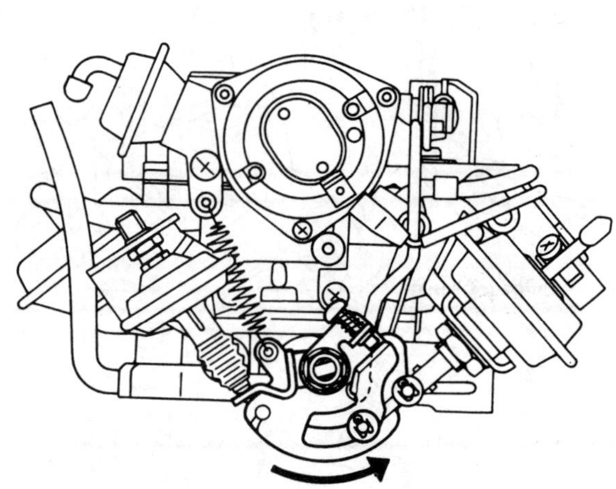

Fig. 3.27 Setting the fast idle cam on the cam second step – California models (Sec 12)

Idle speed control actuator

6 With the engine at normal operating temperature run it at idle speed.
7 Disconnect and plug the vacuum sensor vacuum hose and vacuum switch vacuum hose (Fig. 3.28).
8 Disconnect and plug the vacuum cut solenoid valve hose on the air filter side.
9 Check that the idle speed is 2800 to 3200 rpm for manual transmission models and 3200 to 3600 rpm for automatic transmission models (N selected). If not, turn the adjustment screw (Fig. 3.30).
10 Reconnect the hoses then turn the diagnosis start switch on.

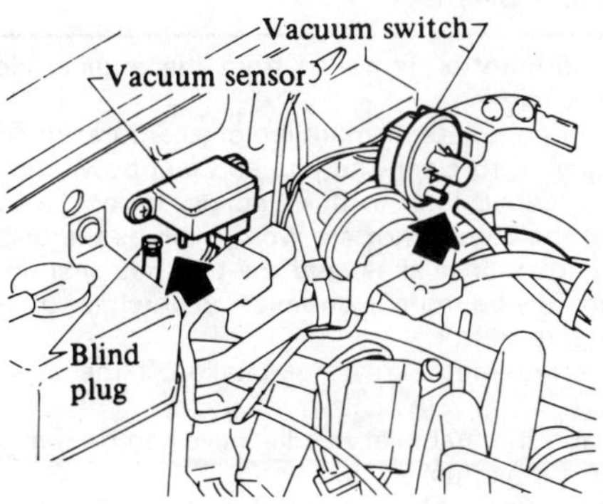

Fig. 3.28 Vacuum switch and sensor location (Sec 12)

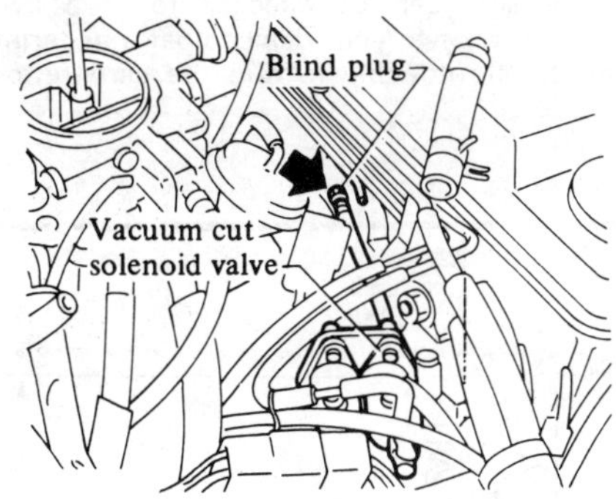

Fig. 3.29 Checking idle speed control actuator (Sec 12)

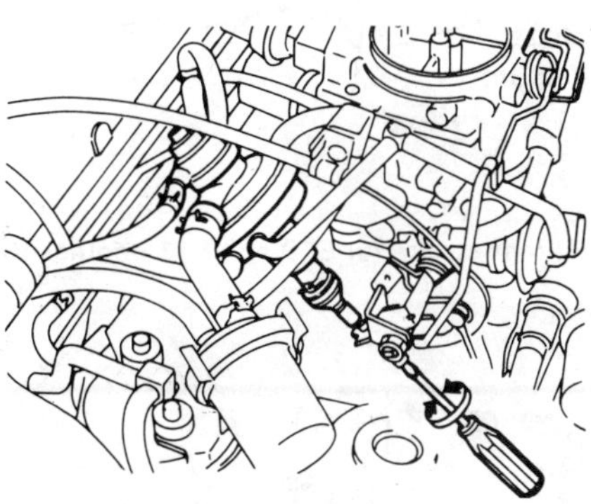

Fig. 3.30 Adjusting idle speed control actuator (Sec 12)

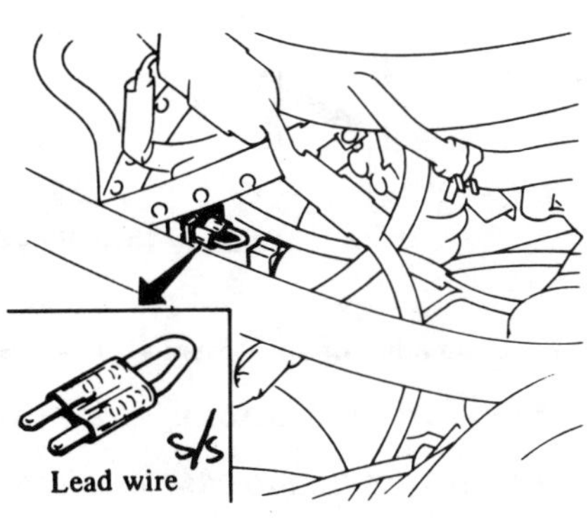

Fig. 3.31 Bridging water temperature sensor (Sec 12)

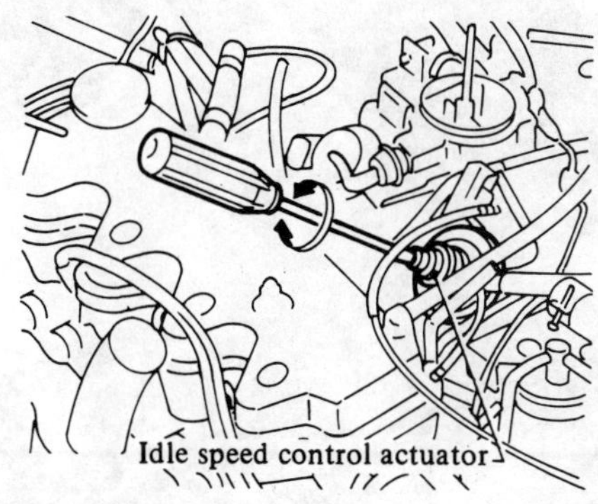

Fig. 3.32 Adjusting idle speed control actuator (Sec 12)

11 Disconnect the water temperature sensor harness connector and bridge the two terminals.
12 Check that the idle speed is now 1500 to 1900 rpm. If not turn the adjustment screw on the top of the idle speed control actuator.
13 Switch off the diagnosis start unit and reconnect the wiring.

Dashpot
14 Where fitted the procedure is identical to that given in Section 11.

13 Carburettor – removal and refitting

1 Remove the air cleaner.
2 Disconnect the accelerator control cable from the carburettor.
3 Disconnect the choke control cable (photo) or the electrical lead (automatic choke).
4 Disconnect the lead from the fuel cut-off solenoid valve.
5 Where fitted (see Section 21) disconnect the leads from the terminals of the mixture heating system.
6 Disconnect and plug the fuel hoses (photo).
7 Unbolt and remove the carburettor from the intake manifold. Take care not to drop the mounting nuts into the carburettor intake.
8 Refitting is a reversal of removal, but always use a new gasket at the manifold joint.

13.3 Manual choke cable at carburettor

13.6 Disconnecting the carburettor fuel hose

14 Carburettor – overhaul

1 With the carburettor removed from the engine, clean away all external dirt and grease.
2 The need for complete dismantling of a carburettor seldom occurs. The usual reason is to clean the jets and fuel bowl and to check the adjustments described later in this Section. In fact, where the major components of the carburettor are worn, such as the throttle or choke valve plate spindles or bushes, and the unit has seen long service, it will almost certainly be more economical to purchase a new or factory-reconditioned carburettor.
3 Extract the retaining screws and take off the top cover (choke chamber).
4 The float and the fuel inlet needle valve can be removed once the float arm pivot pin is pushed out.
5 Clean out the float chamber and clean the jets and bleed holes by applying air pressure from a tyre pump. If the jets are badly clogged or are suspected of being the wrong ones they should be unscrewed. *Never probe a jet with wire,* but if air pressure fails to clear it use a nylon bristle. The jets can be checked for size by quoting the carburettor index number to your dealer's parts department.
6 To remove the throttle block from the carburettor main body,

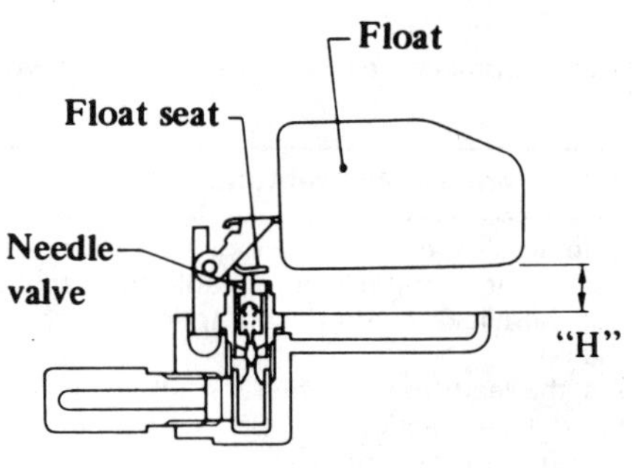

Fig. 3.33 Float setting diagram (Sec 14)

H = 15.0 mm (0.59 in) for E10 and E13 engine
H = 12.0 mm (0.47 in) for E15 and E16 engine

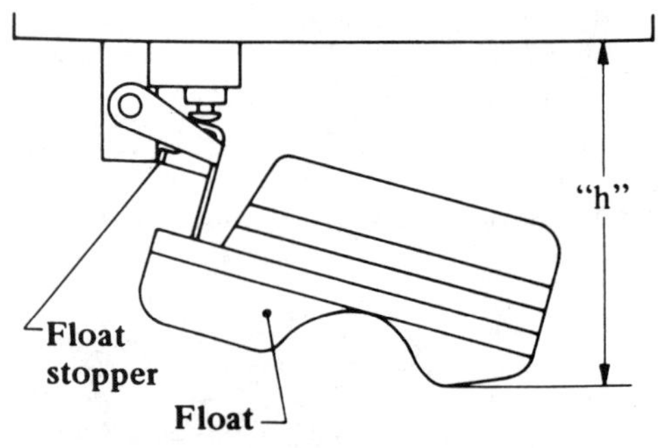

Fig. 3.35 Float stroke for E10 and E13 engines (Sec 14)

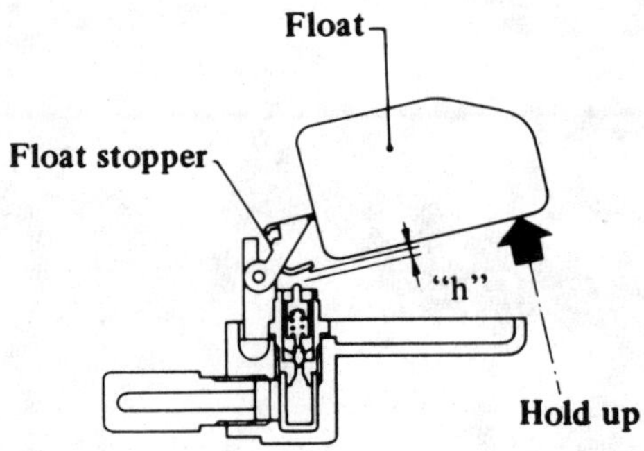

Fig. 3.34 Float-to-needle clearance for E15 and E16 engines (Sec 14)

h = 1.3 to 1.7 mm (0.051 to 0.067 in)

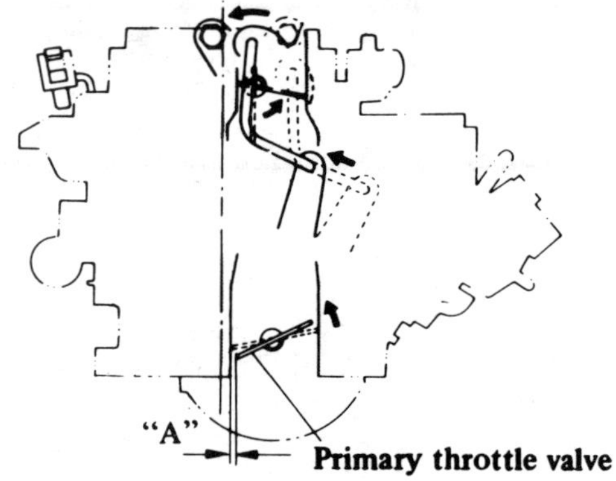

Fig. 3.36 Fast idle adjustment – manual choke (Sec 14)

A = 1.37 to 1.51 mm (0.0539 to 0.0595 in) for E10 and E13 engine
A = 1.42 to 1.56 mm (0.0559 to 0.0614 in) for E15 and E16 engine

disconnect the choke linkage, invert the carburettor and extract the fixing screws.
7 Obtain a repair kit for your carburettor which will contain all the necessary gaskets, seals and other renewable items.
8 Reassemble by reversing the dismantling operations, but as work progresses check the following adjustments and settings. A twist drill is useful when measuring valve plate clearances, as the precise diameter will serve as a gauge.

Float setting
9 Invert the carburettor cover so that the float arm rests on the fuel inlet needle valve under its own weight. Measure the distance H between the surface of the float and the face of the top cover (see Fig. 3.33).
10 If this is not as specified, carefully bend the float seat.
11 On E15 and E16 engined models, check the clearance H between the float seat and the end of the fuel inlet needle valve (see Fig. 3.34). If it is not as specified, bend the float stop.
12 On E10 and E13 engined models the float stroke should be measured (see Fig. 3.35). If not as specified, bend the float stop.
13 With the carburettor in the vehicle, the fuel level can be checked through the sight glass on the fuel bowl.

Fast idle (manual choke)
14 Close the choke valve plate and then check the gap A between the edge of the primary throttle valve plate and the wall of the carburettor (see Fig. 3.36). Adjust the clearance to specification by turning the fast idle screw.

Accelerator pump
15 To check the volume of fuel ejected from the fuel pump, pour fuel into the float chamber and then turn the throttle lever from fully closed to fully open ten times. Hold the lever in the fully open position for a period of three seconds between strokes.
16 Catch the ejected fuel in a measuring glass and then divide the total volume by ten to obtain the volume of fuel ejected per stroke. If this is not as specified, check the pump linkage for wear or distortion.
17 On North American models a pump stroke limiter is fitted. To check the setting, actuate the accelerator pump rod until the pump lever just comes into contact with the piston pin. Now measure the gap between the edge of the primary throttle valve plate and the carburettor throat wall (see Fig. 3.37). If it is not as given in Specifications, bend the stroke limiter.
18 If the solenoid fuel cut-off valve was removed, screw it in, using a new sealing washer, to the specified torque setting.

15 Electronically controlled carburettor (ECC) – description

1 The electrically controlled carburettor is fitted to certain North American models. It incorporates solenoid operated orifices to control the fuel mixture instead of a choke valve as used in a conventional carburettor. A vacuum controlled idle speed actuator controls the idle speed together with an electrically controlled modulator.
2 A micro-computer control unit is used to monitor engine speed, load and temperature and it is this which determines the carburettor settings. The same control unit controls the ignition timing.
3 An exhaust gas sensor fitted in the exhaust manifold senses the quantity of oxygen in the exhaust gas and feeds the information to the control unit.
4 If the system develops a fault check that the wiring and connectors are intact and secure.
5 Note that the battery should be disconnected before disconnecting the ECC wiring otherwise the control unit may be damaged.

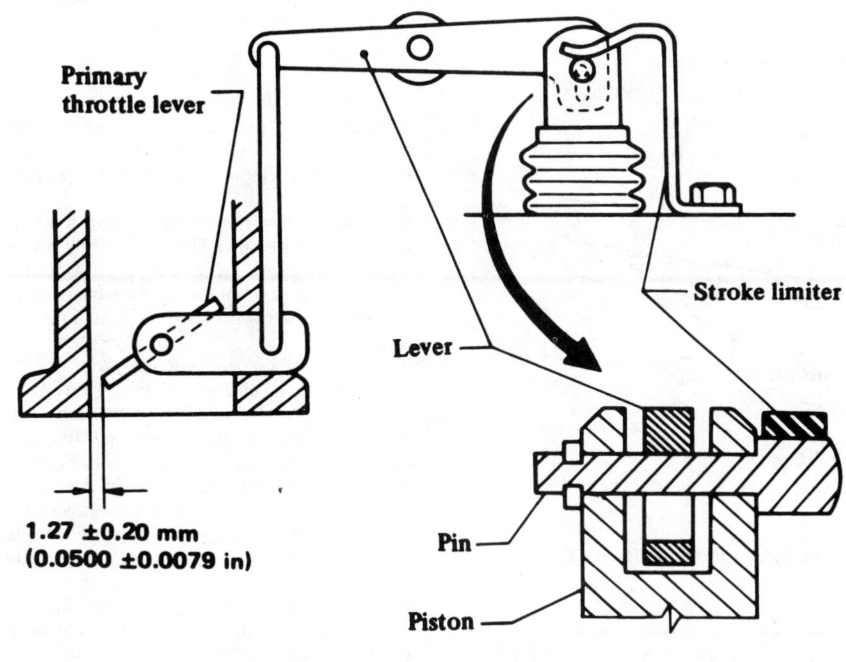

Fig. 3.37 Accelerator pump – North American models (Sec 14)

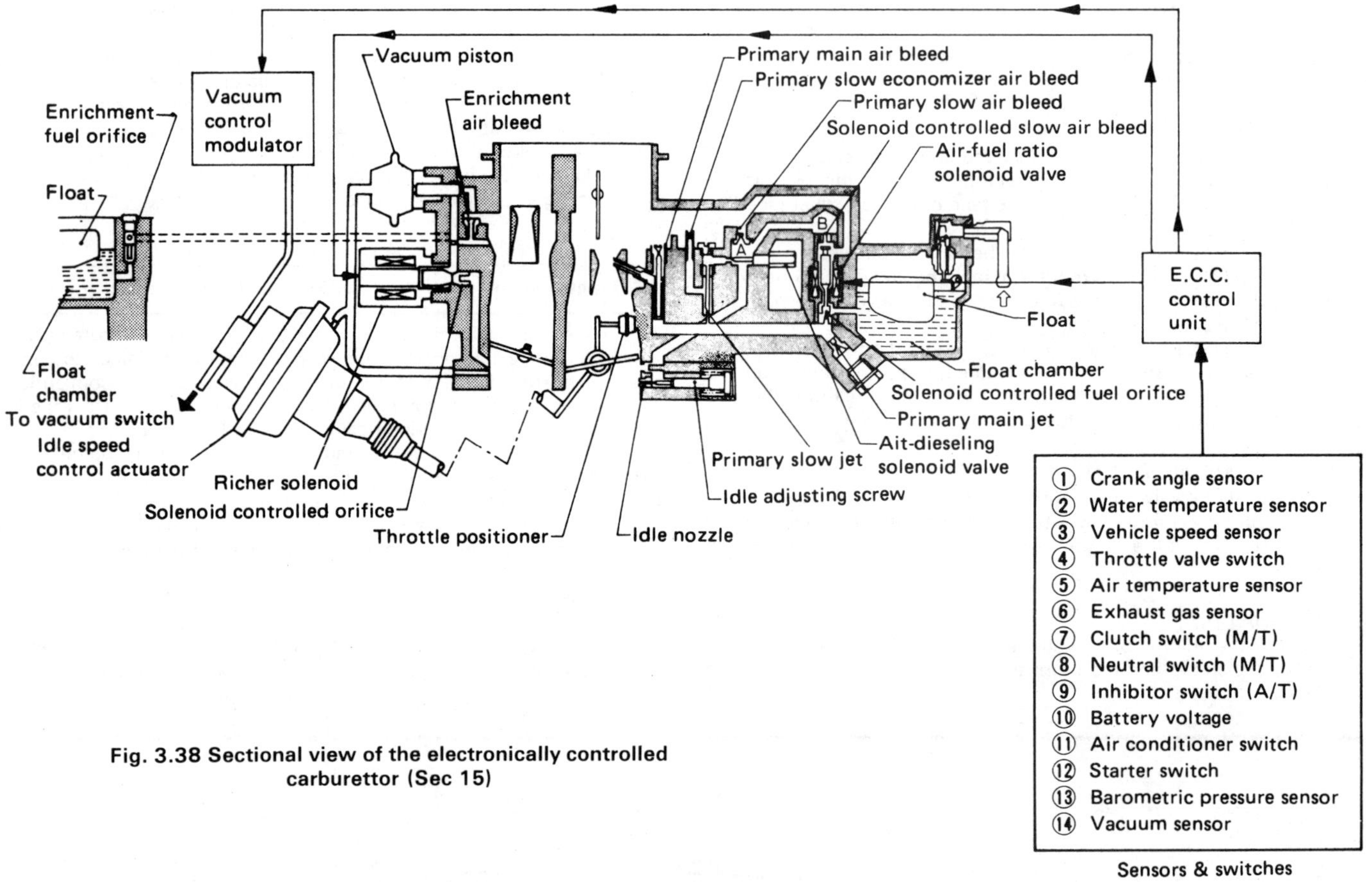

Fig. 3.38 Sectional view of the electronically controlled carburettor (Sec 15)

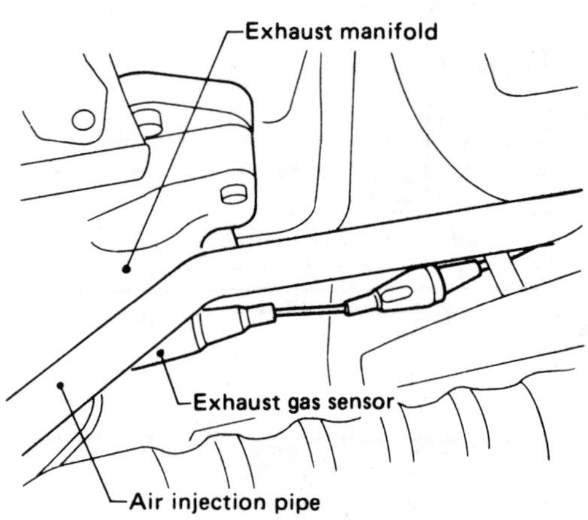

Fig. 3.39 Exhaust gas sensor location (Sec 15)

16 Electronic concentrated engine control system (ECCS) – description

1 The electronic concentrated engine control system is fitted to Turbo models (E15ET engine). The system components are shown in Fig. 3.40. A micro-computer control unit is used to control fuel injection, ignition timing, idle speed, fuel pump operation and mixture ratio feedback.

2 The roller type fuel pump is located in the fuel tank and is integral with the fuel transmitter unit.

3 The injectors operate on the solenoid principle, the quantity of fuel injected being proportional to the length of time they are energised.

4 The throttle valve switch is located on the throttle chamber and incorporates an idle contact which is closed when the throttle is at the idle position.

5 The air flow meter incorporates a flap, a bypass circuit, and an air temperature sensor. The unit sends an electrical signal to the control unit which varies according to the quantity of air passing through it.

6 The idle control valve stabilizes the engine idling speed when the electrical load on the alternator is altered, such as when turning on the headlamps.

7 The exhaust gas sensor monitors the amount of oxygen in the exhaust gases, and sends the information to the control unit.

8 The detonation sensor is located on the rear of the cylinder block. It detects the onset of engine knocking and the control unit then adjusts the ignition timing accordingly.

9 The pressure regulator maintains the fuel pressure 36.3 lbf/in^2 (2.6 bar) above the vacuum (negative pressure) within the intake manifold. This is necessary to ensure accurate metering of fuel by the injectors.

10 The air regulator bypasses the throttle valve and provides extra air for starting when the engine is cold.

11 The turbocharger utilizes the flow of exhaust gases to rotate a compressor turbine in the air intake system. The unit incorporates a waste gate to prevent excessive inlet pressures, and in addition the intake manifold is fitted with a relief valve.

12 The ECCS system incorporates micro-computer circuitry, and if a fault occurs, a set self-diagnosis procedure must be followed in order to detect the area or component which is malfunctioning. The control unit is fitted with a red and green Light Emitting Diode (LED) which blink in coded fashion to indicate the area malfunctioning. Therefore it is recommended that the system is checked by a qualified person in the event of a fault. However the work outlined in Sections 17 and 18 can be carried out by the home mechanic.

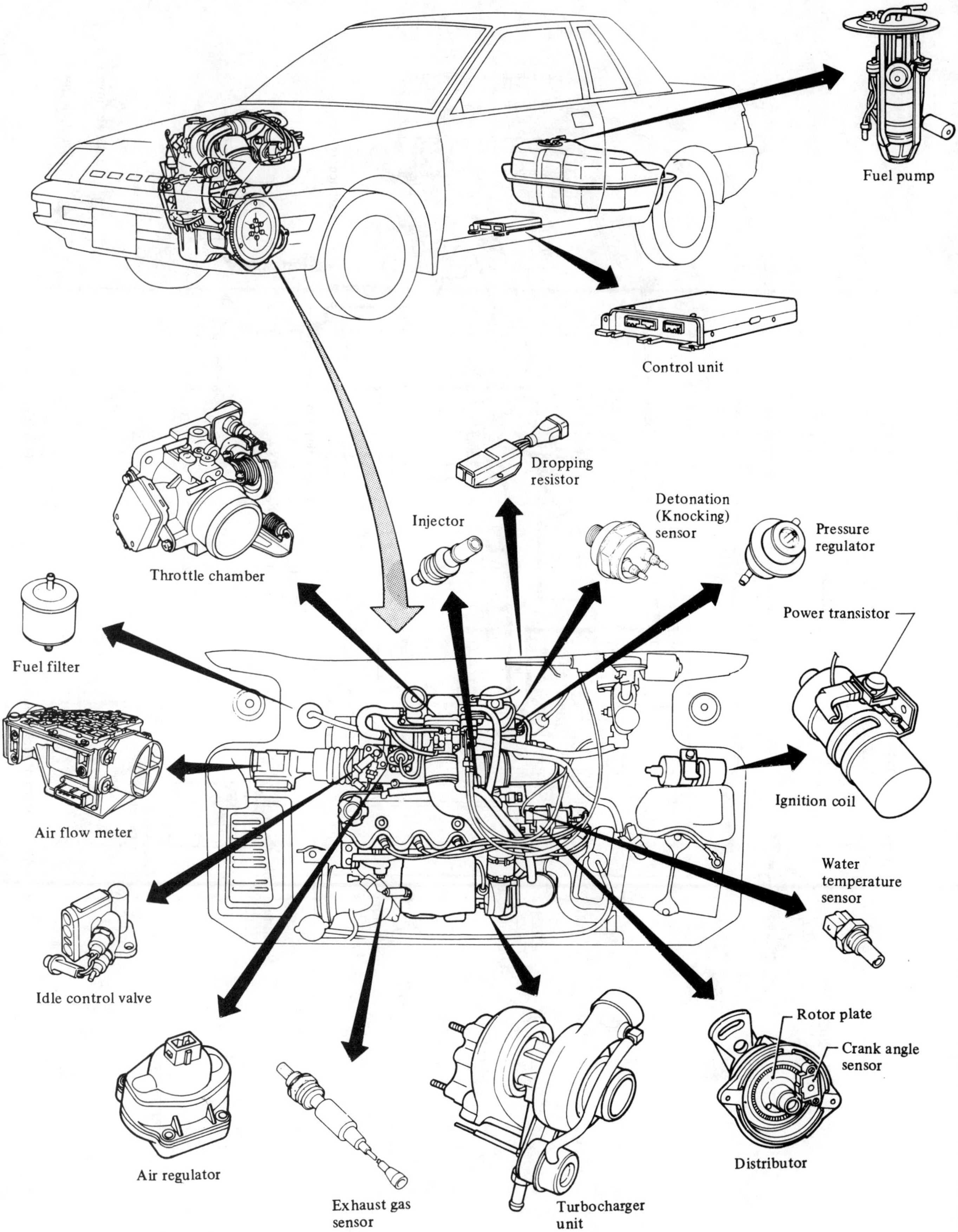

Fig. 3.40 Electronic concentrated engine control system components (Sec 16)

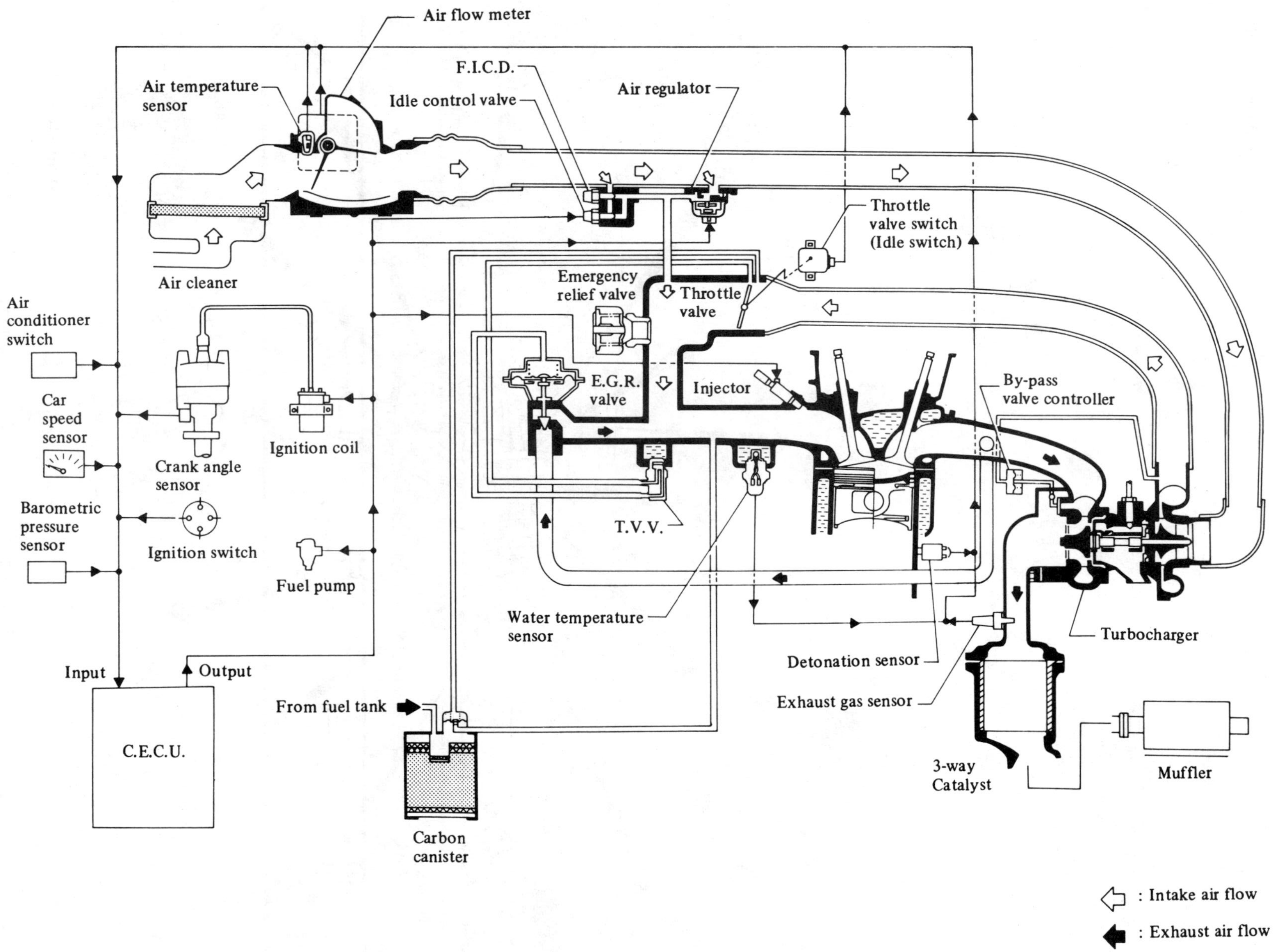

Fig. 3.41 Diagram of electronic concentrated engine control system (Sec 16)

17 Electronic concentrated engine control system – adjustments

Throttle valve switch

1 Disconnect the wiring plug from the switch.
2 Using an ohmmeter check that continuity exists between the two terminals in the plug.
3 Reconnect the plug then run the engine at 1100 rpm and adjust the switch position so that the internal contacts just separate.
4 Switch off the engine.

Idle control valve

5 The idle control valve adjustment is preset by the factory and will not normally require further adjustment. However if the adjustment has been turned by mistake, proceed as follows.
6 Run the engine to normal operating temperature then let it idle.
7 Disconnect the wiring plug from the valve and check that the engine idles at the normal idling speed.
8 Now, using suitable leads, apply battery voltage to the wiring plug terminals and check that the engine speed increases by 90 to 120 rpm. If not, prise out the rubber plug and turn the adjusting screw A (Fig. 3.43) as necessary.
9 Stop the engine, refit the rubber plug, and reconnect the plug.

Idle speed and mixture

10 Adjustment of the idle mixture setting requires the use of a resistor, therefore this work should be entrusted to a Nissan dealer.
11 Idle speed adjustment is not normally necessary, however the adjusting screw is located on the idle control valve (Fig. 3.43).

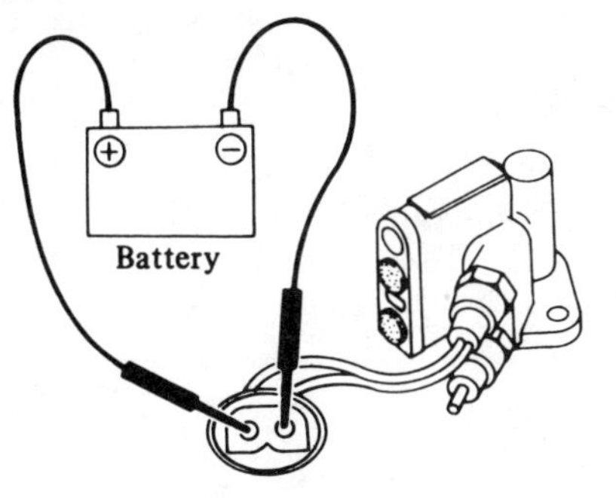

Fig. 3.42 Energising the ECCS idle control valve (Sec 17)

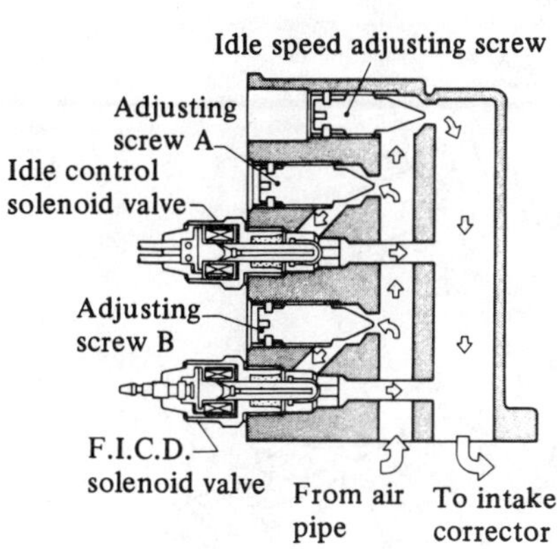

Fig. 3.43 Cross section of the ECCS idle control valve (Sec 17)

18 ECCS components – removal and refitting

Fuel pressure regulator

1 Release any fuel pressure in the system by following the preliminary procedure given in Section 4.

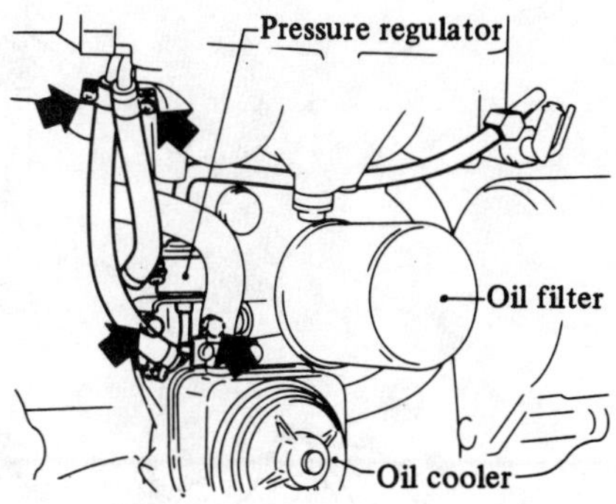

Fig. 3.44 ECCS fuel pressure regulator removal (Sec 18)

2 Disconnect the intake manifold vacuum hose from the regulator.
3 Unbolt the pressure regulator then disconnect the fuel hose.
4 Refitting is a reversal of removal.

Turbocharger (UK models)

5 Disconnect and remove the blow-by hose, air inlet pipe, air pipe, and air duct hose.
6 Unbolt and remove the heat insulator and temperature sensor.
7 Unscrew the union and disconnect the oil delivery pipe.
8 Disconnect the oil drain hose.
9 Unscrew the nuts from the exhaust downpipe flange.
10 Unbolt the support bracket then unscrew the manifold nuts and remove the manifold together with the turbocharger.
11 Unscrew the nuts and separate the turbocharger from the manifold.
12 Refitting is a reversal of removal, but use new gaskets.

Turbocharger (North American models)

13 Disconnect and remove the heat insulator, inlet tube, air duct, suction air pipe, and temperature sensor.
14 Disconnect the wiring from the exhaust gas sensor.
15 Disconnect the front hose and oil drain pipe.
16 Unscrew the union and disconnect the oil delivery pipe.
17 Unbolt the catalytic converter support.
18 Unscrew the nuts from the exhaust downpipe flange.
19 Unscrew the nuts and remove the exhaust outlet and catalytic converter assembly.
20 Unscrew the nuts and remove the exhaust manifold and turbocharger.
21 Separate the manifold from the turbocharger by unscrewing the nuts.
22 Refitting is a reversal of removal, but use new gaskets.

Injectors and fuel pipe

23 Release any fuel pressure in the system by following the preliminary procedure given in Section 4.
24 Remove the air inlet pipe and hose.
25 Disconnect the throttle and where applicable the automatic transmission kick-down cables.
26 Remove the throttle chamber.
27 Remove the crankcase ventilation valve and hose.
28 Disconnect the air pipe, also the wiring plugs from the idle control valve and air regulator.
29 Disconnect the wiring plugs from the injectors.
30 Disconnect and remove the fuel hoses.
31 Unscrew the fuel pipe mounting bolts, also the injector mounting bolts.
32 Pull out the fuel pipe and injectors.
33 Loosen the hose clips and remove the injectors from the fuel pipe.
34 Refitting is a reversal of removal, but check that the rubber insulators are serviceable. If the injector hose is renewed, wet the inner surface before pushing it onto the injector.

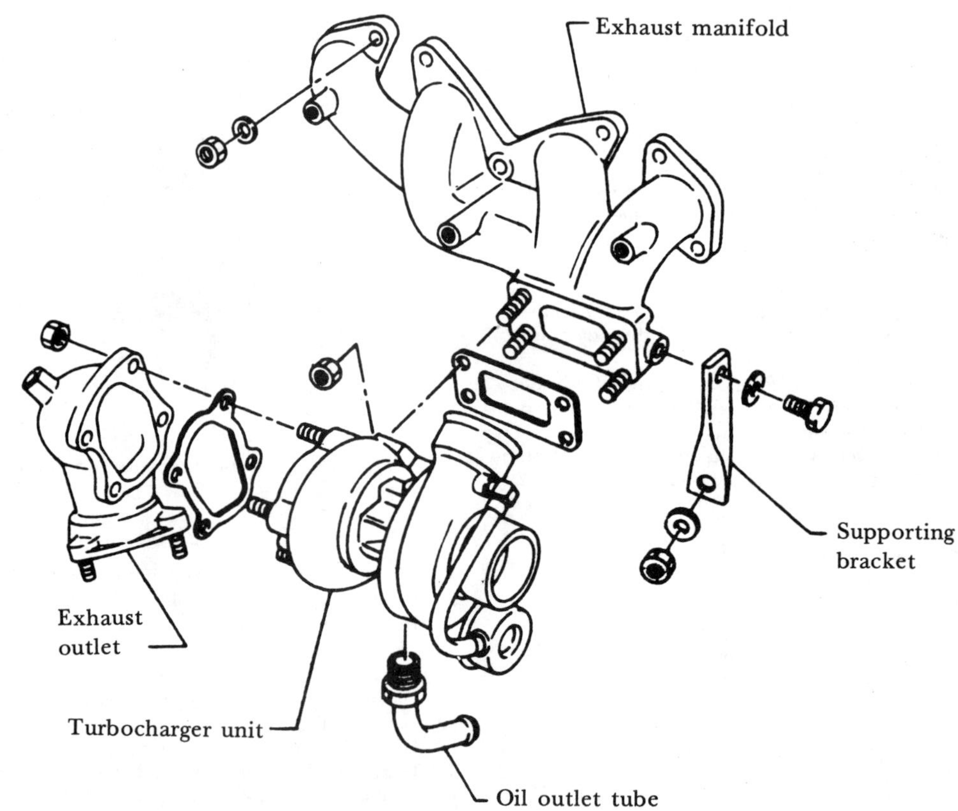

Fig. 3.45 Turbocharger components for UK models (Sec 18)

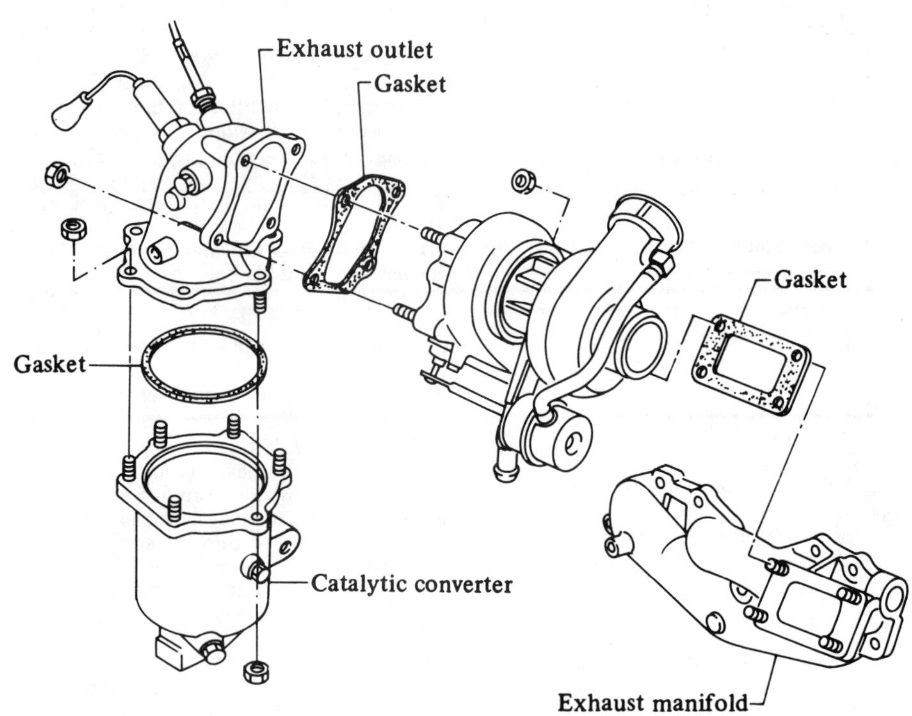

Fig. 3.46 Turbocharger components for North American models (Sec 18)

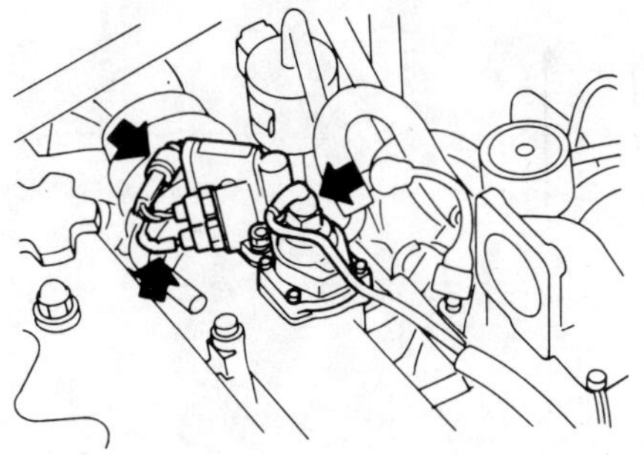

Fig. 3.47 ECCS idle control valve and air regulator wiring plug locations (Sec 18)

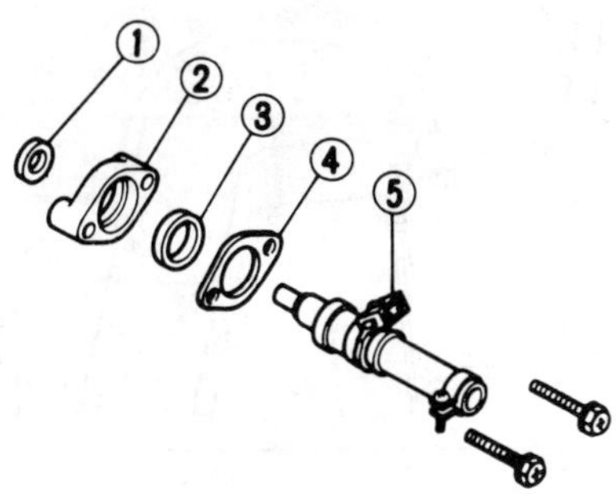

Fig. 3.48 Injector components (Sec 18)

1. *Injector lower rubber insulator*
2. *Injector lower holder*
3. *Injector upper rubber insulator*
4. *Injector upper holder*
5. *Injector*

19 Emission control systems – general

All models are equipped with some kind of emission control system. On North American versions, more sophisticated and complex systems are used.

It must be appreciated that keeping the ignition and fuel systems correctly set and the valve clearances precisely adjusted all contribute to the general reduction in the emission of noxious exhaust gases.

20 Emission control systems (UK models) – description and maintenance

Crankcase ventilation system (PCV)
1 This is described in Chapter 1.

Temperature controlled air cleaner
2 This is described in Section 3 of this Chapter.

21 Emission control systems (North American models) – description and maintenance

Crankcase emission control system
1 This is described in Chapter 1.

Temperature controlled air cleaner
2 This is described in Section 3 of this Chapter.

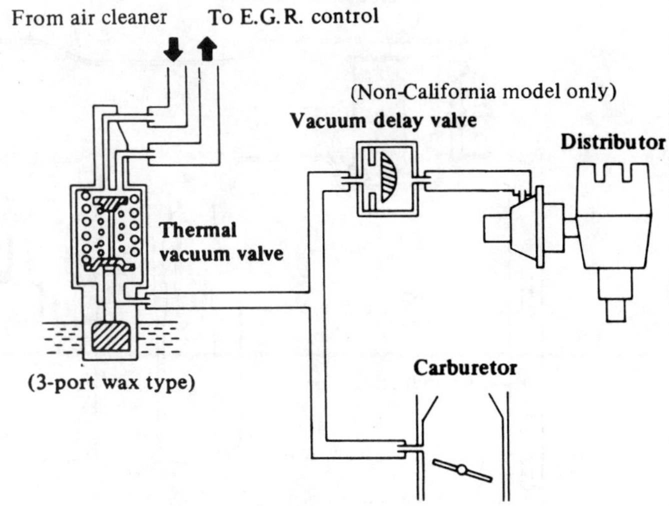

Fig. 3.49 Spark timing control system (Sec 21)

Spark timing control system
3 This is a supplementary vacuum advance system which varies the ignition advance according to prevailing engine conditions as a means of reducing the emission of noxious gases from the exhaust.
4 The system includes the following components:

Thermal vacuum valve – see EGR system

Vacuum delay valve
This delays vacuum to the distributor.
5 Check the system hoses regularly for security and condition. Connect a stroboscope to the engine when it is cold and check that the following conditions are met.

Coolant temperature	Thermal vacuum valve	Ignition spark
Below 10°C (50°F)	*Closed*	*Full advance*
Between 10° and 50°C (50° to 122°F)	*Open*	*No advance*
Above 50°C (122°F)	*Closed*	*Full advance*

6 If the advance does not operate as specified and indicated by the stroboscopic timing lamp, check the thermal vacuum valve.
7 Drain some coolant from the cooling system, remove the valve and blow into the top port while blocking the middle port. Immerse the valve in water and check that air comes out of the bottom port in accordance with the operating chart. Do not allow water to enter the valve.
6 On models with computerised ignition, spark timing control is operated by the control unit.

Air induction system (AIS)
7 This system, which is not used on Californian models, is designed to inject air into the exhaust manifold in order to dilute the CO and HC content of the exhaust gas.
8 Pressure within the exhaust manifold changes to partial vacuum at regular intervals due to the opening and closing of the exhaust valves. The volume of injected air is directly proportional to the vacuum pressure created.
9 Maintenance consists of periodically checking the security and condition of the hoses and renewing the air induction valve filter.
10 An anti-backfire (AB) valve is fitted between the air cleaner and inlet manifold to prevent after-burning in the exhaust system during the initial period of deceleration.
11 To check the AB valve, run the engine to normal operating temperature, then remove the air cleaner cover and place a finger over the valve inlet. Increase the engine speed to 3000 rpm, then release the throttle and check that brief vacuum can be felt. If not, the valve is faulty.
12 On models with a catalytic converter the system is different (Fig. 3.51) and only operates when the engine is cold.

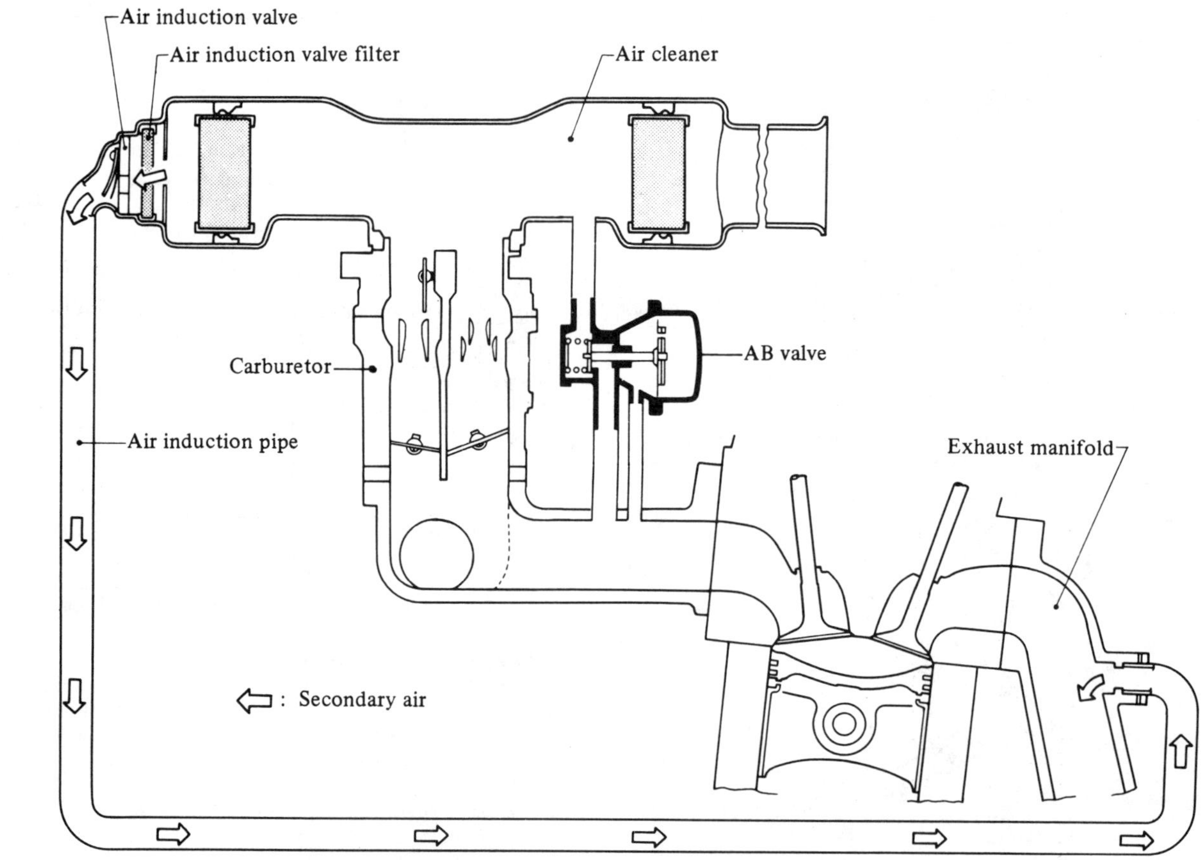

Fig. 3.50 Air induction system – non-catalytic converter models (Sec 21)

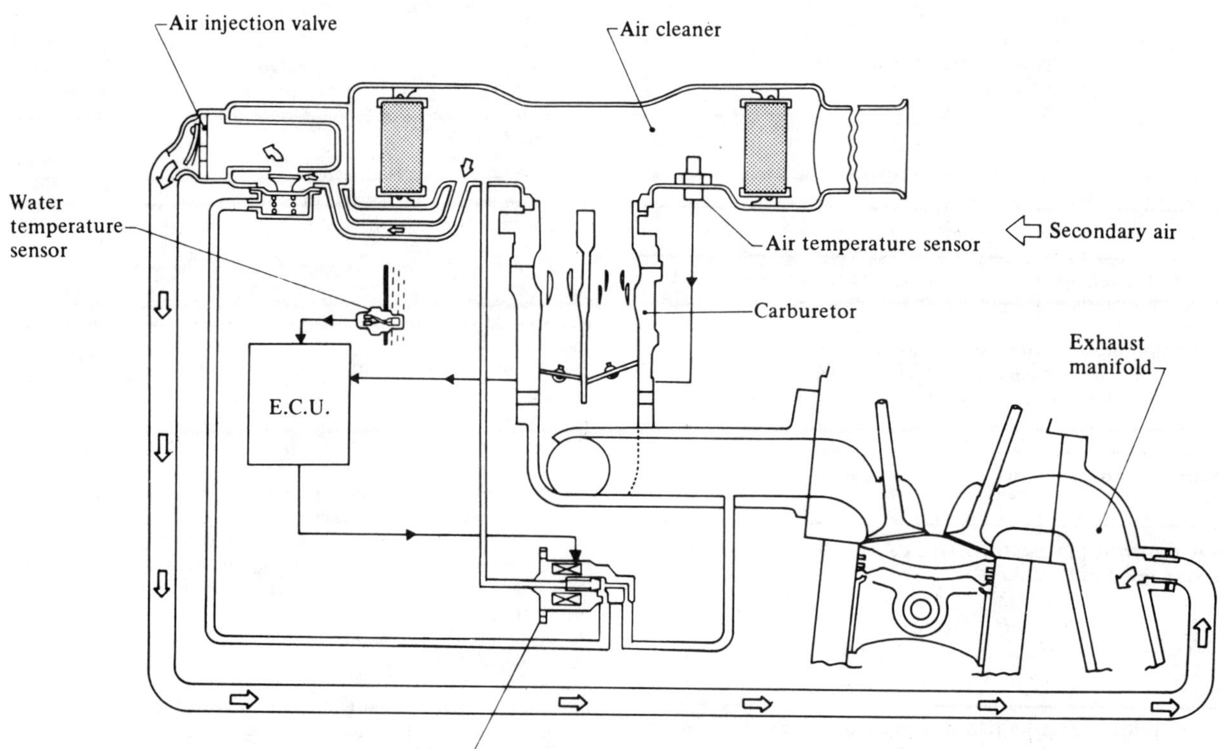

Fig. 3.51 Air induction system – catalytic converter models (Sec 21)

Exhaust gas recirculation system (EGR)

13 This system arranges the return of a proportion of the exhaust gas to the combustion chamber in order to reduce the flame temperature during combustion. This reduces the level of nitrogen oxide in the exhaust gas.

14 The system varies, as shown in the diagram, according to vehicle operating temperature (see Fig. 3.52).

15 Components of the system include the following:

Control valve – controls the volume of exhaust gas admitted to the intake manifold by responding to the degree of vacuum caused by the position of the carburettor throttle valve plate.

Venturi vacuum transducer (VVT) – monitors exhaust pressure and carburettor venturi vacuum which in turn influences the EGR control valve setting according to engine conditions and load.

Thermal vacuum valve (three port type) – sensitive to engine coolant temperatures, it controls the opening of the air passage from the air cleaner. This valve also serves the catalyst warm-up system and the fuel evaporative emission control system.

Thermal vacuum valve (two port type) – mounted on the intake manifold, it is sensitive to engine coolant temperature and controls the opening of the vacuum passage in the thermal vacuum valve.

Vacuum delay valve (VDV) – fitted in the vacuum control line to the EGR valve, it reduces the rate of vacuum change when the throttle valve is opened rapidly.

Vacuum switching valve – controlled by current from the vacuum switch, it supplies vacuum from the reservoir tank to the EGR control valve.

Vacuum switch – actuated when there is an increase in intake manifold vacuum during deceleration and interrupts the electrical signal to the vacuum switching valve.

One-way valve – designed to retain the vacuum in the reservoir tank.

BPT valve – monitors exhaust pressure to control throttle chamber vacuum applied to the EGR valve.

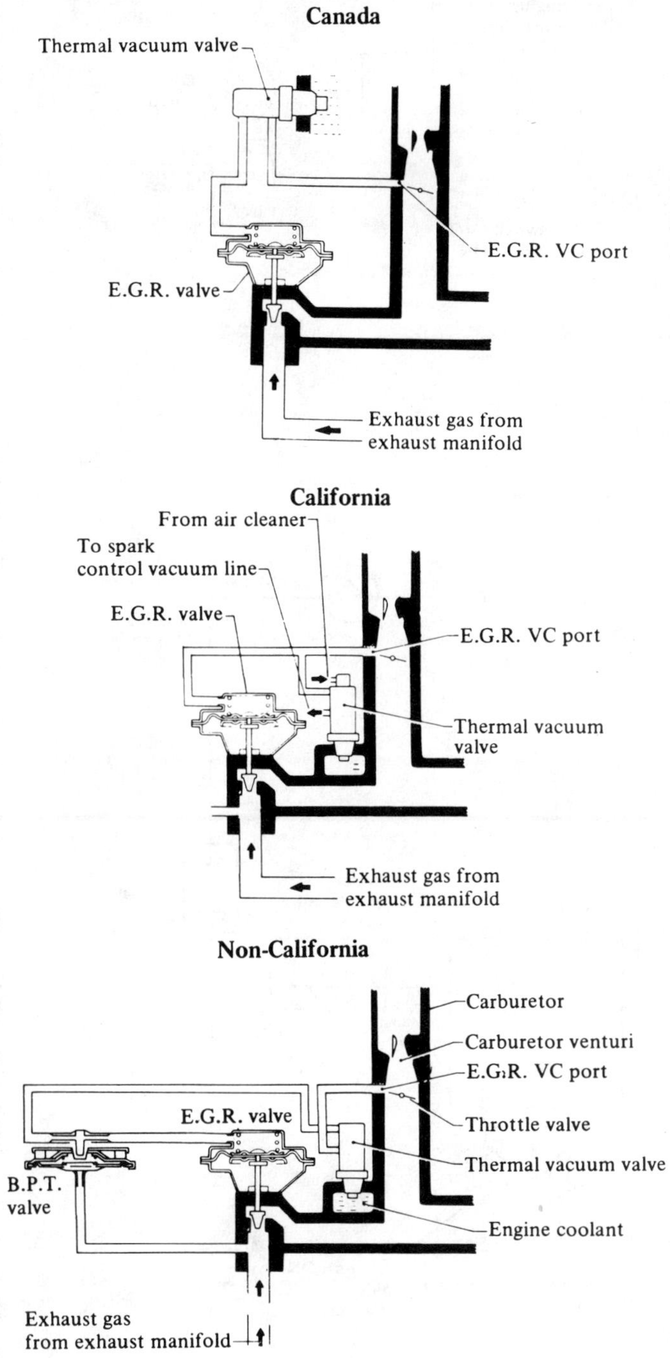

Fig. 3.52 Exhaust gas recirculation system (Sec 21)

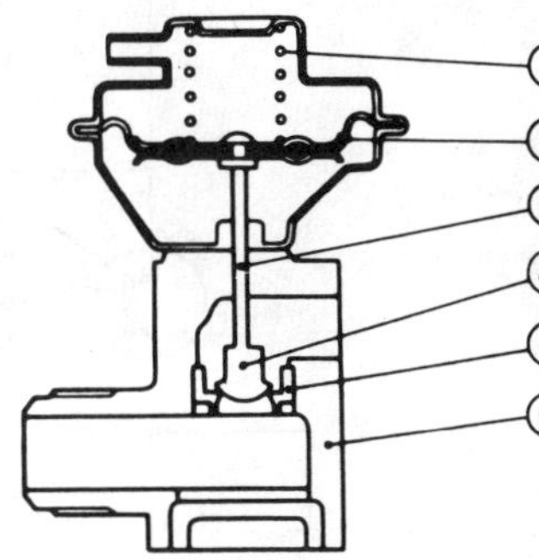

Fig. 3.53 EGR valve components (Sec 21)

1	*Diaphragm spring*	*4*	*Valve*
2	*Diaphragm*	*5*	*Valve seat*
3	*Valve shaft*	*6*	*Valve chamber*

16 Check the security of the system pipe connections at regular intervals. After a high mileage, the EGR control valve should be removed for cleaning of deposits or renewal. To do this, disconnect the hose and remove the mounting nuts. Take the opportunity to clean deposits from the connecting pipe and manifold port. Use a new gasket when refitting the valve.

17 If the thermal vacuum valve must be removed for any reason, treat it carefully as it is of plastic construction. Drain about 1.0 litre (1.76 Imp pts, 1.057 US qts) of coolant before attempting to unscrew the valve.

Air/fuel ratio control system

18 This system operates through a closed-loop or open-loop control, depending upon engine conditions such as coolant temperature, speed and the signal from the exhaust gas sensor (see paragraph 20). The purpose of the system is to precisely control the air/fuel ratio to enable the catalyst in the catalytic converter to reduce the emissions of CO, HC and NOx to the lowest possible levels.

19 Main components of the system include a control unit and a solenoid valve. If the solenoid valve must be removed during overhaul of the carburettor, remove the top cover from the carburettor and invert it. Extract the enrichment jet and retaining screw, and withdraw the solenoid valve.

Exhaust gas sensor

20 This is fitted into the exhaust manifold and monitors the oxygen content in the exhaust gas. A warning lamp is fitted which illuminates after 48 000 km (30 000 miles) as a reminder to inspect aand possibly renew the sensor.

Fuel cut-off system (except Californian models)

21 The system is designed to cut off the fuel supply during deceleration at high speeds and also to prevent running on when the ignition is switched off.

22 Components of the system include:

Solenoid valve. The valve is open to allow fuel to flow into the carburettor idling circuit only when the ignition is switched on.

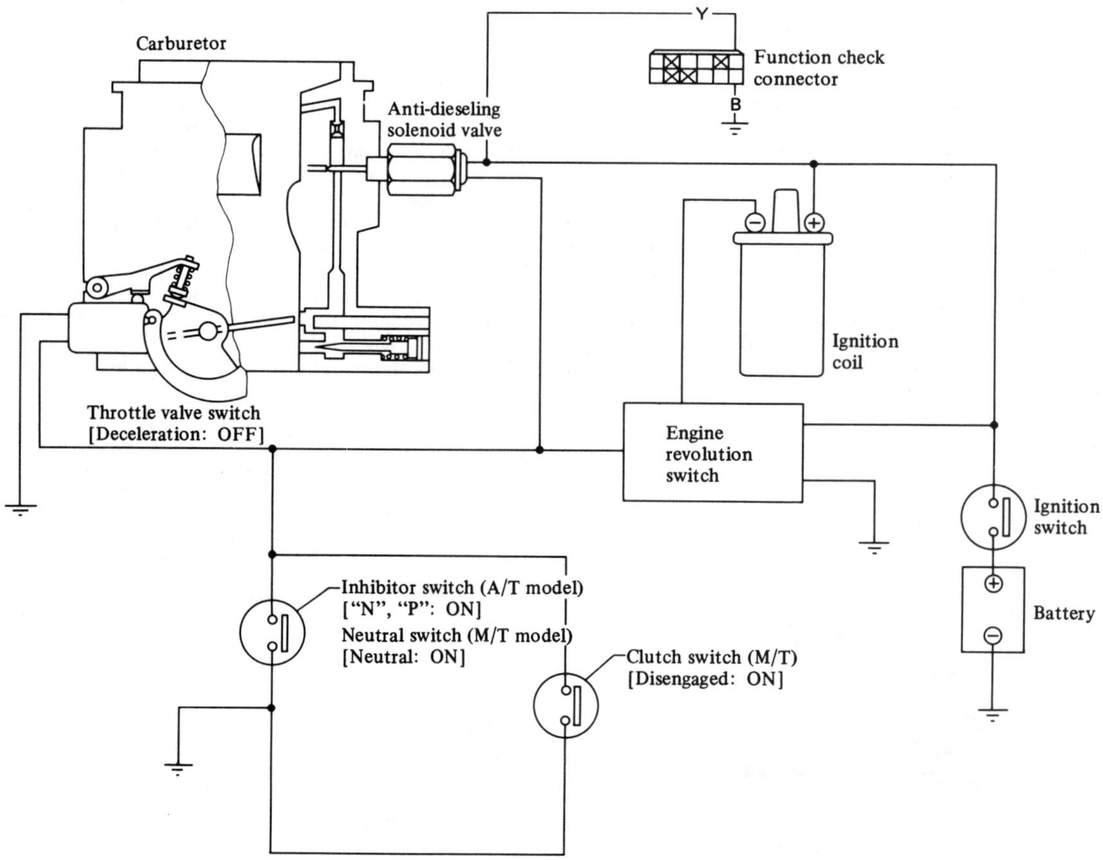

Fig. 3.54 Fuel cut-off system – early models (Sec 21)

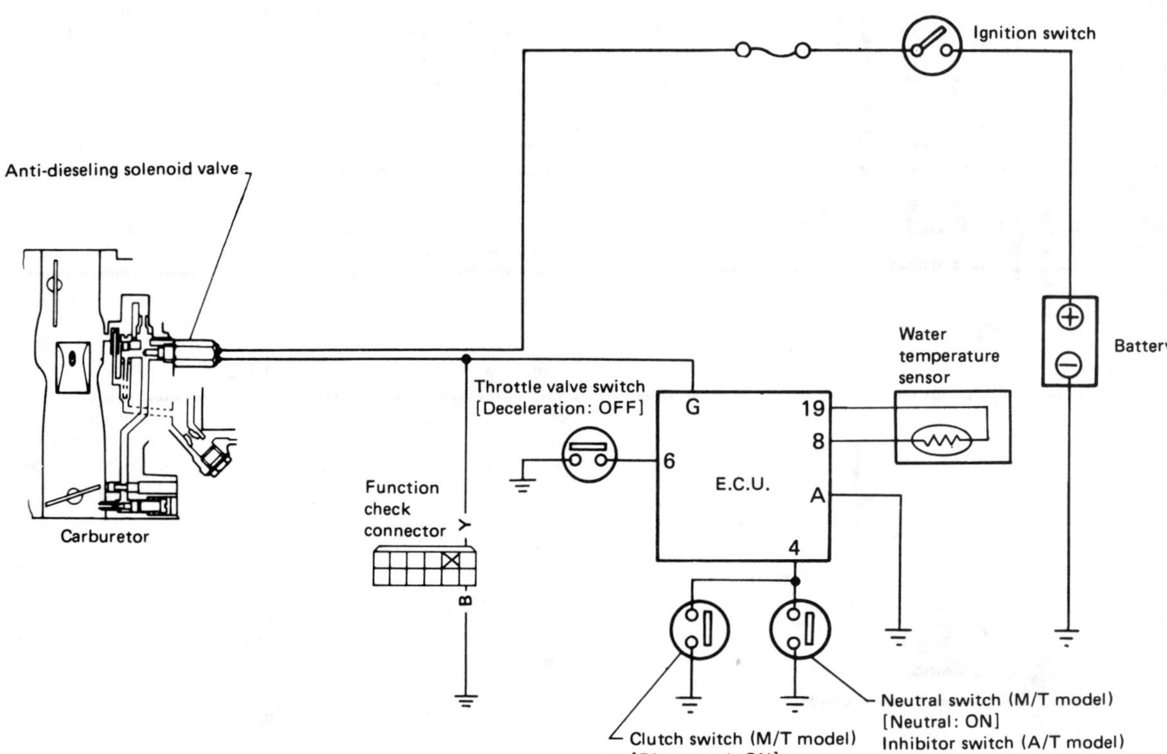

Fig. 3.55 Fuel cut-off system – late models (Sec 21)

Neutral switch. This energises the vacuum switch when the transmission is in neutral.
Clutch switch. This energises the vacuum switch when the clutch pedal is depressed.
Engine speed switch. This detects the engine speed when the fuel shut-off system comes into operation.

Fuel cut-off system (Californian models)

23 This system differs from that described in earlier paragraphs in respect of some of the components used.
Throttle valve switch. This is connected to the electronic control unit (ECU), it operates according to the position of the throttle valve plate.
Vacuum switch. When intake manifold vacuum increases during deceleration, the switch actuates to interrupt the electrical signal to the vacuum switching valve.

Mixture heating system

24 This is a method of heating the carburettor beneath the throttle block by an electric element during the warm-up period. The heater is switched off by the coolant temperature switch as engine temperature rises.

Throttle opener control system (TOCS) – Canada

25 Designed to open the throttle valve plate slightly during deceleration to prevent excessive emissions of unburned HC.

Evaporative control system

26 This system prevents fuel vapour being released to atmosphere from the fuel tank.

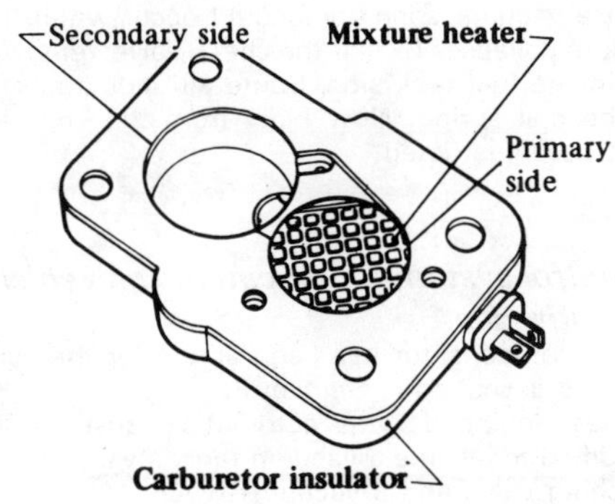

Fig. 3.56 Mixture heating element (Sec 21)

27 When the engine is not running, vapour from the fuel tank passes through a vent line to a canister filled with activated charcoal where it is stored until the engine is started.

28 Once the engine is started and running, a purge control valve opens and the canister and vent line are cleared of vapour which is drawn into the intake manifold and then burned during the normal engine combustion process.

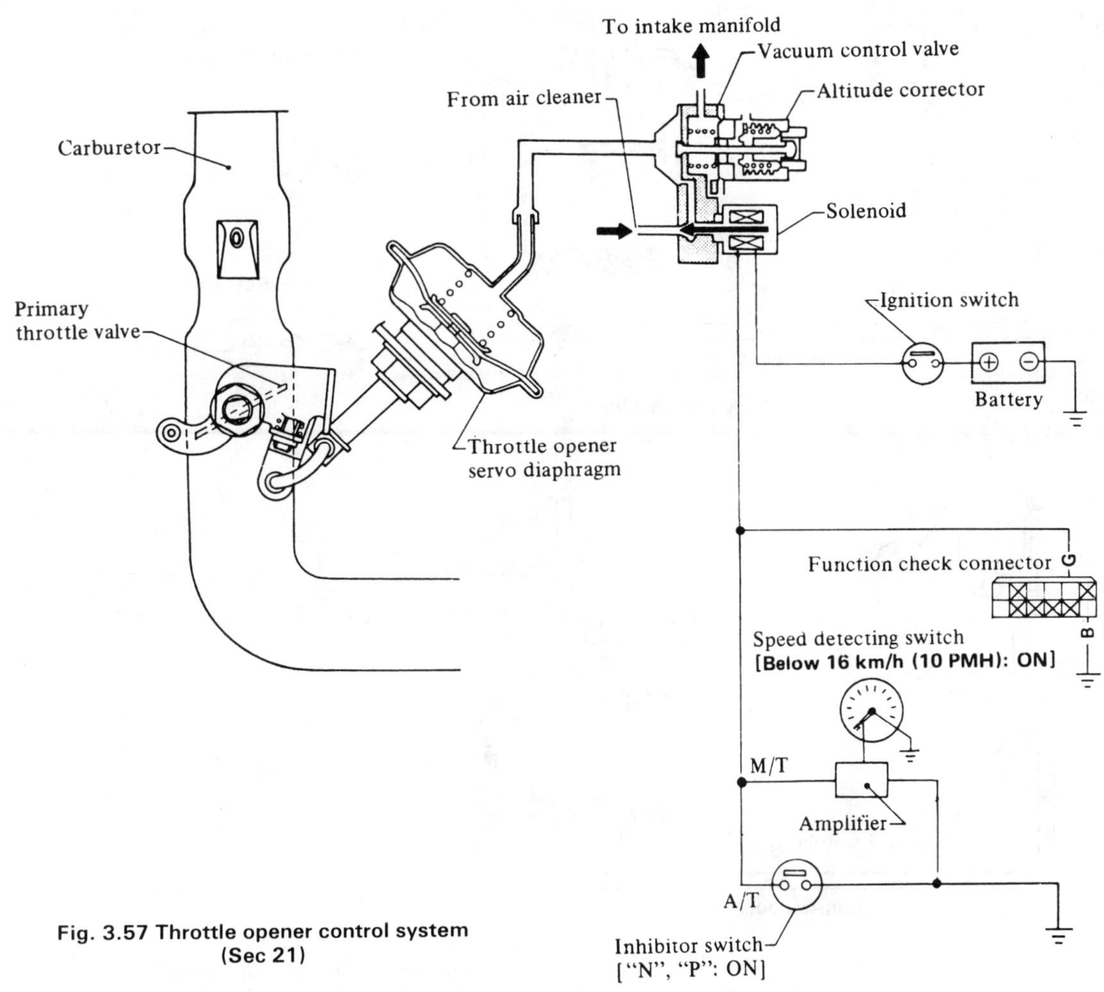

Fig. 3.57 Throttle opener control system (Sec 21)

29 As the fuel tank is of sealed type, it is important that the vent hose is never crushed and that the check (non-return) valve is kept in good order, otherwise vacuum conditions could occur within the tank and cause the tank to collapse. To test the check valve, remove it and blow through it from the fuel tank side. Some air should pass against the pressure of the coil spring. Now blow from the engine side – the airflow should be unrestricted.

Emission control system modifications on vehicles operating at high altitudes

30 An altitude compensator has been added for the carburettor, its purpose being as described in Section 8.
31 On vehicles equipped with automatic transmission, the EGR system is modified to include a vacuum tank, a vacuum delay valve, a vacuum switching valve and a vacuum switch.

22 Catalytic converter – description and precautions

1 This device is fitted into the exhaust system of North American vehicles. It is basically a container for a catalyst which, as exhaust gases pass through it, becomes oxidised and changes the gas to harmless water and carbon dioxide.
2 If the exhaust system is being dismantled, take care not to damage the catalytic converter when removing its shield or flange connecting bolts.
3 On vehicles fitted with this device, certain precautions should be taken:

Do not allow the engine to idle for excessive lengths of time
Use only unleaded fuel
Avoid running out of fuel
Do not park over long grass or other combustible material

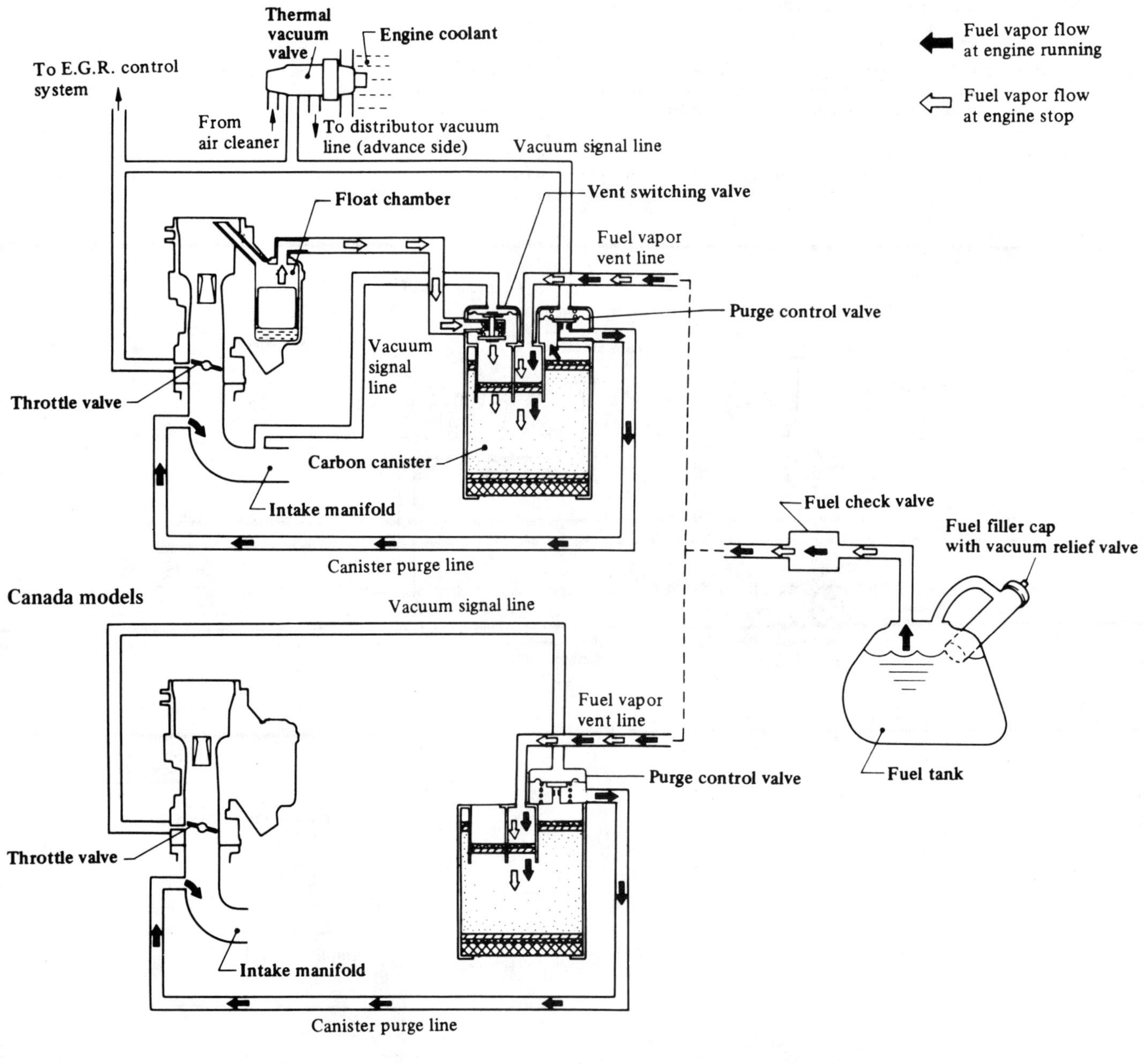

Fig. 3.58 Fuel evaporative emission control system – non-Turbo models (Sec 21)

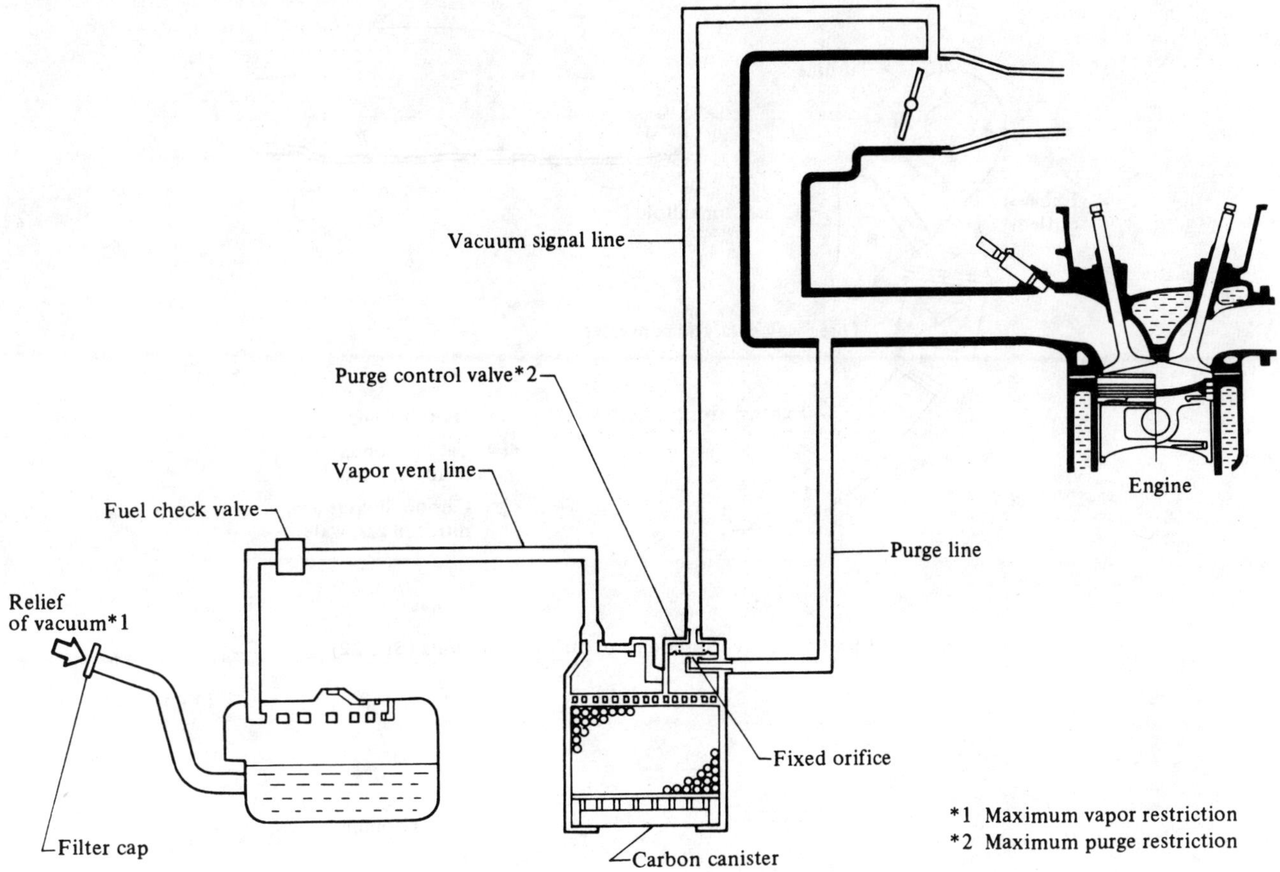

Fig. 3.59 Fuel evaporative emission control system – Turbo models (Sec 21)

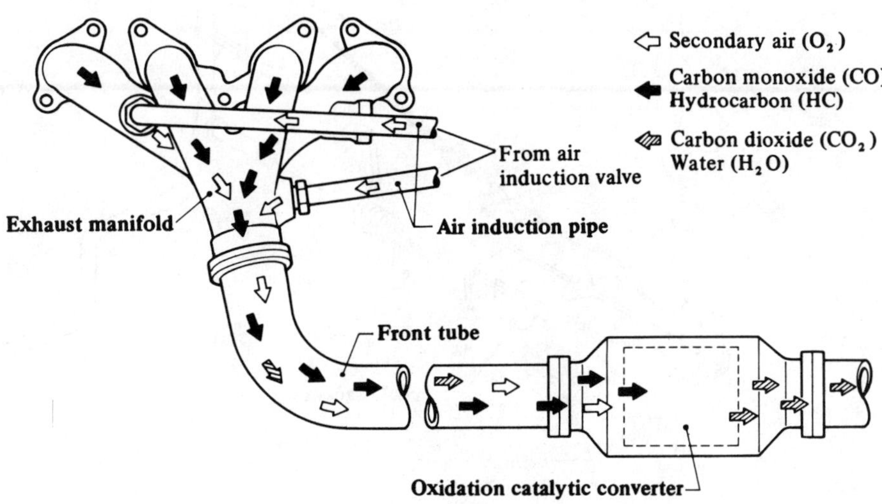

Fig. 3.60 Catalytic converter location on non-Turbo models (Sec 22)

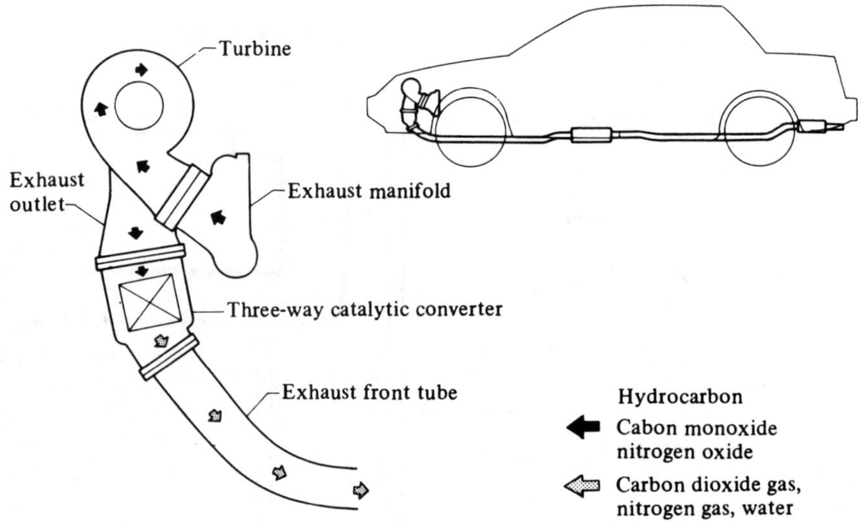

Fig. 3.61 Catalytic converter location on Turbo models (Sec 22)

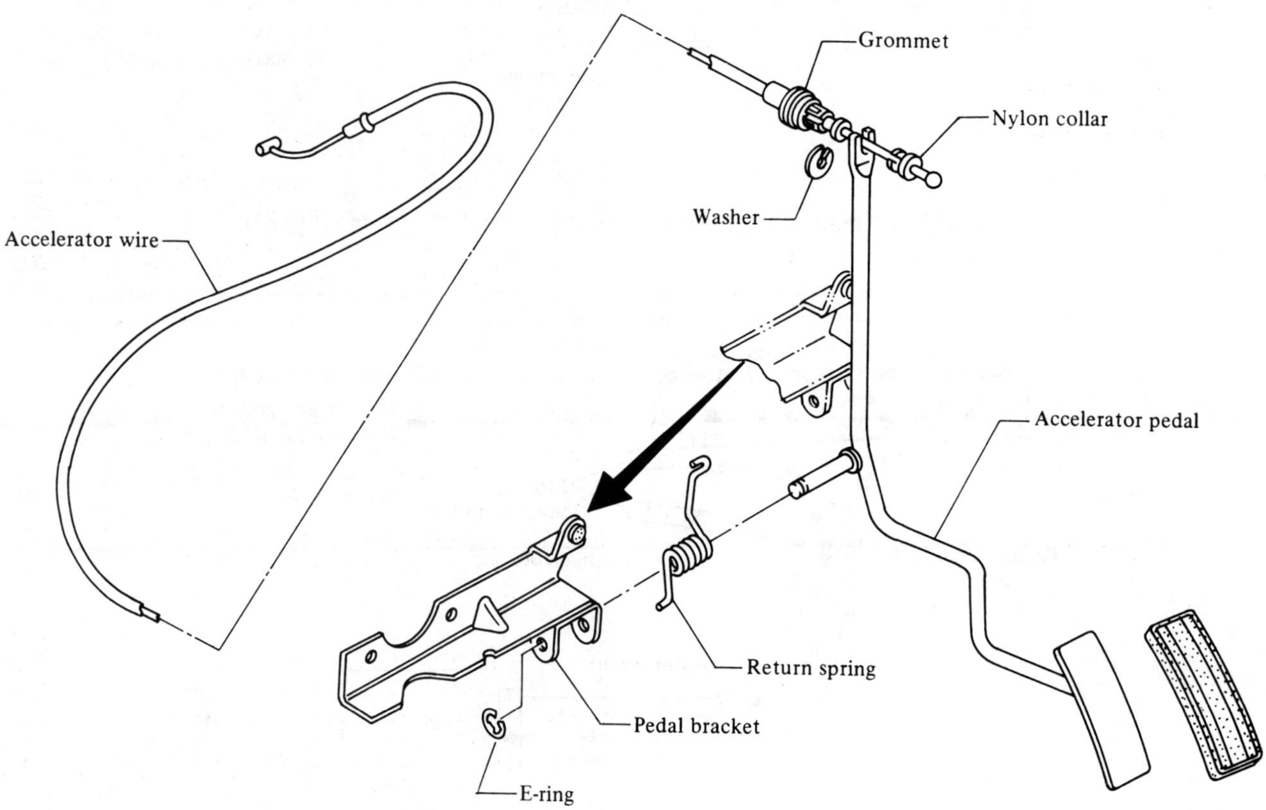

Fig. 3.62 Typical accelerator control (Sec 23)

23 Accelerator pedal and cable – removal, refitting and adjustment

Pedal

1 Release the cable from the top of the pedal arm.
2 Extract the E-clip from the pedal pivot shaft, disengage the return spring and remove the pedal.
3 Refitting is a reversal of removal, but apply grease to the moving parts.

Cable

4 Disconnect the cable from the top of the pedal arm.
5 Release the plastic grommet at the engine compartment rear bulkhead.
6 Disconnect the cable at the throttle and then withdraw it into the engine compartment.
7 Fit the new cable by reversing the removal operations. Adjust it by moving the position of the outer cable clamp to give the correct free movement at the pedal pad (see Fig. 3.63). Make sure that the automatic choke valve plate is fully open before adjusting the cable (if applicable).
8 On vehicles equipped with automatic transmission, check that the pedal can be depressed fully into the kickdown position (also refer to Chapter 7).

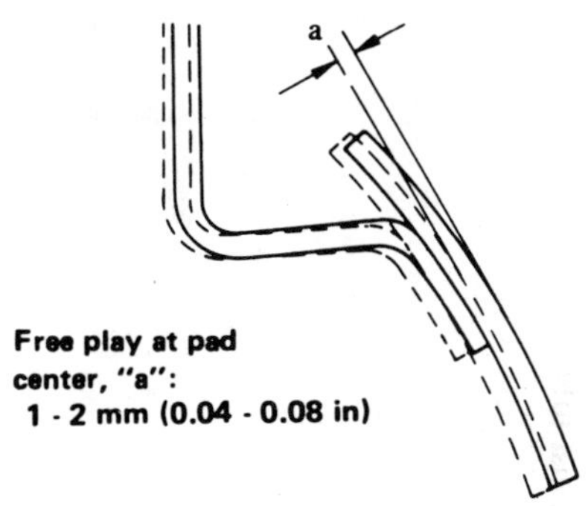

Fig. 3.63 Accelerator pedal free play (Sec 23)

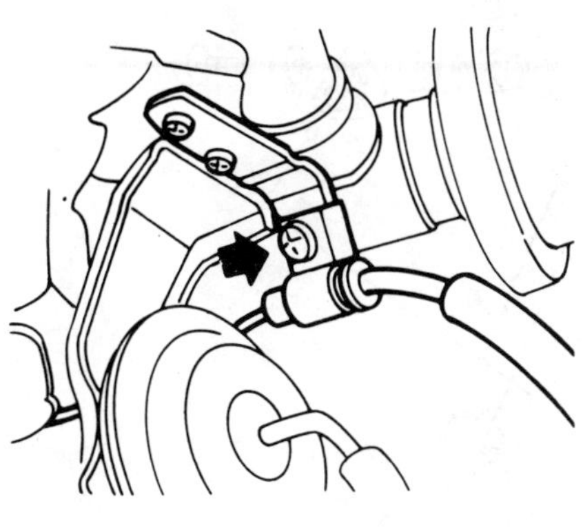

Fig. 3.64 Accelerator cable conduit clamp (Sec 23)

24 Choke control cable – removal and refitting

1 Remove the air cleaner from the carburettor, release the cable pinch screw on the trunnion at the carburettor and the cable clamp screw.
2 Working inside the vehicle at the facia panel, unscrew the choke control knob bezel nut and withdraw the cable into the vehicle interior. If a choke warning lamp switch is fitted, disconnect the electrical leads.
3 Refitting is a reversal of removal. Make sure that the bulkhead cable grommet makes a good seal.
4 With the cable in position, do not tighten the pinch screw until the choke control knob has been pulled out by about 3.0 mm (0.12 in) and the choke valve plate has been checked as being in the fully open position.
5 Finally check that with the control pulled right out, the valve plate is fully closed. Adjust the position of the outer cable in the clamp if necessary to achieve this.

25 Manifolds and exhaust system – general

1 The manifolds are located on opposite sides of the engine. As the intake manifold is coolant heated, the cooling system must be drained before it can be removed (photos).

25.1A Removing the intake manifold and carburettor

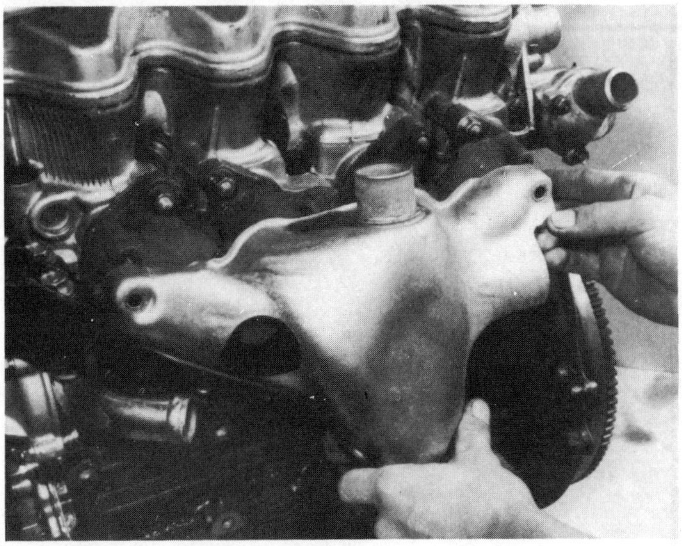

25.1B Removing the exhaust manifold hot air shroud

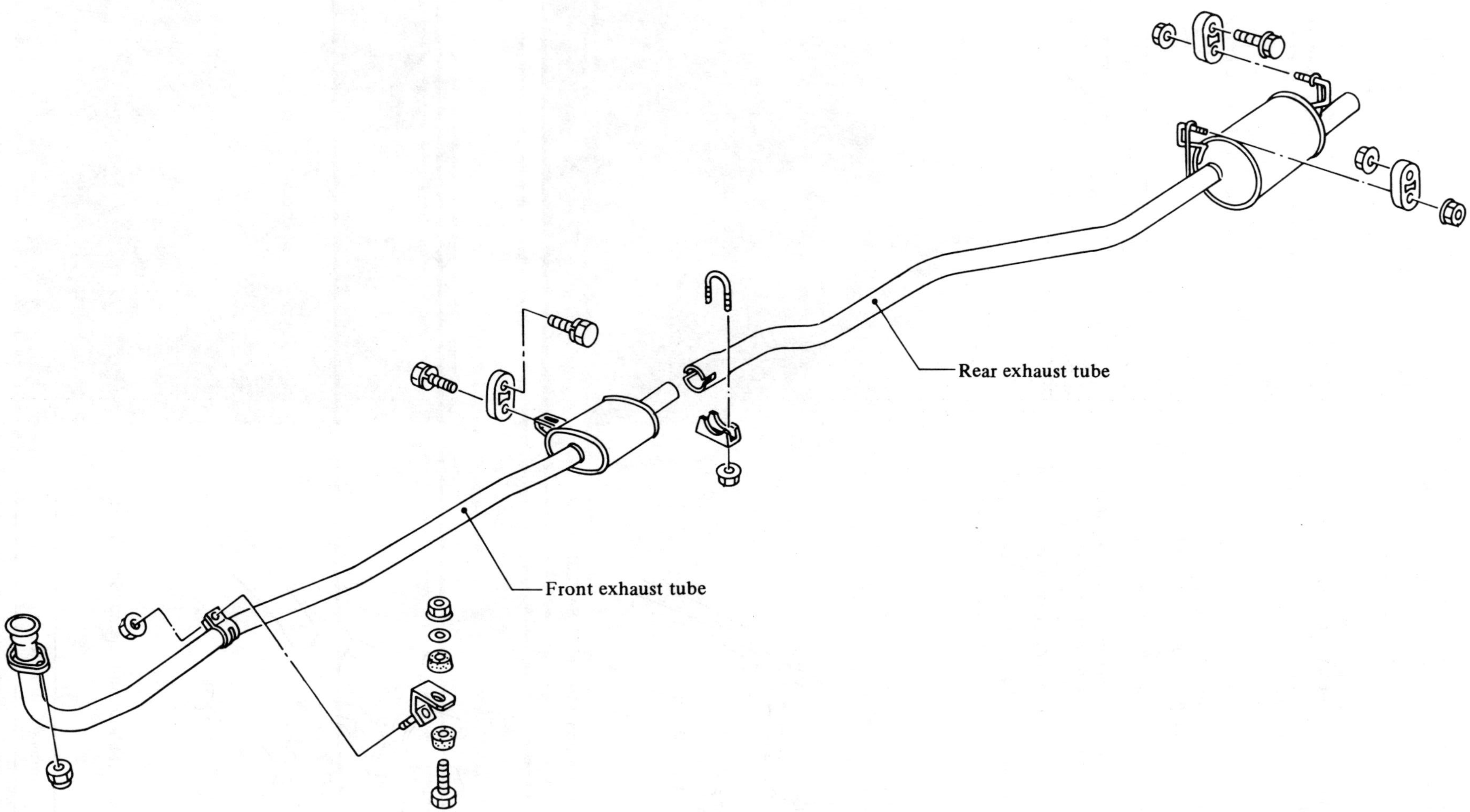

Fig. 3.65 Exhaust system for UK models (Sec 25)

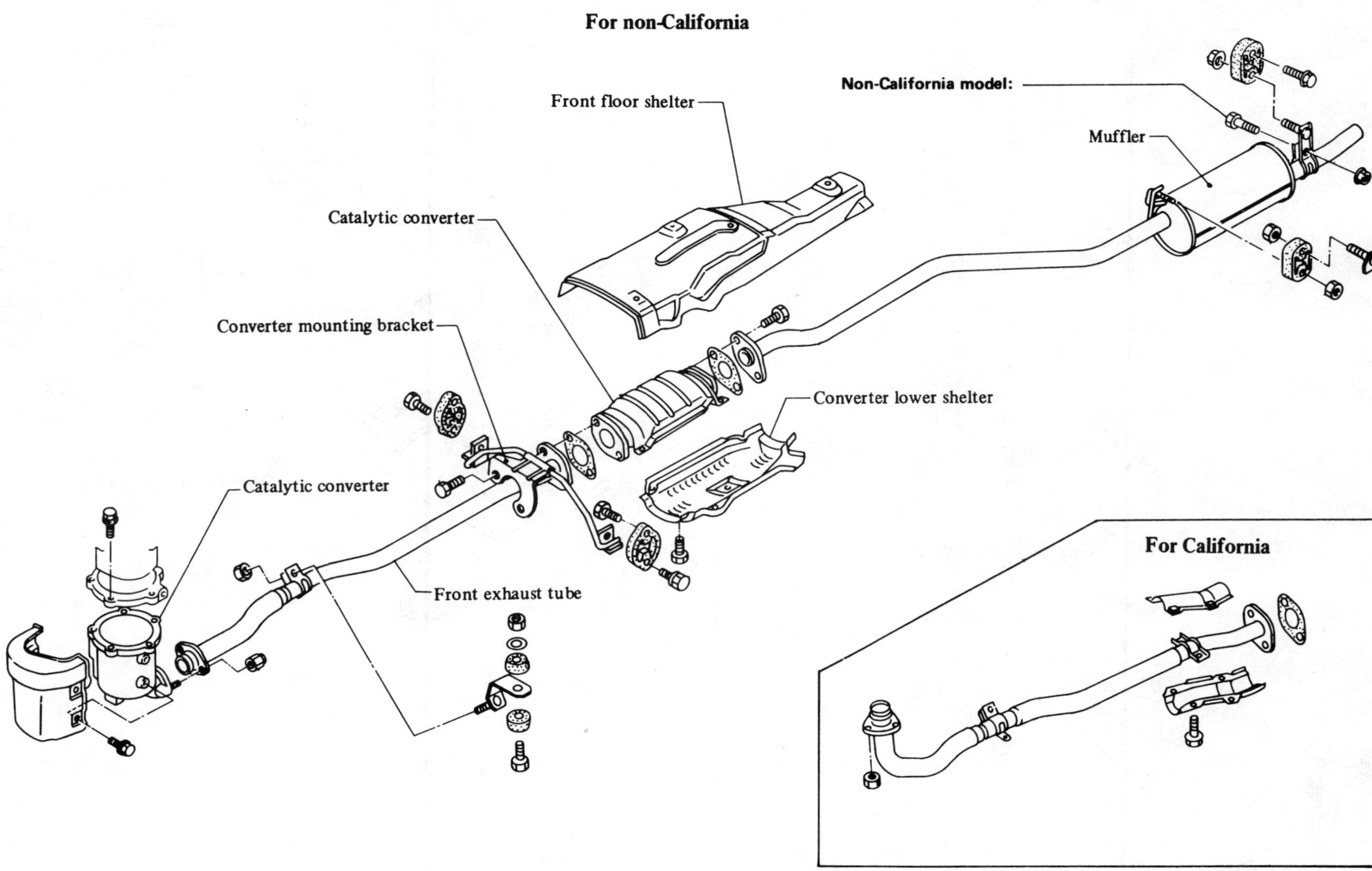

Fig. 3.66 Exhaust system for USA models (Sec 25)

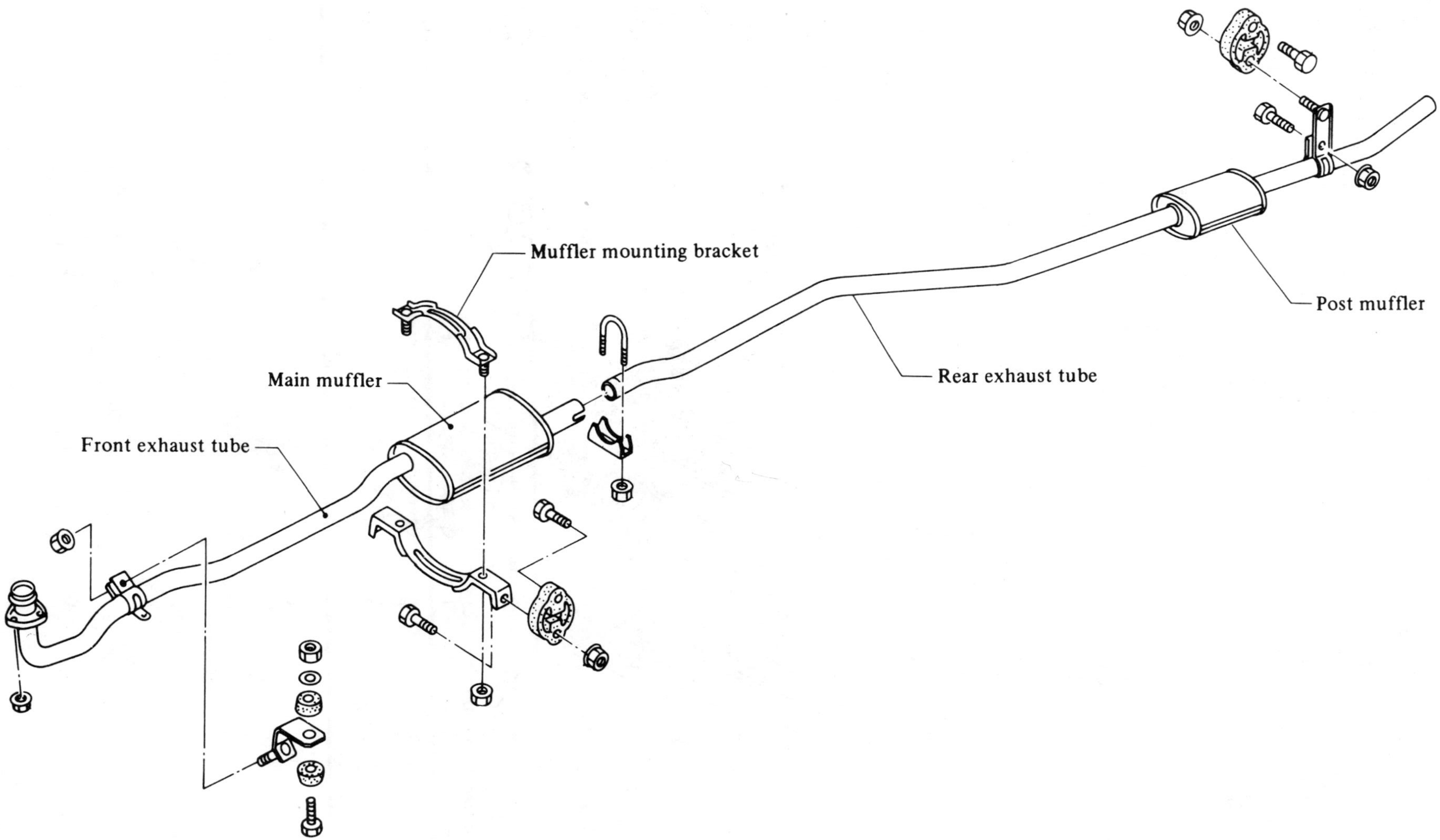

Fig. 3.67 Exhaust system for Canada non-Turbo models (Sec 25)

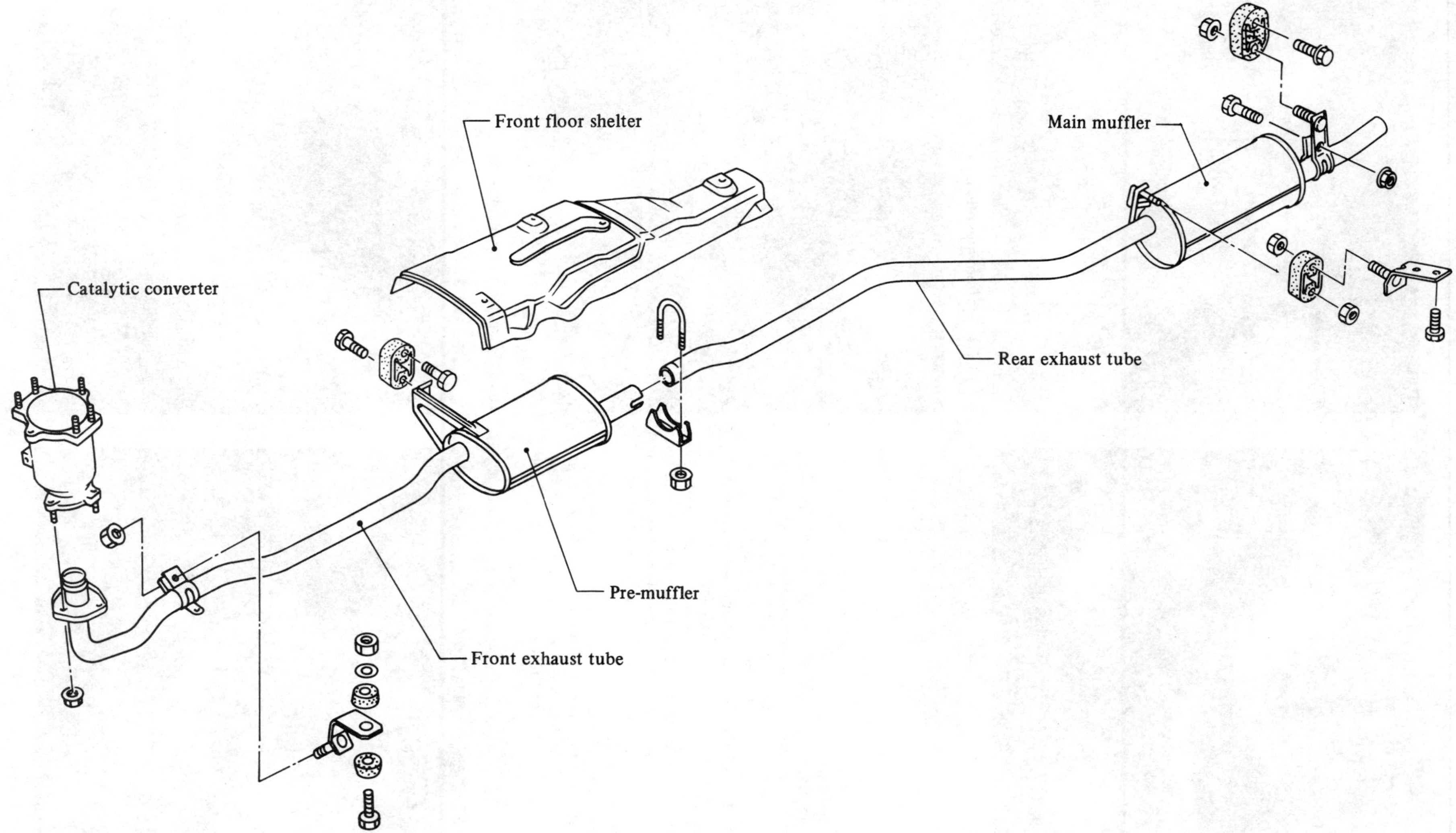

Fig. 3.68 Exhaust system for Canada Turbo models (Sec 25)

25.2 Exhaust manifold gasket

25.3B Exhaust front mounting

2 Always use new gaskets when refitting the manifolds and tighten all nuts and bolts to the specified torques (photo).

3 The exhaust system may have one or two downpipes. On models without a catalytic converter, the system incorporates an expansion box and a silencer (photos).

4 The exhaust system is suspended on flexible mountings (photos).

5 When any one section of the exhaust system needs renewal it often follows that the whole system is best replaced.

6 It is most important when fitting exhaust systems that the bends and contours are carefully followed and that each connecting joint overlaps the correct distance. Any stresses or strain imparted in order to force the system to fit the hanger rubbers, will result in early fractures and failures.

7 When fitting a new part of a complete system it is well worth removing all the system from the car and cleaning up all the joints so that they fit together easily. The time spent struggling with obstinate joints whilst flat on your back under the car is eliminated and the

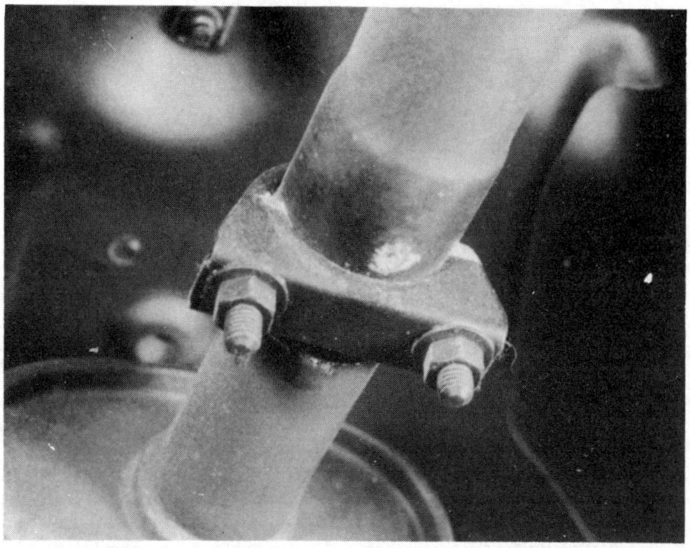

25.3C Exhaust intermediate clamp

25.3A Exhaust downpipe flange and collar

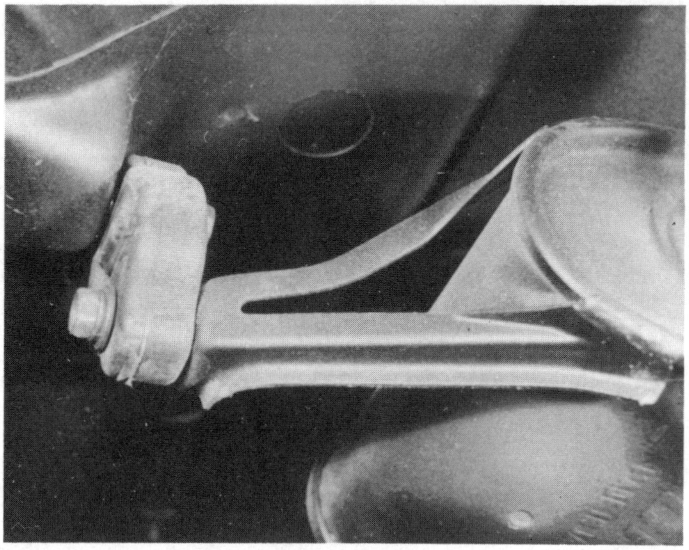

25.4A Exhaust intermediate flexible mounting

25.4B Exhaust rear flexible mounting

25.4C Exhaust silencer mounting arm

likelihood of distorting or even breaking a section is greatly reduced. Do not waste a lot of time trying to undo rusted and corroded clamps and bolts. Cut them off. New ones will be required anyway if they are that bad.

8 Use an exhaust pipe joint sealant when assembling pipe sections to ensure that the respective joints are free from leaks.

9 When fitting the new system, only semi-tighten the retainers initially until the complete system is fitted, then when you have checked it for satisfactory location, tighten the securing bolts/nuts. If the hangers are stretched, perished or broken they must be renewed, otherwise the system will vibrate, leading to leaks, premature wear or even fractures.

26 Fault diagnosis – fuel, exhaust and emission control systems

Unsatisfactory engine performance and excessive fuel consumption are not necessarily the fault of the fuel system or carburettor. In fact they more commonly occur as a result of ignition and timing faults. Before acting on the following it is necessary to check the ignition system first. Even though a fault may lie in the fuel system it will be difficult to trace unless the ignition is correct. The faults below, therefore, assume that this has been attended to first (where appropriate).

Symptom	Reason(s)
Smell of fuel when engine is stopped	Leaking fuel lines or unions Leaking fuel tank
Smell of fuel when engine is idling	Leaking fuel line unions between pump and carburettor injectors Overflow of fuel from float chamber due to wrong level setting, ineffective needle valve or punctured float (carburettor models)
Excessive fuel consumption for reasons not covered by leaks or float chamber faults	Worn jets Over-rich setting Sticking mechanism Dirty air cleaner element Sticking air cleaner thermostatic mechanism
Difficult starting, uneven running, lack of power, cutting out	One or more jets blocked or restricted Float chamber fuel level too low or needle valve sticking Fuel pump not delivering sufficient fuel Faulty solenoid fuel shut-off valve (if fitted) Induction leak
Difficult starting when cold	Choke control or automatic choke maladjusted Automatic choke not cocked before starting
Difficult starting when hot	Automatic choke malfunction Accelerator pedal pumped before starting Vapour lock (especially in hot weather or at high altitude)
Engine does not respond properly to throttle	Faulty accelerator pump Blocked jet(s) Slack in accelerator cable

Symptom	Reason(s)
Engine idle speed drops when hot	Defective temperature compensator Overheated fuel pump
Emission control system	
Excessive HC or CO in exhaust gas	Air cleaner clogged Float level too high Faulty spark control system Faulty throttle opener control system Leaking intake manifold gasket
Excessive HC, CO and NOx in exhaust gas	Worn piston rings Incorrect valve clearances Faulty thermostat Blown cylinder head gasket Clogged PCV valve Incorrect idle mixture Clogged fuel filter Faulty idle compensator Choke not fully off Incorrect ignition settings Malfunction of emission control system component

HC Hydrocarbons
CO Carbon monoxide
NOx Nitrogen oxide

Chapter 4 Ignition system

Contents

Specifications

System types	Conventional with coil and contact breaker points, transistorized with distributor magnetic control and coil, or computerised ECCS and coil

Distributor

Firing order	1–3–4–2 (No 1 cylinder at timing belt end)
Rotor rotation	Anti-clockwise
Cap carbon brush minimum length	10.0 mm (0.39 in)
Contact points gap (conventional system)	0.45 to 0.55 mm (0.018 to 0.022 in)
Contact points dwell angle (conventional system)	49° to 55°
Air gap (transistorized system)	0.3 to 0.5 mm (0.012 to 0.020 in)
HT lead maximum resistance	30 k ohm

Coil

Primary resistance:	
Conventional system	1.35 to 1.65 ohm
Transistorized system	1.04 to 1.27 ohm
Computerised system	0.84 to 1.02 ohm
Secondary resistance:	
Conventional system	6.8 to 10.2 k ohm
Transistorized system	7.3 to 11.0 k ohm
Computerised system	8.2 to 12.4 k ohm

Ignition timing

UK models

E10 and E13 engines	2 ± 2° BTDC at idle speed with distributor vacuum hose disconnected and plugged
E15 engine	4 ± 2° BTDC at idle speed with distributor vacuum hose disconnected and plugged
E15 ET engine	15 ± 2° BTDC at idle speed

North American models

E16 engine:	
1983 model	5 ± 2° ATDC at idle speed with distributor vacuum hose disconnected and plugged
1984/5 model, non-California, manual	15 ± 2° BTDC at idle speed
1984/5 model, non-California, automatic	8 ± 2° BTDC at idle speed
1984/5 model, California	5 ± 2° ATDC at idle speed with distributor vacuum hose disconnected and plugged
1984/5 model, Canada	5 ± 2° ATDC at idle speed with distributor vacuum hose disconnected and plugged
E15 ET engine:	
1984 model, Canada	15 ± 2° BTDC at idle speed with idle speed control valve harness disconnected

Spark plugs

UK models

E15 ET engine	NGK BPR6ES
E10, E13, E15 engines	NGK BPR5ES
Gap	0.8 to 0.9 mm (0.031 to 0.035 in)

USA models

California	NGK BPR5ES-11
Gap	1.0 to 1.1 mm (0.039 to 0.043 in)
Non-California	NGK BPR5ES-11
Gap	1.0 to 1.1 mm (0.039 to 0.043 in)
Non-California alternative	NGK BPR5ES
Gap	0.8 to 0.9 mm (0.031 to 0.035 in)
Canada	NGK BPR5ES
Gap	0.8 to 0.9 mm (0.031 to 0.035 in)

Torque wrench setting

	lbf ft	Nm
Spark plug	14 to 22	20 to 29

1 General description

The ignition system may be one of three different types depending on the engine type. The conventional system incorporates contact breaker points to switch the coil primary circuit on and off, but on the transistorized system the circuit is switched by an electronic unit located in the distributor incorporating a magnetic stator, reluctor and transistor control. The coil primary circuit on engines equipped with the electronic concentrated engine control system (ECCS) is controlled by a micro-computer in conjunction with a power transistor.

In order that the engine can run correctly it is necessary for an

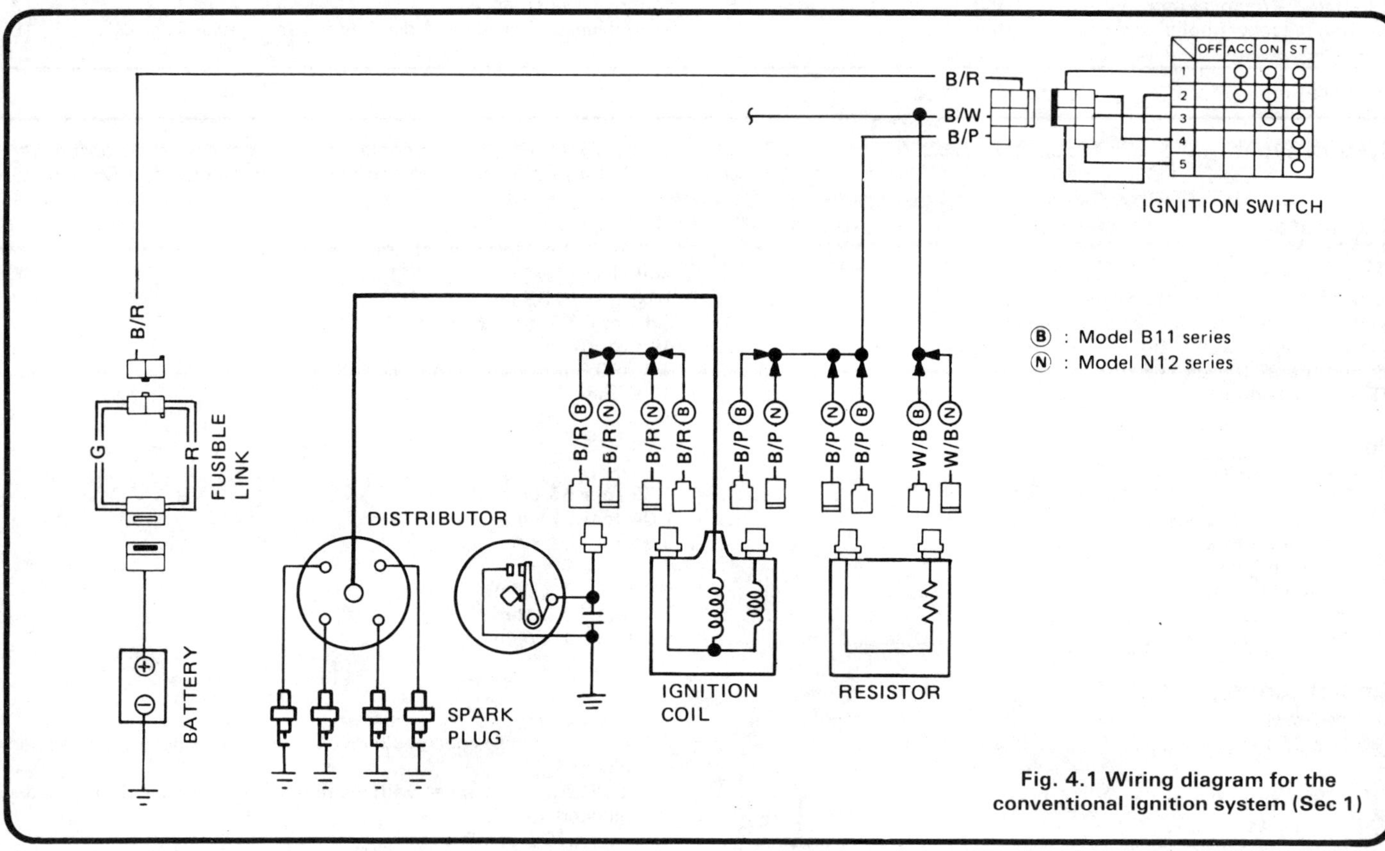

Fig. 4.1 Wiring diagram for the conventional ignition system (Sec 1)

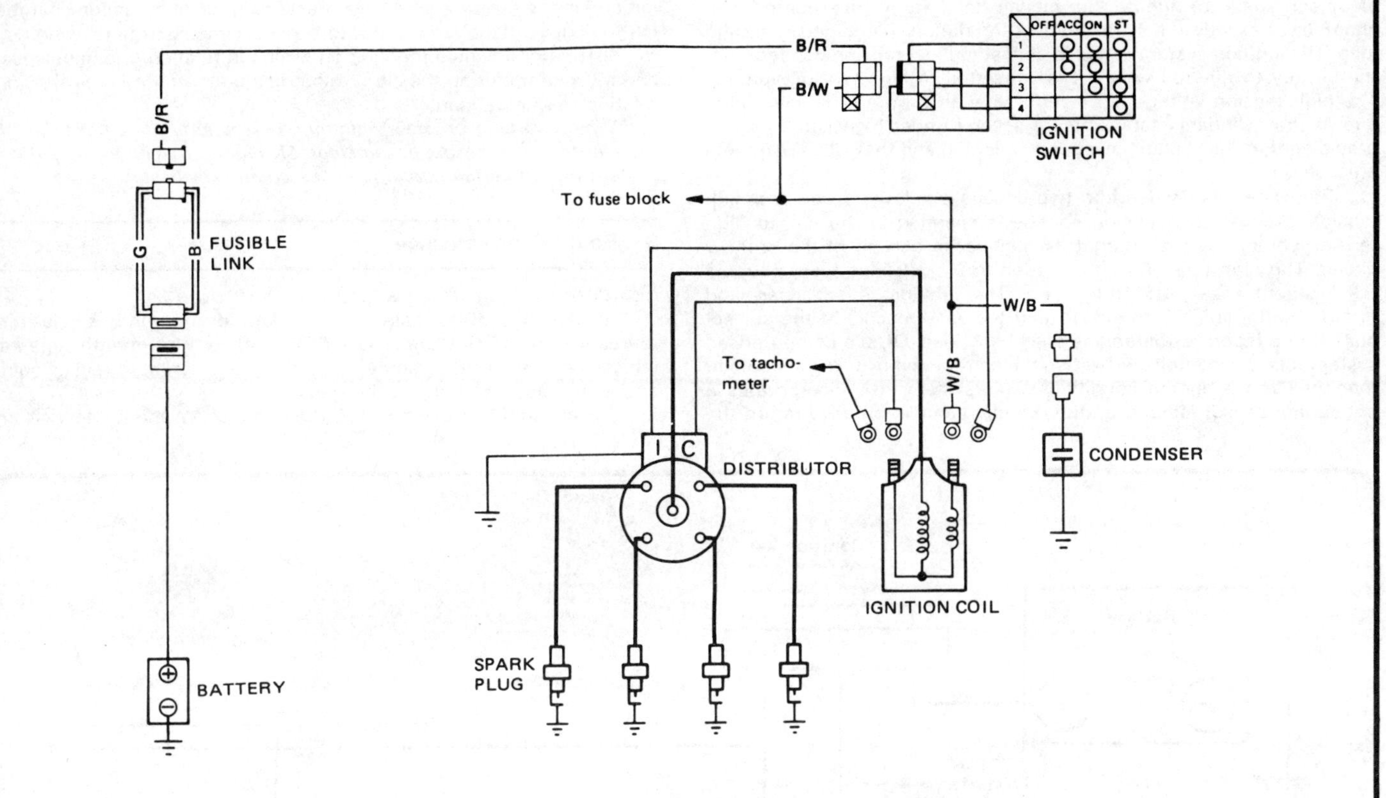

Fig. 4.2 Wiring diagram for the transistorized ignition system (Sec 1)

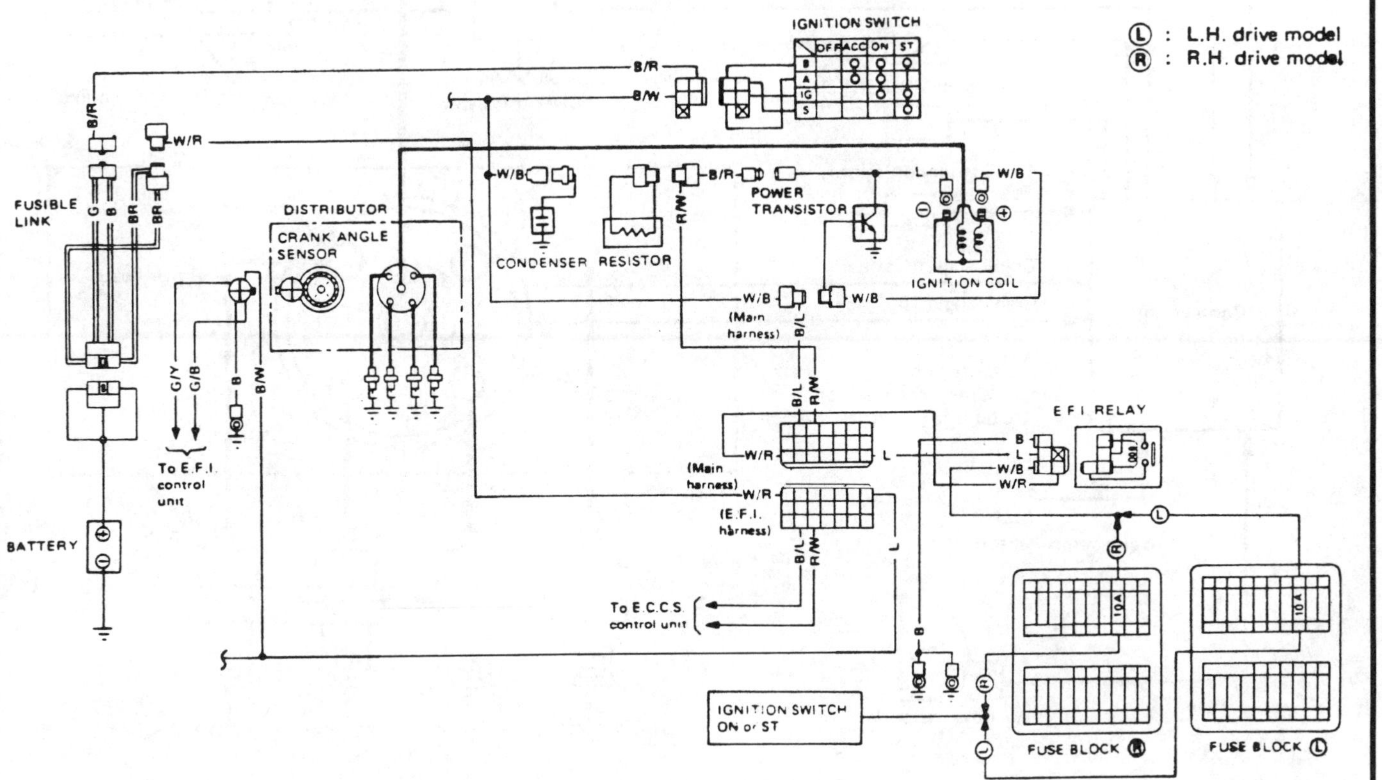

Fig. 4.3 Wiring diagram for the computerised ignition system fitted to UK Turbo models (Sec 1)

electrical spark to ignite the fuel/air mixture in the combustion chamber at exactly the right moment in relation to engine speed and load. The ignition system is based on feeding low tension voltage from the battery to the coil where it is converted to high tension voltage. The high tension voltage is powerful enough to jump the spark plug gap in the cylinders many times a second under high compression, providing that the system is in good condition and that all adjustments are correct.

The system is divided into two circuits; the low tension and high tension circuits. Low tension voltage is changed in the coil to high tension voltage by the alternate switching on and off of the primary circuit. The high tension voltage is fed to the relevant spark plug via the distributor cap and rotor arm. The ignition is advanced and retarded automatically to ensure that the spark occurs at the correct instant in relation to the engine speed and load. On the computerised system this is accomplished within the control unit by using sensors to monitor various engine conditions. This system also incorporates a detonation sensor located on the cylinder block in order to retard the ignition in the event of excessive pressure in the combustion chamber. On the conventional and transistorized systems centrifugal weights in the distributor advance the ignition timing in relation to engine speed, and a vacuum unit on the side of the distributor controls the timing in relation to engine load.

When working on electronic ignition systems remember that the high tension voltage can be considerably higher than on a conventional system and in certain circumstances could prove fatal.

2 Routine maintenance

Conventional ignition system (UK models)

1 At the first 6000 mile (10 000 km) or 6 month service and thereafter at 12 000 mile (20 000 km) or 12 month intervals whichever comes first, remove and clean the contact breaker points and spark plugs and adjust the gaps.

2 At the first 12 000 mile (20 000 km) or 12 month service and

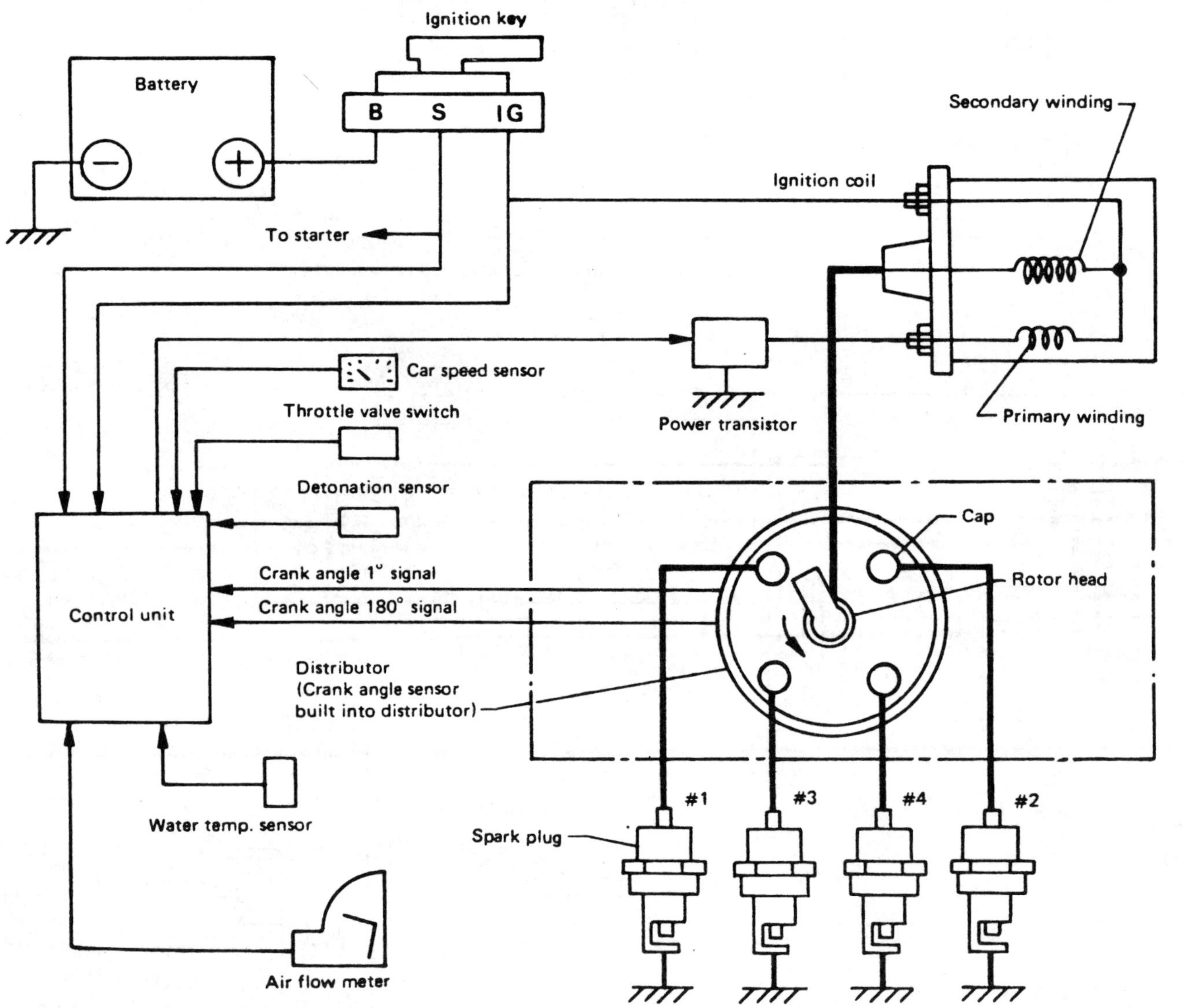

Fig 4.4 Diagram of the computerised ignition system fitted to some North American models (Sec 1)

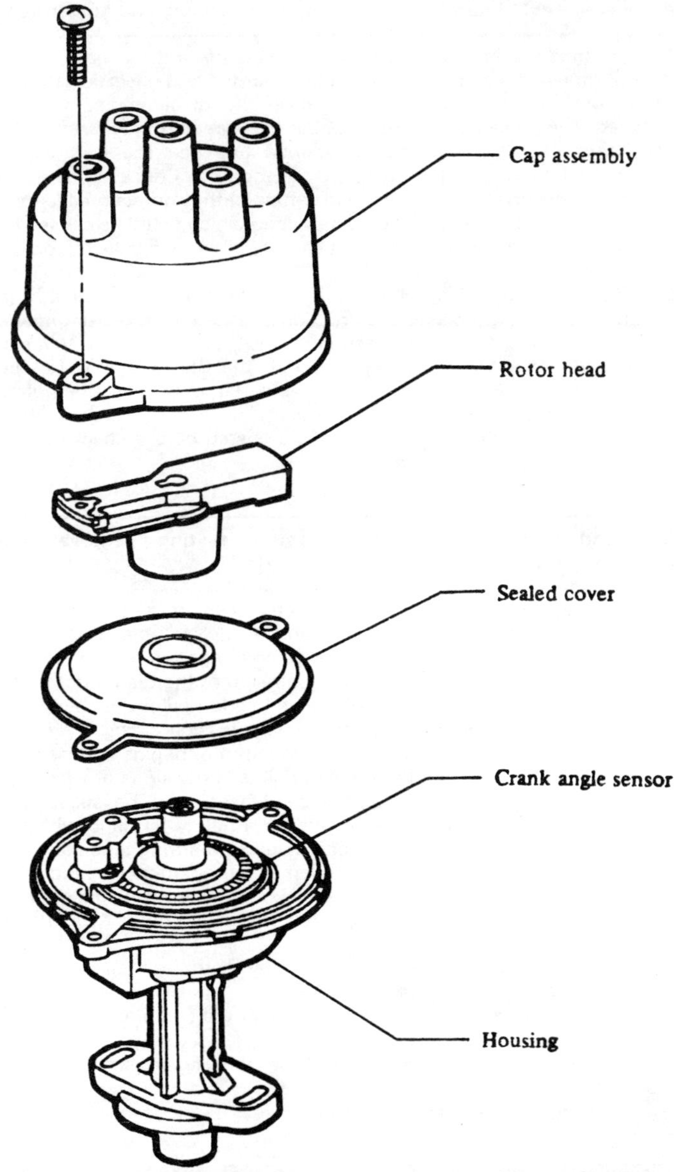

Fig. 4.5 Distributor components on the computerised ignition system (Sec 1)

thereafter at 12 000 mile (20 000 km) or 12 month intervals whichever comes first, renew the contact breaker points and spark plugs and adjust the gaps.
3 Every 6000 miles (10 000 km) or 6 months whichever comes first check and if necessary adjust the ignition timing.
4 Every 24 000 miles (40 000 km) or every 24 months whichever comes first check the ignition LT and HT wiring for condition and security.

Computerised ignition system (UK models)

5 At the first 6000 mile (10 000 km) or 6 month service and thereafter at 12 000 mile (20 000 km) or 12 month intervals whichever comes first, remove and clean the spark plugs and adjust the gaps.
6 At the first 12 000 mile (20 000 km) or 12 month service and thereafter at 12 000 mile (20 000 km) or 12 month intervals whichever comes first, renew the spark plugs and adjust the gaps.
7 Every 6000 miles (10 000 km) or 6 months whichever comes first check and if necessary adjust the ignition timing.
8 Every 24 000 miles (40 000 km) or every 24 months whichever comes first check the ignition wiring including the HT wires for condition and security.

Transistorized and computerised ignition systems (North American models)

9 Every 15 000 miles (24 000 km) check and if necessary adjust the ignition timing.
10 Every 30 000 miles (48 000 km) on all USA models and Canadian Turbo models, and every 15 000 miles (24 000 km) on Canadian non-Turbo models, renew the spark plugs and adjust the gaps.
11 Every 2 years check the ignition wiring including the HT wires for condition and security.

3 Contact breaker points (conventional system) – servicing and adjustment

1 At the intervals specified in Routine Maintenance open the bonnet, prise back the clips which secure the distributor cap and move the cap complete with HT leads to one side.
2 Pull off the rotor.
3 Open the contact points by prising the spring contact arm with the thumb nail. If the faces of the points are dirty or severely eroded they must be removed.
4 To remove the contact breaker, disconnect the LT (negative) lead from the terminal on the distributor body (photo).
5 Release, *but do not remove,* the screws which hold the contact breaker to the distributor baseplate. Slide the contact breaker out from under the screws.

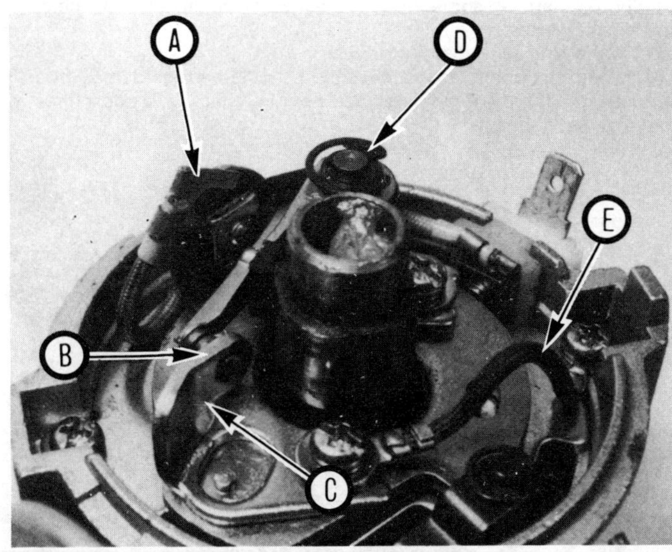

3.4 Distributor with cap removed (conventional system)

A LT lead
B Contact points
C Fixed contact point
D C-clip
E Earth lead

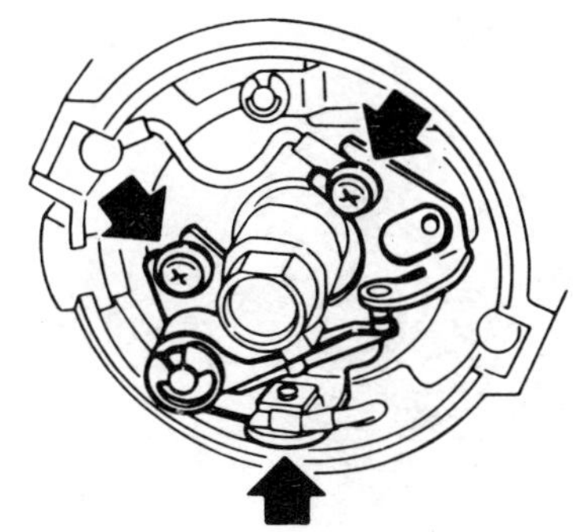

Fig. 4.6 Contact breaker securing screws and LT terminal (Sec 3)

6 To remove the spring contact arm, extract the C-clip and slide the arm off the pivot post.
7 If only slight surface pitting or build-up is evident on the faces of the contact points this may be removed by using abrasive paper or rubbing squarely on an oilstone.
8 If the points faces are severely eroded, or there is a large pip on one face, do not attempt to dress them as an excessive amount of metal will have to be removed. Renew them instead.
9 Wipe the faces of new contact points with solvent to remove any protective grease.
10 Apply a drop of oil to the pivot, fit the spring contact arm and the C-clip.
11 Fit the contact breaker assembly to the distributor baseplate, but leave the screws just finger tight. Connect the earth lead.
12 The crankshaft must now be turned until the plastic heel of the spring contact arm is positioned on one of the cam peaks of the distributor shaft.
13 Adjust the fixed contact arm until, using feeler blades, the points gap is as specified – with the feeler blade a stiff sliding fit (photo). A cut-out is provided in the fixed contact arm with a fulcrum pin so that a small screwdriver can be used to lever the arm and control its movement.
14 Once the gap is correct, tighten the fixed contact screws. Recheck the points gap.
15 Apply a smear of high melting-point grease to the distributor shaft cams.
16 Apply two drops of oil to the felt pad in the recess at the top of the distributor shaft. On some models, a felt pad is not used, in which case just oil the screw.
17 Fit the rotor and distributor cap.
18 Setting the points as just described should be regarded as a basic setting only, the dwell angle should now be checked, as described in the next Section.

3.13 Checking contact breaker points gap (conventional system)

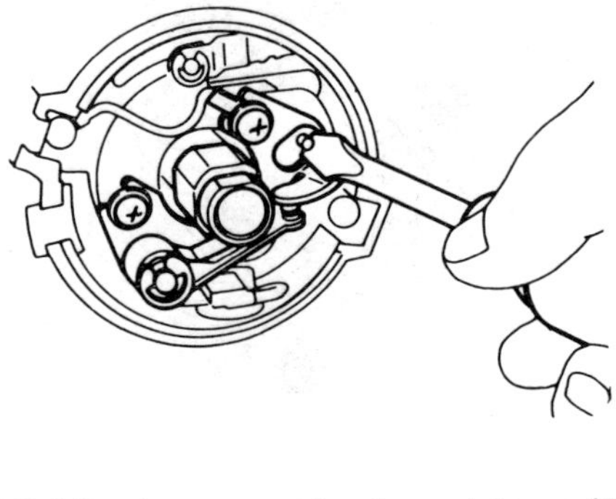

Fig. 4.7 Adjusting contact breaker points gap (Sec 3)

4 Dwell angle (conventional system) – checking and adjustment

1 On modern engines, setting the contact breaker gap in the distributor using feeler gauges must be regarded as a basic adjustment only. For optimum engine performance, the dwell angle must be checked. The dwell angle is the number of degrees through which the distributor cam turns during the period between the instance of closure and opening of the contact breaker points. Checking the dwell angle not only gives a more accurate setting of the contact breaker gap but also evens out any variations in the gap which could be caused by wear in the distributor shaft or its bushes, or difference in height of any of the cam peaks.
2 The angle should be checked with a dwell meter connected in accordance with the maker's instructions. If the check is being made with the engine running, first disconnect and plug the vacuum advance hose at the distributor capsule. Refer to the Specifications for the correct dwell angle. If the dwell angle is too large, increase the points gap, if too small, reduce the points gap.
3 The dwell angle should always be adjusted before checking and adjusting the ignition timing.

5 Condenser (conventional system) – testing, removal and refitting

1 The purpose of the condenser (sometimes known as the capacitor) is to ensure that when the contact breaker points open there is no sparking across them which would waste voltage and cause wear.
2 The condenser is mounted on the outside of the distributor body. If it develops a short-circuit it will cause ignition failure as the points will be prevented from interrupting the low tension circuit.
3 If the engine becomes very difficult to start or begins to miss after several miles running and the breaker points show signs of excessive burning, then the condition of the condenser must be suspect. A further test can be made by separating the points by hand with the ignition switched on. If this is accompanied by a strong blue flash it is indicative that the condenser has failed in the open circuit mode.
4 Without special equipment the only sure way to diagnose condenser trouble is to replace a suspected unit with a new one and note if there is any improvement.
5 To remove the condenser from the distributor take off the distributor cap and rotor arm.
6 Release the terminal nut and disconnect the condenser lead from the terminal.
7 Remove the screw and withdraw the condenser from the distributor.
8 Refitting is a reversal of removal.

6 Distributor (conventional system) – overhaul

1 With the distributor removed from the engine (Section 8), clean away external dirt.
2 Prise off the spring clips and take off the cap.
3 Pull off the rotor.
4 Extract the fixing screws and remove the vacuum advance capsule. It will need tilting in order to release the link rod from the pivot on the baseplate once the E-clip has been prised off the pivot.
5 Extract the two screws and remove the baseplate.
6 Remove the contact breaker from the baseplate.
7 If the contact breaker fixing screws are removed, take care that the anti-friction balls are not lost as the baseplate upper and lower sections separate.
8 Mount the shaft collar in a vice then drive out the roll pin and remove the collar and washer.
9 Withdraw the shaft from the distributor body.
10 Extract the felt lubrication pad from the recess in the top of the shaft and remove the screw which is exposed.
11 Mark the relationship of cam assembly to shaft and separate them.
12 If the cam counter weights and springs must be dismantled, make quite sure that the springs are marked with a dab of quick-drying paint so that they can be reconnected in their original positions.
13 With the distributor dismantled, clean and examine all components for wear. If the shaft bushes and other items are worn, it may well be more economical to purchase a new distributor complete.

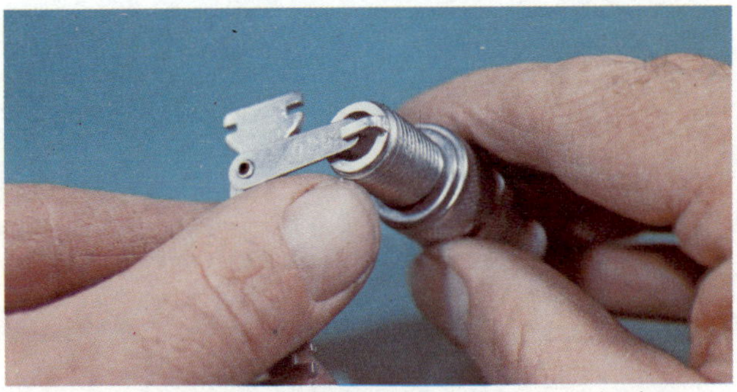

Measuring plug gap. A feeler gauge of the correct size (see ignition system specifications) should have a slight 'drag' when slid between the electrodes. Adjust gap if necessary

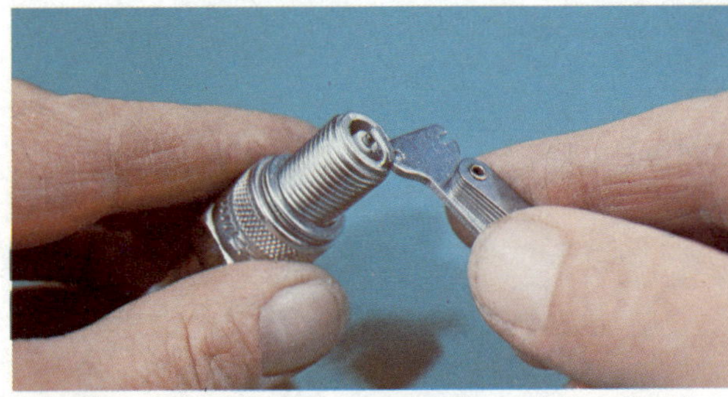

Adjusting plug gap. The plug gap is adjusted by bending the earth electrode inwards, or outwards, as necessary until the correct clearance is obtained. Note the use of the correct tool

Normal. Grey-brown deposits, lightly coated core nose. Gap increasing by around 0.001 in (0.025 mm) per 1000 miles (1600 km). Plugs ideally suited to engine, and engine in good condition

Carbon fouling. Dry, black, sooty deposits. Will cause weak spark and eventually misfire. Fault: over-rich fuel mixture. Check: carburettor mixture settings, float level and jet sizes; choke operation and cleanliness of air filter. Plugs can be re-used after cleaning

Oil fouling. Wet, oily deposits. Will cause weak spark and eventually misfire. Fault: worn bores/piston rings or valve guides; sometimes occurs (temporarily) during running-in period. Plugs can be re-used after thorough cleaning

Overheating. Electrodes have glazed appearance, core nose very white – few deposits. Fault: plug overheating. Check: plug value, ignition timing, fuel octane rating (too low) and fuel mixture (too weak). Discard plugs and cure fault immediately

Electrode damage. Electrodes burned away; core nose has burned, glazed appearance. Fault: pre-ignition. Check: as for 'Overheating' but may be more severe. Discard plugs and remedy fault before piston or valve damage occurs

Split core nose (may appear initially as a crack). Damage is self-evident, but cracks will only show after cleaning. Fault: pre-ignition or wrong gap-setting technique. Check: ignition timing, cooling system, fuel octane rating (too low) and fuel mixture (too weak). Discard plugs, rectify fault immediately

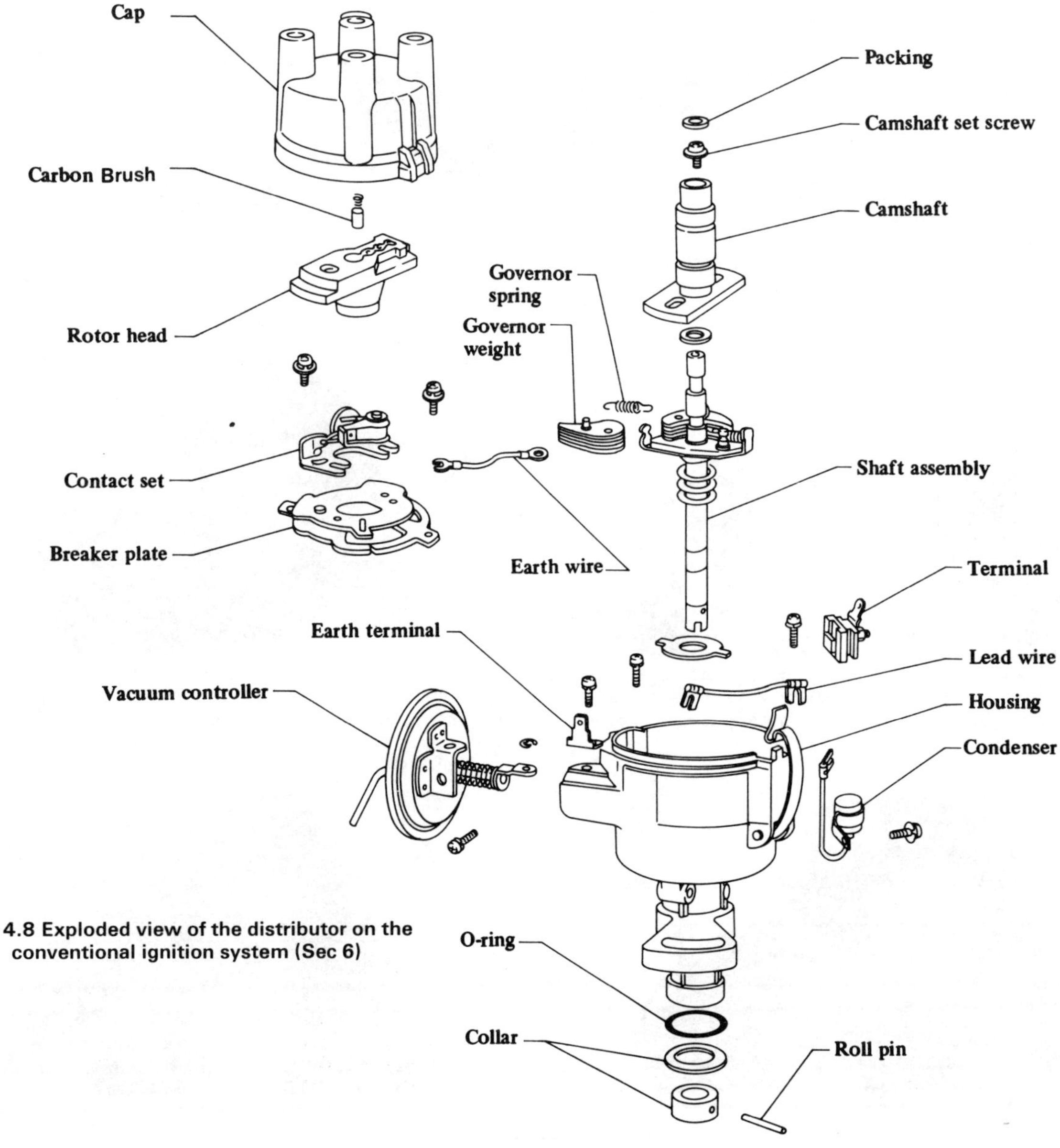

Fig. 4.8 Exploded view of the distributor on the conventional ignition system (Sec 6)

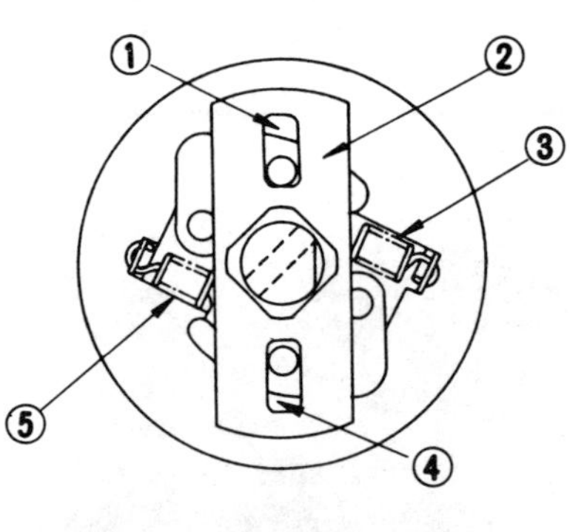

Fig. 4.9 Counterweight spring locations (Sec 6)

1. *Short rectangular hole*
2. *Cam shaft*
3. *Governor spring B (Rectangular hook type)*
4. *Long rectangular hole*
5. *Governor spring A (Circular hook type)*

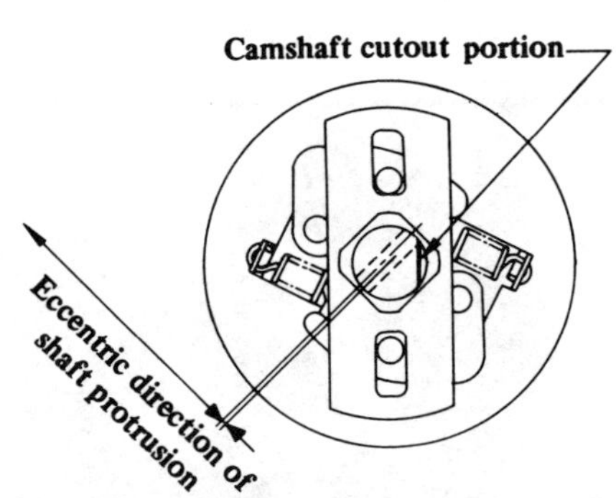

Fig. 4.10 Distributor shaft cut-out relation to offset groove (Sec 6)

6.14 Showing carbon brush location in distributor cap

14 Take the opportunity to carefully examine the cap and rotor for tiny cracks. These can cause conductance paths and prevent starting or be responsible for erratic running. Renew these components if necessary, particularly the rotor if the metal contact is eroded or the carbon brush inside the cap is worn to the specified minimum (photo).

15 Reassembly is a reversal of dismantling, apply light grease to all components as work proceeds and make quite sure that they are aligned with each other in their original positions.

7 Distributor (transistorized system) – overhaul

1 With the distributor removed from the engine (Section 8), clean away external dirt.

2 Remove the cap.

3 Pull off the rotor.

4 Extract the fixing screws and remove the vacuum advance capsule.

5 Disconnect the wiring harness by sliding the insulator block out of the cut-out in the distributor body.

6 Using two screwdrivers as levers, prise the reluctor carefully from the distributor shaft and recover the roll pin.

7 Extract the fixing screws and remove the baseplate.

8 Extract the screws and remove the pick-up unit and spacers.

9 Remove the magnet and stator from the baseplate.

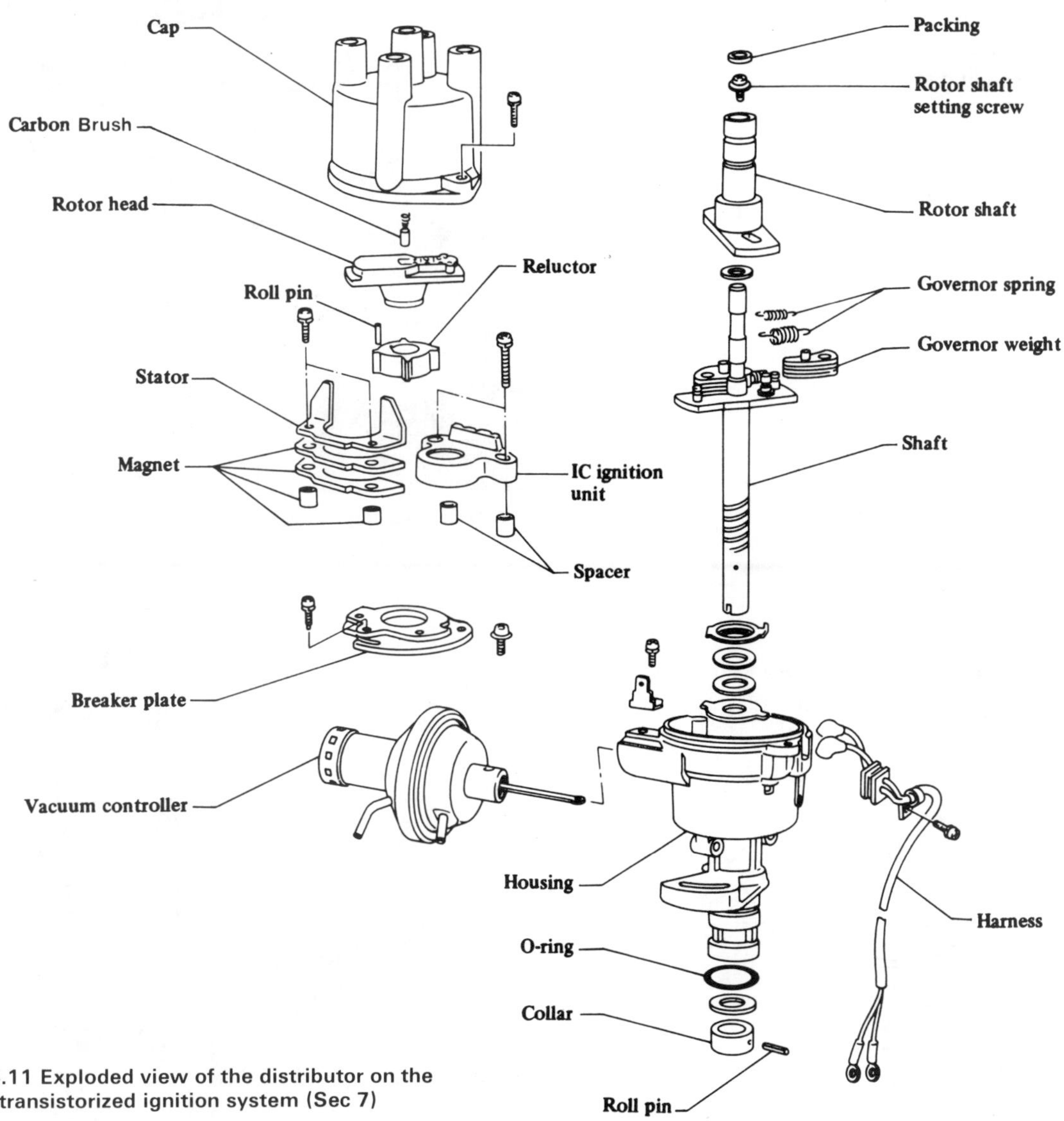

Fig. 4.11 Exploded view of the distributor on the transistorized ignition system (Sec 7)

10 Mount the shaft collar in a vice then drive out the roll pin and remove the collar and washer.
11 Withdraw the shaft from the distributor body.
12 Extract the felt lubrication pad from the recess in the top of the shaft and remove the screw now exposed. Separate the rotor shaft from the mainshaft.
13 If necessary, disconnect the springs and remove the counterweights, but mark the location of the springs with a dab of quick-drying paint.
14 Clean and inspect all components. The reluctor and stator should be unscratched and not distorted, otherwise renew them. If the carbon brush is worn down to its specified minimum length, renew the cap.
15 Apply grease to all friction and bearing surfaces as work proceeds.
16 Reassembly is a reversal of dismantling, but observe the following points.
17 When reconnecting the counterweight springs, fit the smaller diameter spring first.
18 Set the relative position of the rotor shaft cut-out to the mainshaft offset groove as shown in Fig. 4.12.
19 Use a new roll pin to fix the drive collar to the shaft.
20 When fitting the baseplate make sure that its mark is in alignment with the one on the distributor body.
21 Fit a new roll pin to the reluctor.
22 Make sure that the wiring harness spade terminals are correctly located (Fig. 4.13).
23 Finally, set the reluctor air gap (reluctor peak to stator prong edge) to the specified dimension using a feeler blade (Fig. 4.14). Adjustment is made by loosening the screws and re-positioning the stator.

8 Distributor – removal and refitting

1 The distributor is driven from the flywheel end of the camshaft. It is located in a recess in the thermostat housing.
2 First make a mark on the slotted flange of the distributor body in line with the mounting stud so that the ignition timing can be reset approximately when the distributor is refitted.

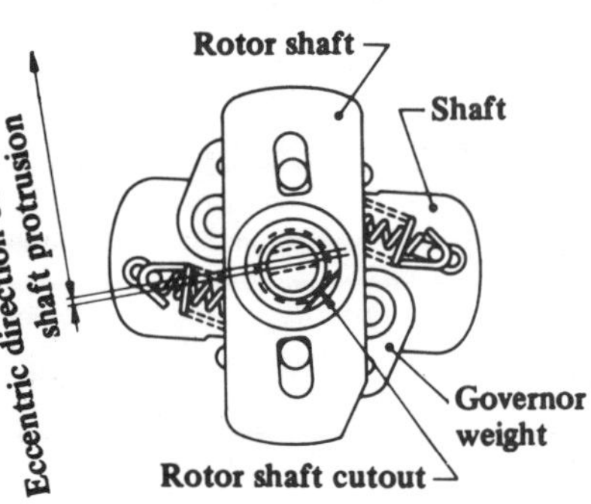

Fig. 4.12 Correct relation of rotor shaft to mainshaft (Sec 7)

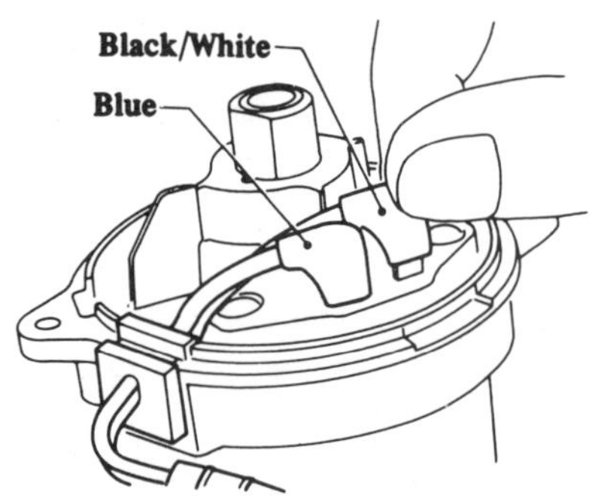

Fig. 4.13 Correct location of sensor leads (Sec 7)

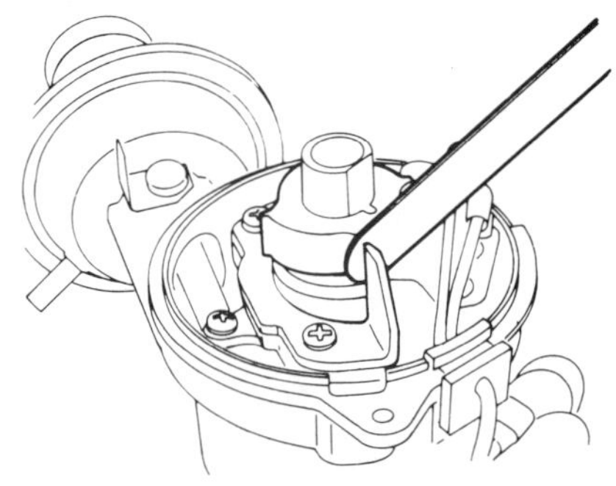

Fig. 4.14 Checking the reluctor air gap (Sec 7)

3 The distributor shaft incorporates an offset slot so there is no need to note the position of the rotor arm. Simply disconnect the wiring and the vacuum hose (where applicable), then unscrew the mounting nut and withdraw the distributor from the cylinder head (photos).
4 Refitting is a reversal of removal, but align the offset slot with the

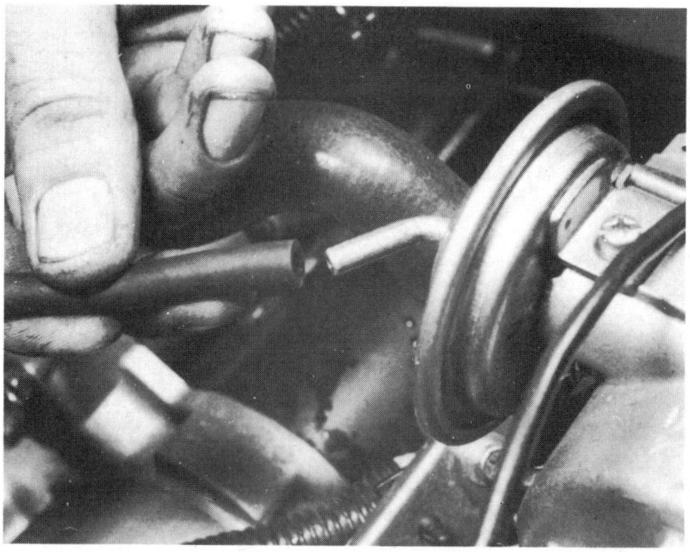

8.3A Disconnecting the distributor vacuum hose

8.3B Removing the distributor

dog on the camshaft before inserting the distributor. Turn the rotor arm if necessary until the slot engages. If the ignition timing setting has been lost, it can be reset approximately by aligning the crankshaft timing marks (Section 9) with No 1 piston on compression, then turning the distributor within the adjustment slot until the contact points just open (conventional system) or the reluctor peak is aligned with the stator post (transistorized system). After refitting the distributor check and adjust the ignition timing as described in Section 9.

9 Ignition timing – adjustment

1 The following procedure applies to vehicles equipped with either a mechanical or electronic type ignition system.
2 Run the engine to normal operating temperature then switch it off and connect a stroboscopic timing light in accordance with the manufacturer's instructions.
3 Disconnect and plug the vacuum advance hose at the distributor except on models equipped with the computerised ECCS ignition.
4 With the engine idling, point the timing light at the index above the crankshaft pulley. The notch in the pulley rim should be in alignment with the mark on the index which applies to your particular vehicle (see Specifications).
5 If the marks are not in alignment, release the distributor clamp nut and turn the distributor in either direction as necessary to align the marks.
6 Tighten the nut, switch off the engine, reconnect the vacuum hose, where applicable, and remove the timing light.
7 Any difficulty experienced in seeing the timing marks clearly can be overcome by applying a spot of white paint to the pulley notch and the specified timing mark on the index.

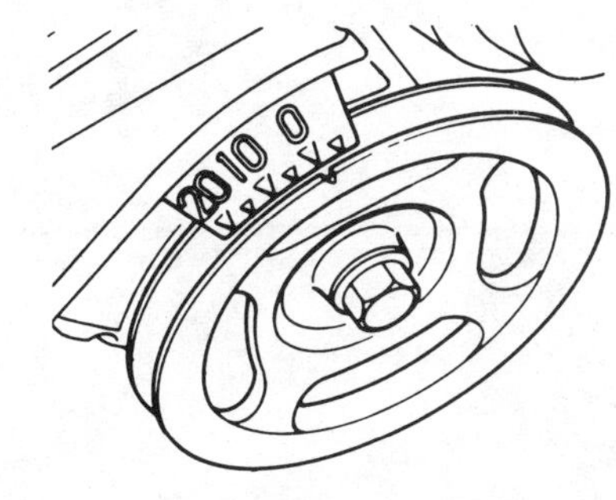

Fig. 4.15 Ignition timing marks and notch in crankshaft pulley (Sec 9)

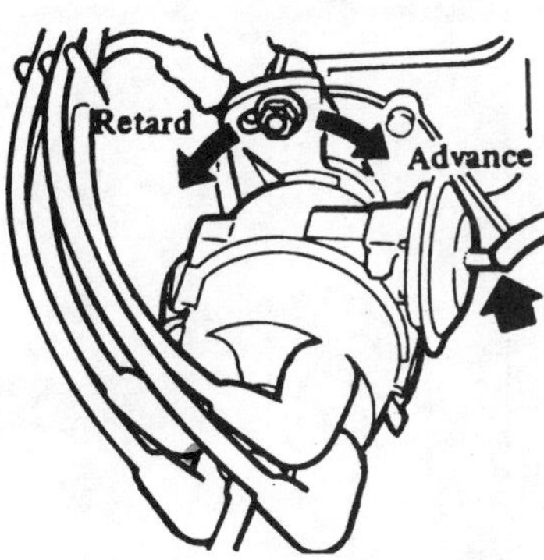

Fig. 4.16 Ignition timing adjustment (Sec 9)

10 Spark plugs, HT leads and distributor cap – general

1 The correct functioning of the spark plugs is vital for economical running and efficiency of the engine.
2 At the intervals specified in Routine Maintenance the plugs should be removed, cleaned and re-gapped.

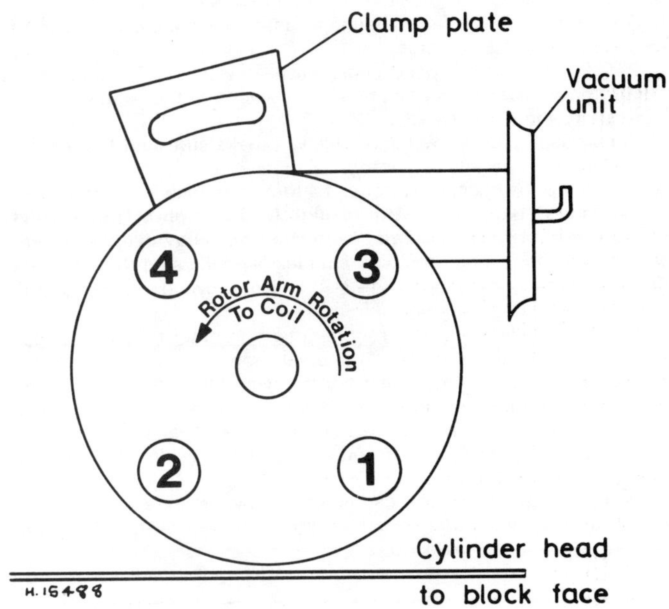

Fig. 4.17 Plug HT lead locations on the distributor cap (Sec 10)

3 To remove the plugs, first open the bonnet, remove the air cleaner (where applicable), and pull off the HT leads. Grip the rubber end fitting not the lead, otherwise the lead connection may be fractured.
4 Brush out any accumulated dirt or grit from the spark plug recess in the cylinder head otherwise it may drop into the combustion chamber when the plug is removed.
5 Unscrew the spark plugs with a deep socket or a box spanner. Do not allow the tool to tilt, otherwise the ceramic insulator may be cracked or broken.
6 Examination of the spark plugs will give a good indication of the condition of the engine.
7 If the insulator nose of the spark plug is clean and white, with no deposits, this is indicative of a weak mixture, or too hot a plug (a hot plug transfers heat away from the electrode slowly, a cold plug transfers heat away quickly).
8 The plugs fitted as standard are specified at the beginning of this Chapter. If the top and insulator nose are covered with hard black-looking deposits, then this is indicative that the mixture is too rich. Should the plug be black and oily, then it is likely that the engine is fairly worn, as well as the mixture being too rich.
9 If the insulator nose is covered with light tan to greyish brown deposits, then the mixture is correct and it is likely that the engine is in good condition.
10 If there are any traces of long brown tapering stains on the outside of the white portion of the plug, the plug will have to be renewed, as this shows that there is a faulty joint between the plug body and the insulator, and compression is being allowed to leak away.
11 Before cleaning a spark plug, wash it in a suitable solvent to remove oily deposits.
12 Although a wire brush can be used to clean the electrode end of the spark plug this method can cause metal conductance paths across the nose of the insulator and it is therefore to be preferred that an abrasive powder cleaning machine is used. Such machines are available quite cheaply from motor accessory stores or you may prefer to take the plugs to your dealer who will not only be able to clean them but also to check the sparking efficiency of each plug under compression.
13 The spark plug gap is of considerable importance, as, if it is too large or too small, the size of the spark and its efficiency will be seriously impaired. For the best results the spark plug gap should be set in accordance with the Specifications at the beginning of this Chapter.
14 To set it, measure the gap with a feeler gauge, and then bend open, or close, the outer electrode until the correct gap is achieved.

The centre electrode should never be bent as this may crack the insulation and cause plug failure if nothing worse.
15 Special spark plug electrode gap adjusting tools are available from most motor accessory stores.
16 Before refitting the spark plugs, wash each one thoroughly again in order to remove all trace of abrasive powder and then apply a trace of grease to the plug threads.
17 Screw each plug in by hand. This will make sure that there is no chance of cross threading (photo).
18 Tighten to the specified torque. If a torque wrench is not available, just nip up each plug. **It is better to slightly undertighten rather than overdo it and strip the threads from the light alloy cylinder head.**
19 When reconnecting the spark plug leads, make sure that they are refitted in their correct order, 1–3–4–2, No 1 cylinder being at the timing belt end of the engine.
20 The plug leads require no routine attention other than being kept clean and wiped over regularly. At intervals however, pull each lead off the plug in turn and remove it from the distributor. Water can seep down into the joints giving rise to a white corrosive deposit which must be carefully removed from the end of each cable. A smear of petroleum jelly applied to the end fitting of the cables will help to eliminate this problem.
21 Whenever the distributor cap is removed wipe it clean and check for thin lines between the electrodes. If evident, renew the cap as the lines are the result of tracking.

11 Coil – description and testing

1 The coil is located on the left-hand side of the engine compartment. It should be kept clean at all times to prevent possible arcing across the high tension tower.
2 To ensure the correct HT polarity at the spark plugs, the LT coil leads must always be connected correctly. Refer to the wiring diagrams at the end of the manual for the lead colour coding.
3 The coil may be tested by using an ohmmeter. Connect the ohmmeter across the two LT terminals on the coil to check the primary winding resistance, and across the negative LT terminal and central HT tower to check the secondary winding resistance. If the readings are not as given in the Specifications, renew the coil. All wiring should be disconnected while checking the coil.
4 A ballast resistor is mounted next to the coil on the conventional ignition system (photo), and its purpose is to supply battery voltage to the coil during starting. For normal running the coil is designed to run on a reduced voltage.

12 Fault diagnosis – ignition system

1 If the engine fails to start but was running normally when it was last used, first check that there is fuel in the fuel tank. If the engine turns over normally on the starter motor and the battery is evidently well charged first check the HT circuit.
2 One of the commonest reasons for bad starting is wet or damp spark plug leads and distributor cap. Check both items and wipe dry, if necessary.
3 If the engine still fails to start, disconnect an HT lead from any spark plug and hold the end of the cable approximately 5.0 mm (0.2 in) away from the cylinder head using well insulated pliers. If the HT lead terminal is concealed use a suitable large nail as an extension. While an assistant spins the engine on the starter motor check that a regular blue spark occurs. If so, then the spark plugs are probably the cause of the engine not starting and they should be removed and re-gapped or renewed.
4 If no spark occurs, disconnect the main feed HT lead from the distributor cap and check for a spark as in paragraph 3. If sparks now occur, check the distributor cap, rotor arm, and HT leads as described in Section 10 and renew them as necessary.

Conventional ignition system

5 Using a 12 volt voltmeter or 12 volt test lamp, test between the LT wire on the battery side of the coil and earth with the ignition switched on and the points open. No reading indicates a break in the supply from the ignition switch.
6 Test between the LT terminal on the distributor side of the coil and earth. If there is no reading suspect the condenser or coil.

Transistorized ignition system

7 Remove the distributor cap and disconnect the sensor *feed* wire. Connect a voltmeter between the wire and earth and check that battery voltage is registered with the ignition switched on. If not, there is a break in the supply from the ignition switch.
8 Remove the coil HT lead from the distributor cap and connect it to earth with a suitable lead.
9 Spin the engine on the starter motor and check that the reading on the voltmeter is between 8.6 volt and 1.0 volt below battery voltage. If not, check the supply wiring.
10 Connect the voltmeter between the battery negative terminal and the earth terminal on the distributor then spin the engine on the starter motor. If more than 0.5 volt is registered on the voltmeter check the earth connections for the distributor, engine and battery. If less than 0.5 volt is registered, a faulty IC ignition unit within the distributor is indicated and it should therefore be renewed.
11 After making the check remove the voltmeter, reconnect the wires and refit the distributor cap.

Computerised (ECCS) ignition system

12 The ECCS Control Unit incorporates a micro-computer to control the ignition timing, fuel injection, idle speed and the fuel pump, and the computer has a self-diagnosis facility. The method for checking the various circuits which includes the ignition system is described in Chapter 3, but the work should be entrusted to a Nissan dealer.

10.17 Fitting a spark plug

11.4 Showing the ignition coil and ballast resistor (conventional system)

Chapter 5 Clutch

Contents

Specifications

Type	Single dry plate, with diaphragm spring, pressure plate, cable-operated	
Driven plate		
Diameter:		
E10 engine	160.0 mm (6.30 in)	
E13, E15 and E16 engines	180.0 mm (7.1 in)	
E15ET engine	200.0 mm (7.9 in)	
Minimum wear limit – lining surface to rivet	0.3 mm (0.012 in)	
Maximum run-out:		
E15ET engine	0.7 mm (0.028 in)	
Except E15ET engine	0.5 mm (0.020 in)	
Hub spline play (measured at outer edge of plate)	0.4 mm (0.016 in)	
Clutch cover		
Diaphragm spring finger maximum height variation	0.5 mm (0.020 in)	
Pedal		
Height dimension:		
UK models	201.0 to 207.0 mm (7.91 to 8.15 in)	
North American models:		
1983/1984	194.0 to 204.0 mm (7.64 to 8.03 in)	
1985	209.0 to 214.0 mm (8.23 to 8.43 in)	
Free play:		
All models except 1985 North American	11.0 to 21.0 mm (0.43 to 0.83 in)	
1985 North American models	12.5 to 17.5 mm (0.492 to 0.689 in)	
Release lever free play:		
All models except 1985 North American	2.0 to 4.0 mm (0.08 to 0.16 in)	
1985 North American models	2.5 to 3.5 mm (0.098 to 0.138 in)	
Torque wrench settings	**lbf ft**	**Nm**
Pedal stop bolt locknut	12 to 16	16 to 22
Cable adjustment locknut	2.2 to 2.9	3 to 4
Clutch cover bolts	12 to 15	16 to 21

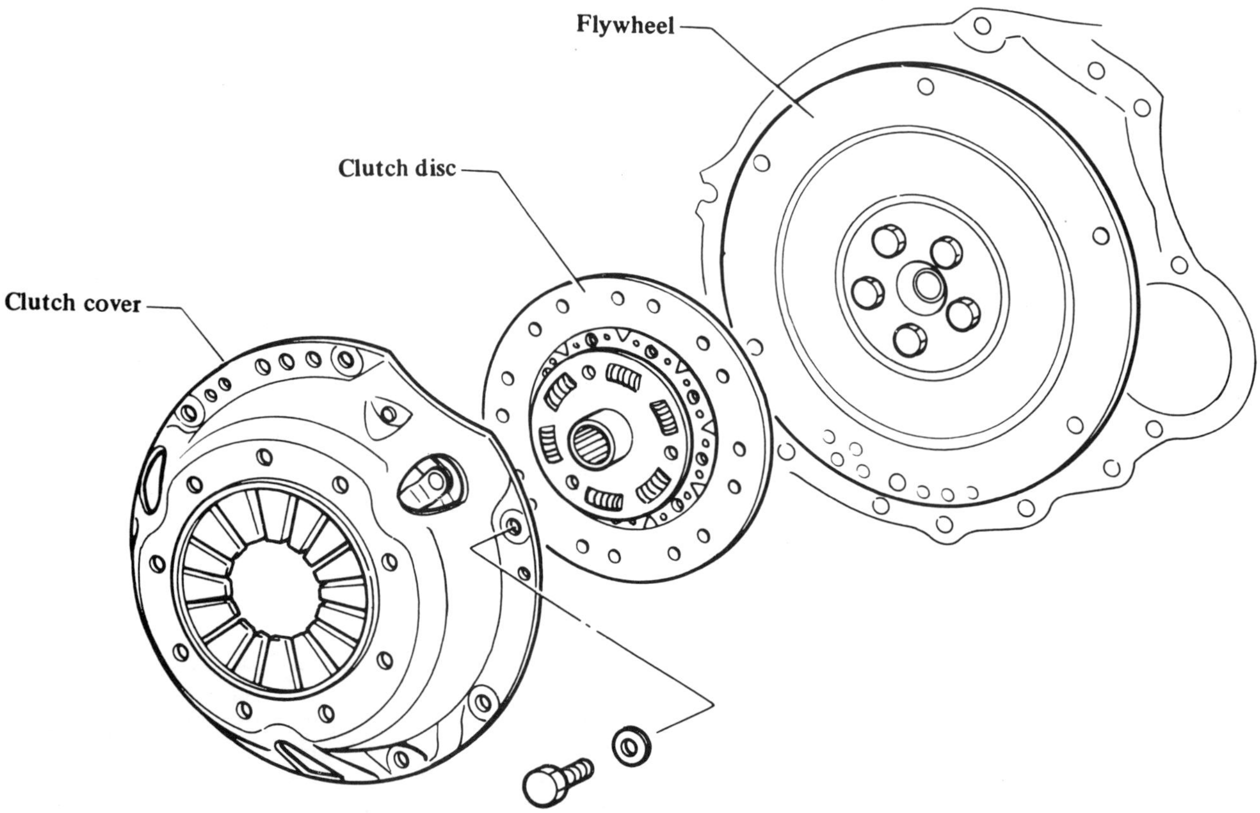

Fig. 5.1 Clutch components (Sec 1)

1 General description

The clutch is of single dry plate type and is cable operated from a pendant pedal.

When the pedal is depressed, the cable pulls the release lever which forces the ball bearing type release bearing against the diaphragm spring fingers of the cover assembly. This releases the pressure plate from the linings of the driven plate and there is then no drive between the engine and transmission.

When the pedal is released, the pressure plate is forced against the driven plate which in turn is forced against the flywheel. Drive is then transmitted from the engine, through the driven plate, and into the transmission by the input shaft.

2 Routine maintenance

1 Every 6000 miles (10 000 km) on UK models or 15 000 miles (24 000 km) on North American models check the clutch pedal height and cable free play, and adjust if necessary. Check that the pedal moves smoothly, and lightly oil the pivot bushes.

3 Clutch – adjustment

1 Working inside the car measure the distance from the upper surface of the clutch pedal pad to the floor pan or steering hole cover as shown in Figs. 5.2 or 5.3.

2 If adjustment is required loosen the locknuts on the pedal stop bolt or switch (as applicable) then tighten the locknuts.

3 Working in the engine compartment check that the free play at the end of the clutch release lever is as given in Specifications. If not, loosen the locknut on the cable end fitting and turn the knurled adjuster as necessary (photo). Tighten the locknut on completion.

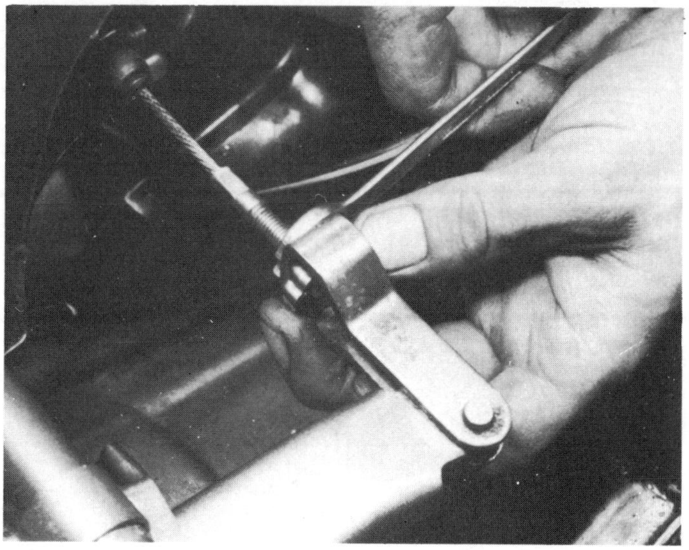

3.3 Adjusting the clutch cable

4 Fully depress the clutch pedal several times and recheck the release lever free play.

5 Finally check that the clutch pedal free play at the pad is as specified.

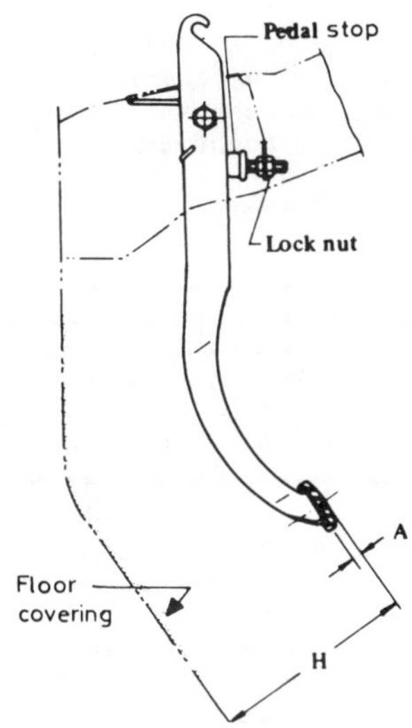

Fig. 5.2 Clutch pedal adjustment on UK models (Sec 3)

A Free play *H Height*

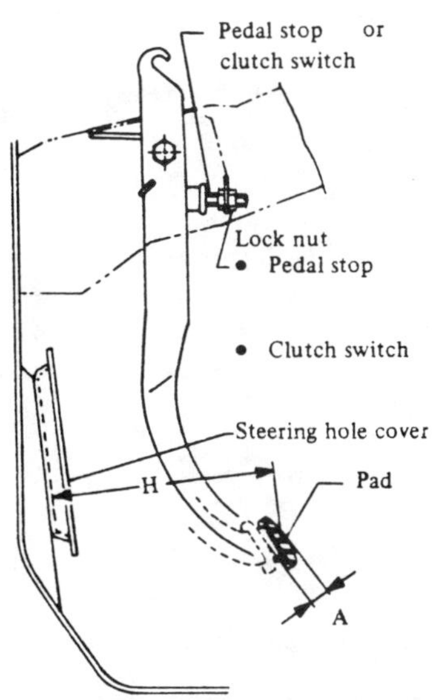

Fig. 5.3 Clutch pedal adjustment on North American models (Sec 3)

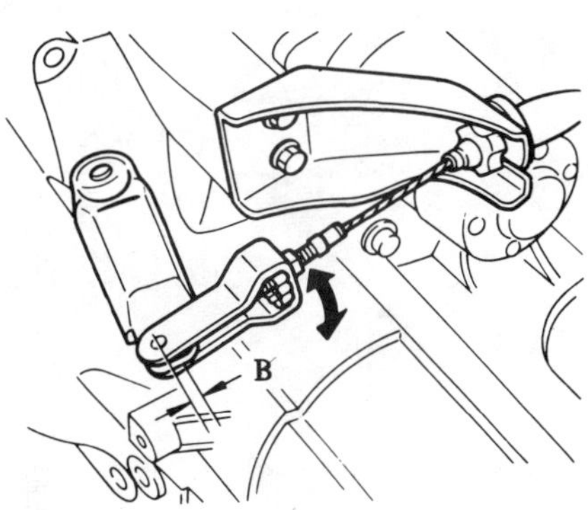

Fig. 5.4 Release lever free play adjustment (Sec 3)

B Free play

4 Clutch cable – renewal

1 Remove the cover from under the facia panel.
2 Release the nuts at the end of the cable at the release lever and disconnect the cable from the lever (photo).
3 Working inside the vehicle, remove the clevis pin and disconnect the cable from the clutch pedal.
4 Unscrew the two nuts which retain the cable grommet to the bulkhead and withdraw the cable from the vehicle.
5 Refit the new cable by reversing the removal operations, and then carry out the adjustments described in Section 3.

4.2 Disconnecting the clutch cable from the release lever

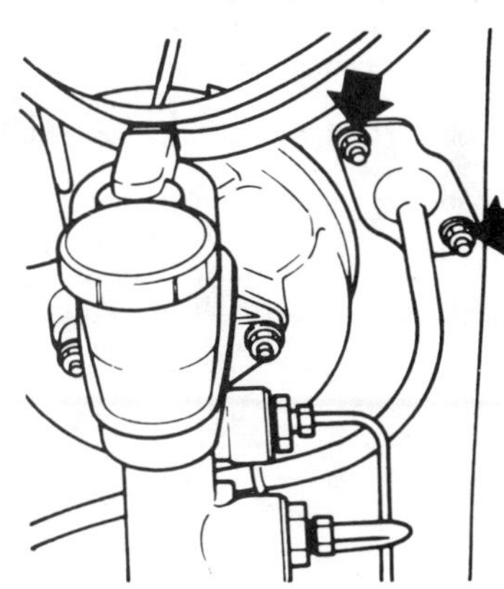

Fig. 5.5 Clutch cable mounting on the bulkhead (Sec 4)

5 Clutch pedal – removal and refitting

1 Remove the cover from under the facia panel.
2 Disconnect the cable from the pedal arm.
3 Extract the E-clip from the end of the pivot shaft.
4 Unhook the pedal return spring and slide the pedal off the pivot shaft.
5 The pivot bushes may be renewed.
6 Refitting is a reversal of removal, apply grease to the pivot shaft and bushes.
7 Adjust the pedal height and free movement, as described in Section 3.

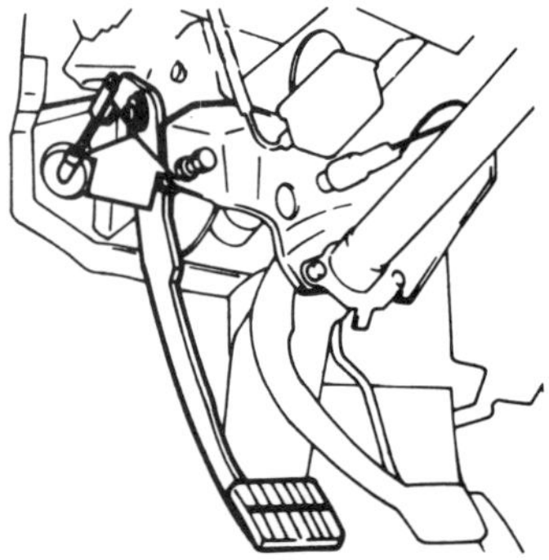

Fig. 5.6 Clutch pedal E-clip location (Sec 5)

6 Clutch – removal

1 Access to the clutch is obtained by removing the transmission, as described in Chapter 6.
2 Unscrew each of the clutch cover bolts progressively a turn at a time until the pressure of the diaphragm spring is relieved. If the flywheel tends to turn as the bolts are unscrewed, jam the teeth of the starter ring gear using a large screwdriver or similar blade.
3 Remove the clutch cover, catching the driven plate as it is released from the flywheel.
4 The clutch cover bolts are offset so it can only be refitted in one position.

7 Clutch – inspection

1 Examine the driven plate. If the friction linings are worn down to, or nearly down to the rivet heads, the plate should be renewed. Do not attempt to re-line the plate yourself, it is unlikely to prove satisfactory.
2 If the linings are good for further service, check the torsion springs and the hub for cracks.
3 Examine the splines in the driven plate hub for wear.
4 Check the clutch pressure plate cover assembly. If any parts show evidence of cracking or severe rusting, or if the fingers of the diaphragm spring are worn or stepped by contact with the release bearing, renew the assembly. Do not attempt to dismantle the pressure plate cover.
5 Check the friction surfaces of both the pressure plate and the flywheel. If grooved, or showing signs of very fine surface cracking, the flywheel may be machined – subject to certain limitations described in Chapter 1 – but the pressure plate assembly will have to be renewed complete.
6 If the clutch components are oil-stained or the inside of the clutch bellhousing is covered in oil, suspect a faulty oil seal on the transmission input shaft or crankshaft rear end. Renew the defective oil seal before installing new components.
7 A pilot bearing is not fitted in the centre of the crankshaft rear mounting flange for the flywheel.

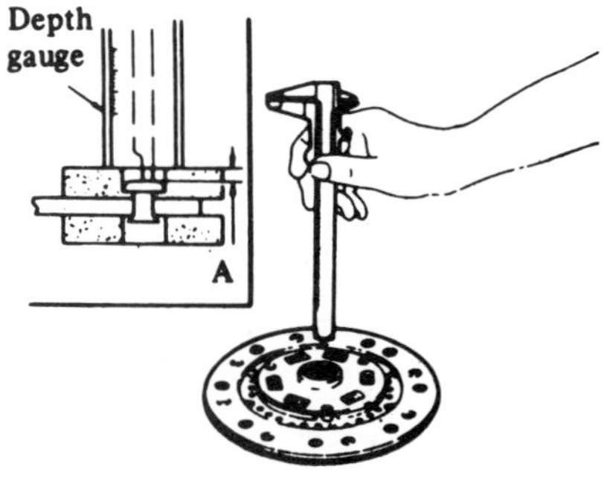

Fig. 5.7 Checking the driven plate lining wear (Sec 7)

A Checking dimension

8 Clutch release bearing – removal, inspection and refitting

1 Whenever the transmission is removed for clutch overhaul, check the release components and renew the bearing as a matter of routine.
2 Disconnect the springs from the release bearing and slide the release bearing off the bearing mounting sleeve (photo).
3 Check that the release control shaft bushes are not worn. If they are, the fork-to-cross-shaft retaining pins will have to be driven out. The return spring is released as the shaft is withdrawn. The pins can only be driven out if the cross-shaft is turned to allow the pins to pass out into the small cavities in the bellhousing.
4 Reassembly is a reversal of dismantling, apply grease to the cross-shaft bushes and to the release bearing recess (photo).

8.2 Removing the release bearing

8.4 Inserting a release fork roll pin

8.5 Release bearing fully entered in spring clips

5 Smear a little molybdenum disulphide grease onto the release bearing mounting sleeve and the input shaft splines. Make sure that the release bearing is pushed fully home until the retaining spring clips are heard to click (photo).

6 When correctly reassembled, the numbers moulded onto the release fork should be visible.

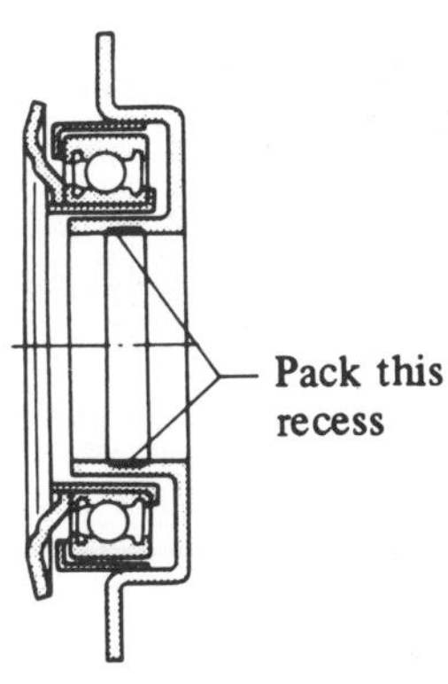

Fig. 5.8 Grease recess in clutch release bearing (Sec 8)

Fig. 5.9 Release bearing showing retaining spring clip (Sec 8)

9 Clutch – refitting

1 Make sure that the flywheel and pressure plate friction surfaces are clean, and free from protective grease.

2 Place the driven plate against the flywheel so that its hub has the greater projecting side away from the flywheel (photo).

3 Offer up the clutch cover (photo) and screw in the bolts finger tight. The mounting dowels are offset so the cover can only be fitted one way.

4 The driven plate must now be centralised. To do this, a conventional clutch alignment tool can be used, or alternatively the splined hub of the driven plate can be centralised within the tips of the diaphragm spring fingers visually or, more accurately, by cutting a cardboard disc of outside diameter equal to the outside diameter of the clutch cover aperture. Now punch a hole centrally in the cardboard disc and then push a bar or rod through the hole in the disc and engage it in the splined hub of the driven plate (photo). The diameter of the bar or rod should be such that it provides a sliding fit in the hub. A bar of incorrect diameter can be adjusted to suit by winding tape around it.

5 The driven plate should now be moved in the appropriate direction to centralise the cardboard disc (photo). Tighten the cover bolts and remove the tool.

6 Refit the transmission with reference to Chapter 6.

9.2 Locating the driven plate

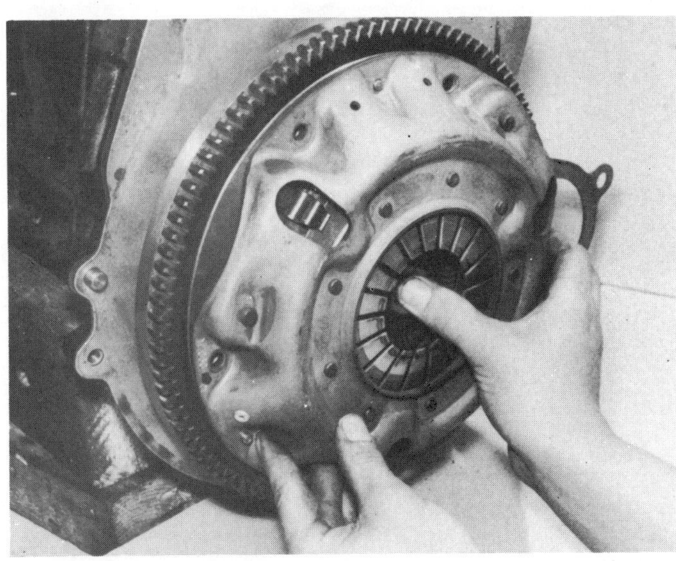

9.3 Locating the clutch cover

9.4 Improvised alignment tool

9.5 Centralising the clutch driven plate

10 Fault diagnosis – clutch

Symptom	Reason(s)
Judder when taking up drive	Loose engine or gearbox mountings Badly worn friction linings or contaminated with oil Worn splines on gearbox input shaft or driven plate hub
Clutch drag (or failure to disengage) so that gears cannot be meshed	Incorrect adjustment Rust on splines (may occur after vehicle standing idle for long periods) Damaged or misaligned pressure plate assembly Cable stretched or broken
Clutch slip (increase in engine speed does not result in increase in car speed – especially on hills)	Incorrect adjustment Friction linings worn out or oil contaminated
Noise from clutch	Worn release bearing Worn or loose components of pressure plate or driven plate

Chapter 6 Manual transmission

Contents

Specifications

Type	Transversely mounted with four or five forward speeds and reverse. Synchromesh on all forward gears, floor-mounted gearchange
Designation	
Four-speed	RN4F30A
Five-speed	RS5F30A or RS5F31A
Ratios	
Four-speed transmission:	
1st	3.333:1
2nd	1.955:1
3rd	1.286:1
4th	0.902:1
Reverse	3.417:1
Five-speed transmission (pre 1985 models):	
1st	3.333:1
2nd	1.955:1
3rd	1.286:1
4th	0.902:1
5th	0.733:1
Reverse	3.417:1
Five-speed transmission (1985 on models):	
1st	3.063:1
2nd	1.826:1
3rd	1.207:1
4th	0.902:1
5th	0.733:1
Reverse	3.417:1
Final drive ratios:	
E10 engine	4.471:1
E13 engine	3.895:1
E15 engine	3.550:1, 3.789:1 or 3.895:1
E16 engine	3.650:1 or 3.895:1
Oil capacity	
Four-speed	4.0 Imp pt; 2.4 US qt; 2.3 litre
Five-speed	4.8 Imp pt; 2.8 US qt; 2.7 litre
Gear endfloat	
1st gear	0.18 to 0.31 mm (0.0071 to 0.0122 in)
2nd, 3rd and 4th gear	0.20 to 0.40 mm (0.0079 to 0.0157 in)
5th gear	0.18 to 0.41 mm (0.0071 to 0.0161 in)
Differential side gear to pinion gear endfloat:	
UK models	0.5 mm (0.020 in)
North American models	0 to 0.3 mm (0 to 0.012 in)

Torque wrench settings

	lbf ft	Nm
Clutch bellhousing to engine	30	40
Clutch bellhousing to casing	15	20
Circular cover to transmission casing	6	8
Bearing retainer screws	15	20
5th/Reverse detent plug	18	25
Crownwheel bolts	59	80
Oil filler/level plug	22	30
Oil drain plug	22	30
Reverse lamp switch	18	25
Neutral switch (seat belt)	18	25
Left-hand mounting bracket	26	35
Rear mounting to bodyframe	26	35
Rear mounting to transmission	26	35
Gearchange support rod to transmission	9	12
Gearchange control rod to transmission	6	8
Front mounting bracket to transmission	15	20
Front suspension strut top mounting nuts	17	23
Suspension lower balljoint to track control arm	45	61
Caliper mounting bolts	45	61
Roadwheel nuts	70	95

1 General description

The manual transmission is of four or five-speed type, depending upon the model. On both transmissions top gear is of overdrive type.

Synchromesh is provided on all forward gears and gear selection is by means of a floor-mounted control lever.

The transmission is mounted transversely in line with the engine. Power is transmitted from the clutch through an input shaft and mainshaft to the final drive/differential which is incorporated within the transmission casing.

The four and five-speed units are so similar that their overhaul is not described separately in this Chapter as it is assumed that owners of four-speed models will be able to ignore reference to a 5th gear in the operations listed.

2 Routine maintenance

1 Every 6000 miles (10 000 km) or 6 months whichever comes first on UK models, or every 15 000 miles (24 000 km) or 12 months, whichever comes first on North American models, check the transmission oil level and top up if necessary. Unscrew and remove the combined filler/level plug (photo). The oil should be level with the bottom of the hole and just starting to dribble out. If not, top up as necessary.

2 If the car is operated under severe conditions, renew the transmission oil every 24 000 miles (40 000 km) on UK models, every 30 000 miles (48 000 km) on North American models, or every 24 months whichever comes first.

2.1 Using an adaptor to remove the filler/level plug

3 Gearchange lever and rods – removal, refitting and adjustment

1 Twist off the gearchange lever knob. It may be necessary to protect the knob and then use grips to remove it.

2 Remove the nuts and separate the control and support rods from the transmission.

3 Unbolt the control rod bracket and lift away the control assembly.

4 Reconnect by reversing the operations just described.

5 Release, but do not remove, the selector stop plate bolts.

6 Move the control lever to 1st gear position.

7 Move the stop plate until a 1.0 mm (0.039 in) gap is visible between the stop and the gearchange lever.

8 Tighten the stop plate bolts. Check the movement and selection of all gears.

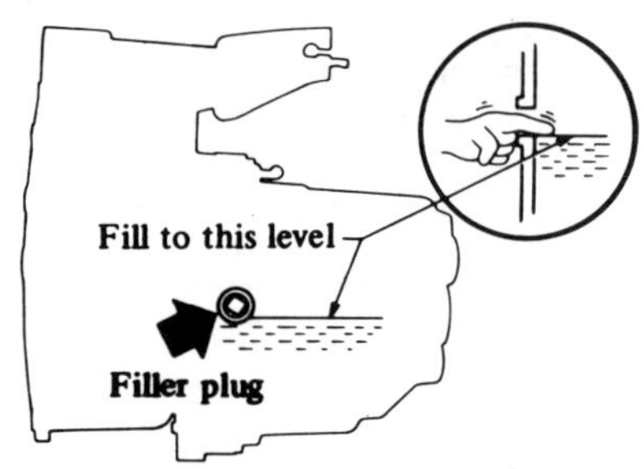

Fig. 6.1 Showing correct transmission oil level (Sec 2)

4 Gearchange control rods – dismantling and reassembly

1 Disconnect the control and support rods, as described in the preceding Section.

2 Remove the control lever bracket, the set spring ring, the bearing set seat and the bearing set spring.

3 Remove the flexible mountings, the dust cover and control lever bearing (photo).

4 The control rod flexible bushes may be drawn out using a bolt, nut, washers and a distance piece. Fit new bushes after having smeared them with hydraulic fluid.

5 Reassembly is a reversal of dismantling, apply grease to the pivot and friction surfaces.

6 Always use a new cotter pin at the end of the control rod and make sure that its ends are bent round as shown (Fig. 6.7).

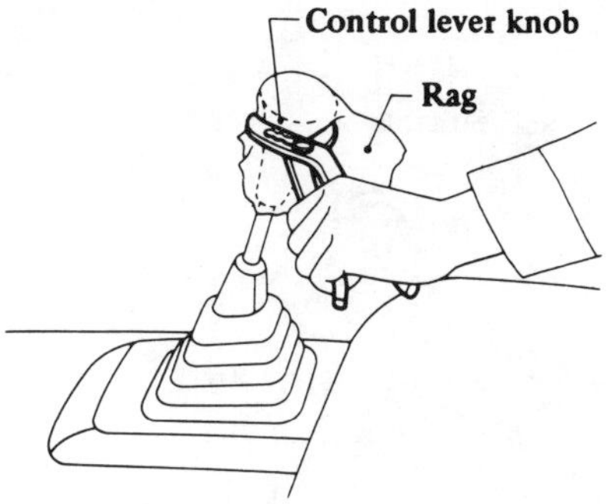

Fig. 6.2 Twisting off gearchange lever knob (Sec 3)

Fig. 6.3 Gearchange control and support rod securing nuts (Sec 3)

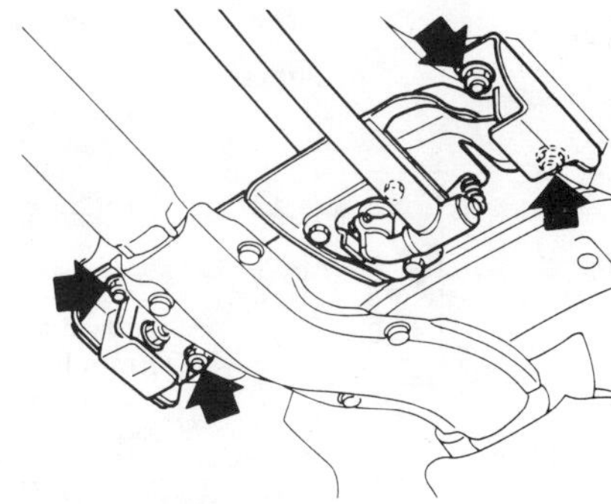

Fig. 6.4 Gearchange control rod bracket retaining nuts (Sec 3)

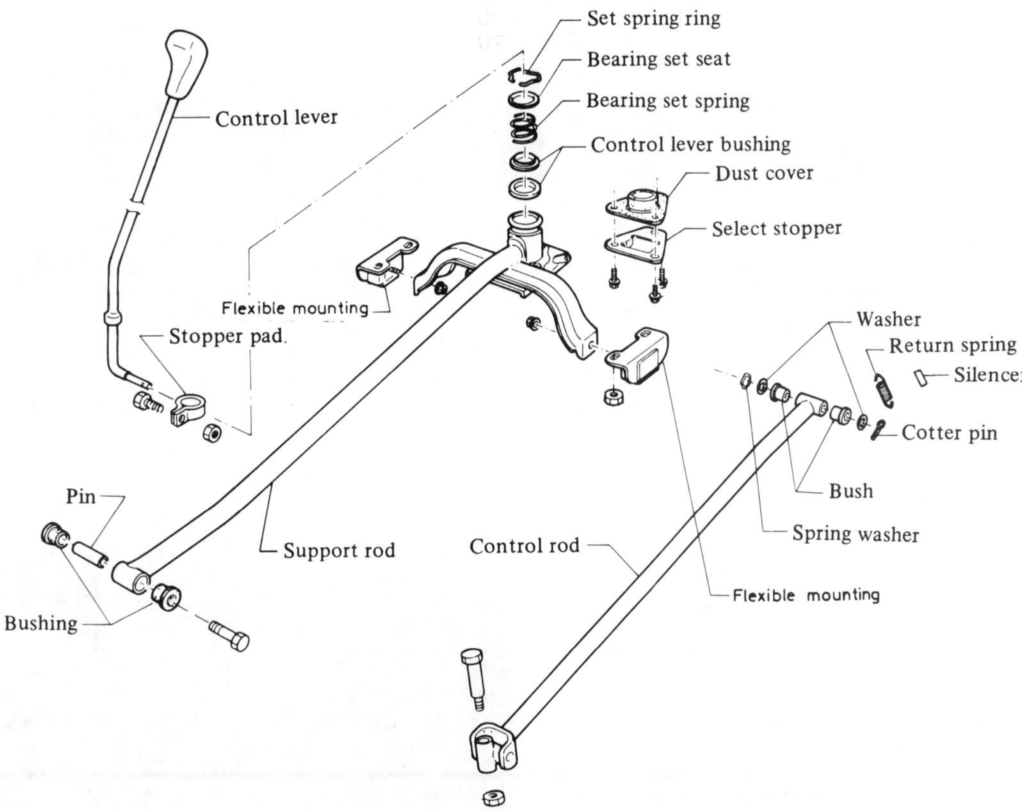

Fig. 6.5 Gearchange control components (Sec 4)

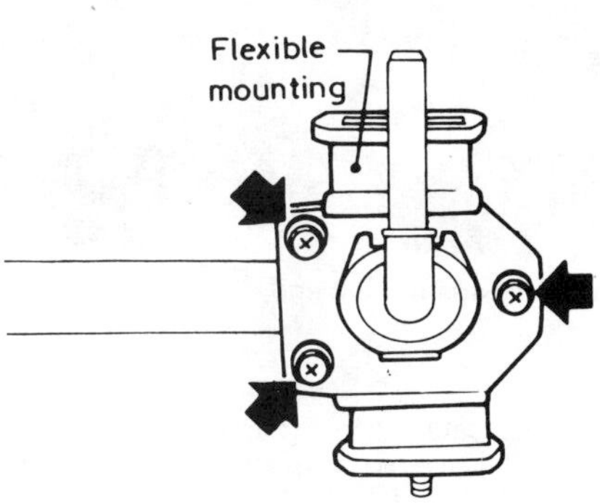

Fig. 6.6 Gearchange control flexible mounting (Sec 4)

4.3 Control lever bearing at lever end

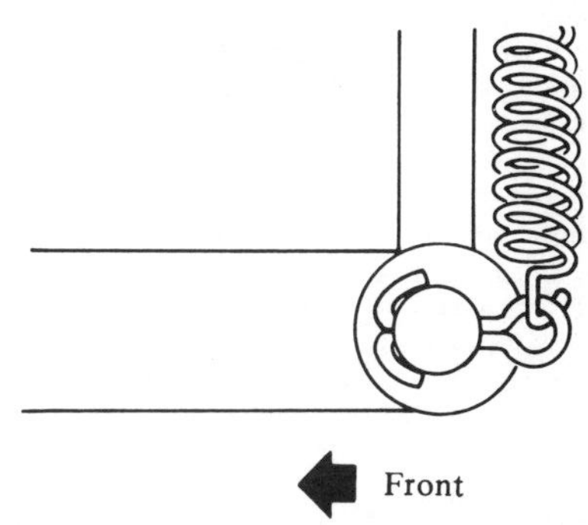

Fig. 6.7 Gearchange control rod cotter pin (Sec 4)

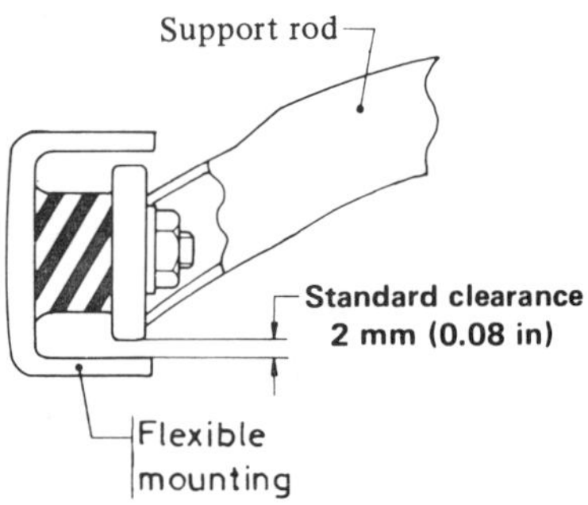

Fig. 6.8 Gearchange control flexible mounting clearance (Sec 4)

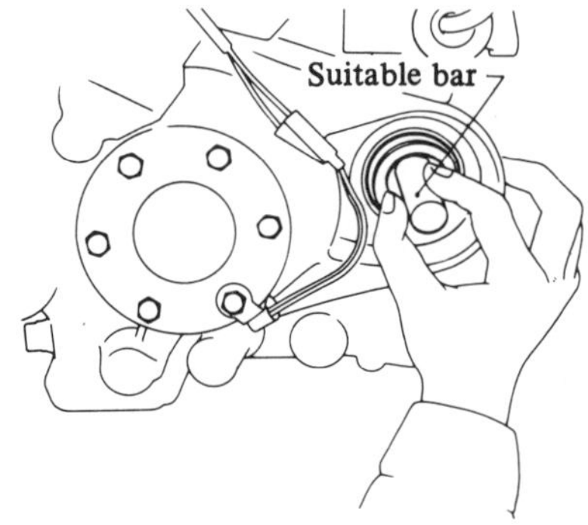

Fig. 6.10 Fitting temporary bar into side gear (Sec 5)

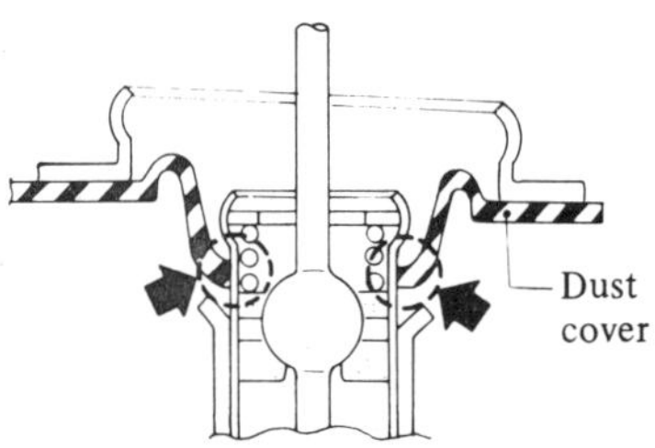

Fig. 6.9 Gearchange lever dust cover (Sec 4)

Fig. 6.11 Front wing inner panel fixing bolts (Sec 5)

7 Adjust the clearance between the support rod and the flexible mounting backplate (see Fig. 6.8).
8 Fit the dust cover securely.

5 Transmission – removal and refitting

1 Open the bonnet and remove the battery and its retaining clamp where necessary.
2 Release the cooling system expansion tank from its retainer and move it out of the way where necessary.
3 Drain the transmission oil.
4 Disconnect the driveshafts from the transmission using the following procedure.
5 Support the vehicle under its side-members and then disconnect the front suspension lower balljoints. Do this by unscrewing the three nuts which hold the balljoint to the suspension arm. It is recommended that new nuts are used at reassembly.
6 Remove the front roadwheels.
7 Unbolt the disc calipers and tie them up out of the way; it is not necessary to disconnect the hydraulic lines.
8 Unscrew, but do not remove, the nuts at the suspension strut top mountings. This is to provide flexibility of the struts when the driveshafts are withdrawn from the transmission.
9 To disconnect a driveshaft, insert a large screwdriver or other suitable lever behind the inboard joint flange and then prise to overcome the resistance of the joint circlip.
10 Take care not to damage the transmission oil seal and do not pull on the outer end of the driveshaft or the joints may come apart.
11 Once the driveshafts have been disconnected, insert a short bar into each of the side gears as it is possible for the side gears to rotate and to fall into the differential case. Alternatively, a U-shaped piece of wire may be used.
12 Unbolt and remove the front wing inner protective panel.
13 Disconnect the gearchange control and support rods from the transmission (photo).
14 Disconnect the clutch operating cable from the release lever.

5.13 Gearchange control rods at transmission end

15 Disconnect the speedometer cable from the transmission.
16 Disconnect the leads from the reverse lamp switch and the cover plate earth terminal where fitted.
17 Support the engine and the transmission on jacks with blocks of wood as insulators. Disconnect the engine front mounting.
18 Disconnect the transmission mountings.
19 Unscrew and remove the bolts which hold the transmission bellhousing to the engine. Note the location of the coolant tube, lower

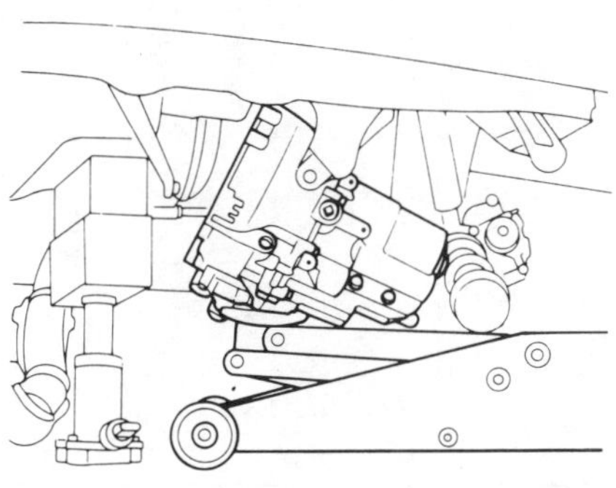

Fig. 6.12 Removing the transmission (Sec 5)

protective shield and the sump-to-transmission reinforcement tube held by some of the bolts.

20 Withdraw the transmission from under the vehicle. It will be found easier to do this if a trolley jack is used.

21 Refitting is a reversal of removal, but observe the following points. Apply a smear of molybdenum disulphide grease to the input shaft splines. If the clutch has been dismantled, make sure that the driven plate has been centralised (Chapter 5).

22 Tighten all bolts and nuts to the specified torques, and reconnect the driveshafts, as described in Chapter 8.

23 Check the clutch adjustment and refill the transmission with oil.

6 Transmission – dismantling

1 With the transmission removed from the vehicle, clean away external dirt using a water-soluble solvent, or paraffin, and a stiff brush. Remove the mounting brackets, noting their location.

2 Drain the transmission oil if not drained previously.

3 With the unit standing on the flange of the clutch bellhousing, unscrew the casing-to-bellhousing bolts and withdraw the casing from the bellhousing. On five-speed units, tilt the casing slightly as it is withdrawn to prevent the selector fork jamming inside the casing. If the casing is stuck, tap it off carefully to break the joint using a plastic hammer.

Transmission casing

4 Unscrew and remove the reverse lamp switch.

5 Remove the oil trough.

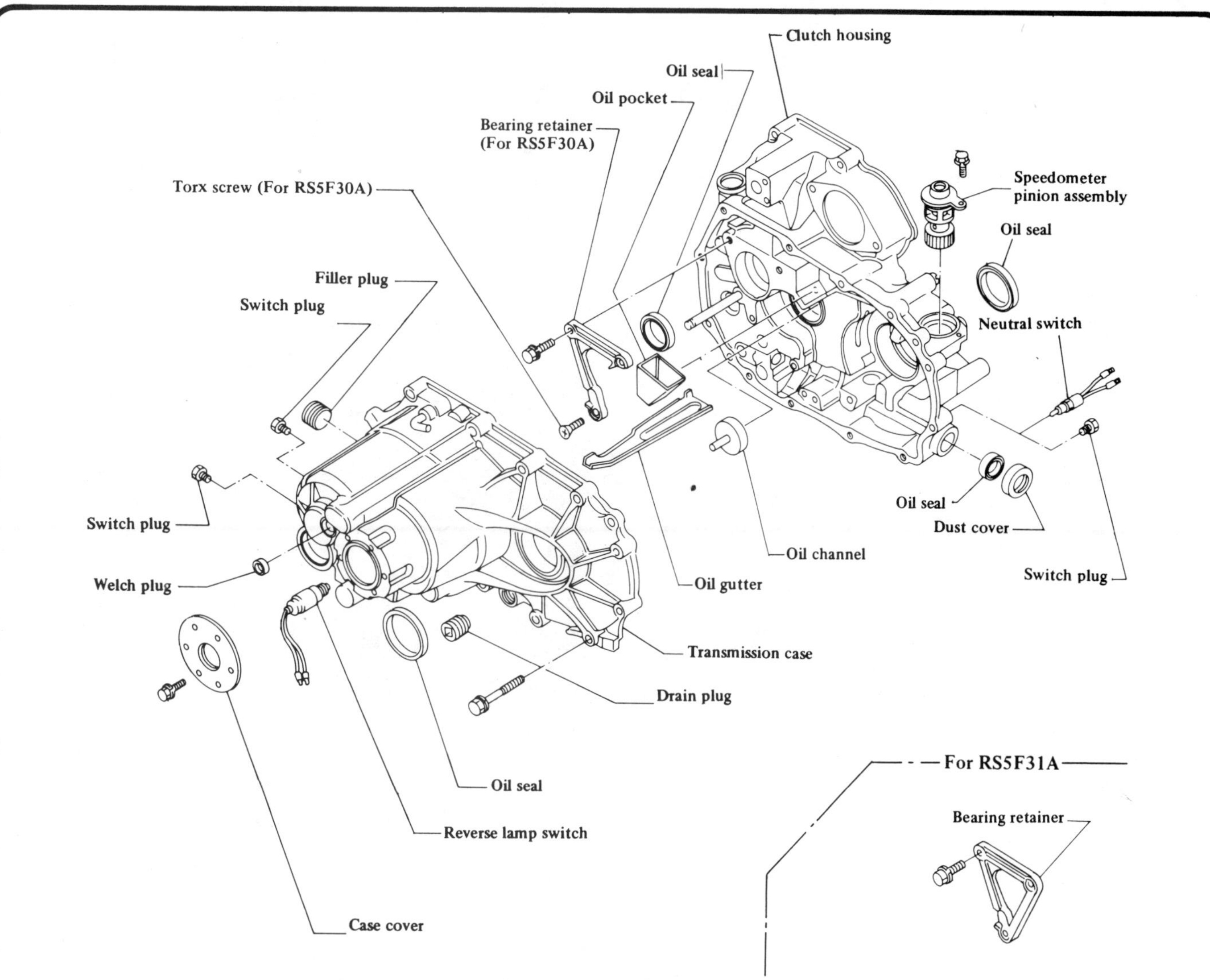

Fig. 6.13 Transmission casing/housing components (Sec 6)

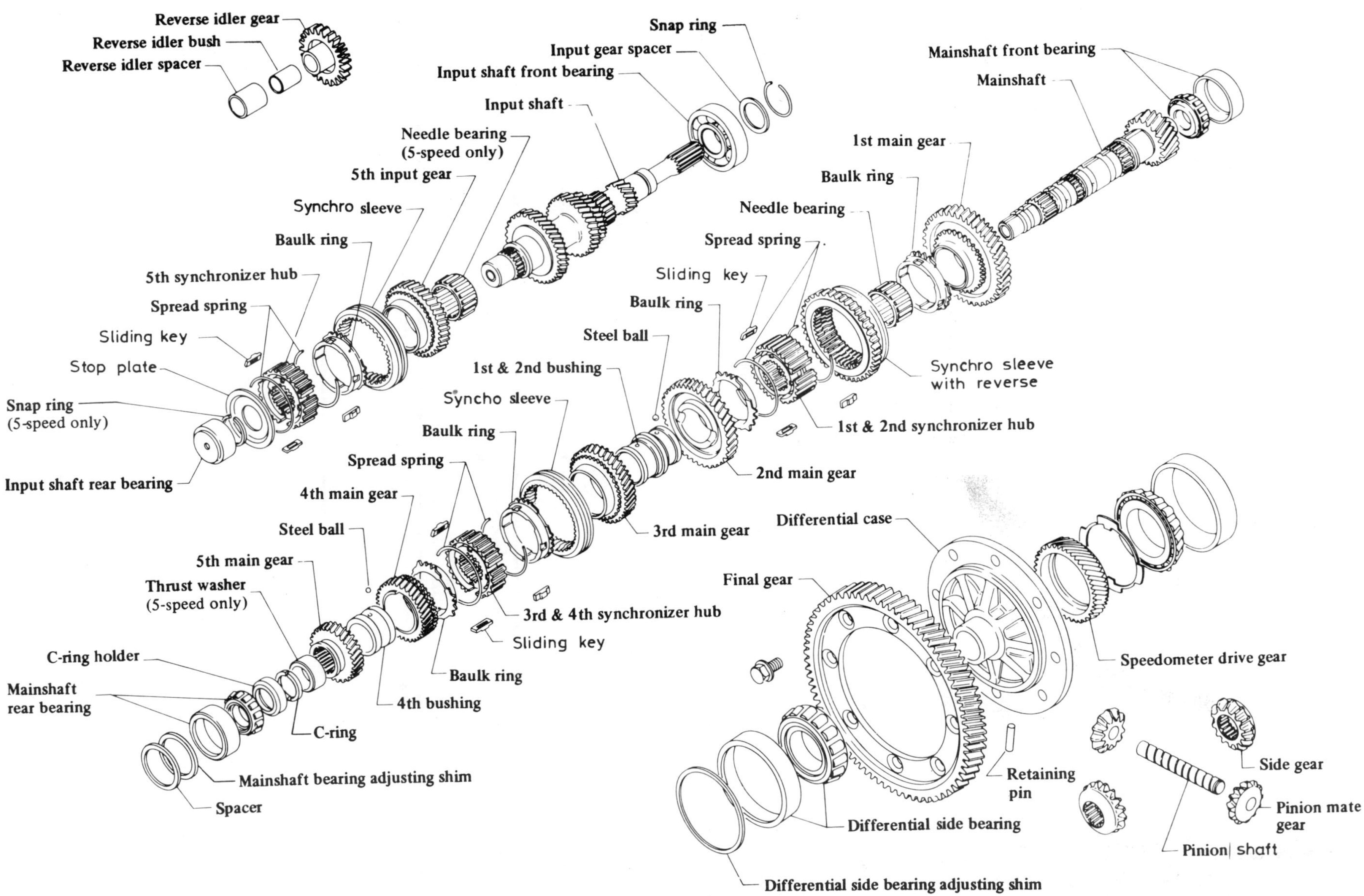

Fig. 6.14 Geartrain components (Sec 6)

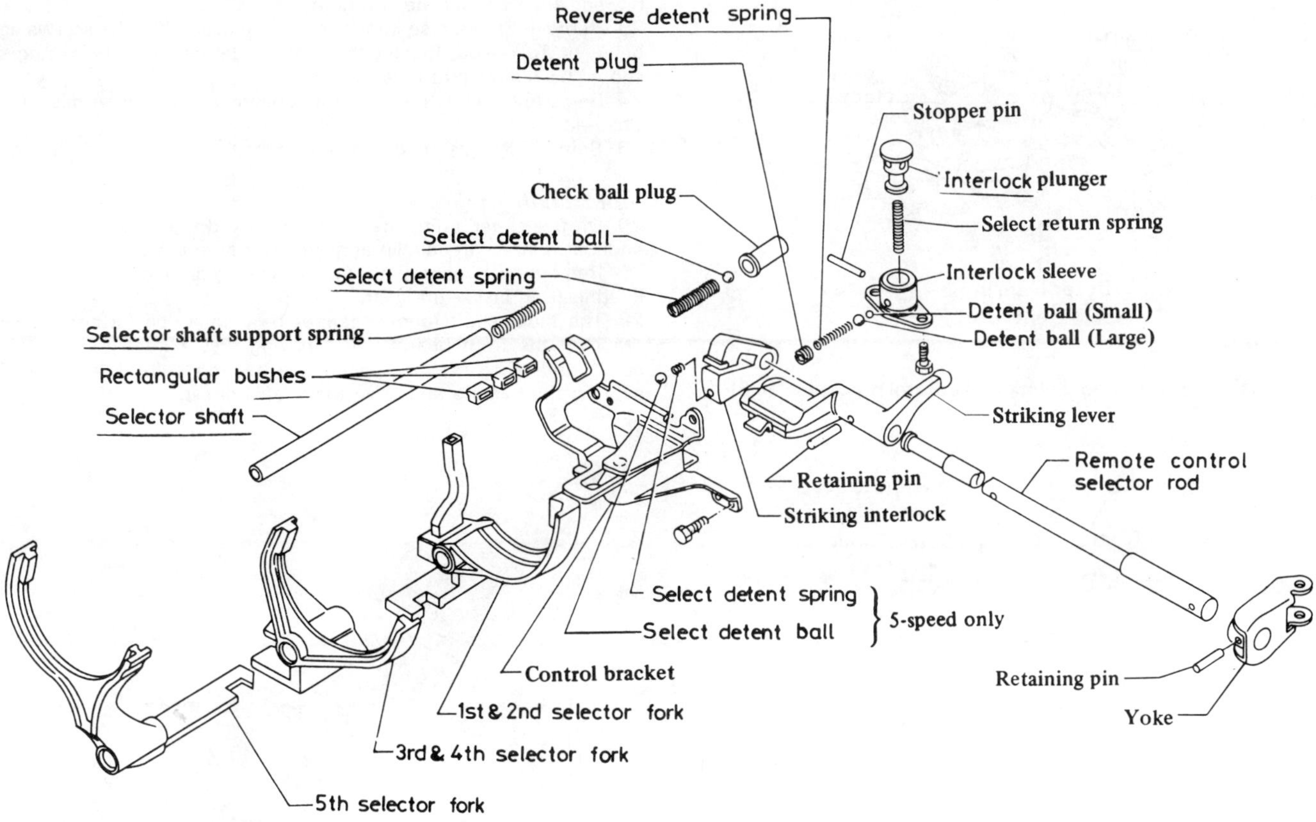

Fig. 6.15 Selector components (Sec 6)

6 If the input shaft rear bearing is to be renewed, remove the very small welch plug from the transmission casing. Do this by drilling a hole in the plug and then screw in a self-tapping screw. The screw will probably force out the plug or its head can be used to lever it out.
7 Unbolt the circular cover from the casing and take out the spacer and the mainshaft bearing adjusting shim. If the mainshaft bearing is to be renewed, drive out the old outer track and fit the new one.
8 If the differential side bearings are to be renewed, drive the bearing outer track from the transmission casing. A new oil seal will be required.

Clutch housing

9 The clutch housing will have been left standing with the geartrains projecting from it when the transmission casing was drawn off.
10 Withdraw the selector shaft out of the 3rd, 4th and 5th selector forks. Extract the coil spring from the end of the shaft.
11 Remove the 5th, 3rd and 4th selector forks. Retain the plastic slides from the forks. Do not lose the rectangular bushes located in the fork arm cut-outs.
12 Remove the control bracket with the 1st/2nd selector fork. Take care not to lose the small 5th speed detent ball and spring. Extract the larger coil spring and ball from the remote control selector rod hole.

Types RS530A and RN4F30A

13 Remove the screws from the triangular shaped bearing retainer. One of these screws is of Torx type and will require a special bit to unscrew it. Hold the reverse idler gear up while the screw is undone. Remove the spacer from the reverse idler shaft.
14 Turn the clutch housing on its side and remove the mainshaft assembly. Remove the input shaft assembly by tapping the end of the shaft with a plastic-faced or copper hammer. Reverse idler gear will come off its shaft as the input shaft is released, but mark the idler gear as to which way up it is fitted.
15 Take out the final drive/differential.

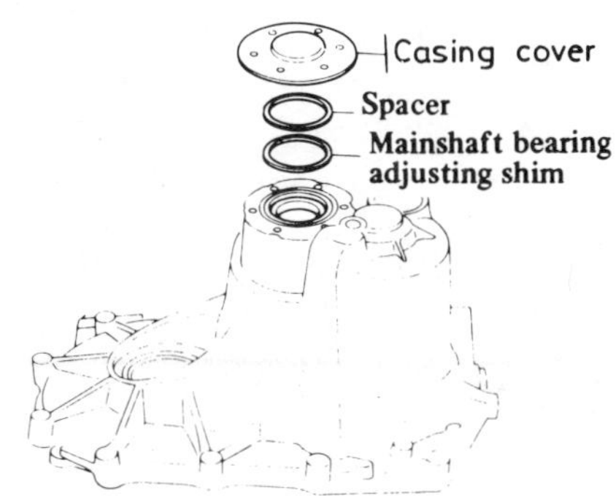

Fig. 6.16 Transmission casing cover components (Sec 6)

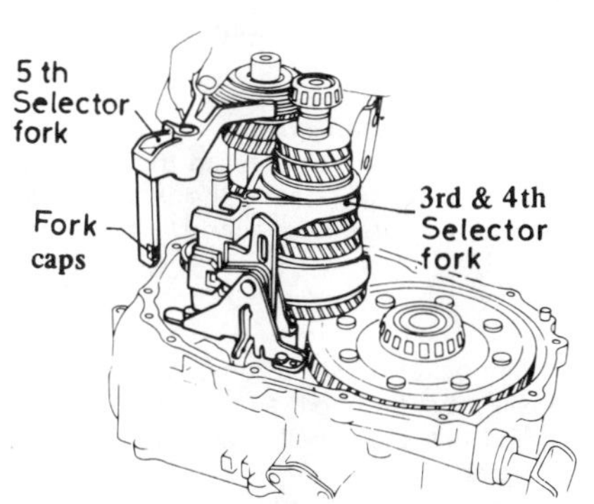

Fig. 6.17 3rd/4th and 5th gear selector forks (Sec 6)

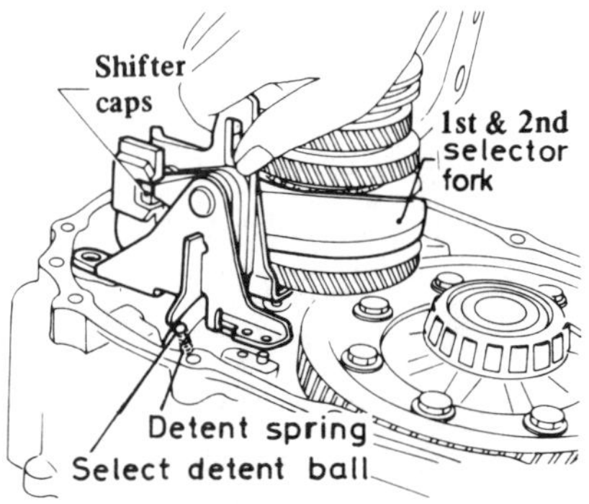

Fig. 6.18 1st/2nd selector fork and control bracket (Sec 6)

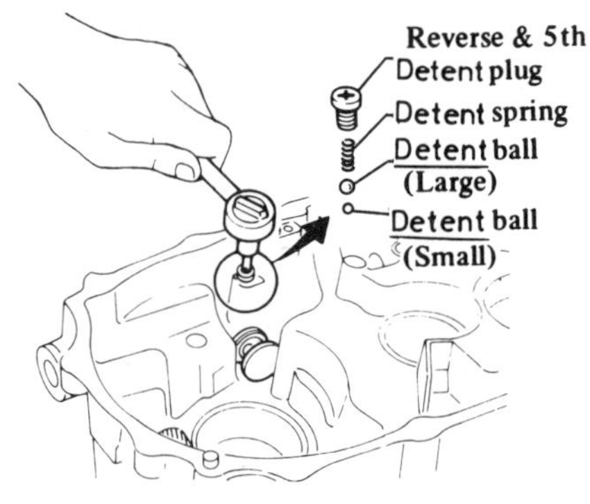

Fig. 6.19 5th/reverse detent components (Sec 6)

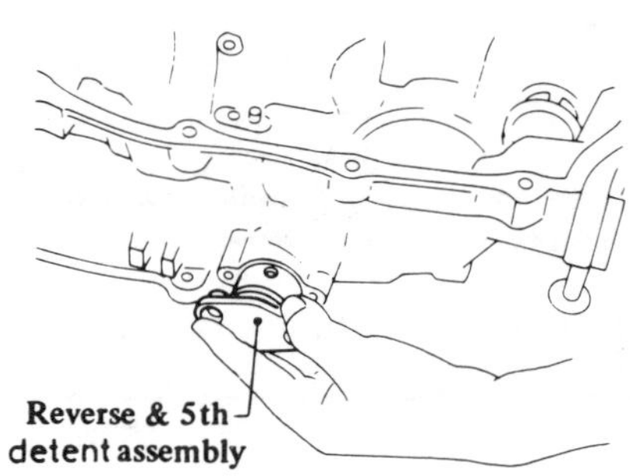

Fig. 6.20 5th/reverse interlock assembly (Sec 6)

Type RS5F31A

16 Lift out the mainshaft assembly at the same time releasing it from the input shaft gears.
17 Take out the final drive/differential.
18 Remove the bearing retainer bolts.
19 With the clutch housing on its side tap the end of the input shaft with a soft-faced hammer and remove the input shaft together with the bearing retainer and reverse idler gear. Do not remove the reverse idler shaft.

All types

20 If the plastic oil pocket must be removed then the bearing outer track which retains it must first be drawn out using a suitable extractor with thin claws. Extract the small retaining bolt and remove the speedometer drivegear.
21 Drive the roll pin from the selector rod dog then withdraw the rod, dog and interlock. When removing the rod, take care not to damage the oil seal lips.
22 Unscrew 5th/reverse detent plug, which will require a Torx type bit, and then extráct the spring and balls.
23 Remove 5th/reverse interlock plunger assembly, the screws again being of Torx type. Extract the smaller detent ball. The O-ring seal should be renewed at reassembly.
24 Remove the clutch release shaft, bearing and lever, as described in Chapter 5.
25 Remove the plastic oil channel from the transmission casing.

Input shaft

26 On five-speed units, measure and record the input shaft 5th gear endfloat. Extract the circlip and 5th gear stop plate.
27 Remove 5th gear with the synchroniser and the split needle bearing from inside the gear.
28 The input shaft (photo) cannot be dismantled further except to draw off the front bearing after having first extracted the retaining circlip and taken off the spacer. If a bearing puller is not available, support the bearing and drive the shaft from it.

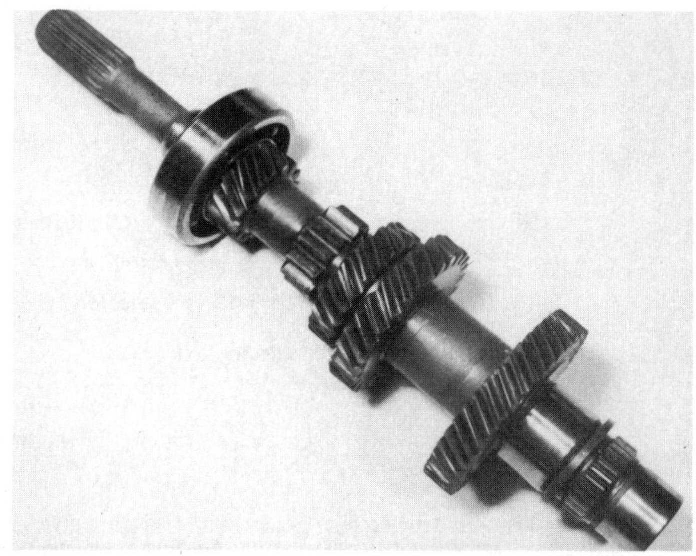

6.28 Input shaft with 5th gear and synchroniser removed

Mainshaft

29 Before dismantling the mainshaft, compare the endfloat of the gears with the specified tolerances. Inspect the components with endfloat which is outside the limits very carefully before reassembly.
30 Remove the bearing inner races from the front and rear ends of the shaft. Use either a two-legged puller or press the shaft out of the bearings.
31 On five-speed units, remove the C-ring retainer, the C-rings and the thrust washer. Remove 5th gear, a puller will be required for this.
32 Remove 4th gear, the gear bush and steel locking ball.
33 Remove the baulk ring.
34 Remove 3rd/4th synchro unit.
35 Remove 3rd gear.
36 Remove 2nd and 3rd gear bush.
37 Remove the steel locking ball.
38 Remove 2nd gear.
39 Remove the baulk ring.
40 Remove 1st/2nd synchro unit with reverse gear (straight-cut teeth on synchro sleeve) together with 1st gear as an assembly. The synchro-hub is tight on the shaft and the best way to remove the assembly is to support under 1st gear and drive the shaft downwards, using a copper-faced hammer.
41 Remove 1st gear split needle bearing.

Differential/final drive

42 Unbolt the crownwheel from the differential case.
43 Using a punch, drive out the pinion shaft lockpin and withdraw the shaft.

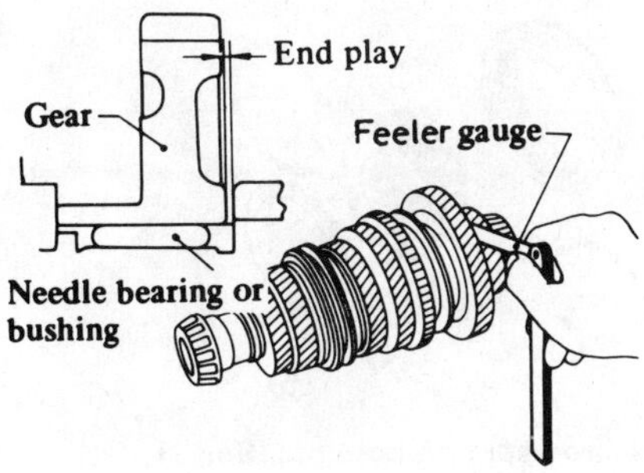

Fig. 6.21 Checking input shaft 5th gear endfloat (Sec 6)

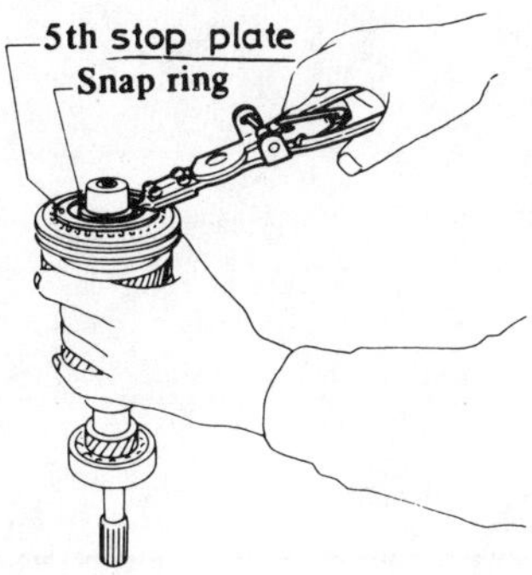

Fig. 6.22 Input shaft 5th gear circlip and stop plate (Sec 6)

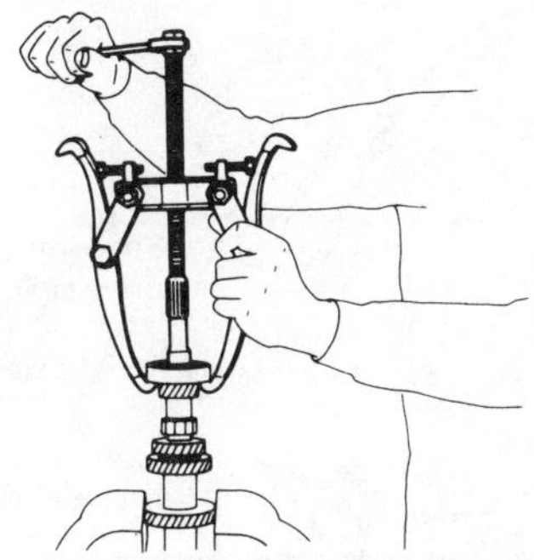

Fig. 6.23 Removing input shaft front bearing (Sec 6)

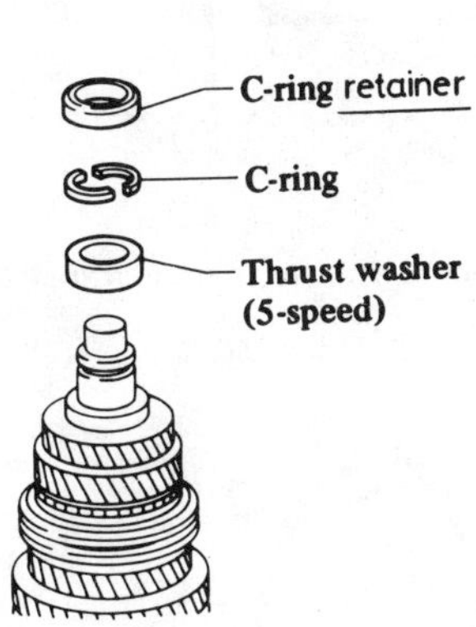

Fig. 6.24 5th gear thrust components on mainshaft (Sec 6)

Fig. 6.25 Mainshaft 4th gear components (Sec 6)

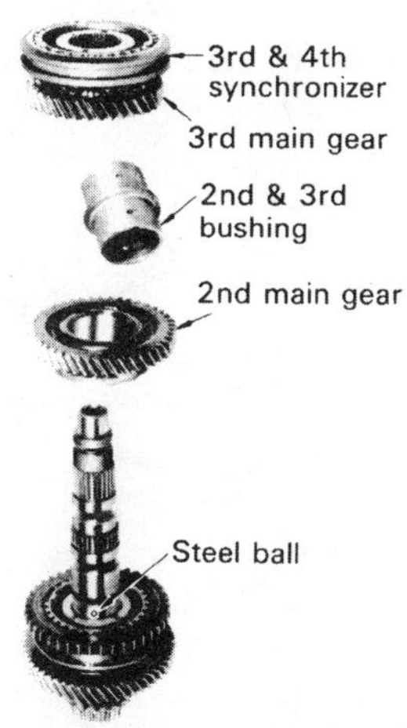

Fig. 6.26 2nd and 3rd gear components on mainshaft (Sec 6)

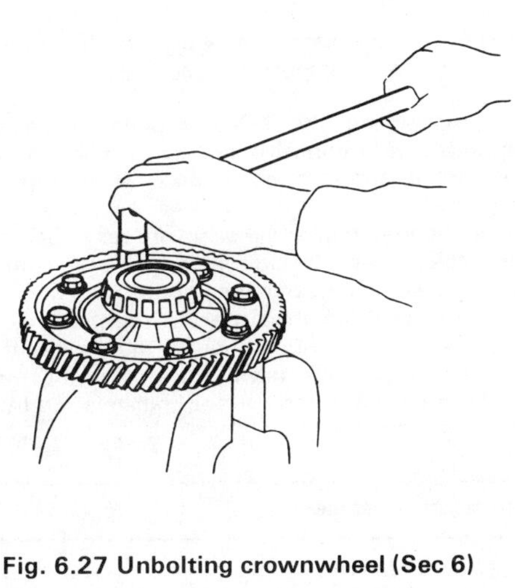

Fig. 6.27 Unbolting crownwheel (Sec 6)

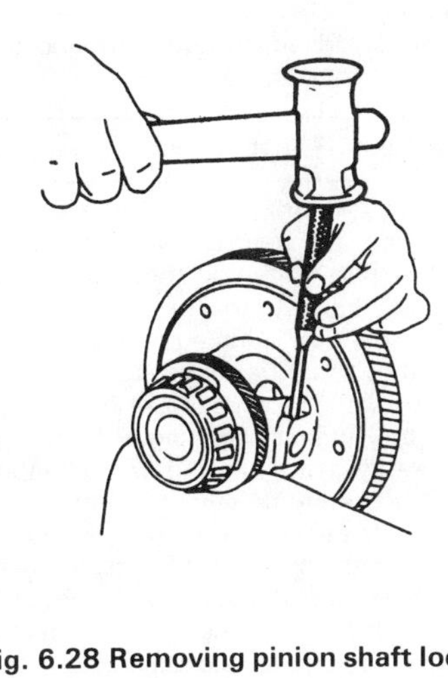

Fig. 6.28 Removing pinion shaft lockpin (Sec 6)

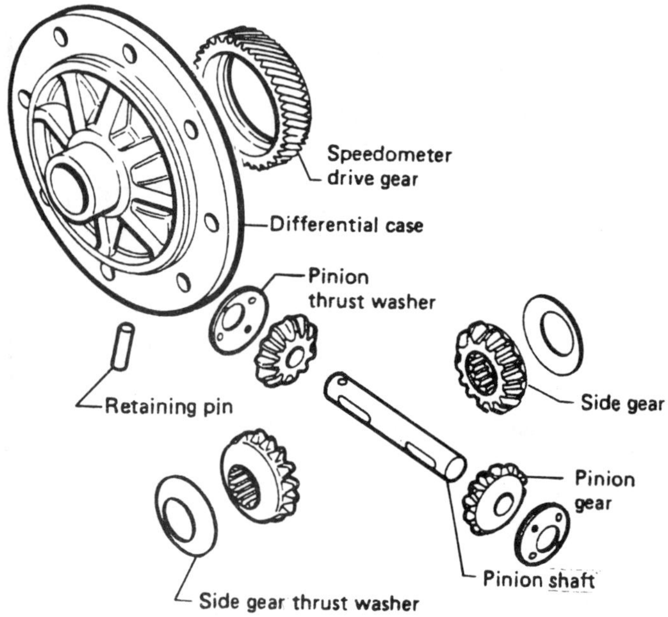

Fig. 6.29 Later type differential fitted with thrust washers (Sec 6)

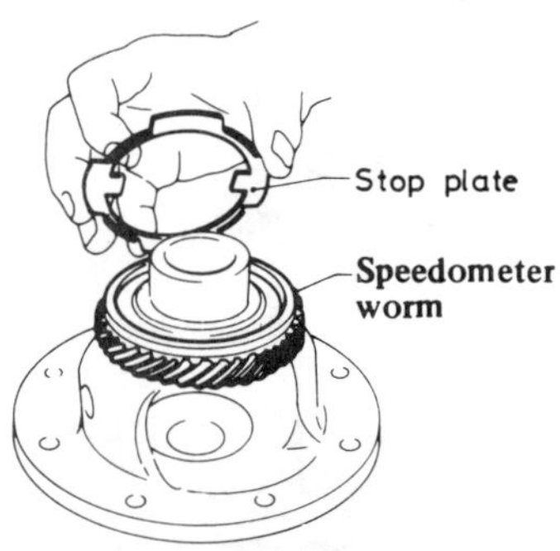

Fig. 6.30 Speedometer worm gear and stop plate (Sec 6)

44 Remove the pinion and side gears together with the thrust washers if fitted.
45 Draw off the differential side bearing races, noting exactly how the taper of the rollers is set.
46 Remove the speedometer drivegear stop plate and the gear.

7 Transmission components – inspection

1 With the transmission completely dismantled, clean all components and inspect them for wear or damage.
2 Check the gears for chipped teeth and their bushes for wear.
3 Check the shafts for scoring or grooving
4 Check the bearings for wear by spinning them with the fingers. If they shake or rattle then they must be renewed.
5 Wear in the synchronisers will usually be known about before dismantling, as a result of noisy gear changing or by the fact that the synchro action could be easily beaten during gear changing.
6 Even if the synchro is operating quietly, it is worthwhile checking the units in the following way at time of major overhaul.
7 Refer to Fig. 6.31. Extract the spreader springs, remove the sliding keys, and then push the hub from the sleeve, but not before having marked the components with quick-drying paint to ensure that their relative positions are maintained at reassembly.
8 Check the synchro components for wear or deformation. Place the baulk ring on its cone and twist it to ensure good contact between the

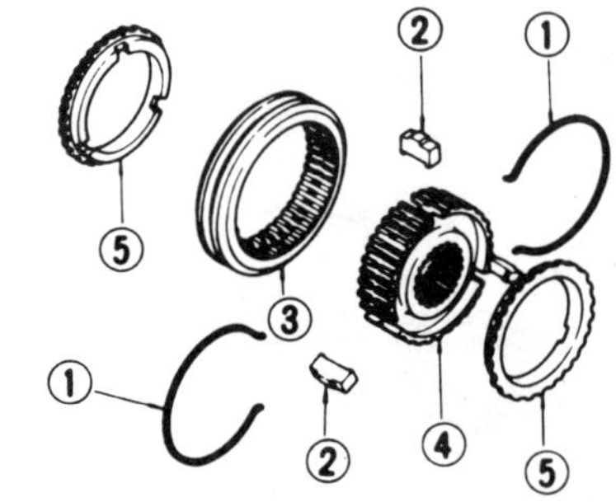

Fig. 6.31 Synchroniser components (Sec 7)

1 Spreader spring
2 Sliding key
3 Sleeve
4 Hub
5 Baulk ring

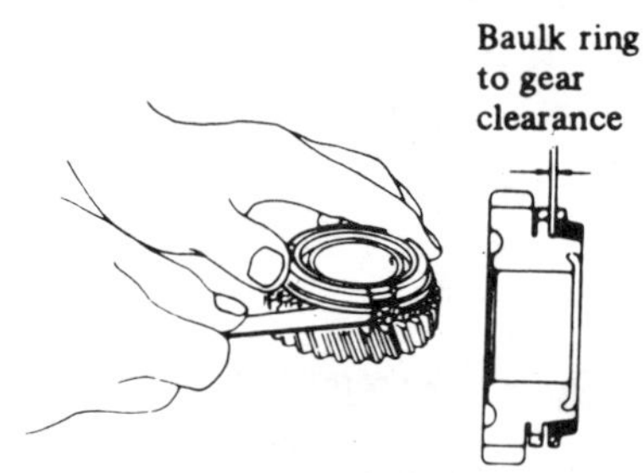

Fig. 6.32 Checking baulk ring-to-gear clearance (Sec 7)

Clearance = 0.7 mm (0.028 in)

Fig. 6.33 Synchro spring arrangement (Sec 7)

surfaces. Using a feeler blade, check that the gap between gear and baulk ring is not less than specified (see Fig. 6.32). If it is, renew the baulk ring.
9 When reassembling the synchro units, make sure that the spreader springs run in opposite directions, when viewed from each side of the synchro, and that the spring ends are not engaged in the same sliding key.
10 It is recommended that all oil seals are renewed at time of major overhaul (photos). These include those for the clutch cross-shaft, differential side bearings, gearchange control rod and the input shaft.
11 Should anything more than the the slightest seepage of oil be observed during the normal operation of the vehicle the oil seals for the differential side bearings and the gearchange control rod can be renewed without having to remove the transmission from the vehicle. Refer to Chapter 7 Section 10.

8 Transmission – reassembly

Differential/final drive

1 Fit the speedometer worm drivegear and its stop plate to the differential case (photo).

7.10A Interior of transmission casing

7.10B Differential bearing outer track and oil seal in transmission casing

7.10C Differential bearing outer track and oil seal in clutch bellhousing

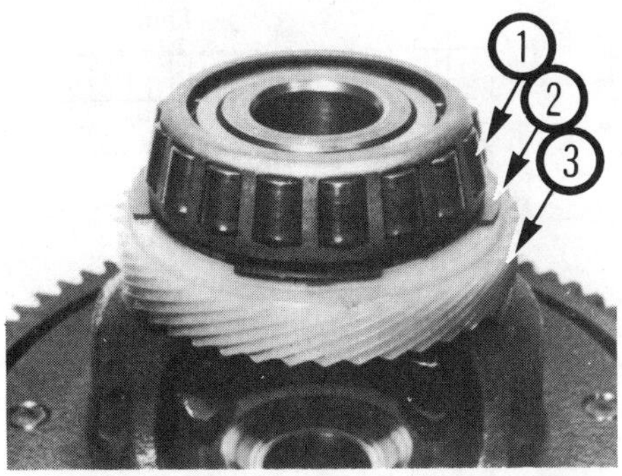

8.1 Differential side bearing (1), stop plate (2), and speedometer worm drive gear (3)

8.2 Method of fitting differential side bearing

8.3 Fitting differential gears and shaft

2 Press or drive on the differential side bearing inner races (photo).
3 Into the differential case fit the pinion and side gears with thrust washers (later models) and the pinion shaft (photo). Where thrust washers are fitted use a dial gauge to check that the side gear-to-pinion gear clearance is as given in the Specifications. If necessary fit different thrust washers.
4 Drive in a new pinion shaft roll pin, making sure that it is flush with the differential case (photo).

8.4 Differential pinion shaft roll pin

5 Clean the threads of the crownwheel bolts and apply thread-locking fluid, then fit the crownwheel, screw them in and tighten to the specified torque.

Mainshaft

6 Oil all components liberally as they are reassembled.
7 Fit 1st gear needle bearing to the mainshaft (photos).
8 Fit 1st gear (photo).
9 Fit 1st gear baulk ring (photo).
10 Fit 1st/2nd synchro unit with reverse (photo). Tap the synchro-hub down the mainshaft using a piece of tubing, but hold the synchro together with the hand in case the vibration makes it fall apart.
11 Locate the steel lock ball in its hole in the shaft. *On no account place the ball in the hole in the shaft groove* (photo).
12 Fit 2nd gear baulk ring (photo).
13 Fit 2nd gear (photo).
14 Fit 2nd/3rd gear bush, turning it slowly to engage its cut-out with the lock ball (photo).
15 Fit 3rd gear (photo).
16 Fit the baulk ring (photo).
17 Fit 3rd/4th synchro unit (photo) so that the engraved dashes in the sleeve are visible.
18 Using thick grease, stick the second (4th gear bush) lock ball in its shaft hole – *not the hole in the shaft groove.*
19 Fit the baulk ring (photo).
20 Fit 4th gear bush, turning it slowly to engage its cut-out with the lock ball (photo).
21 Fit 4th gear (photo).
22 If the transmission is of five-speed type, fit 5th gear (photo). Drive in onto the mainshaft using a piece of tubing (photo).
23 On five-speed models only, fit the thrust washer (photo).
24 Fit the C-ring (photo). These are supplied in various thicknesses to correct gear endfloat.
25 Fit the C-ring retainer (photo).

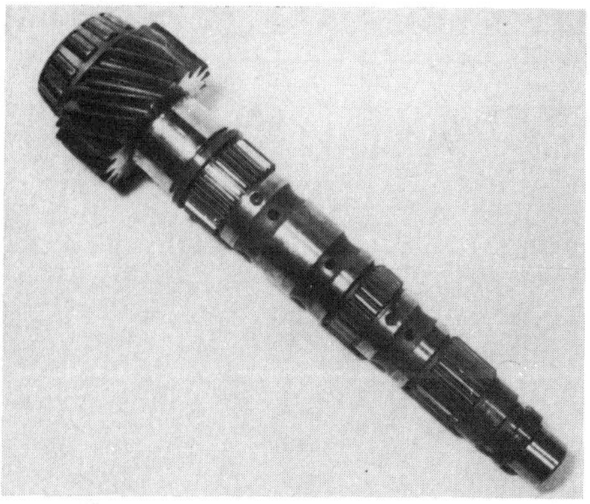

8.7A Mainshaft dismantled

8.7B 1st gear split needle bearing

8.8 Fitting 1st gear

8.9 1st gear baulk ring

8.10 1st/2nd synchro with reverse gear

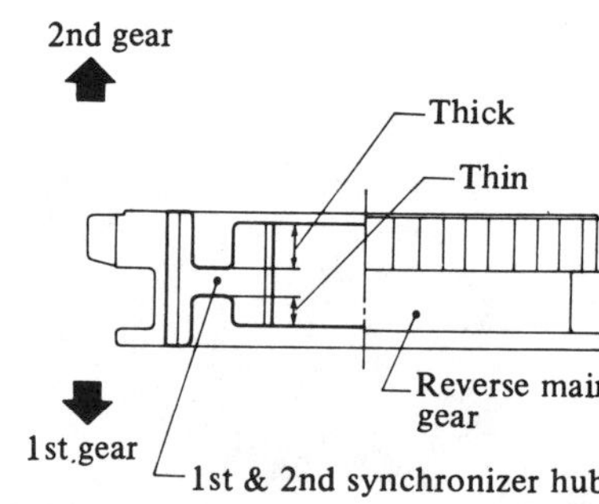

Fig. 6.34 1st/2nd synchro-hub to sleeve relationship (Sec 8)

8.11 Steel lock ball for 2nd/3rd gear bush in correct hole

8.12 2nd gear baulk ring

8.13 2nd gear

8.14 2nd/3rd gear bush

8.15 3rd gear

8.16 Baulk ring

8.17 3rd/4th synchro unit

8.19 4th gear baulk ring showing lock ball location

8.20 4th gear bush

8.21 4th gear

8.22A 5th gear

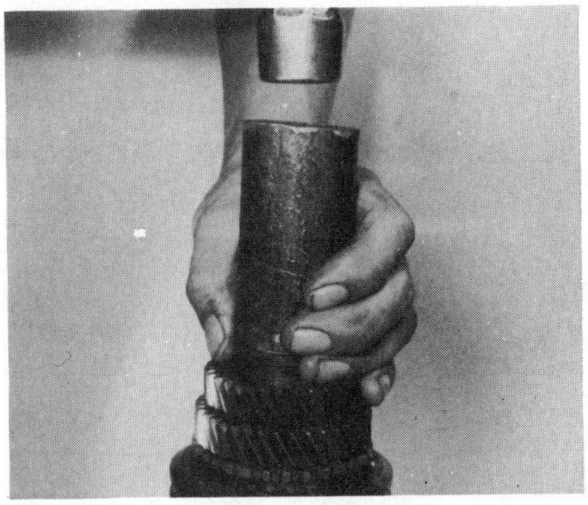
8.22B Driving 5th gear onto mainshaft

8.23 5th gear thrust washer

8.24 C-rings

8.25 C-ring retainer

8.26 Mainshaft bearing

8.27 Checking mainshaft gear endfloat

8.28 Input shaft split type needle roller bearing

26 Press on new bearing inner races to both ends of the mainshaft (photo).
27 Using feeler blades check that the gear endfloat is within the specified tolerances (photo). If any endfloat measurements are outside the specified limits the shaft should be dismantled and the components re-examined.

Input shaft

28 Fit the split type needle roller bearing (photo).
29 Fit 5th gear (photo).
30 Fit the baulk ring (photo).
31 Fit the synchro unit (photo) so that the engraved dashes on the sleeve are visible.
32 To the synchro unit fit the stop plate and the circlip (photos). The circlips are available in various thicknesses to eliminate endfloat.
33 Press on a new shaft front bearing, fit the spacer and use a new circlip (photo).

Clutch housing

34 Fit a new oil channel so that its relieved area is towards the oil pocket when installed (photo).
35 Press or drive the differential and mainshaft bearing outer tracks into their seats.
36 Remember that the mainshaft bearing outer track retains the oil pocket, so align the pocket correctly before fitting the bearing track.
37 Reassemble the clutch release mechanism (Chapter 5).
38 Refit the interlock plunger assembly (photo). Fit reverse/5th detent balls (small one first) the spring and plug (photos).
39 The force of the reverse detent should now be checked using a spring balance. On four-speed models the torque required to move

8.29 Input shaft 5th gear

8.30 5th gear baulk ring

8.31 5th gear synchro on input shaft

8.32A Input shaft synchro unit stop plate

8.32B Input shaft stop plate circlip

8.33 Input shaft front bearing circlip

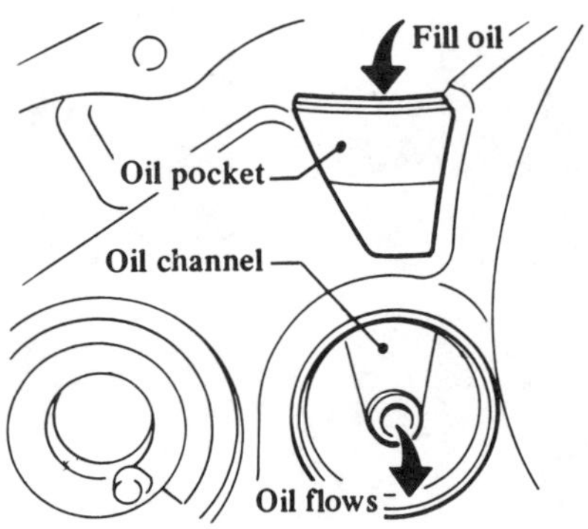

Fig. 6.35 Oil pocket/oil channel arrangement (Sec 8)

8.34 Oil channel

8.38A Using a magnet to fit the small reverse/5th detent ball

8.38B Using a magnet to fit the larger reverse/5th detent ball

8.38C Using a magnet to fit reverse/5th detent spring

8.38D Reverse/5th detent plug

8.38E Interlock plunger assembly

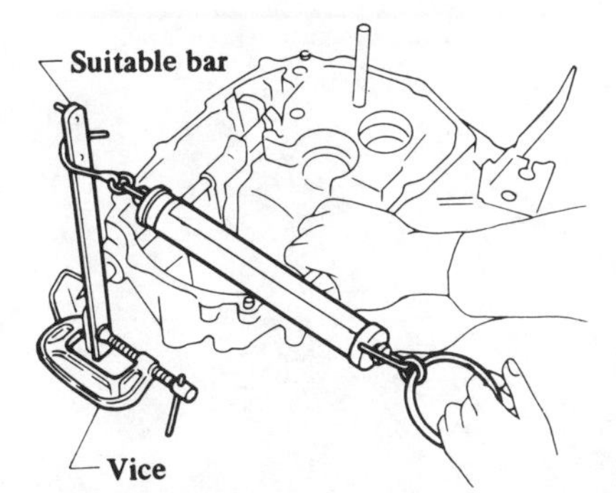

Fig. 6.36 Checking reverse detent resistance/torque (Sec 8)

against the detent should be between 15.7 and 22.6 Nm (139 and 200 lbf in). On pre 1985 five-speed models the torque should be between 22.1 and 27.0 Nm (195 and 239 lbf in), and on 1985 on five-speed models between 10.3 and 12.7 Nm (91 and 113 lbf in). Use a bar clamped to the reverse detent and determine the correct pull for the spring balance according to the distance from the fulcrum point.

40 The detent force may be increased by changing the detent plug for one of greater length.

8.41A Assembling remote control rod, striking lever and interlock

8.42 Speedometer drivegear

8.41B Striking lever roll pin (outer)

8.43 Method of retaining differential side gears

8.41C Striking lever roll pin (inner)

41 Using new double roll pins, refit the remote control rod, striking lever and interlock (photos). The notch in the control rod should be downwards.
42 Refit the speedometer drive gear and screw in the lockbolt (photo).
43 Retain the differential side gears with rods or a wire clip bent into a U-shape (photo). Some later models have side gear thrust washers.

Types RS5F30A and RN4F30A

44 Lower the differential/final drive into position (photo).
45 Fit the input shaft and reverse idler gear simultaneously (photo). The idler gear (marked before removal) should be refitted in its original position. Use a plastic-faced or copper hammer to tap the input shaft fully home in the clutch housing.
46 Fit the spacer to the reverse idler shaft (photo).
47 Fit the triangular shaped bearing retainer (photo). Apply thread locking fluid to the screw threads and tighten them as tightly as possible (photo). Note the Torx type screw next to the idler.
48 Fit the mainshaft (photo) carefully meshing the gearteeth with those of the input shaft as the operation proceeds, push both synchro sleeves downwards and hold the reverse idler gear upwards.

8.44 Lowering differential/final drive into clutch housing

8.45 Fitting input shaft with reverse idler gear

8.46 Reverse idler gear spacer

8.47A Fitting triangular-shaped bearing retainer

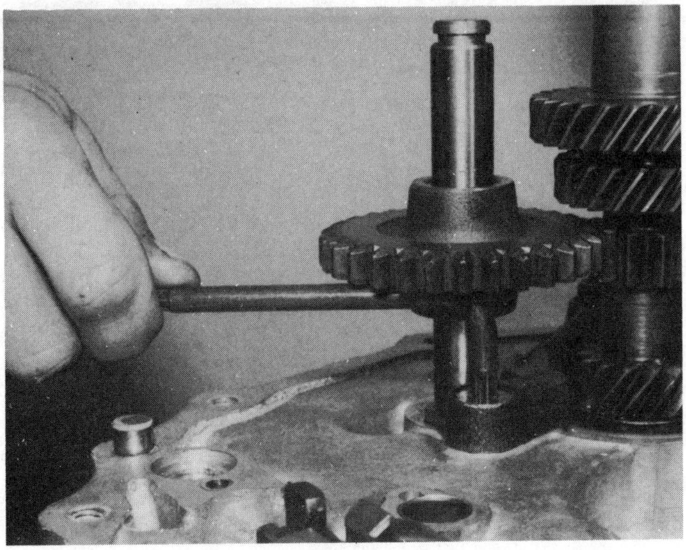

8.47B Tightening bearing retainer screw

8.48 Fitting mainshaft/geartrain

Type RS5F31A

49 Fit the input shaft and reverse idler gear simultaneously, at the same time locating the gear over the shaft and the bearing retainer onto the housing. Tap the input shaft with a soft-faced hammer.
50 Insert the bearing retainer bolts and tighten them to the specified torque.
51 Fit the mainshaft while meshing the gear teeth with those of the input shaft.
52 Lower the differential/final drive into position.

All types

53 Fit the bush, the ball and the large coil spring to the hole in the remote control rod housing (photos).
54 Fit the control bracket which incorporates reverse selector fork. Make sure that 1st/2nd selector fork is located under the bracket and 5th speed detent spring and ball are placed into the hole in the remote control interlock as the assembly operations progress (photos). Tighten the control bracket screws.
55 Locate the 3rd/4th and 5th selector forks (photo). Make sure that the plastic slides are in position in the fork arm cut-outs (photo) also the rectangular shaped metal bushes are in the selector dog cut-outs (photo).
56 Pass the selector shaft through the forks (photo), making sure that the coil spring is located in the recess in the lower end of the shaft (photo) using grease.

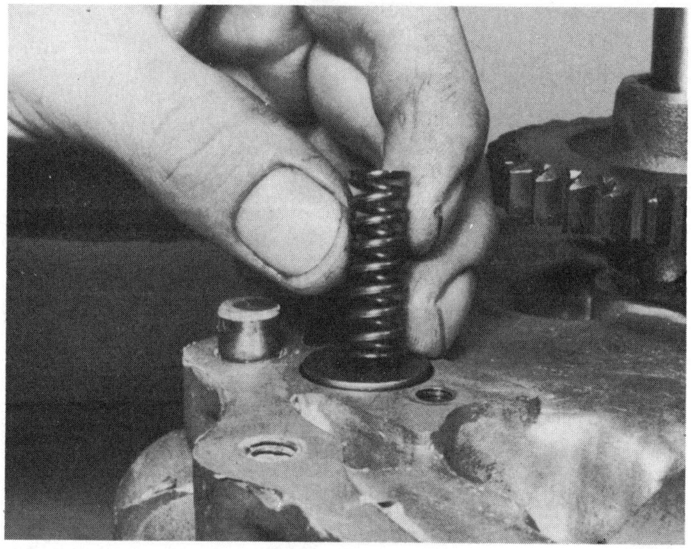

8.53C Remote control housing spring

8.53A Remote control housing bush

8.54A Detent spring (remote control interlock)

8.53B Remote control housing ball

8.54B Detent ball (remote control interlock)

8.54C Remote control bracket (screws arrowed)

8.54D Remote control interlock detent ball correctly located (arrowed)

8.55A Selector forks

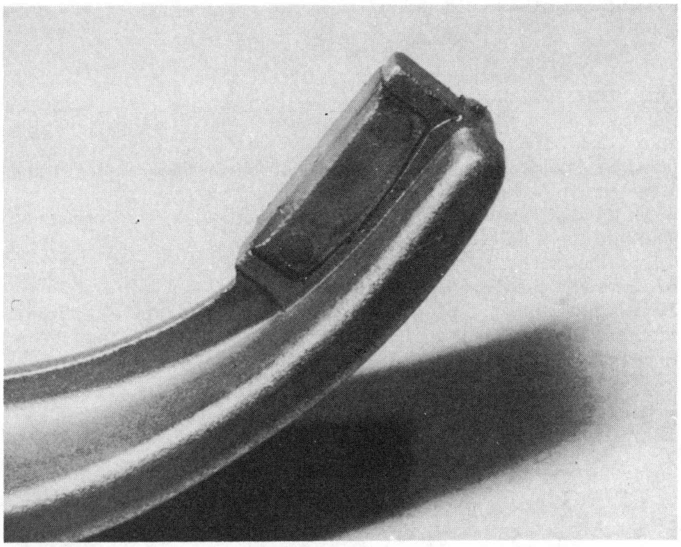

8.55B Selector fork plastic slide

8.55C Selector dog metal bushes

8.56A Fitting selector shaft into forks

8.56B Selector shaft coil spring

Transmission casing

57 If the differential side bearings were renewed, fit the new bearing outer track now with a new oil seal.
58 If the mainshaft bearing was renewed, fit the new track into the casing now.
59 If the input shaft rear bearing was renewed, tap a new small welch plug into the hole in the casing.
60 Fit the plastic oil trough.
61 Screw in the reverse lamp switch
62 With the clutch housing standing on the bench with the geartrains vertical, applying jointing compound to the mating faces of the transmission casing and clutch housing.
63 Lower the casing into position over the geartrains (photo). On five-speed units, tilt the casing as necessary to clear the selector fork.
64 Fit the connecting bolts and tighten to the specified torque (photo).
65 If the mainshaft bearing has not been changed, fit the original adjusting shim and spacer (photos). If a new bearing has been fitted, refer to Section 10 for details of mainshaft bearing adjustment.
66 Apply jointing compound to the edges of the circular cover and bolt it into position on the transmission casing (photo).

9 Final drive – adjustment

1 If any of the following components of the transmission have been renewed during overhaul then you will have to take the assembly to your dealer for the final drive to be adjusted to ensure correct crownwheel-to-pinion meshing and the specified bearing preload.

Differential casing
Differential side bearing
Clutch housing
Transmission casing

2 Owing to the need for special tools, this work is not within the scope of the home mechanic.

10 Mainshaft bearing preload – adjustment

1 If any of the following components have been renewed during overhaul, then the mainshaft bearing preload must be checked and adjusted.

Mainshaft
Mainshaft bearings
Clutch housing
Transmission casing

2 Remove the circular cover from the transmission.
3 To carry out the adjustment, measure between the machined face of the transmission casing and the surface of the spacer. A shim

8.63 Fitting transmission casing

8.64 Tightening casing bolts

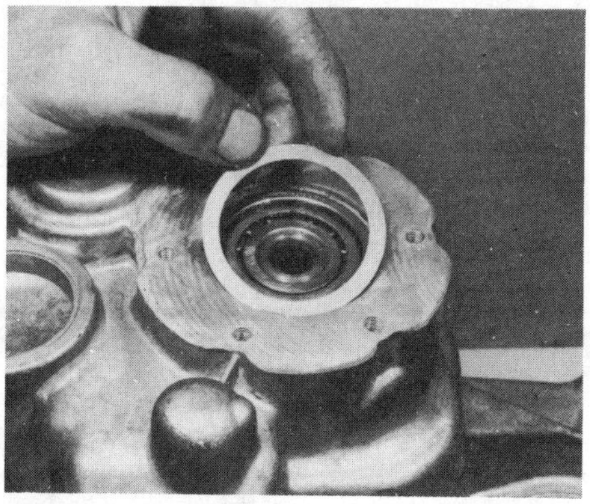

8.65A Mainshaft bearing adjusting shim

8.65B Mainshaft bearing spacer

8.66 Transmission casing circular cover

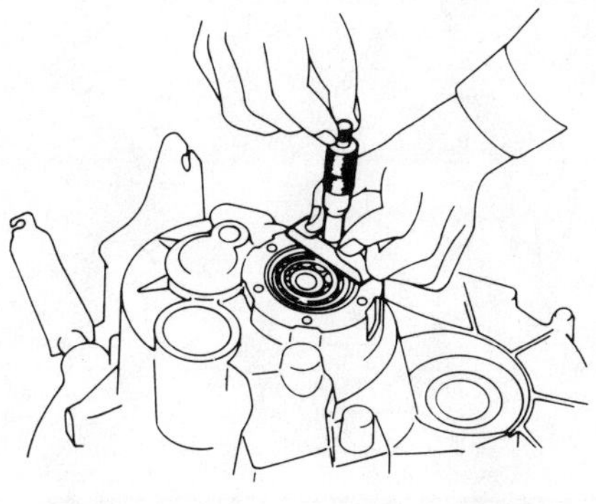

Fig. 6.37 Determining the mainshaft bearing preload shim (Sec 10)

should now be selected which is 0.2 mm (0.008 in) thicker than the dimension just taken.

4 Fit the spacer, the selected shim and the cover and check that the input shaft turns smoothly with 4th gear selected. A special tool is available from dealers (KV38105900) which engages in the side gears and gives a torque reading for rotation of the final drive when 4th gear is selected.

5 The correct turning torque should be between 7.4 and 10.8 Nm (5.4 to 7.9 lbf ft).

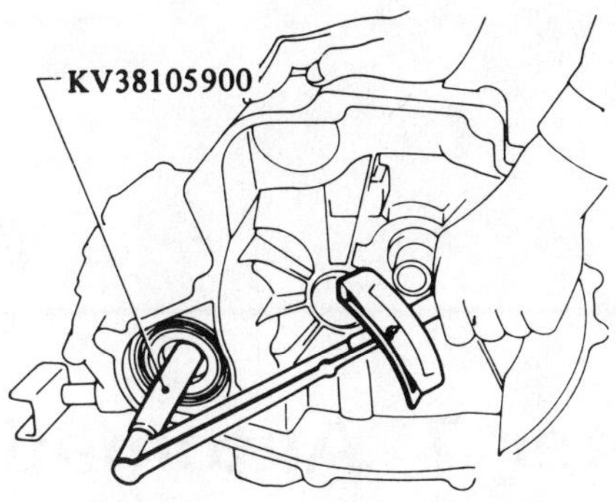

Fig. 6.38 Using special tool to check final drive turning torque (Sec 10)

11 Fault diagnosis – manual transmission

Symptom	Reason(s)
Weak or ineffective synchromesh	Synchro baulk rings worn, split or damaged Synchromesh units worn or damaged
Jumps out of gear	Gearchange mechanism worn Synchromesh units badly worn Selector fork badly worn
Excessive noise	Incorrect grade of oil in gearbox or oil level too low Gear teeth excessively worn or damaged Shaft thrust washers worn allowing excessive end play Worn bearings
Difficulty in engaging gears	Clutch pedal adjustment incorrect
Noise when cornering	Wheel bearing or driveshaft fault Differential fault

Note: *It is sometimes difficult to decide whether it is worthwhile removing and dismantling the gearbox for a fault which may be nothing more than a minor irritant. Gearboxes which howl, or where the synchromesh can be beaten by a quick gearchange, may continue to perform for a long time in this state. A worn gearbox usually needs a complete rebuild to eliminate noise because the various gears, if re-aligned on new bearings, will continue to howl when different wearing surfaces are presented to each other. The decision to overhaul, therefore, must be considered with regard to time and money available, relative to the degree of noise or malfunction that the driver has to suffer.*

Chapter 7 Automatic transmission

Contents

Specifications

Type .... RL3FO1A fully automatic, with three element torque converter and two planetary geartrains. Three forward speeds and reverse. Final drive is integral

Ratios

1st	2.826:1
2nd	1.543:1
3rd	1.000:1
Reverse	2.364:1

Final drive ratios

UK models	3.364:1
North American models:	
Except Turbo	3.476:1
Turbo	3.167:1

Fluid capacity

10.6 Imp pt; 6.3 US qt; 6.0 litre

Torque wrench settings

Torque wrench settings	lbf ft	Nm
Drive plate to torque converter	44	60
Torque converter housing to engine	15	20
Sump pan bolts	5	7
Control valve body to transmission casing	7	9
Governor valve body to shaft	5	7
Oil cooler union nut at transmission casing	35	48
Reinforcement strut to engine crankcase	26	35
Reinforcement strut to transmission casing	15	20
Front suspension strut top mounting nuts	17	23
Suspension lower balljoints to track control arm	45	61
Caliper mounting bolts	45	61
Roadwheel nuts	70	95

1 General description

The automatic transmission incorporates a torque converter with planetary geartrains and the final drive/differential unit.

Six speed selector control lever positions are used:

P *Park – to lock up the transmission mechanically*
R *Reverse gear*
N *Neutral*
D *Forward speed – changing automatically up and down between 1st, 2nd and 3rd gear ratios*
2 *Second gear hold, will change between 1st and 2nd gears only*
1 *First (low) gear hold*

Kickdown in D is used for rapid acceleration during overtaking, it changes down to 2nd or 1st gear, depending upon roadspeed, when accelerator is fully depressed.

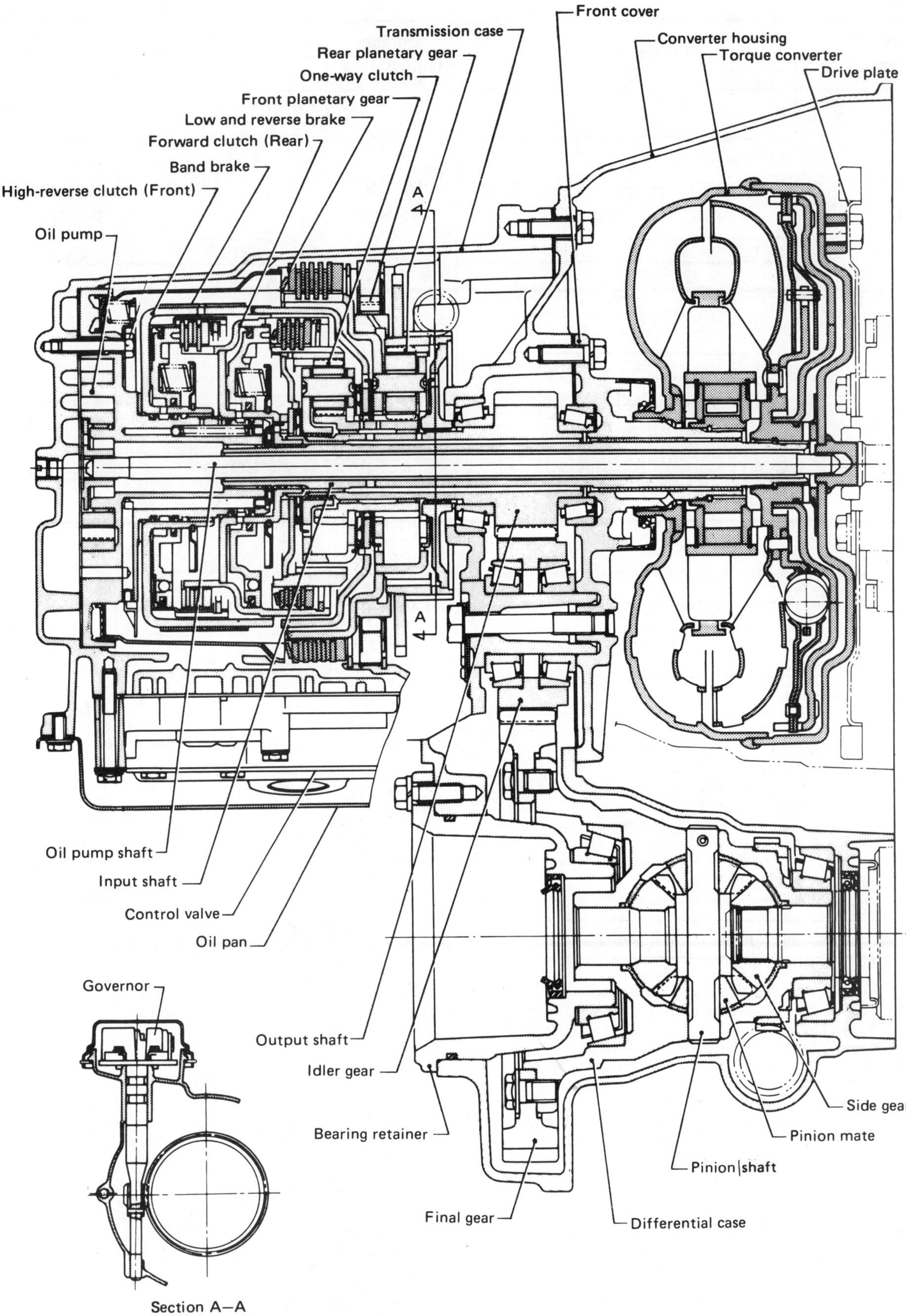

Fig. 7.1 Sectional view of the automatic transmission (Sec 1)

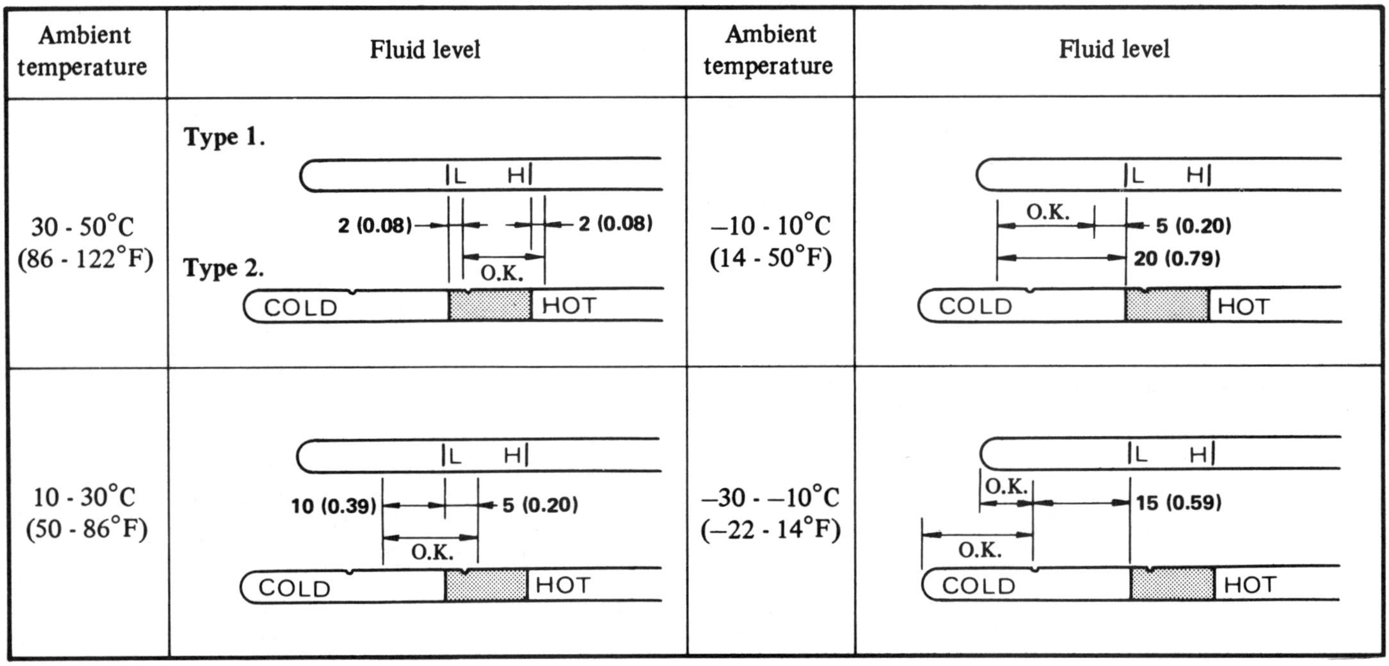

Fig. 7.2 Transmission fluid dipstick level ranges at various ambient temperatures (Sec 3)

2 Routine maintenance

1 Every 6000 miles (10 000 km) or 6 months whichever comes first on UK models, or every 15 000 miles (24 000 km) or 12 months whichever comes first on North American models check the transmission fluid level and top up if necessary. Also check the fluid for contamination or signs of internal wear.
2 If the car is operated under severe conditions, renew the transmission fluid every 24 000 miles (40 000 km) on UK models, every 30 000 miles (48 000 km) on North American models, or every 24 months whichever comes first.

3 Fluid level – checking, topping up and changing

1 Check the fluid level at the intervals given in Section 2.
2 The precise level of the fluid will depend on the ambient temperature (see Fig. 7.2). First park the car on a level surface and apply the handbrake.
3 Run the engine for ten minutes and allow it to idle.
4 Move the selector lever slowly through each position ending in Park (P).
5 With the engine idling, withdraw the transmission dipstick, wipe it clean, re-insert it and then withdraw it again and read off the fluid level.
6 If it is not as indicated in the illustrations, top up. On no account overfill the automatic transmission or run it with too low a fluid level.
7 Renewal of the automatic transmission fluid is only specified by the manufacturers if the vehicle is operated under arduous conditions – such as trailer towing. However, it would seem to make sense to change the fluid on all vehicles after a reasonably high mileage in order to remove any impurities from the system. The additives in the fluid will almost certainly have lost some of their characteristics by this time as well.
8 Before draining the fluid, have it at normal operating temperature by running on the road for a distance of at least five miles (eight kilometres).
9 Unscrew and remove the transmission drain plug and catch the fluid in a container.

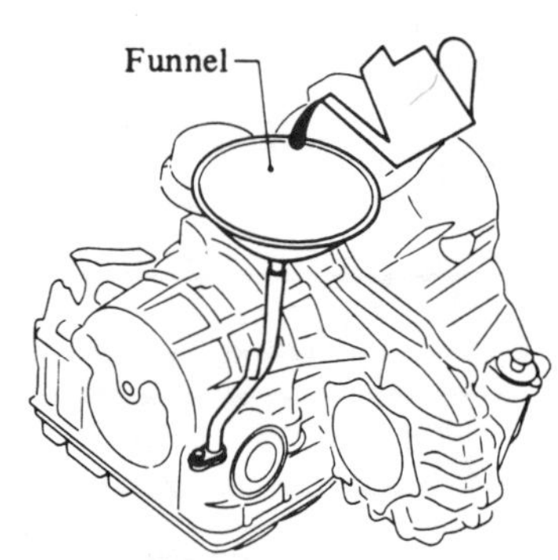

Fig. 7.3 Use a clean funnel to top up the transmission fluid level (Sec 3)

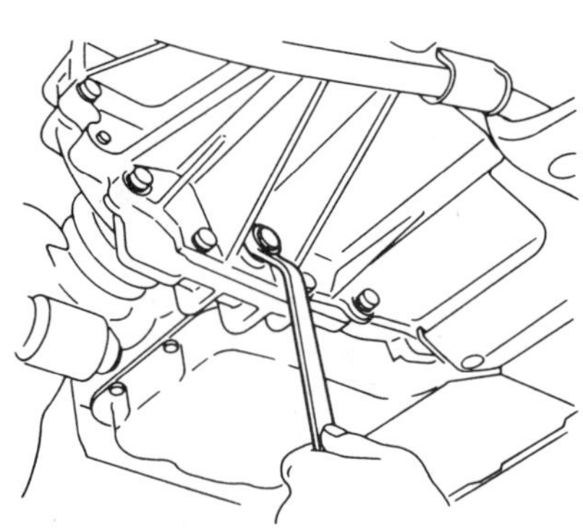

Fig. 7.4 Removing the transmission drain plug (Sec 3)

10 The condition of the fluid is an indication of the serviceability of the transmission. If it is very dark or nearly black and smells of burning, suspect worn friction components within the transmission. If there is no odour then the discolouration may be due to a small leak of coolant coming from the fluid cooler within the radiator.
11 If the fluid is an opaque pink in colour this will be due to a coolant leak or flood water contamination.
12 If the fluid is dark brown in colour and sticky, this will probably be due to overheating by under or over filling.
13 Refit the drain plug,withdraw the dipstick and pour the fresh fluid into the transmission through the dipstick guide tube.
14 Check the fluid level, as previously described.

4 Overhaul and adjustment – general

1 Owing to the need for special tools and equipment, operations to the automatic transmission should be limited to the in-vehicle work described in the following Sections.
2 Where more extensive overhaul is required, it is best to leave this to your dealer, or, where necessary, remove the transmission for professional repair or renewal, as described in Section 11.

5 Speed selector cable – adjustment

1 When the hand control lever is moved to all positions on the index, the individual detents should be positively felt. If this is not so, adjust in the following way.
2 Set the control lever to P.
3 Working at the selector lever on the transmission casing, disconnect the cable.
4 With the fingers, move the selector lever positively into its P detent.
5 Using the cable end fitting locknuts, adjust the cable until it applies no tension in either direction to the selector lever on the transmission. Tighten the locknuts.
6 Check that all selector positions are positively obtained.

6 Inhibitor switch – adjustment

1 The inhibitor switch controls the reversing lamps when R is selected, and also prevents operation of the starter when the hand control lever is in any position but P or N.
2 If the inhibitor switch does not operate correctly, adjust in the following way.
3 Loosen, but do not remove, the switch screws.
4 Set the hand control lever to N.
5 Push a 2.5 mm (0.098 in) diameter pin through the switch lever and switch body holes to align them. Hold the pin and tighten the switch screws. Remove the pin.

7 Kickdown cable – adjustment and renewal

1 Release the cable locknuts at the carburettor.
2 With the throttle cable pulley held in the full throttle position move the cable end fitting in the direction T (see Fig. 7.7). Tighten nut B to eliminate any free movement.
3 Unscrew nut B between one and one and a half turns and secure it in this position by tightening nut A (see Fig. 7.8).
4 Check that the throttle cable end fitting movement L is within the specified tolerances (see Fig. 7.9).
5 To renew the cable, first remove the fluid sump pan and control valve assembly, as described in the next Section.
6 Disconnect the kickdown cable from the carburettor.
7 Disconnect the other end of the cable from the lever and then release the cable conduit from the casing by flattening the lockplate tab and unscrewing the nut.
8 Fit the new cable by reversing the removal operations. Bend up the locktab around the nut.
9 Adjust as previously described, and finally check that the rubber bellows (if fitted) on the inner cable at the carburettor end is not twisted.

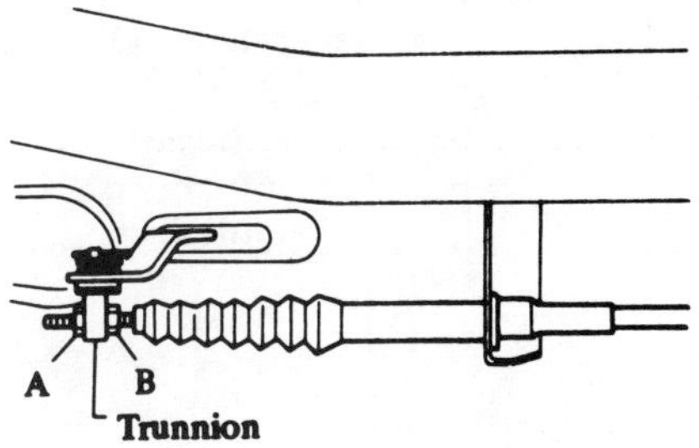

Fig. 7.5 Selector cable at transmission end (Sec 5)

A Locknut *B Locknut*

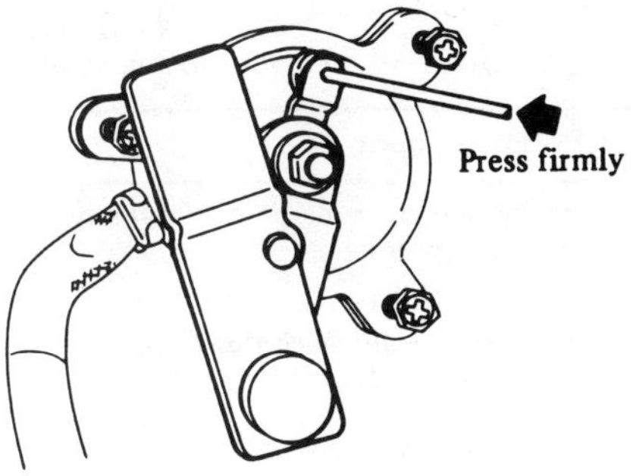

Fig. 7.6 Adjusting the inhibitor switch (Sec 6)

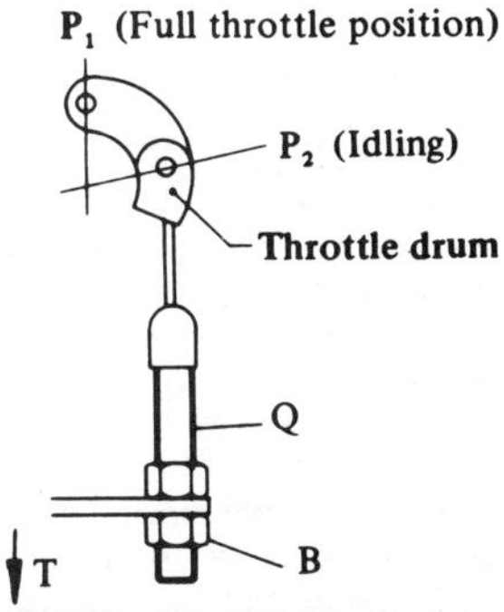

Fig. 7.7 Kickdown (throttle) cable end fitting at carburettor (Sec 7)

B Locknut
P1 Full throttle position
P2 Idle position
Q Cable end fitting
T Adjustment movement direction

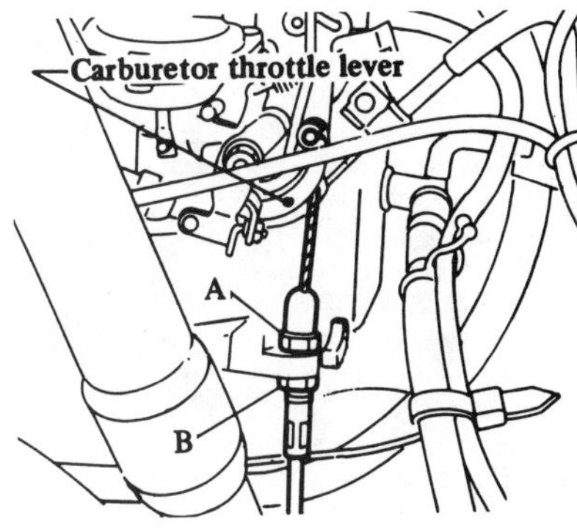

Fig. 7.8 Kickdown (throttle) cable locknuts A and B at carburettor end (Sec 7)

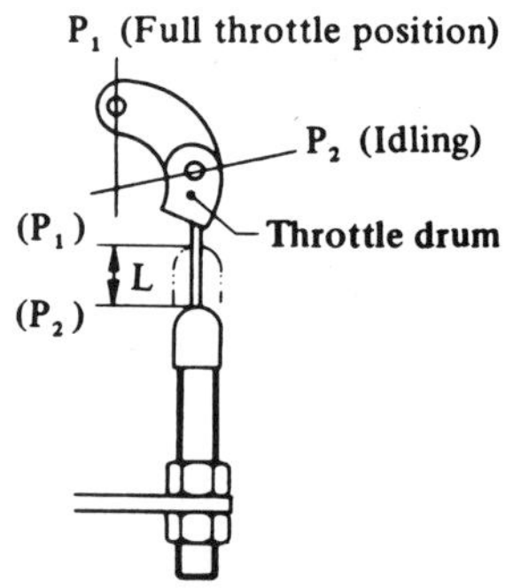

Fig. 7.9 Kickdown (throttle) cable movement diagram (Sec 7)

L = 27.4 to 31.4 mm (1.079 to 1.236 in)
P1 Full throttle position
P2 Idle position

8 Brake band – adjustment

1 This will normally only be required if a fault develops, indicated by one of the following symptoms.

No change from 1st to 2nd
Speed changes direct from 1st to 3rd
Severe jerk on 1st to 2nd upshift
Poor acceleration
Maximum speed not obtained
No 3rd to 2nd downshift
No kickdown when in 3rd gear
Slip in 3rd to 2nd downshift
No manual 3rd to 2nd downshift
Transmission overheats

2 Drain the transmission fluid.
3 Remove the sump pan shield, the sump pan and the gasket from the transmission.
4 Unscrew the bolts evenly and progressively and withdraw the control valve assembly.
5 Release the brake band anchor pin locknut and then, using a torque wrench, tighten the pin to between 4.0 and 6.0 Nm (3 and 4 lbf ft).
6 Now unscrew the anchor pin through $2\frac{1}{2}$ complete turns. Hold the anchor pin stationary and tighten the locknut to between 16.0 and 22.0 Nm (12 and 16 lbf ft).
7 Refit the control valve assembly and tighten the bolts. Make sure that the manual and detent valves are correctly engaged – the manual valve should be set at neutral, and the groove on the detent valve should face forward.
8 Refit the sump pan together with a new gasket, followed by the shield. Fill the transmission with the specified fluid then check the level as described in Section 3.

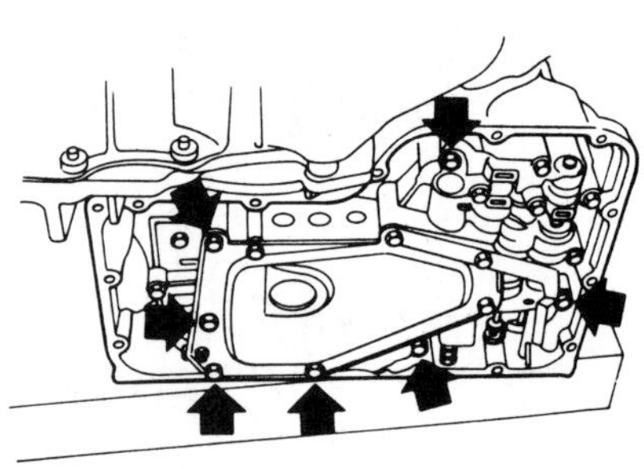

Fig. 7.10 Control valve assembly bolt locations (Sec 8)

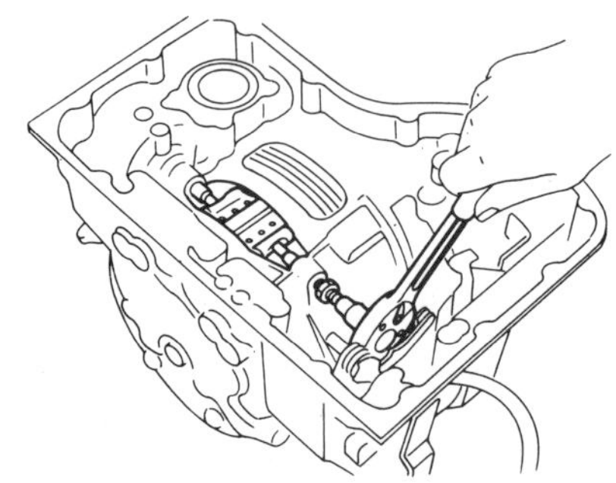

Fig. 7.11 Adjusting the brake band anchor pin (Sec 8)

9 Governor shaft – removal and refitting

1 On pre 1984 models disconnect the battery, remove the cooling system expansion tank, and remove the battery and its support bracket.
2 On 1984 on models remove the distributor cap.
3 Release the governor cap snap retainer, the cap with breather hose, and the sealing ring.
4 Unscrew the governor shaft lockbolt.
5 Withdraw the governor shaft.
6 The governor body may be unbolted from the shaft, and scratched or worn components renewed. The worm may be removed from the governor shaft after driving out the securing pin.
7 Refitting is a reversal of removal, but make sure that the cap is located on the case protrusion correctly.

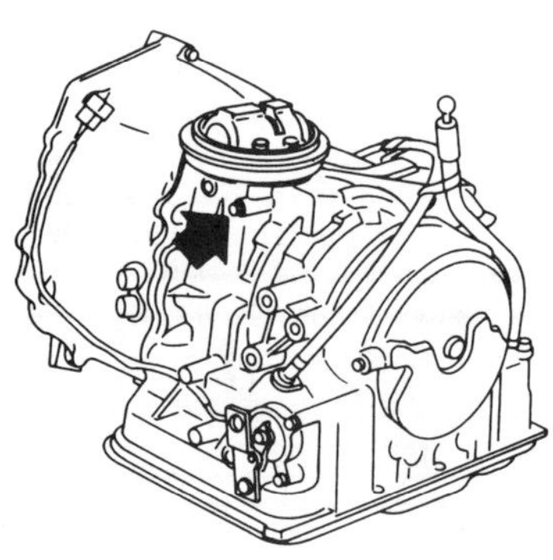

Fig. 7.12 Governor shaft lockbolt location (Sec 9)

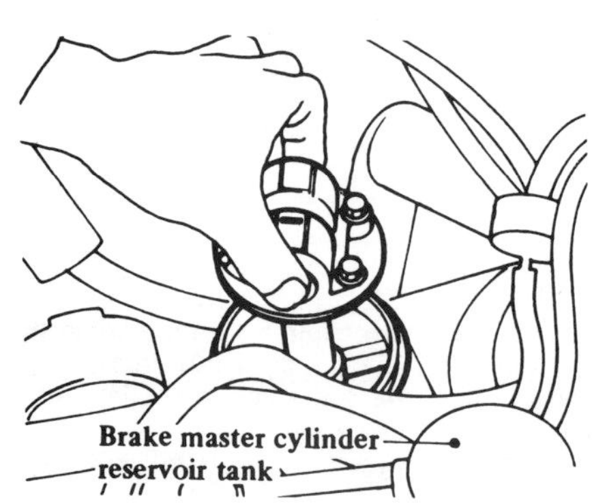

Fig. 7.13 Removing the governor shaft (Sec 9)

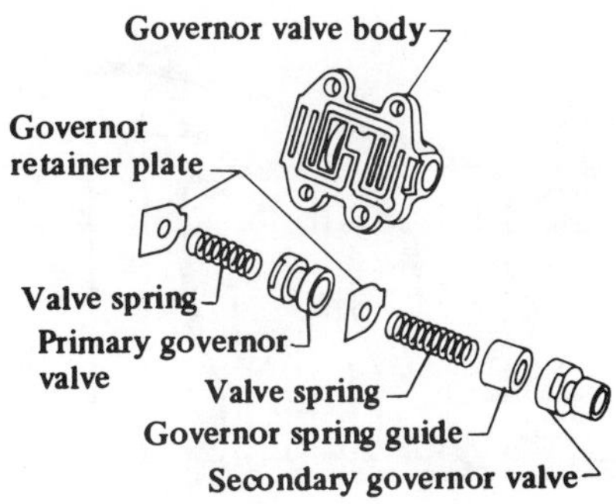

Fig. 7.14 Governor body components (Sec 9)

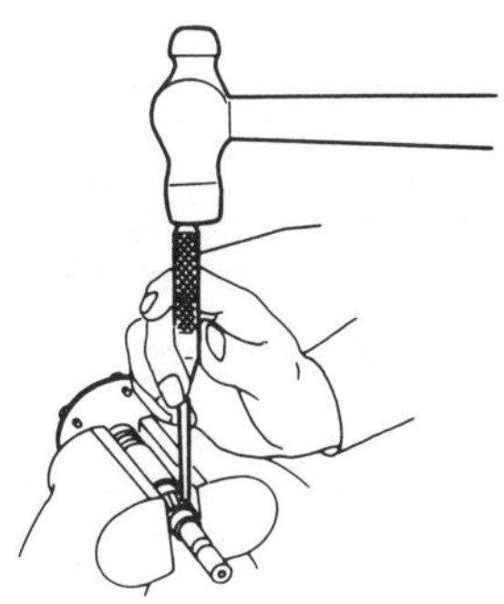

Fig. 7.15 Driving out the governor shaft worm pin (Sec 9)

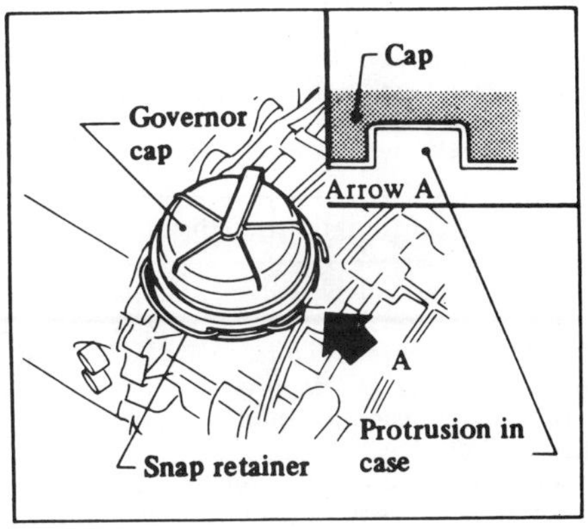

Fig. 7.16 Showing correct location of governor cap (Sec 9)

10 Differential bearing oil seals – renewal

1 In the event of oil leakage from the differential bearing oil seals, it is necessary to ascertain whether the cause is simply worn oil seals or additionally, excessive wear of the side gears resulting in lateral movement in the oil seal. If the latter is the case, new oil seals will not necessarily cure the problem and the transmission should be renewed.
2 Jack up the front of the car and support on axle stands. Apply the handbrake.
3 Using a dial gauge at right angles to the driveshaft bellows clip (Fig. 7.17), push the driveshaft joint fully towards the transmission then lightly move the joint up and down. If the play is more than 1.0 mm (0.039 in) the side gears are worn excessively, and renewal of the transmission should be considered.
4 If the side gear wear is not excessive remove the driveshafts as described in Chapter 8 and extract the oil seals using a two-legged puller.
5 Apply transmission fluid to the lips of the new oil seals before fitting them. A piece of tubing applied to the outer rim of the oil seal can be used to drive them into their seats.
6 Refit the driveshafts with reference to Chapter 8.

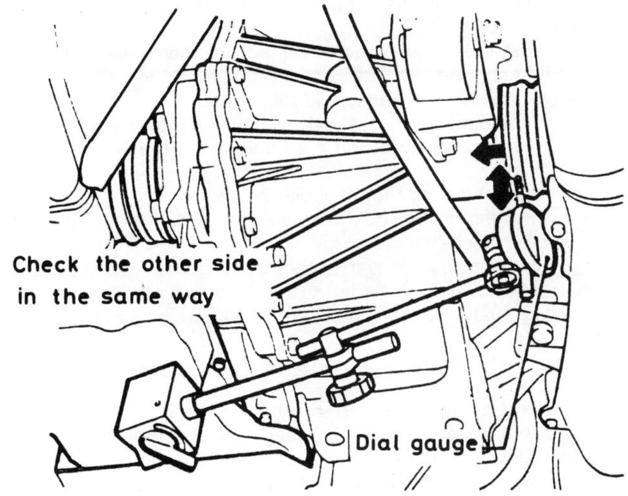

Fig. 7.17 Using a dial gauge to check the differential side gear wear (Sec 10)

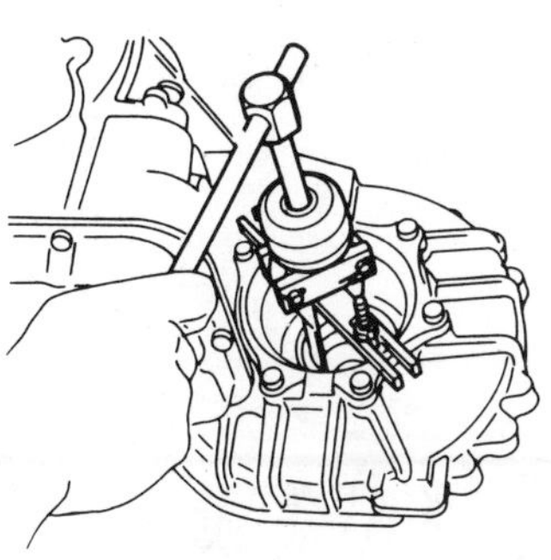

Fig. 7.18 Using a puller to remove a differential bearing oil seal (Sec 10)

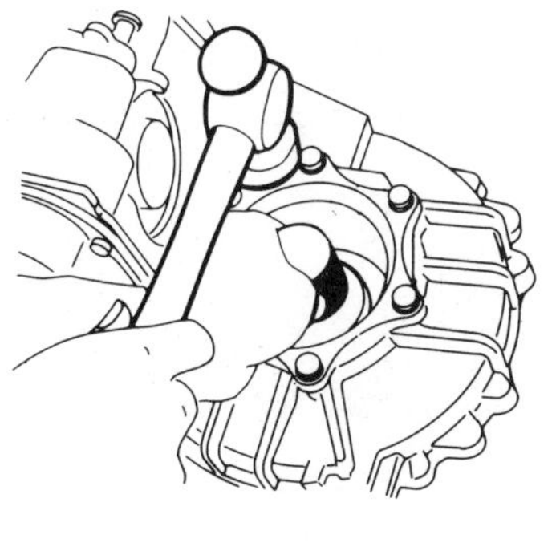

Fig. 7.19 Driving in a new differential bearing oil seal (Sec 10)

11 Transmission – removal and refitting

1 Place the vehicle over an inspection pit or raise the front end and support it securely on axle stands positioned under the side-members.
2 Disconnect the battery.
3 Remove the left-hand roadwheel.
4 Drain the transmission fluid.
5 Remove the left-hand plastic wing shield.
6 Disconnect both driveshafts, as described in Chapter 6, Section 5, paragraphs 4 to 11.
7 Disconnect the speedometer cable from the transmission.
8 Disconnect the throttle cable from the carburettor and the leads from the inhibitor switch.
9 Disconnect the speed selector cable from the transmission lever, and the cable support bracket from the transmission casing.
10 Remove the dipstick guide/fluid filler tube.
11 Support the engine on a jack with a block of wood as an insulator.
12 Support the transmission on a second jack – preferably of trolley type.
13 Disconnect and plug the oil cooler pipes.
14 Mark the relationship of the torque converter to the driveplate using a dab of quick-drying paint.
15 Unscrew the torque converter-to-driveplate connecting bolts. The crankshaft will have to be turned to bring each bolt into view within the cut-out of the torque converter housing before a spanner or socket wrench can be used.
16 Withdraw the automatic transmission flexible mounting pivot bolts.
17 Unbolt and remove the starter motor.
18 Unscrew and remove the torque converter housing-to-engine connecting bolts. Record the location of the coolant tube and mounting brackets held by some of these bolts. Unbolt the engine-to-transmission reinforcement strut.
19 Withdraw the transmission from under the front wing, having an assistant hold the torque converter in full engagement with the oil pump driveshaft to prevent loss of fluid.
20 If the transmission is being replaced with a new or rebuilt unit, check what is fitted to the new unit before parting with the original transmission. The parts not supplied can then be removed from the old unit.
21 Before offering up the transmission to the engine, check that the

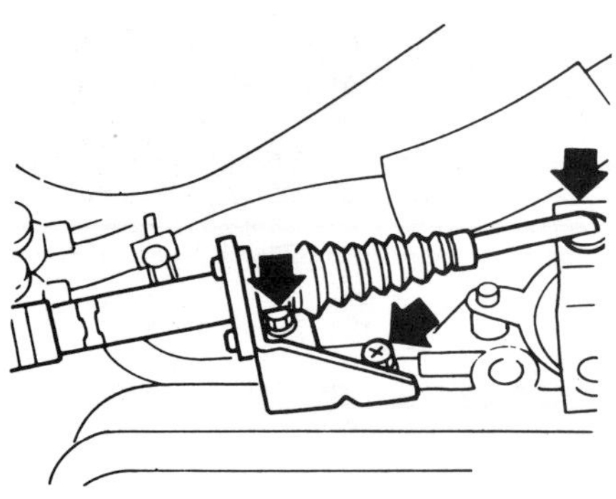

Fig. 7.20 Speed selector cable fixings at transmission end (Sec 11)

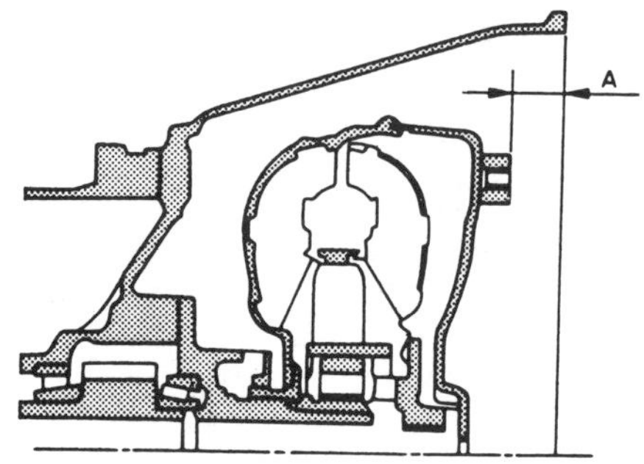

Fig. 7.21 Diagram showing torque converter fully installed dimension (Sec 11)

A = Not less than 21.1 mm (0.831 in)

converter is pushed fully home. This can be determined if dimension A is not less than that specified (see Fig. 7.21).
22 Align the marks on the driveplate and torque converter (made before dismantling), apply thread locking fluid to clean bolt threads, and screw in and tighten the bolts to the specified torque.
23 Bolt on the starter motor.
24 Fit the engine-to-transmission connecting bolts, making sure to locate the coolant tube and mounting brackets under their correct bolts. Refit the mounting pivot bolts. Refit the reinforcement strut.
25 Reconnect the speed selector control cable, and adjust if necessary.
26 Reconnect the inhibitor switch leads.
27 Reconnect the fluid cooler hoses.
28 Refit the wing protective shield.
29 Refit the dipstick guide/fluid filler tube.
30 Reconnect the throttle cable.
31 Reconnect the speedometer drive cable.
32 Reconnect the driveshafts, as described in Chapter 8.
33 Refit the left-hand roadwheel.
34 Reconnect the battery.
35 Fill the transmission with the specified fluid then check the level as described in Section 3.

12 Fault diagnosis – automatic transmission

1 As has been mentioned elsewhere in this Chapter, no service repair work should be considered by anyone without the specialist knowledge and equipment required to undertake this work. This is also relevant to fault diagnosis. If a fault is evident, carry out the various adjustments previously described, and if the fault still exists consult the local garage or specialist.
2 Before removing the automatic transmission for repair, make sure that the repairer does not require to perform diagnostic tests with the transmission installed.
3 Most minor faults will be due to incorrect fluid level, incorrectly adjusted selector control or throttle cables and the internal brake band being out of adjustment (refer to Section 7).

Chapter 8 Driveshafts

Contents

Specifications

Type ... Open shafts with spider or ball and cage type joints at each end

Torque wrench settings

	lbf ft	Nm
Lower balljoint to arm	40 to 47	54 to 64
Knuckle to strut:		
Except Pulsar 1984/85	51 to 65	69 to 88
Pulsar 1984/85	72 to 87	98 to 118
Driveshaft nut:		
UK except E15	58 to 116	78 to 157
UK E15	87 to 145	118 to 196
North America	87 to 145	118 to 196

1 General description

The driveshafts are of open type, transmitting power from the final drive/differential within the transmission to the front roadwheels.

At the roadwheel end of the shaft, a Rzeppa (spider) or Birfield (ball and cage) joint is fitted. At the inboard end of the shaft either a spider or double offset ball type joint is fitted, depending upon the model.

None of the joints can be repaired or dismantled, but can only be renewed as assemblies after detaching them from the shaft.

2 Routine maintenance

1 Every 12 000 miles (20 000 km) on UK models or 15 000 miles (24 000 km) on North American models check the driveshaft rubber bellows for damage, cracks, leakage and security (photos). Where the car is operating under extreme conditions, reduce the service interval mileage by half or carry out the check every 6 months, whichever comes first.

2.1A Driveshaft outer rubber bellows (1985 on models)

2.1B Driveshaft inner rubber bellows (1985 on models)

3 Driveshaft – checking, removal and refitting

1 Jack up the front of the car and support on axle stands. Apply the handbrake.

2 Grip the driveshaft and attempt to turn it in alternate directions against the rotation of the roadwheel. If possible, have an assistant depress the brake pedal during the check. If any play is evident, the joint must be renewed (Section 4 or 5) or the complete driveshaft replaced.

3 To remove the driveshaft first remove the roadwheel.

3.4 Removing the nut retainer

4 Extract the split pin, take off the nut retainer (photo), and loosen the driveshaft-to-hub retaining nut. In order to hold the hub against rotation, either refit the roadwheel and lower the vehicle to the ground, have an assistant apply the brakes, or use a length of steel rod or bar placed between two roadwheel studs as a lever. Take steps to prevent damage to the stud threads by screwing on the nuts.
5 Unbolt the brake caliper and tie it up out of the way (photo). There is no need to disconnect the hydraulic line.
6 Extract the split pin, unscrew the nut, and detach the steering tie-rod from the steering knuckle using a separator tool.
7 Disconnect the front suspension lower balljoint. Do this by unscrewing the nuts which hold the balljoint to the suspension arm. It is recommended that new nuts are used at reassembly.
8 Place a suitable container beneath the transmission to catch any spilled oil.
9 Insert a large screwdriver or other lever between the transmission casing and the inboard joint flange of the driveshaft. On the right-hand side of the casing an aperture is provided for inserting the lever. Prise the joint to overcome resistance of the retaining circlip. Take care not to damage the transmission oil seal and do not pull on the outer end of the driveshaft as the joints may come apart.
10 Unscrew and remove the bolts holding the knuckle to the strut, then withdraw the driveshaft from the transmission together with the hub and knuckle (photo).
11 Once the inboard end of the shaft has been disconnected, insert a short bar into the differential side gear as it is possible for the side gear to rotate and to fall into the differential case.
12 Working at the outboard end of the shaft, remove the driveshaft nut and washer and push the driveshaft out of the hub (photo). If it is tight, use a two or three-legged puller to push it out.
13 It is recommended that a new oil seal is fitted to the transmission whenever the driveshaft is removed. Refer to Chapter 6 for the procedure.
14 Refitting is a reversal of removal, but use a new circlip when fitting the inboard end of the driveshaft (photo). Push the shaft fully home in the side gear, after having first removed the temporary gear retaining bar and having applied grease to the oil seal lips. Pull the joint cover to check that the circlip is positively engaged. The flange of the joint casing may be tapped with a plastic-faced hammer if necessary to drive the shaft fully home. Tighten all nuts and bolts to the specified torque and fit a new split pin to the hub nut (photos). Top up the transmission oil level as necessary, and finally depress the footbrake pedal several times to set the disc pads in their normal position.

3.5 Front brake caliper tied to coil spring

3.10 Removing the driveshaft from the transmission

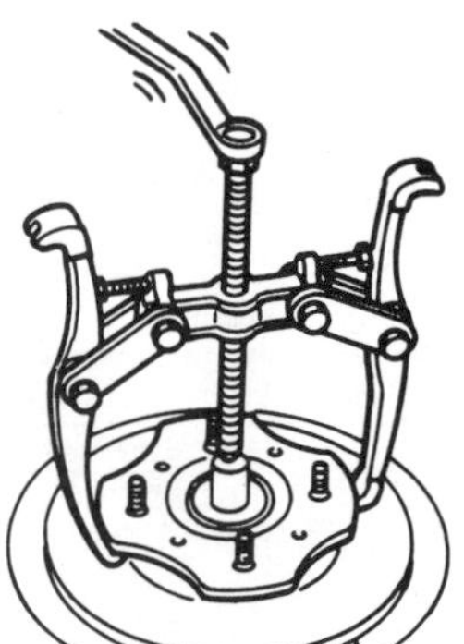

Fig. 8.1 Using a two-legged puller to push the driveshaft from the hub (Sec 3)

3.12A Removing the driveshaft nut ...

3.12B ... and washer

3.14A Circlip at driveshaft inboard end

3.14B Tightening the driveshaft nut

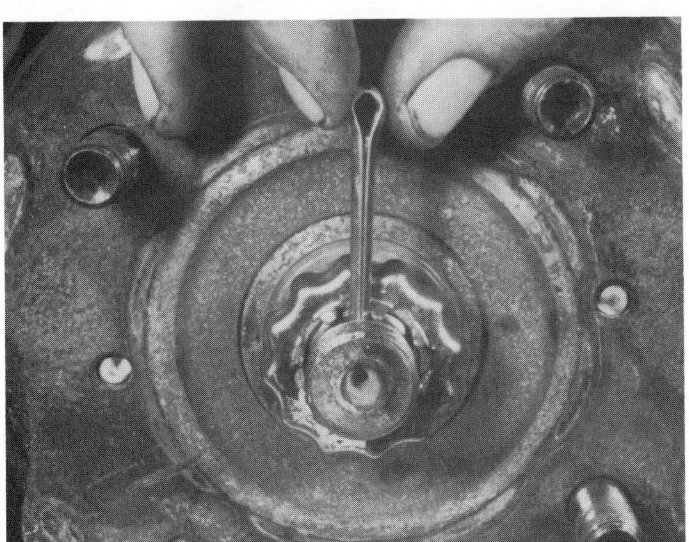

3.14C Insert a new split pin ...

3.14D ... and bend up the legs

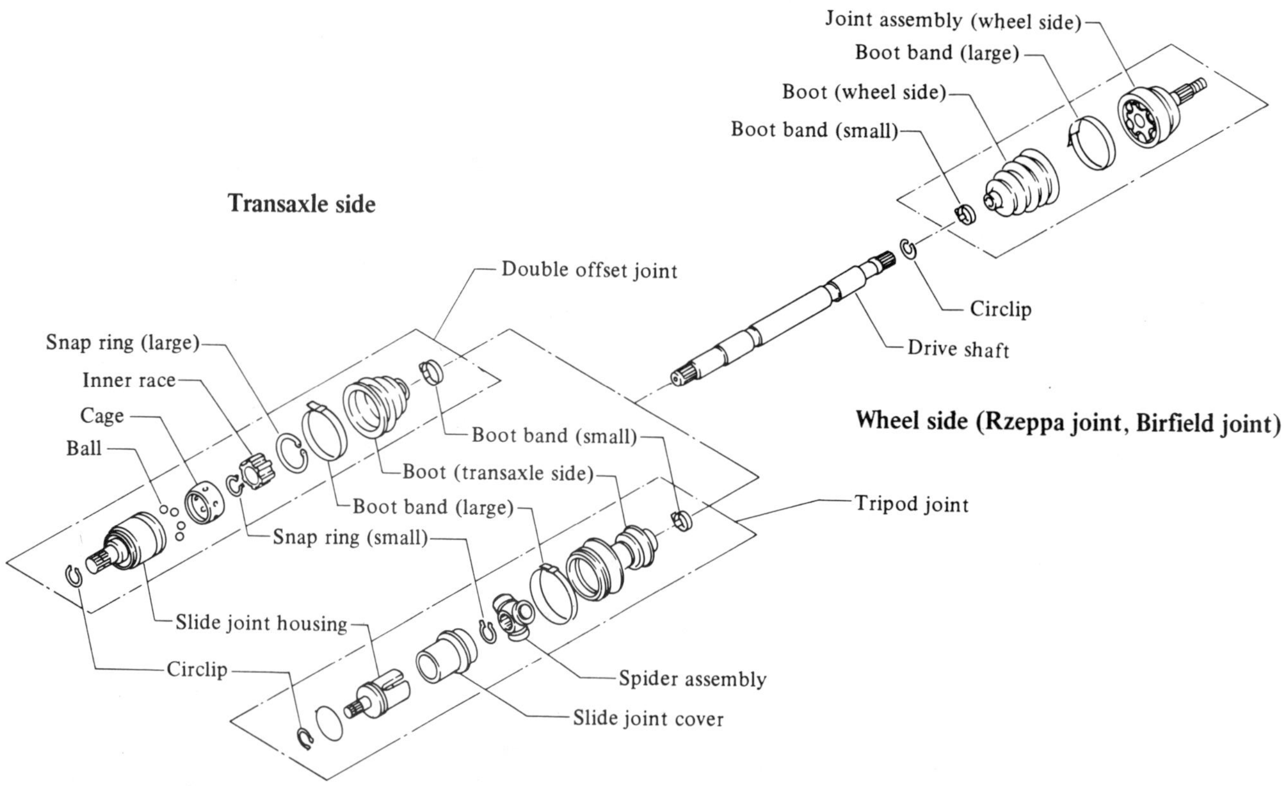

Fig. 8.2 Exploded view of the driveshaft (Secs 4 and 5)

4 Driveshaft outboard joint – removal and refitting

1 With the driveshaft removed, as described in the preceding Section, remove and discard the bellows securing bands.
2 Using quick-drying paint, put alignment marks on the shaft and joint.
3 Using a plastic or copper-faced hammer tap the outboard joint assembly off the shaft against the resistance of the retaining circlip (photo). Quite heavy blows will be required to release it.
4 Withdraw the flexible bellows from the shaft.
5 If the joint is worn it can only be renewed as an assembly.
6 Commence refitting by sliding the bellows onto the shaft and fitting a new securing band to the narrower diameter.
7 If the original joint is being fitted, use a new retaining circlip and, having wiped away as much lubricant as possible, tap the joint onto the shaft until the circlip snaps home (photo). The marks made before removal should be in alignment.
8 If a new joint is being fitted, and it is of spider type, make sure that its spider is offset to the one at the opposite end of the shaft if applicable. This can be felt through the bellows.
9 Peel back the bellows and apply the specified quantity of the

4.3 Removing the driveshaft outboard joint

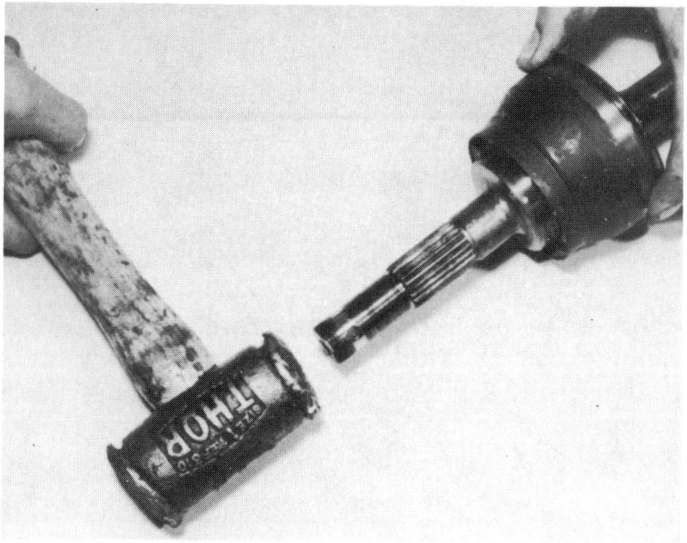

4.7 Tapping the outboard joint onto the driveshaft

special grease which is supplied with each new joint. If the original joint is being refitted use a suitable molybdenum disulphide based grease.

Rzeppa joint 110g (3.88 oz)
Birfield joint (E10 engine) 90g (3.17 oz)
Birfield joint (except E10 engine) 100g (3.53 oz)

Note that on new Rzeppa type joints, the inboard and outboard grease packs are different.

10 Set the bellows length (Fig. 8.3) then fit the large securing band. Tighten both bands using pliers and a screwdriver, bend over the tabs, then bend back the remaining end and cut off the excess.

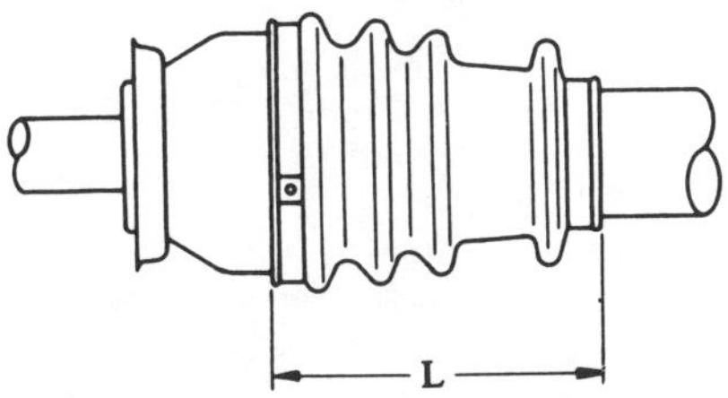

Fig. 8.3 Outboard joint bellows setting dimension (Sec 4)

Dimension L
Rzeppa joint = 100 mm (3.94 in)
Birfield joint – E10 engine = 85 mm (3.85 in)
Birfield joint – E13, E15 and E16 engines = 90 mm (3.54 in)

Note: *The bellows contour differs on some late models*

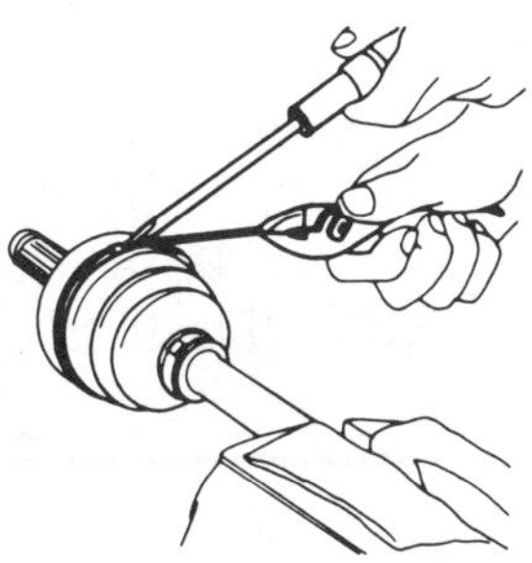

Fig. 8.4 Tightening the bellows securing bands (Sec 4)

5 Driveshaft inboard joint – removal and refitting

Spider type joint

1 With the driveshaft removed from the vehicle, grip it in a vice fitted with jaw protectors.

2 Remove the bellows retaining bands and discard them. Pull off the joint housing (photo).

3 Extract the joint circlip (photo).

4 If the original joint is to be refitted, mark the spider in relation to the shaft with a dab of quick-drying paint.

5 Support the spider and press the shaft from it. If a press is not available, the spider can be removed using a backing plate and a two or three-legged puller.

6 Slide the bellows from the shaft.

7 The slide joint cover can be removed by cutting slots in it with a hacksaw and bending the areas between the cuts outward.

8 Remove the cover O-ring and renew both the cover and O-ring at reassembly.

5.2 Removing the driveshaft inboard joint housing

5.3 Driveshaft inboard joint spider and circlip

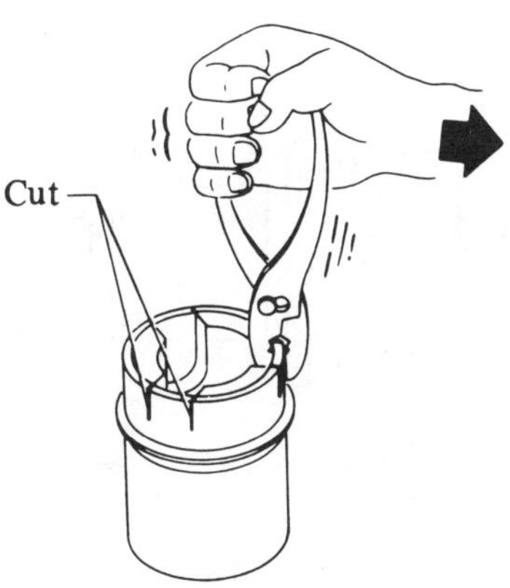

Fig. 8.5 Method of removing the inboard joint slide cover (Sec 5)

9 When reassembling, grease the O-ring and peen the rim of the new slide joint cover over its entire length and then seal the seam with RTV type sealant.
10 Fit the bellows to the shaft and locate a new securing band on the smaller diameter.
11 Fit the spider to the shaft so that the marks made on removal are in alignment (original joint) or, if a new joint is being used, it is offset to the spider at the opposite end of the shaft if applicable. This can be felt through the bellows.
12 Fit the spider to the shaft so that the chamfered face of the spider goes onto the shaft first. Drive the spider onto the shaft using a hammer and a piece of tubing.
13 Fit a new circlip so that its rounded side is towards the spider.
14 Pack the joint with grease. If a new joint is being fitted a grease pack is supplied with the joint, however if the original joint is being refitted use a suitable molybdenum disulphide based grease.

All models except 1985 North American 180g (6.35 oz)
1985 North American models – RH 215g (7.58 oz)
1985 North American models – LH 190g (6.70 oz)

15 Set the bellows length (Figs 8.7 or 8.8) then fit the large securing band. Tighten both bands using pliers and a screwdriver, bend over the tabs, then bend back the remaining end and cut off the excess.

Double offset joint

16 With the driveshaft gripped in the jaws of a vice fitted with jaw protectors, remove the bellows retaining bands and discard them.
17 Pull back the bellows and prise out the large circlip now exposed.
18 Pull off the slide joint housing.
19 Wipe away the grease from the ball cage and remove the balls.
20 Turn the cage through half a turn and pull it from the inner race.
21 Extract the circlip and tap the inner race from the driveshaft.
22 Pull the bellows from the driveshaft.
23 Refitting is a reversal of removal, but fit new circlips and pack the joint with new grease. New joints are supplied with a grease pack, but if the original joint is being refitted use a suitable molybdenum disulphide based grease.

E15 engine models 90g (3.17 oz)
Except E15 engine models 100g (3.53 oz)

24 Set the bellows length (Fig. 8.10) then fit the large securing band. Tighten both bands using pliers and a screwdriver, bend over the tabs, then bend back the remaining end and cut off the excess.

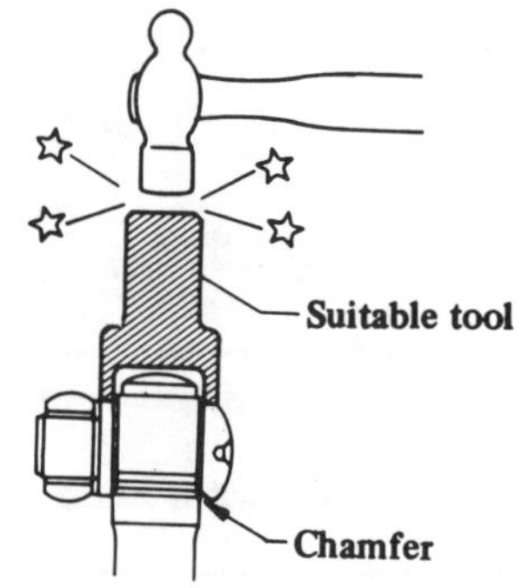

Fig. 8.6 Fitting the inboard joint spider showing the chamfer location (Sec 5)

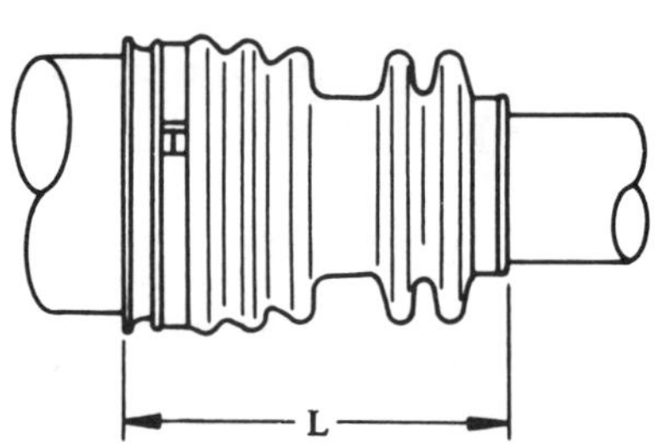

Fig. 8.7 Inboard spider joint bellows setting dimension for pre 1985 models (Sec 5)

Dimension L = 112 mm (4.41 in)

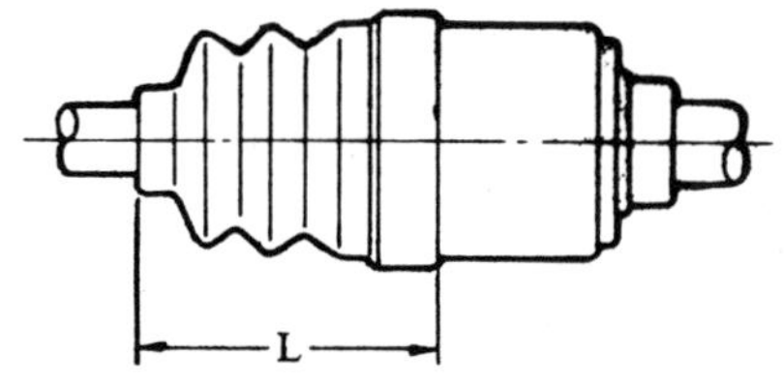

Fig. 8.8 Inboard spider joint bellows setting dimension for 1985 on models (Sec 5)

Dimension L
RH = 110 mm (4.33 in)
LH = 97 mm (3.82 in)

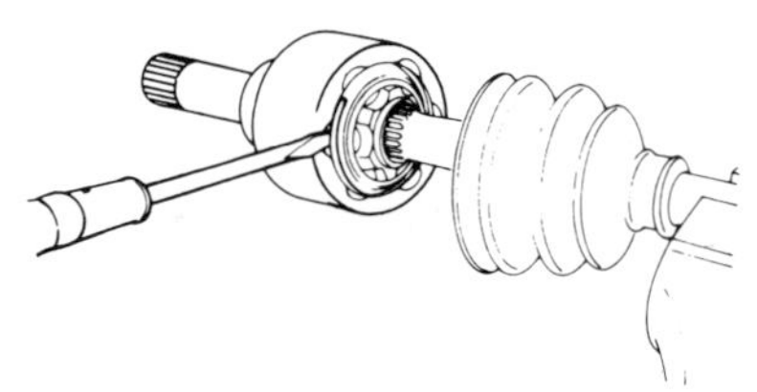

Fig. 8.9 Prising out the circlip from the double offset joint (Sec 5)

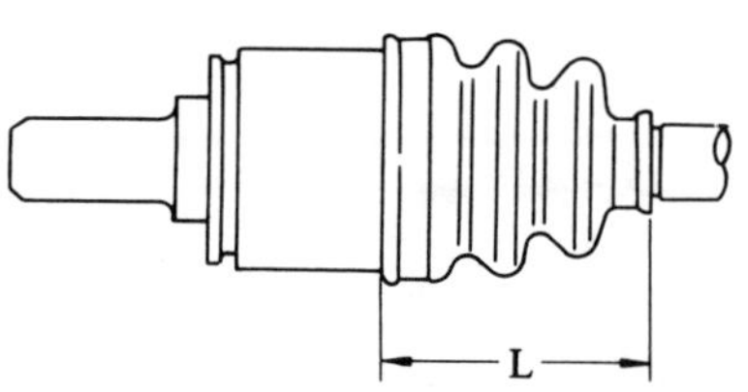

Fig. 8.10 Inboard double offset joint bellows setting dimension (Sec 5)

Dimension L
E10 and E13 engine = 82 mm (3.23 in)
E15 and 1983/84 E16 engine = 84 mm (3.31 in)
1984 E16 engine = 85 mm (3.35 in)

6 Driveshaft bellows – renewal

1 Renewal of either driveshaft bellows is only possible after removal of the relevant joint as described in Section 4 or 5.

7 Fault diagnosis – driveshafts

Symptom	Reason(s)
Vibration	Worn joints Worn wheel or differential bearings
Noise on taking up drive	Worn driveshaft splines Worn joints Loose driveshaft nut

Chapter 9 Braking system

Contents

Specifications

System type	Discs front, drums rear, vacuum servo assistance, dual hydraulic circuit split diagonally, self-adjusting rear brakes, cable operated handbrake on rear brakes, rear brake pressure regulating valve
Front brake discs	
Diameter	240.0 mm (9.45 in)
Maximum run-out:	
UK models, except Turbo	0.12 mm (0.005 in)
UK Turbo models	0.07 mm (0.003 in)
North American models	0.07 mm (0.003 in)
Minimum thickness:	
UK models, except Turbo	11.0 mm (0.433 in)
UK Turbo models	10.0 mm (0.394 in)
North American models	10.0 mm (0.394 in)
Pad friction material minimum thickness	2.0 mm (0.080 in)
Rear brakes	
Drum inner diameter:	
UK models	180.0 mm (7.09 in)
North American models (pre 1985 except Turbo)	180.0 mm (7.09 in)
North American models (Turbo and 1985 on)	203.2 mm (8.0 in)
Drum maximum inner diameter:	
UK models	181.0 mm (7.13 in)
North American models (pre 1984)	181.0 mm (7.13 in)
North American models (1984 on)	204.5 mm (8.05 in)
Drum maximum run-out:	
UK models, except Turbo	0.1 mm (0.004 in)
UK Turbo models	0.05 mm (0.002 in)
North American models	0.05 mm (0.002 in)
Shoe friction material minimum thickness	1.5 mm (0.060 in)
Master cylinder	
Maximum piston-to-cylinder clearance	0.20 mm (0.008 in)

Brake pedal

Free height:	
UK models – manual transmission (except Turbo)	201.0 to 207.0 mm (7.91 to 8.15 in)
UK models – manual transmission (Turbo)	194.0 to 204.0 mm (7.64 to 8.03 in)
UK models – automatic transmission	203.0 to 209.0 mm (7.99 to 8.23 in)
North American models – manual transmission	194.0 to 204.0 mm (7.64 to 8.03 in)
North American models – automatic transmission	197.0 to 207.0 mm (7.76 to 8.15 in)
Stop-lamp switch clearance	0 to 1.0 mm (0 to 0.04 in)
Depressed height (engine running, force of 50 kg/110 lb):	
UK models – except Turbo	More than 85.0 mm (3.35 in)
UK models – Turbo	More than 95.0 mm (3.74 in)
North American models – except Turbo	More than 85.0 mm (3.35 in)
North American models – Turbo	More than 80.0 mm (3.15 in)

Torque wrench settings

	lbf ft	Nm
Bleed screw	5.1 to 6.5	7 to 9
Stop-lamp switch locknut	12 to 16	16 to 22
Servo input locknut	12 to 16	16 to 22
Pedal bracket bolt	5.8 to 8.0	8 to 11
Hydraulic pipe union	11 to 13	15 to 18
Pressure valve	2.9 to 3.6	4 to 5
Master cylinder stop screw:		
Nabco	1.1 to 2.2	1.5 to 2.9
Tokico	1.4 to 2.5	2.0 to 3.4
Caliper bracket	40 to 47	54 to 64
Caliper cylinder to pins	16 to 23	22 to 31
Rear wheel cylinder	4.3 to 5.8	6 to 8
Rear brake backplate	18 to 25	25 to 33
Handbrake lever	5.8 to 8.0	8 to 11
Handbrake rod bush bracket	2.3 to 3.2	3.1 to 4.3
Handbrake rear cable locknut	2.3 to 3.2	3.1 to 4.3
Rear seat belt anchor	17 to 23	24 to 31
Servo to bracket	5.8 to 8.0	8 to 11
Master cylinder to servo	5.8 to 8.0	8 to 11
Disc to hub	18 to 25	25 to 33

1 General description

The braking system is of dual-line four wheel hydraulic type with servo assistance. The hydraulic circuit is split diagonally.

A pressure regulating valve is incorporated in the hydraulic circuit to prevent the rear wheels locking up during heavy applications of the brake pedal.

The handbrake operates mechanically on the rear wheels and incorporates an 'on' warning lamp switch.

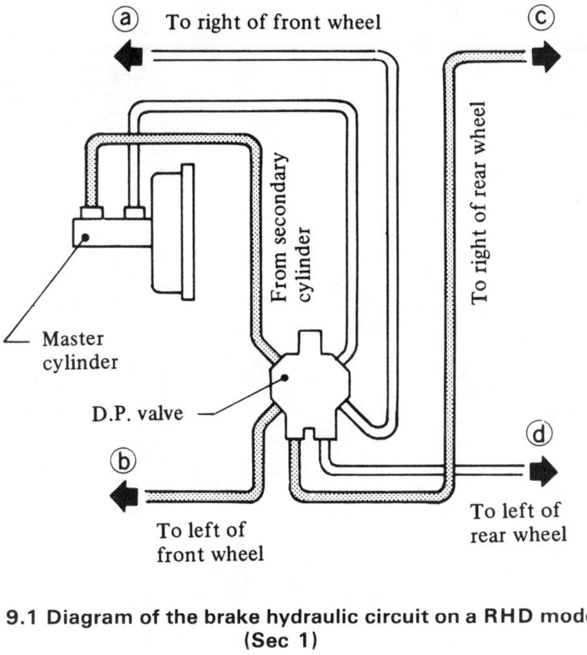

Fig. 9.1 Diagram of the brake hydraulic circuit on a RHD model (Sec 1)

2 Routine maintenance

1 Every 6000 miles (10 000 km) or 6 months whichever comes first on UK models, or 15 000 miles (24 000 km) or 12 months whichever comes first on North American models, check the front brake disc pads for wear and check the condition and security of the discs, calipers and all hydraulic hoses and lines. Check the footbrake pedal and handbrake for wear and correct adjustment. Check that the brake fluid level in the reservoir is at or near the maximum mark, and top up if necessary (photo). Slight variations of level will occur according to the wear of the brake linings, however if the level drops considerably the complete hydraulic system should be checked for leaks.

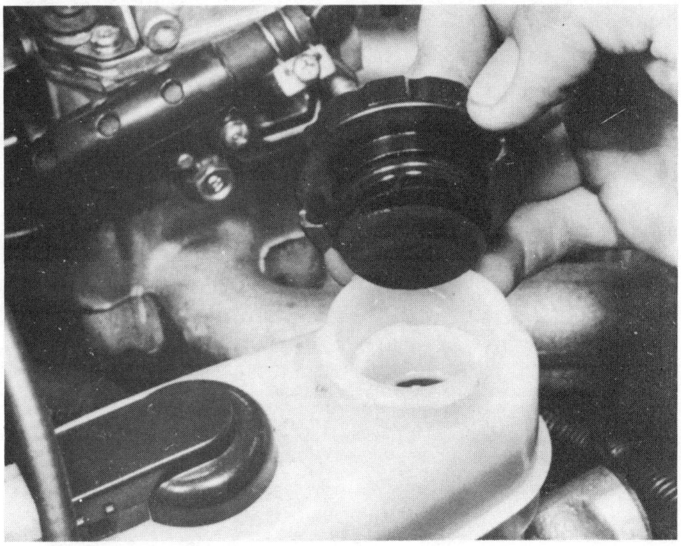

2.1 Brake fluid reservoir and filler cap

2 Every 12 000 miles (20 000 km) or 12 months whichever comes first on UK models, or 15 000 miles (24 000 km) or 12 months whichever comes first on North American models, check the rear brake shoes and drums for wear.
3 Every 12 000 miles (20 000 km) or 12 months whichever comes first on UK models, or 30 000 miles (48 000 km) or 24 months whichever comes first on North American models renew the brake fluid in the hydraulic system.
4 Every 24 000 miles (40 000 km) or 24 months whichever comes first on UK models, or 30 000 miles (48 000 km) or 24 months whichever comes first on North American models, check the servo unit, hoses and check valve for operation, condition and security. If necessary renew the servo air filter.

3 Disc pads – inspection and renewal

1 Raise the front of the vehicle, support it securely, and remove the roadwheels.
2 Check the thickness of the friction material. This must not be less than that specified.
3 If the thickness is less than that figure the pads must be renewed as an axle set (four pads).
4 Unscrew and remove the caliper lower lockpin bolt (photo).
5 Swivel the caliper/cylinder body upwards (photo).
6 Remove the anti-squeal shim (photo).
7 Take out the pads (photo).
8 The retaining springs can be prised out, if required (photos), also the piston shim (photo).
9 Brush away dirt and dust. *Avoid inhaling it as it is injurious to health.* Do not depress the brake pedal while the pads are out of the caliper.
10 Smear a trace of high melting-point grease onto the pad backplates and then locate the pads (friction surface to disc) and the anti-squeal shims.
11 The piston must now be fully depressed into the cylinder in order to accommodate the increased thickness of the new pads. Depressing the piston will cause the fluid level to rise in the master cylinder reservoir, so anticipate this by syphoning out some fluid using a clean battery hydrometer or meat baster.
12 Swivel the caliper/cylinder body downwards then fit and tighten the lower lockpin bolt. Check the tightness of both lockpin and guide pin bolts.
13 Renew the pads on the opposite side, refit the roadwheels and lower the vehicle.
14 Apply the brakes hard to position the pads against the disc.
15 Check the fluid level, and top up the reservoir if necessary.

4 Rear brake lining – inspection and renewal

1 Chock the front wheels. Raise the rear of the vehicle, support it securely and remove the roadwheels. Release the handbrake.
2 Prise off the hub grease cap, extract the cotter pin, take off the nut retainer and unscrew and remove the nut.
3 Pull off the brake drum, catching the outboard bearing which will be displaced. It is possible for the brake drum to be held on the axle due to the brake shoes being locked in grooves which have been worn in the drum. Should this occur, prise the plug from the brake backplate and, using a screwdriver, lift the toggle lever from the automatic adjuster star wheel and turn the star wheel to contract the shoes.
4 Inspect the shoe linings. If their thickness is less than that specified the shoes must be renewed as an axle set (four shoes).
5 It is recommended that new shoes are purchased complete with linings. Attempting to reline old shoes yourself seldom proves satisfactory.
6 If the old shoes appear oil stained, this will be due to a leaking wheel cylinder (defective seal) or to a faulty bearing oil seal.
7 Remove the shoe steady springs. To do this, grip the edges of the spring cup with a pair of pliers (photo), depress it against pressure of the coil spring and turn it through 90°. Release the spring cup and take off the spring (photos).

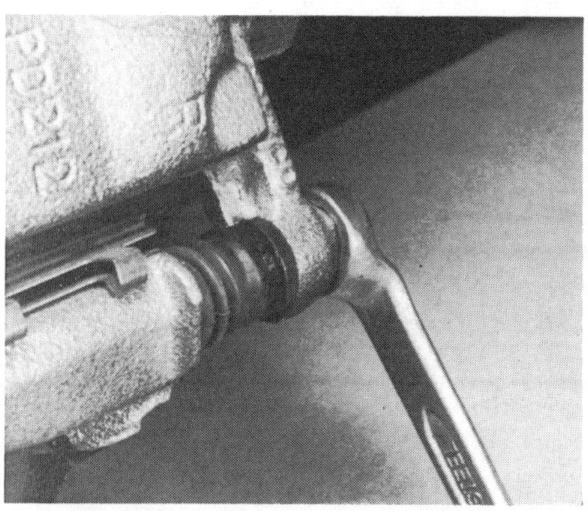

3.4 Unscrewing the caliper lower lockpin bolt

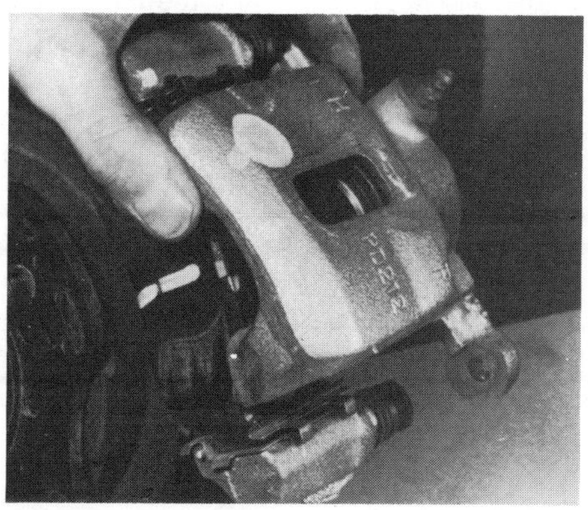

3.5 Raising the caliper to gain access to the disc pads

3.6 Removing the anti-squeal shims ...

3.7 ... and disc pads

3.8A Removing the pad retaining spring ...

3.8B ... and piston shim

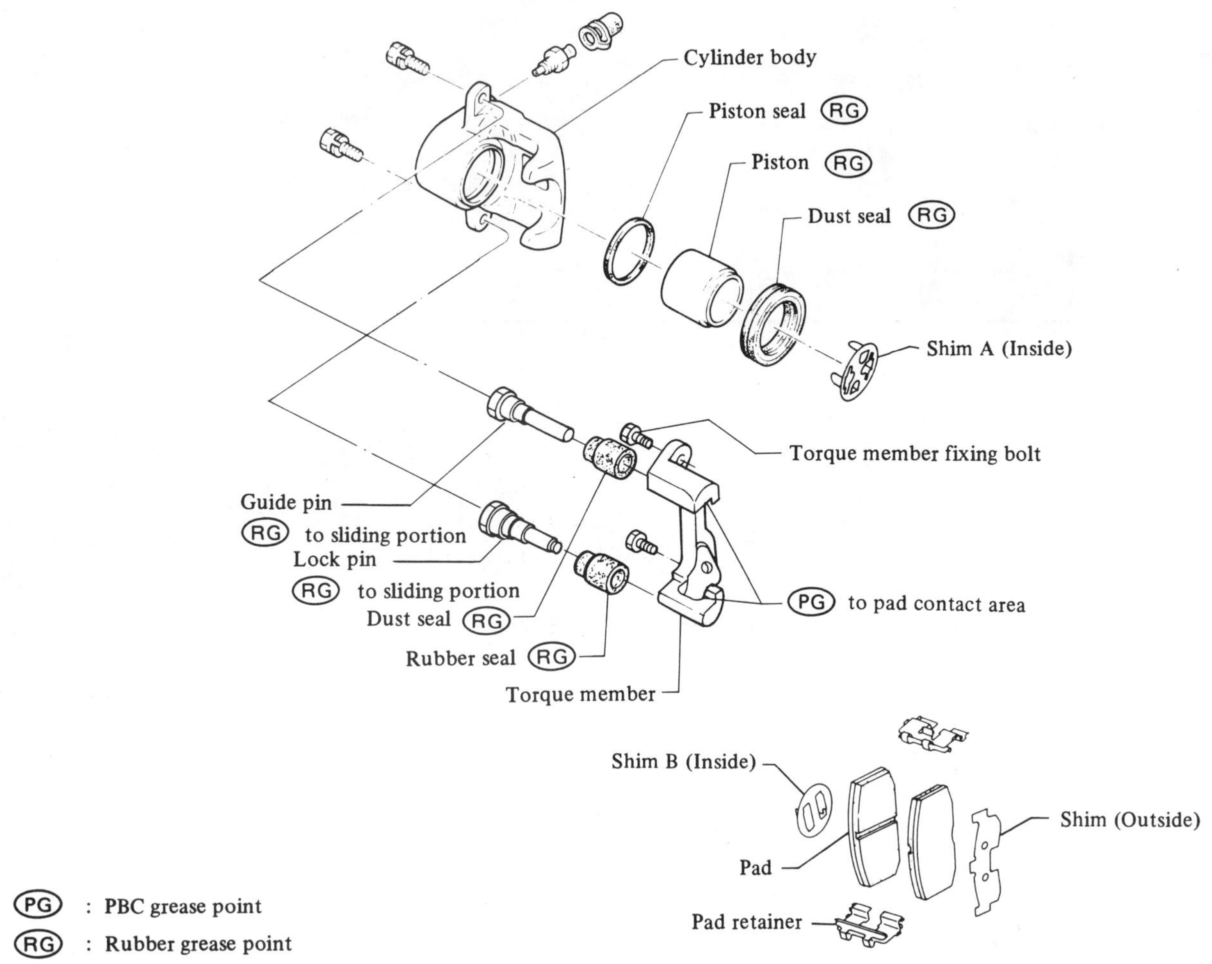

Fig. 9.2 Exploded view of the front disc caliper (Secs 3 and 5)

4.7A Removing a shoe steady spring cup ...

4.7B ... and spring

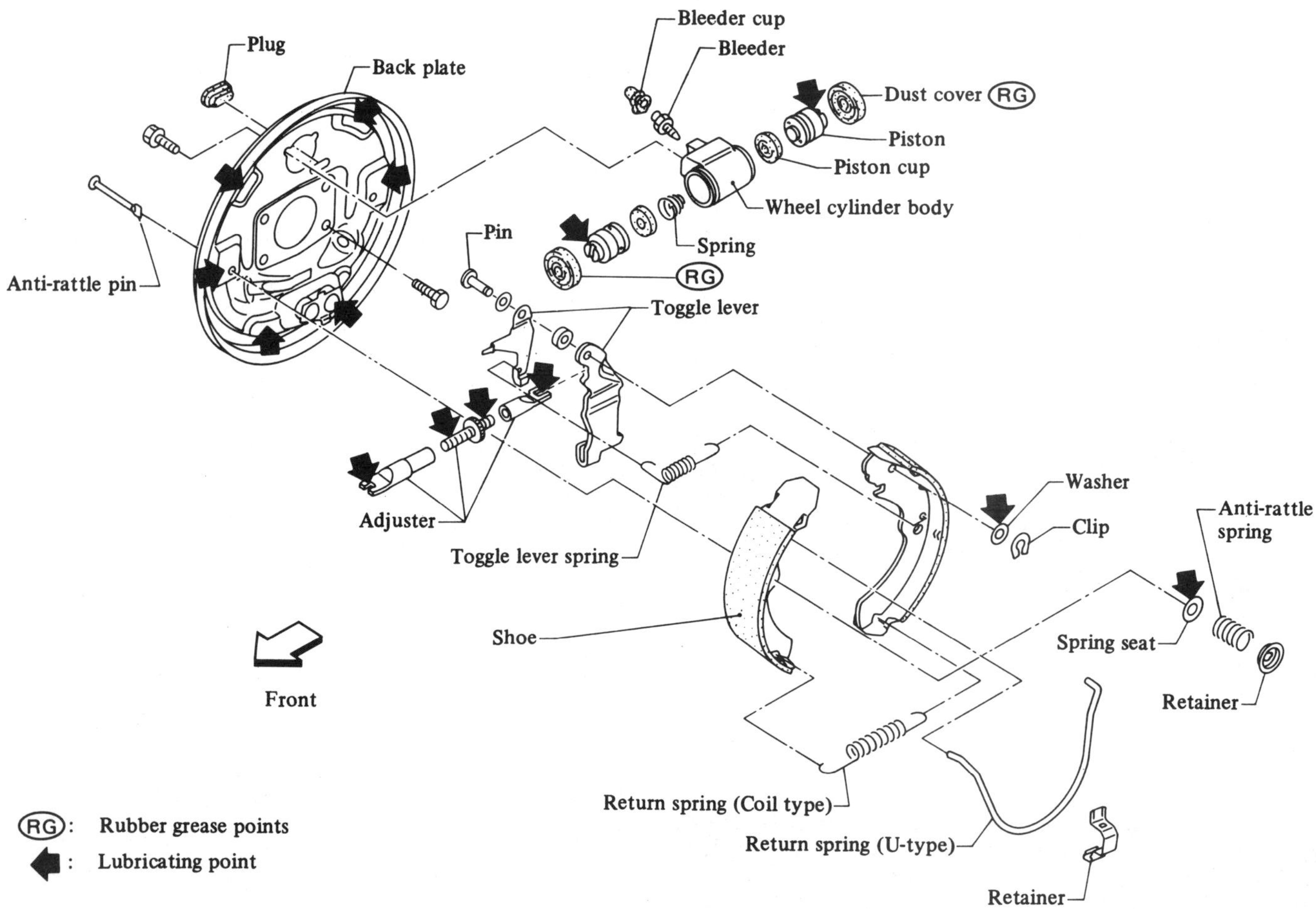

Fig. 9.3 Exploded view of the rear brake (Secs 4 and 7)

Note: *1985 on North American models incorporate a conventional upper shoe return spring*

8 Note the location of the shoes on the backplate with respect to the leading and trailing ends, as the lining material does not cover both ends of the shoes equally. Also note the spring location points on the shoes.

9 On all except 1985 North American models prise off the shoe return spring clip (photo), and then release the shoe return spring. This is a U-shaped spring and one arm should be gripped and levered towards the other arm to release it from the hole in the shoe (photos). It is recommended that a rag is placed over the spring to prevent it flying out accidentally.

10 On 1985 North American models unhook and remove the upper shoe return spring.

11 Pull the upper ends of the shoes apart and remove the adjuster strut (photo).

4.9A Prise off the spring clip ...

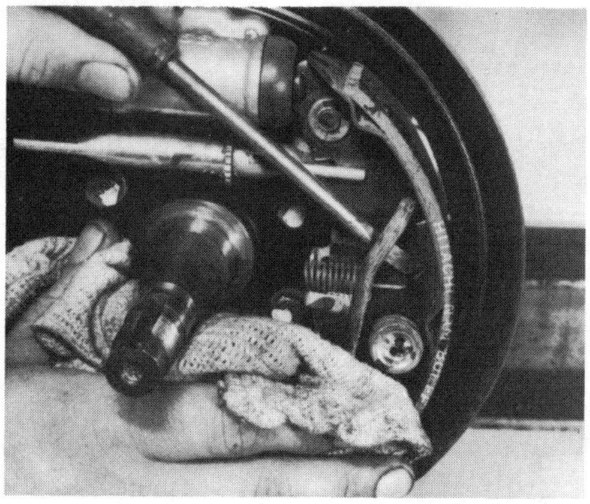

4.9B ... followed by the shoe return spring

4.11 Removing the adjuster strut

12 Unhook and remove the lower return spring and at the same time remove the shoes (photo). As they come away, disconnect the handbrake cable. Do not depress the brake pedal while the shoes are removed.
13 The handbrake and automatic adjuster toggle levers (photo) must be removed from the old shoe and fitted to the new one using the pin, washer, spring and U-shaped clip.

4.12 Releasing the lower return spring when removing the brake shoes

4.13 Showing the handbrake and automatic adjuster toggle lever and spring

14 Apply a smear of grease to the shoe contact high spots on the brake backplate and to the ends of the wheel cylinder pistons.
15 Fit the new shoes by reversing the removal operations, but before fitting the automatic adjuster strut turn the star wheel to contact the strut fully.
16 Refit the brake drum and adjust the bearing preload, as described in Chapter 11.
17 Repeat the operations on the opposite brake.
18 Apply the handbrake several times to actuate the automatic adjuster and to position the shoe linings as close as possible to the drum.
19 Refit the roadwheels and lower the vehicle.

5 Caliper – removal, overhaul and refitting

1 Raise the front of the vehicle and support it securely. Remove the roadwheel (photo). Fit a suitable clamp to the flexible hydraulic hose or alternatively tighten the fluid reservoir filler cap onto a sheet of polythene to prevent the loss of brake fluid, then disconnect the hydraulic hose from the caliper by unscrewing the hollow bolt from the banjo union. Note the copper washers, one each side of the union.
2 Unscrew the caliper mounting bolts and remove the caliper from the brake disc.
3 Clean away external dirt, *avoiding inhaling any dust.* Remove the pads Section 3).
4 Unscrew the guide pin and the lock pin bolts, and separate the cylinder body from the caliper bracket. Remove the pins.
5 Apply air pressure (such as is generated by a foot-operated pump) to the fluid entry hole in the caliper and eject the piston, dust excluder and retaining ring.
6 Inspect the surfaces of the piston and cylinder bore. If pitted or corroded, reassemble the caliper and renew it complete.
7 If the piston and cylinder are in good condition, use a sharp instrument to pick the piston seal out of its groove and discard it.
8 Wash all components in methylated spirit or clean hydraulic fluid and obtain a repair kit which will contain all the necessary new seals and other renewable items.
9 Commence reassembly by manipulating the new piston seal into its groove using the fingers only.
10 Push the piston part way into its bore, having first lubricated it with hydraulic fluid.
11 Fit the dust excluder and its retainer.
12 Smear the guide and lockpins with a little rubber grease and locate them in the caliper bracket together with their dust covers.
13 Connect the cylinder body to the caliper bracket then insert and tighten the guide pin (upper) bolt.
14 Fit the caliper to the stub axle carrier and tighten the bolts.
15 Depress the piston fully then locate the pads, shims and springs, lower the cylinder body and insert and tighten the lock pin (lower) bolt.
16 Bleed the hydraulic system as described in Section 12 then apply the brakes several times to position the pads on the disc.

5.1 Front brake caliper viewed from below

6 Disc – inspection and renovation

1 Whenever the disc pads are inspected for wear, take the opportunity to examine the disc for deep scoring, grooving or cracks. Light scoring is normal.
2 The disc should not run out-of-true by more than the specified tolerance (see Specifications). This may be checked using a dial gauge or feeler blades between the disc and a fixed point as the disc is rotated.

3 Provided the thickness of the brake disc will not be reduced below the specified minimum, a scored disc may be reground for further use.
4 To remove the disc, first remove the hub with reference to Chapter 11, then unbolt the disc from the hub.
5 Clean the mating faces before fitting the disc to the hub and tightening the bolts evenly to the specified torque.
6 Refit the hub with reference to Chapter 11.

7 Rear wheel cylinder – removal, overhaul and refitting

1 Remove the brake shoes, as described in Section 4.
2 Disconnect the hydraulic line from the cylinder.
3 Unbolt the wheel cylinder from the brake backplate (photo).
4 Clean away external dirt and pull off the dust covers.
5 Eject the internal components by tapping the cylinder on a block of wood or by applying air pressure from a foot-operated tyre pump to the fluid entry hole. Note which direction the seal lips face.
6 Examine the surface of the pistons and cylinder bore. If scored or corroded, renew the wheel cylinder complete.
7 If these components are in good condition, clean everything in either methylated spirits or hydraulic fluid, nothing else.
8 Discard the old seals and fit the new ones. These are contained in a repair kit, together with other renewable items.
9 Assemble the cylinder, applying hydraulic fluid as a lubricant as work progresses.
10 Refit the cylinder to the backplate, connect the fluid line and fit the brake shoes.
11 Bleed the brake circuit, as described in Section 12.

7.3 Rear wheel cylinder mounting bolt location

8 Brake drum – inspection and renovation

1 Whenever the brake drum is removed to inspect the wear of the shoe linings, take the opportunity to examine the interior friction surface of the drum.
2 If it is badly scored or grooved it is possible to have it machined, provided the internal diameter will not exceed the maximum specified dimension (see Specifications).

9 Master cylinder – removal, overhaul and refitting

1 Disconnect the fluid lines from the master cylinder and allow the fluid to drain into a suitable container.
2 Unbolt the master cylinder from the front face of the vacuum servo unit (photo). Disconnect the low fluid switch wiring where applicable.

9.2 Master cylinder located on front of servo unit

3 Withdraw the master cylinder, taking care not to spill hydraulic fluid on the paintwork.
4 Clean away external dirt.
5 Prise off the end cap and be prepared for the primary piston to be ejected.
6 Where applicable slightly depress the secondary piston and unscrew and remove the stop screw.
7 Shake out the secondary piston. Note the direction in which all seal lips face.
8 Inspect the surfaces of the pistons and the cylinder bore. If they are scored or corroded, renew the master cylinder complete.
9 If these components are in good condition, clean them in either methylated spirit or hydraulic fluid, nothing else. Discard the seals, and obtain a repair kit which will contain all the necessary new seals and other renewable components. It should be noted that a primary piston of Nabco make is supplied as an assembly.
10 Manipulate the new seals into position with the fingers only.
11 Renew the reservoir seals. Where applicable, the fast-fill valve assembly can be removed after extracting the circlip. Renew the valve O-ring.
12 As reassembly progresses, lubricate the components with clean hydraulic fluid.
13 Insert the secondary piston spring and then the assembled secondary piston into the cylinder.
14 Where applicable, depress the secondary piston slightly with a rod and screw in the stop screw.
15 Fit the primary piston spring and the primary piston assembly and stake a new stop cap into position.
16 Fit the master cylinder to the front face of the vacuum servo unit.
17 Reconnect the fluid pipelines and wiring as applicable.
18 Bleed the complete system, as described in Section 12, however if the fluid reservoir aperture in the master cylinder incorporates a fast-fill valve it is only necessary to bleed the master cylinder using the following procedure.
19 Top up the reservoir and keep it full during the procedure.
20 Disconnect the primary line (nearest the servo), place a container beneath the master cylinder, then fully depress the brake pedal.
21 Release the pedal and wait five seconds.
22 Repeat until clear fluid emerges from the master cylinder, then refit and tighten the primary line.
23 While an assistant depresses the brake pedal loosen the primary line union to bleed the remaining air then tighten the union. If air is still present release the pedal, wait five seconds, and bleed out the remaining air.
24 Bleed the secondary line in a similar manner but instead of waiting five seconds after releasing the pedal, wait twenty seconds.

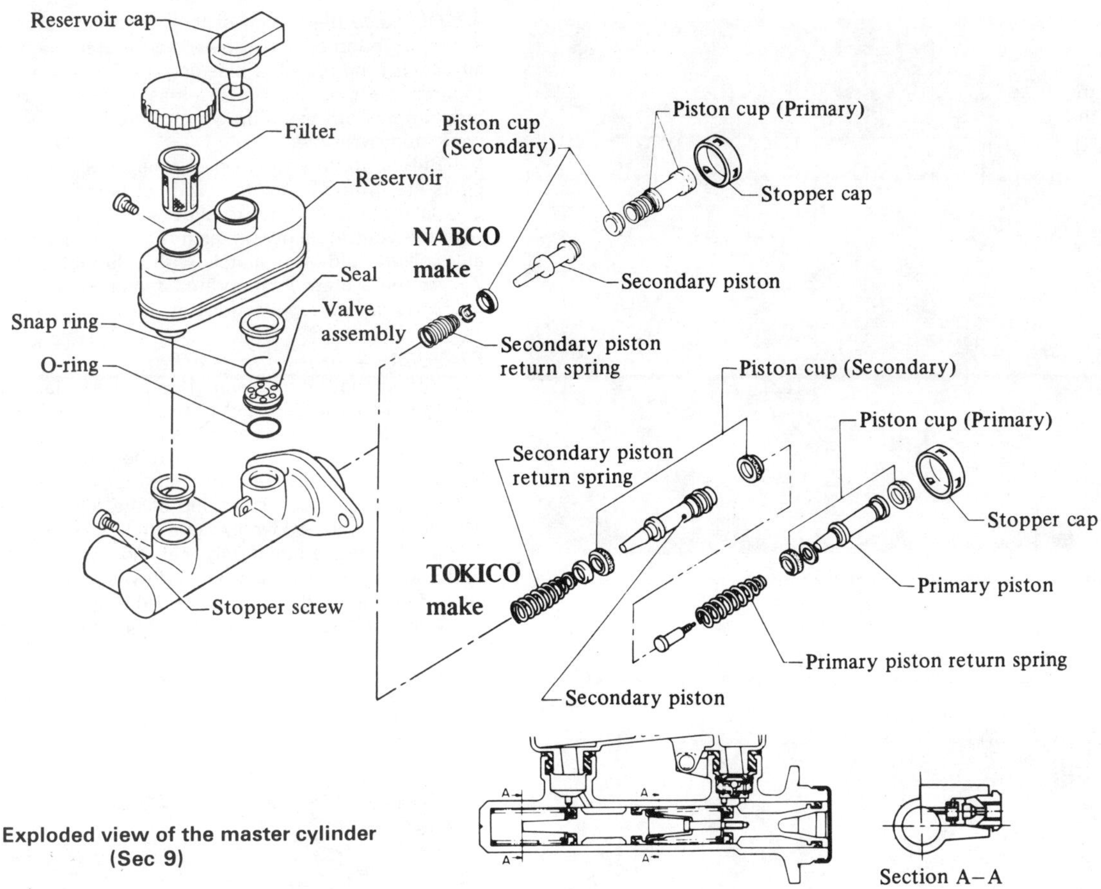

Fig. 9.4 Exploded view of the master cylinder (Sec 9)

25 After bleeding the master cylinder, depress the brake pedal and check that it feels firm. If it feels 'spongy', air must be present in another section of the hydraulic circuit and the complete system should then be bled.

10 Pressure regulating valve – general

1 This valve (also referred to as a dual proportioning valve) is located on the engine compartment rear bulkhead (photo).

10.1 Brake pressure regulating valve

2 Any fault can only be rectified by renewal of the valve, no repair being possible.
3 Removal is simply a matter of disconnecting the fluid lines and the central anchor bolt. Note the reconnection points for the individual pipelines to facilitate refitting.
4 Bleeding the system on completion (Section 12).

11 Hydraulic pipes and hoses – general

1 Periodically inspect the condition of the flexible brake hoses. If they appear swollen, chafed or when bent double with the fingers tiny cracks are visible they must be renewed.
2 Always uncouple the rigid pipe from the flexible hose first, then release the end of the flexible hose from the support bracket (photo). To do this, pull out the lockplate using a pair of pliers.
3 Now unscrew the flexible hose from the caliper or connector. On calipers, a banjo type hose connector is used. When installing the hose, always use a new sealing washer.
4 When installation is complete, check that the flexible hose does not rub against the tyre or other adjacent components. Its attitude may be altered to overcome this by pulling out the clip at the support bracket and twisting the hose in the required direction by not more than one quarter turn.
5 Bleed the hydraulic system (Section 12).
6 At regular intervals wipe the steel brake pipes clean and examine them for signs of rust or denting caused by flying stones.
7 Examine the fit of the pipes in their insulated securing clips and bend the tongues of the clips if necessary to ensure a positive fit (photo).
8 Check that the pipes are not touching any adjacent components or rubbing against any part of the vehicle. Where this is observed, bend the pipe gently away to clear.
9 Any section of pipe which is rusty or chafed should be renewed. Brake pipes are available to the correct length and fitted with end unions from most dealers and they can also be made to pattern by many accessory suppliers. When installing the new pipes use the old

11.2 Flexible hose to rigid pipe connection

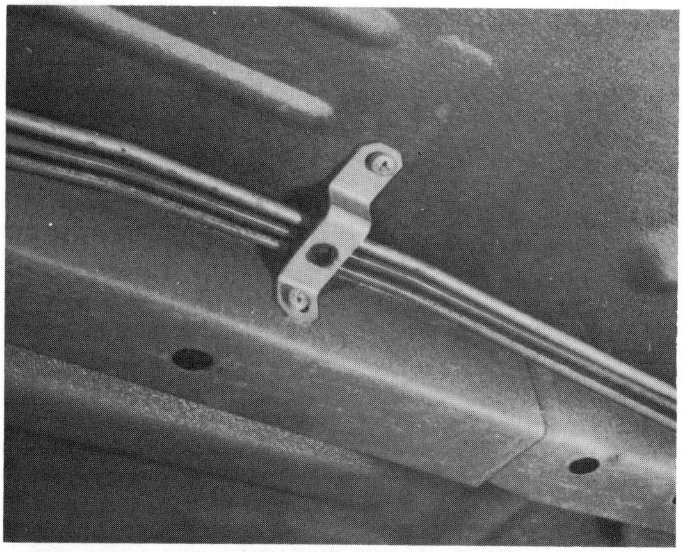

11.7 Retaining clip for brake pipes and fuel pipe

pipes as a guide to bending and do not make any bends sharper than is necessary.
10 The system will of course have to be bled when the circuit has been reconnected.

12 Hydraulic system – bleeding

1 The two independent hydraulic circuits are as follows:

(a) Front right-hand caliper and left rear wheel cylinder
(b) Front left-hand caliper and right rear wheel cylinder

On RHD models (a) is the primary circuit and (b) is the secondary circuit, however on LHD models the circuits are reversed. The *secondary* circuit should be bled first (rear wheel then front wheel), followed by the *primary* circuit (rear wheel then front wheel).
2 If the master cylinder or the pressure regulating valve has been disconnected and reconnected then the complete system (both circuits) must be bled.
3 If the component of only one circuit has been disturbed then only the particular circuit need be bled.
4 Owing to the design of the hydraulic system and pipeline layout, it will be found easier to bleed the system using a pressure bleeding kit. Unless the pressure bleeding method is being used, do not forget to keep the fluid level in the master cylinder reservoir topped-up to prevent air from being drawn into the system which would make any work done worthless.
5 Before commencing operations, check that all system hoses and pipes are in good condition with all unions tight and free from leaks.
6 Take great care not to allow hydraulic fluid to come into contact with the vehicle paintwork as it is an effective paint-stripper. Wash off any spilled fluid immediately with cold water.
7 As the system incorporates a vacuum servo, destroy the vacuum by giving several applications of the brake pedal in quick succession.

Bleeding – two-man method

8 Gather together a clean glass jar and a length of rubber or plastic tubing which will be a tight fit on the brake bleed screws.
9 Engage the help of an assistant.
10 Push one end of the bleed tube onto the first bleed screw and immerse the other end in the glass jar which should contain enough hydraulic fluid to cover the end of the tube (photo).
11 Open the bleed screw one half turn and have your assistant depress the brake pedal fully then slowly release it. Tighten the bleed screw at the end of each pedal downstroke to obviate any chance of air or fluid being drawn back into the system.
12 Wait between 10 and 20 seconds then repeat the operations as many times as is necessary until clean hydraulic fluid, free from air bubbles can be seen coming through into the jar.
13 Tighten the bleed screw at the end of a pedal downstroke and remove the bleed tube. Bleed from the remaining screws in a similar way.

12.10 Bleed tube connected to a front caliper bleed screw

Bleeding – using one-way valve kit

14 There are a number of one-man, one-way brake bleeding kits available from motor accessory shops. It is recommended that one of these kits is used wherever possible, rather than just a tube, as it will greatly simplify the bleeding operation and reduce the risk of air or fluid being drawn back into the system, quite apart from being able to do the work without the help of an assistant.
15 To use the kit, connect the tube to the bleed screw and open the screw one half turn.
16 Depress the brake pedal fully and slowly release it. The one-way valve in the kit will prevent expelled air from returning at the end of each pedal downstroke. Repeat this operation several times to be sure of ejecting all air from the system. Some kits include a translucent container which can be positioned so that the air bubbles can actually be seen being ejected from the system.

17 Tighten the bleed screw, remove the tube and repeat the operations in the remaining brakes.
18 On completion, depress the brake pedal. If it still feels spongy repeat the bleeding operations as air must still be trapped in the system.

Bleeding – using a pressure bleeding kit

19 These kits are available from motor accessory shops and are usually operated by air pressure from the spare tyre.
20 By connecting a pressurised container to the master cylinder fluid reservoir, bleeding is then carried out by simply opening each bleed screw in turn and allowing the fluid to run out, rather like turning on a tap, until no air is visible in the expelled fluid.
21 By using this method, the large reserve of hydraulic fluid provides a safeguard against air being drawn into the system during bleeding which often occurs if the fluid level in the reservoir is not maintained.
22 Pressure bleeding is particularly effective when bleeding 'difficult' systems or when bleeding the complete system at a time of routine fluid renewal.

All methods

23 When bleeding is completed, check and top up the fluid level in the master cylinder reservoir.
24 Check the feel of the brake pedal. If it feels at all spongy, air must still be present in the system and further bleeding is indicated. Failure to bleed satisfactorily after a reasonable repetition of the bleeding operations may be due to worn master cylinder seals.
25 Discard brake fluid which has been expelled. It is almost certain to be contaminated with moisture, air and dirt, making it unsuitable for further use. Clean fluid should always be stored in an airtight container as it absorbs moisture readily (hygroscopic) which lowers its boiling point and could affect braking performance under severe conditions.

13 Vacuum servo unit – description and maintenance

1 The vacuum servo unit is fitted into the brake hydraulic circuit in series with the master cylinder to provide assistance to the driver when the brake pedal is depressed. This reduces the effort required by the driver to operate the brakes under all braking conditions.
2 The unit operates by vacuum obtained from the induction manifold and comprises, basically, a booster diaphragm and check valve. The servo unit and hydraulic master cylinder are connected together so that the servo unit piston rod acts as the master cylinder pushrod. The driver's effort is transmitted through another pushrod to the servo unit piston and its built-in control system. The servo unit piston does not fit tightly into the cylinder, but has a strong diaphragm to keep its edges in constant contact with the cylinder wall, so assuring an airtight seal between two parts. The forward chamber is held under vacuum conditions created in the inlet manifold of the engine and, during periods when the brake pedal is not in use, the controls open a passage to the rear chamber so placing it under vacuum conditions as well. When the brake pedal is depressed, the vacuum passage to the rear chamber is cut off and the chamber opened to atmospheric pressure. The consequent rush of air pushes the servo piston forward in the vacuum chamber and operates the main pushrod to the master cylinder.
3 The controls are designed so that assistance is given under all conditions and when the brakes are not required, vacuum in the rear chamber is established when the brake pedal is released. All air from the atmosphere entering the rear chamber is passed through a small air filter.
4 Under normal operating conditions the vacuum servo unit is very reliable and does not require overhaul except at very high mileages. in this case it is far better to obtain a service exchange unit, rather than repair the original unit.
5 It is emphasised that the servo unit assists in reducing the braking effort required at the foot pedal and, in the even of its failure, the hydraulic braking system is in no way affected except that the need for higher pedal pressure will be noticed.
6 Periodically inspect the condition of the vacuum hose in which is incorporated a non-return valve (photo). Renew the hose if it is split or has hardened.
7 At the specified intervals (or earlier in dusty climates) the air filter should be renewed.

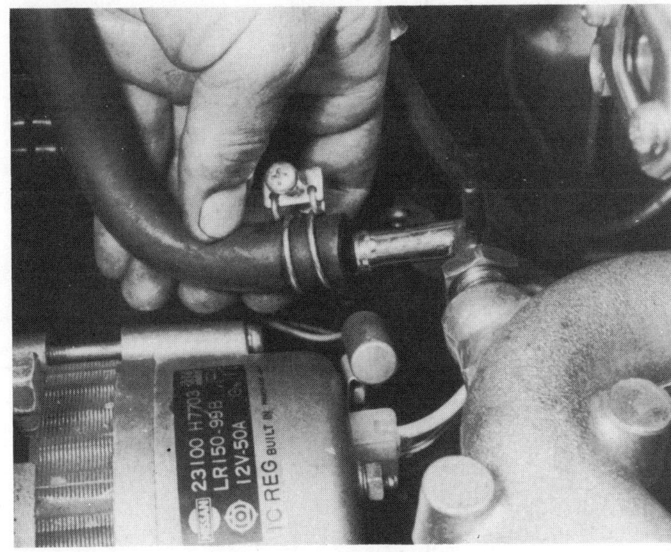

13.6 Vacuum servo hose at the intake manifold

8 To remove it from its location around the servo unit pushrod, peel back the dust excluder.
9 To save disconnecting the pushrod, the old filter may be cut away and removed.
10 Make a diagonal cut from the centre hole to the outside edge of the new filter and install it.
11 Refit the dust excluder.

14 Vacuum servo unit – removal and refitting

1 Remove the master cylinder, as described in Section 9.
2 Disconnect the vacuum hose from the servo unit.
3 Working inside the vehicle, disconnect the pushrod from the brake pedal.
4 Unbolt the brake servo unit from the bulkhead and remove it (photo).
5 Refitting is a reversal of removal, but carry out the following checks before actually installing it.

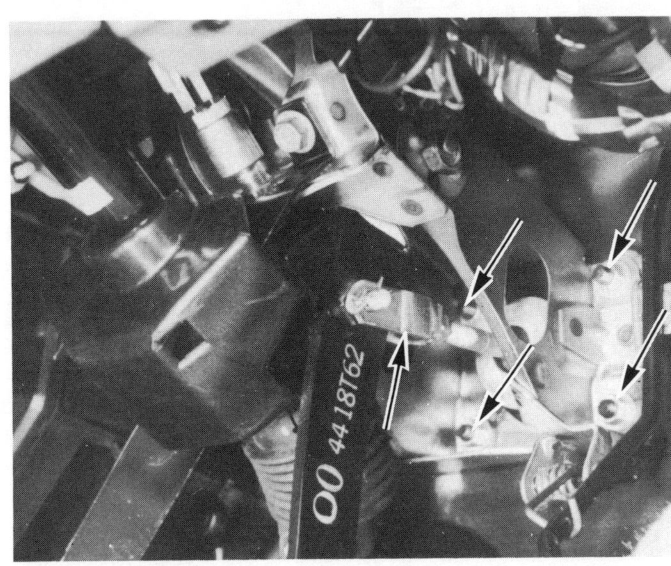

14.4 Showing brake pedal pushrod and vacuum servo unit mounting nuts (arrowed)

6 Check the projection (A) of the pushrod which enters the master cylinder (see Fig. 9.5). The projection is set in production and the threads locked, so if it does not conform to specifications renew the complete unit.
7 Now check the length (B) of the pushrod which connects with the brake pedal (see Fig. 9.6). Adjust if necessary by turning the clevis fork after having released the locknut.

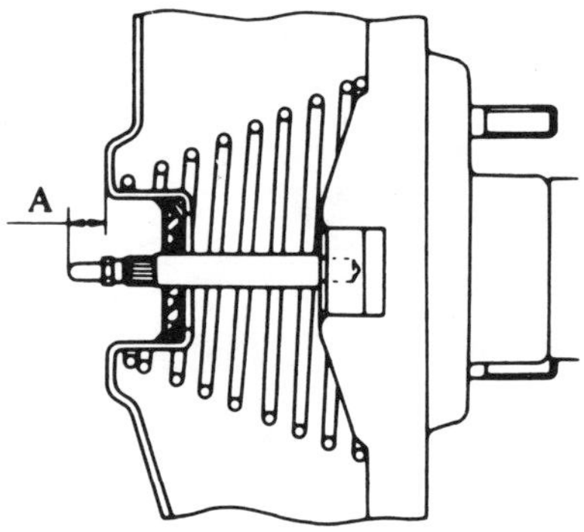

Fig. 9.5 Vacuum servo unit-to-master cylinder pushrod setting dimension (Sec 14)

Non-Turbo models – A = 10.275 to 10.525 mm (0.4045 to 0.4144 in)
Turbo model – A = 10.375 to 10.425 mm (0.4085 to 0.4104 in)

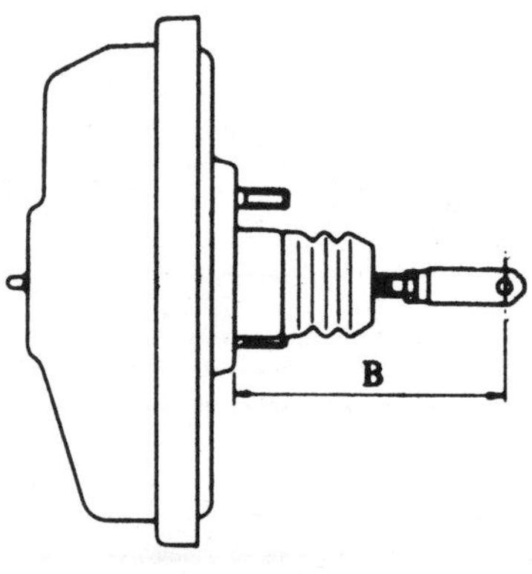

Fig. 9.6 Brake pedal-to-vacuum servo unit pushrod setting dimension (Sec 14)

All models – B = 150.0 mm (5.91 in)

15 Handbrake – adjustment

1 The handbrake is adjusted by the action of the rear shoe automatic adjuster and will require no further attention unless the cable stretches, normally only after a high mileage has been covered.
2 The handbrake should be fully applied with the rear wheels locked if the handbrake control lever is pulled up 6 to 7 notches (clicks) of the ratchet.
3 If the lever moves over an excessive number of notches, release the handbrake and adjust the cable by moving the locknuts at the cable stirrup which is adjacent to the equaliser under the vehicle (photo).
4 Keep the groove in the equaliser well greased at all times.
5 Working inside the vehicle, remove the centre console (see Chapter 12) and bend the handbrake warning switchplate down so that, with the ignition on, the warning lamp comes on when the lever is pulled up one notch.

15.3 Handbrake equaliser and adjustment stirrup (arrowed)

16 Handbrake cables – renewal

Primary cable

1 Disconnect the secondary cable from the connecting stirrup which is adjacent to the equaliser under the vehicle (see photo 15.3). Do this by unscrewing the locknuts and then passing the cable end fitting through the groove in the equaliser.
2 Working inside the vehicle, remove the centre console, as described in Chapter 12.
3 Disconnect the lead from the handbrake warning switch (photo).
4 Remove the seat belt stalks.
5 Remove the hand control lever fixing screws and the cable bush fixing screws.
6 Withdraw the hand control lever with the primary cable attached into the vehicle interior.
7 To separate the primary cable from the control lever, drill out the connecting pin.
8 Refitting is a reversal of removal, a clevis pin and cotter pin are supplied for connecting the new cable to the hand control lever.
9 Adjust, as described in the preceding Section.

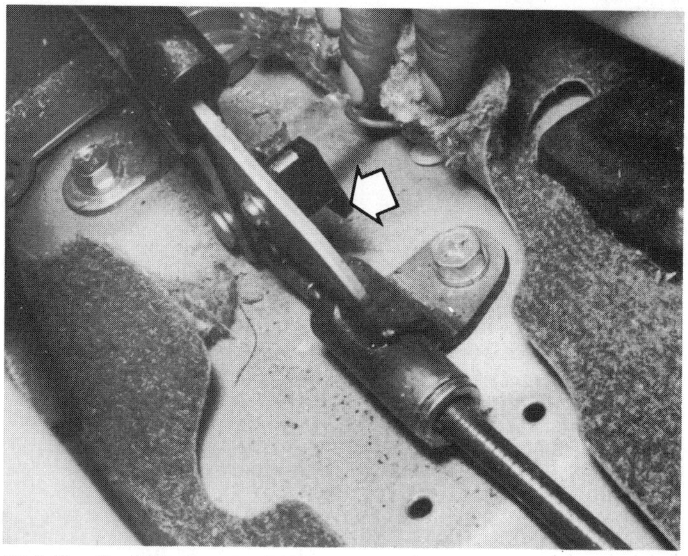

16.3 Handbrake lever warning switch location

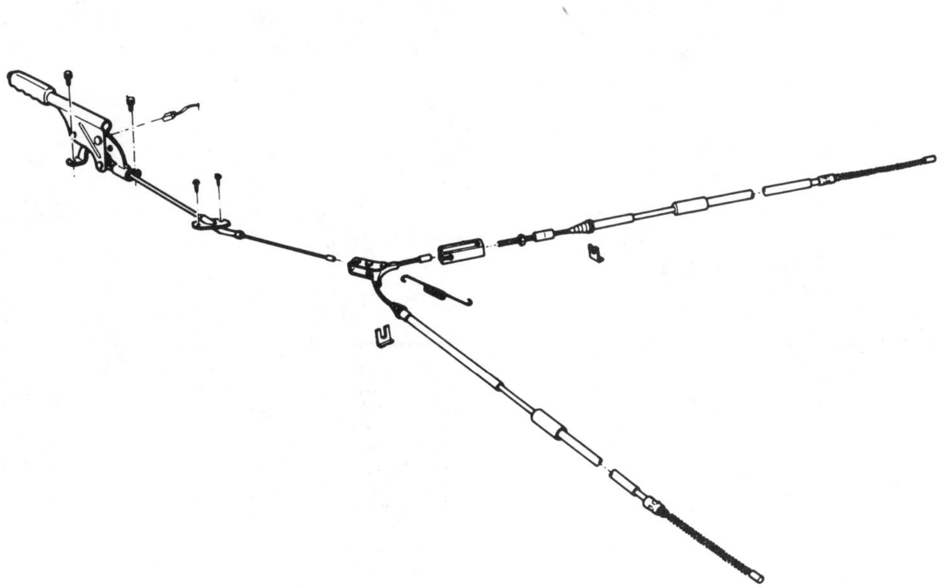

Fig. 9.7 Handbrake lever and cable components (Secs 15 and 16)

Secondary cable

10 Disconnect the cable at the stirrup which is adjacent to the equaliser under the vehicle.
11 Draw the longer cable through the groove in the equaliser.
12 Bend back the cable clips on the rear suspension arms (photo) and release the cables.
13 Raise the rear of the vehicle, support it securely and chock the front wheels. Remove both roadwheels and brake drums.
14 Unhook the handbrake cable and fittings from the handbrake lever on the brake shoes.
15 Remove both cables through the backplates (photo).
16 Fit the new cables by reversing the removal operations. Apply grease to the cable friction surfaces.
17 Refit the brake drums and adjust the bearings, as described in Chapter 11.

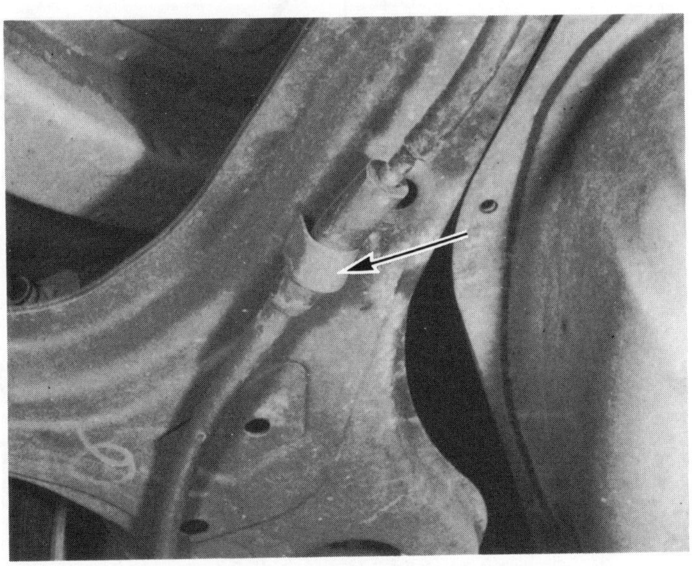

16.12 Handbrake cable clip on rear suspension arm

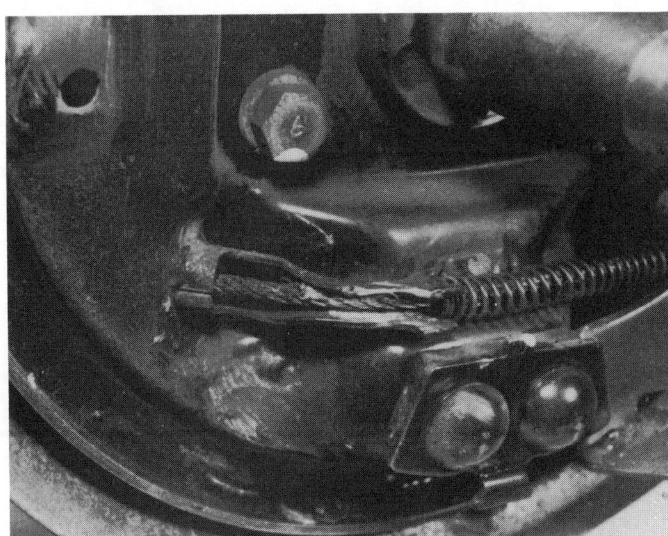

16.15 Handbrake cable guide at the backplate

18 Refit the roadwheels.
19 Adjust the cables, as described in the preceding Section.

17 Brake pedal – removal, refitting and adjustment

1 Working within the vehicle, under the facia panel, disconnect the pushrod from the brake pedal arm. Where necessary the lower facia panel and heater duct must be removed.
2 Prise the clip from the end of the pivot shaft, then unhook the return spring and withdraw the clutch pedal.
3 Slide the pivot shaft from the bracket and lower the brake pedal.
4 The pivot bushes may be renewed.
5 Refitting is a reversal of removal, but apply grease to the pivot shaft and bushes and check the pedal height as follows.

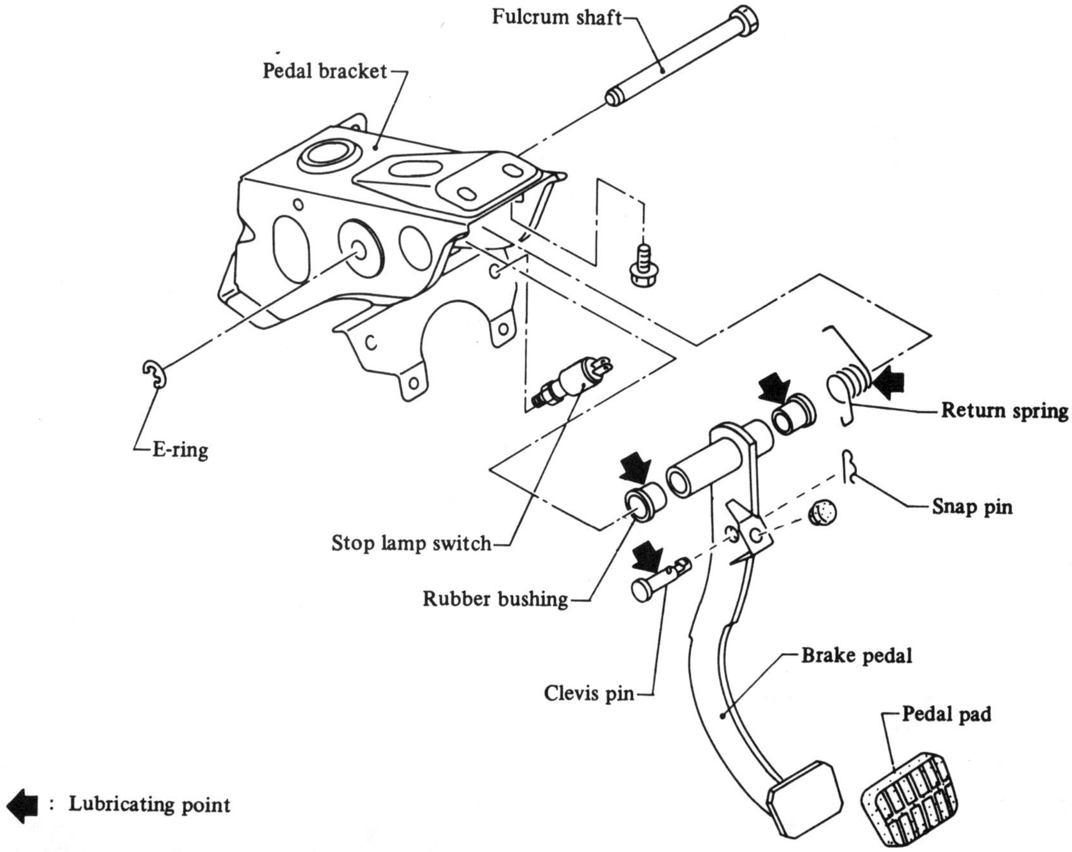

Fig. 9.8 Brake pedal components (Sec 17)

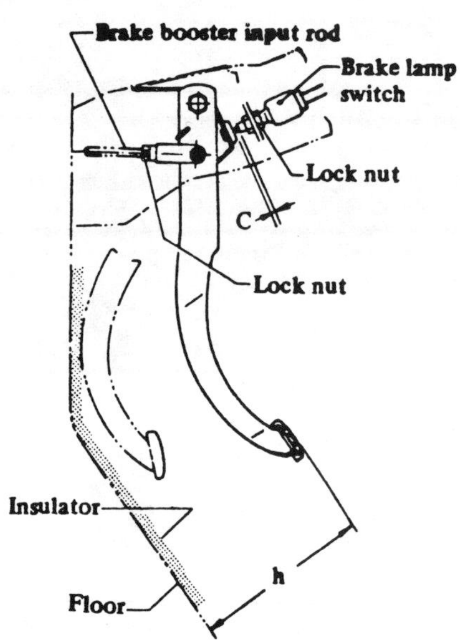

Fig. 9.9 Brake pedal adjustment dimensions for non-Turbo UK models (Sec 17)

h = Height
c = Stoplamp switch clearance } *see Specifications*

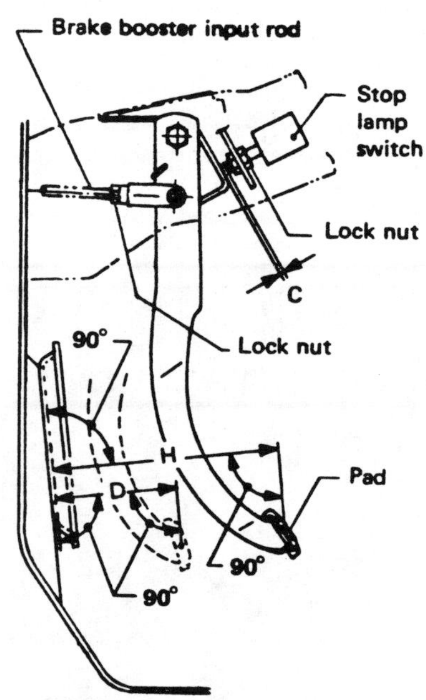

Fig. 9.10 Brake pedal adjustment dimensions for all except non-Turbo UK models (Sec 17)

H = Height (released)
C = Stoplamp switch clearance } *see Specifications*
D = Depressed height

6 Measure the distance from the upper surface of the brake pedal pad to the floor pan or steering hole cover as shown in Figs. 9.9 or 9.10. If the dimension is not as specified, loosen the locknut and turn the pedal pushrod as required making sure that the end of the rod protrudes into the clevis. Tighten the locknut after adjusting the rod.
7 With the pedal fully released check that the specified clearance exists between the pedal and stop-lamp switch. If not, loosen the locknut and adjust the stop-lamp as necessary, then tighten the locknut.
8 Finally run the engine and check that the depressed height of the pedal is as specified. If below the minimum amount check the hydraulic system for leaks or accumulation of air.

18 Fault diagnosis – braking system

Symptom	Reason(s)
Pedal travels a long way before the brakes operate	Incorrect pedal adjustment Brake shoes set too far from the drums (seized adjuster)
Stopping ability poor, even though pedal pressure is firm	Linings, discs or drums badly worn or scored One or more wheel hydraulic cylinders seized, resulting in some brake shoes not pressing against the drums (or pads against disc) Brake linings contaminated with oil Wrong type of linings fitted (too hard) Brake shoes wrongly assembled Servo unit not functioning
Car veers to one side when the brakes are applied	Brake pads or linings on one side are contaminated with oil Hydraulic wheel cylinder on one side partially or fully seized A mixture of lining materials fitted between sides Brake disc not matched Unequal wear between sides caused by partially seized wheel cylinders
Pedal feels spongy when the brakes are applied	Air is present in the hydraulic system
Pedal feels springy when the brakes are applied	Brake linings not bedded in (after fitting new ones) Master cylinder or brake backplate mounting bolts loose Severe wear in brake drums causing distortion when brakes are applied Discs out of true
Pedal travels right down with little or no resistance and brakes are virtually non-operative	Leak in hydraulic system resulting in lack of pressure for operating wheel cylinders If no signs of leakage are apparent the master cylinder internal seals are failing to sustain pressure
Binding, juddering, overheating	One or a combination of reasons given above Shoes installed incorrectly with reference to leading and trailing ends Broken shoe return spring Disc worn Drum distorted Incorrect pedal adjustment
Lack of servo assistance	Vacuum hose disconnect or leaking Non-return valve defective or incorrectly fitted Servo internal defect

Chapter 10 Electrical system

Contents

Specifications

System type	12 volt, negative earth
Battery capacity	
UK models	30, 40, 45 or 60 amp hr
USA models	60 amp hr
Canadian models	65 amp hr
Alternator	
Rating	50 or 60 amp
Output voltage	14.4 to 15.0 volt at 20°C (68°F)
Minimum brush wear limit	7.0 mm (0.28 in)
Starter motor	
Type	Pre-engaged
Minimum brush wear limit	11.0 mm (0.43 in)
Drive pinion-to-stop clearance (manual transmission starter)	0.3 to 2.5 mm (0.012 to 0.098 in)
Difference in idler gear height between solenoid and hand pressure activation (Turbo and automatic transmission starter)	0.3 to 2.5 mm (0.012 to 0.098 in)
Bulbs	**Wattage**
Headlamp:	
UK models	60/55
North American models	65/35
Front direction indicator:	
UK models	23
North American Coupe	23
North American Saloon	27
Front side marker:	
North American Coupe	5
North American Saloon	3.4
Sidelamp:	
UK models	5
North American models	8

Bulbs (continued)	Wattage
Number plate lamp:	
UK models	5
North American models	10
Rear direction indicator	23
Reversing lamp	23
Stop/tail lamp	23/8
Rear side marker:	
Pre 1985 North American models	3.4
1985 on North American models	5
Luggage compartment lamp	5
Interior lamp	10

Fuses (typical North American)*

Circuit number	Circuit protected	Rating (amps)
1	Headlamp (RH)	10
2	Headlamp (LH)	10
3	Stoplamp	10
4	Tail lamp, illumination lamp	10
5	Clock, interior lamp	10
6	Horn, hazard light	10
7	Radio, cassette	10
8	Windscreen wiper	20
9	Heater	20
10	Air conditioner	20
11	Cigarette lighter	10
12	Flasher unit, instrument panel	10
13	Heated rear window	20
14	Engine control	20
15	Fan motor	20
16	Fuel pump	10

* Refer to fusebox cover

Torque wrench settings

	lbf ft	Nm
Alternator bracket bolt	7 to 9	9 to 12
Alternator adjustment bolt	12 to 15	16 to 21
Starter motor tie-bolt	4 to 5	5 to 6
Alternator pulley nut	29 to 43	39 to 59

1 General description

The major components of the 12 volt negative earth system consist of a 12 volt battery, an alternator (driven from the crankshaft pulley), and a starter motor.

The battery supplies a steady amount of current for the ignition, lighting and other electrical circuits and provides a reserve of power when the current consumed by the electrical equipment exceeds that being produced by the alternator.

The alternator has its own regulator which ensures a high output if the battery is in a low state of charge and the demand from the electrical equipment is high, and a low output if the battery is fully charged and there is little demand from the electrical equipment.

When fitting electrical accessories to cars with a negative earth system it is important, if they contain silicon diodes or transistors, that they are connected correctly, otherwise serious damage may result to the components concerned. Items such as radios, tape players, electronic ignition systems, electronic tachometer, automatic dipping etc, should all be checked for correct polarity.

2 Battery – maintenance

Low maintenance type

1 The battery fitted as original equipment on early models is of the low maintenance type. The addition of water is not required with this type of battery.

2 Periodically wipe the top of the battery casing clean and apply petroleum jelly to the battery terminals to prevent build-up of corrosion.

3 The significance of the charge indicator on this type of battery should be understood (photo). If the indicator appears blue in colour, then the battery is fully charged.

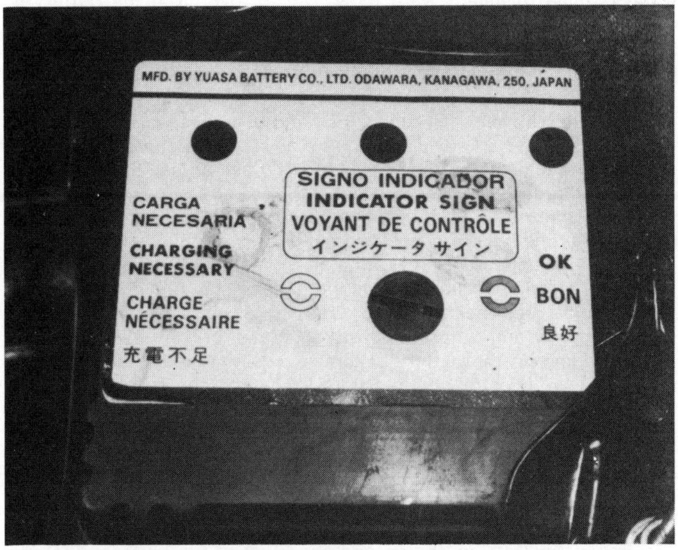

2.3 Top of low-maintenance type battery

4 If the indicator is colourless, then the battery is discharged.

5 Do not re-charge the battery at a rate of more than 10A. If the battery indicator does not change to blue after the charging period, then the battery must be considered to have reached the end of its useful life.

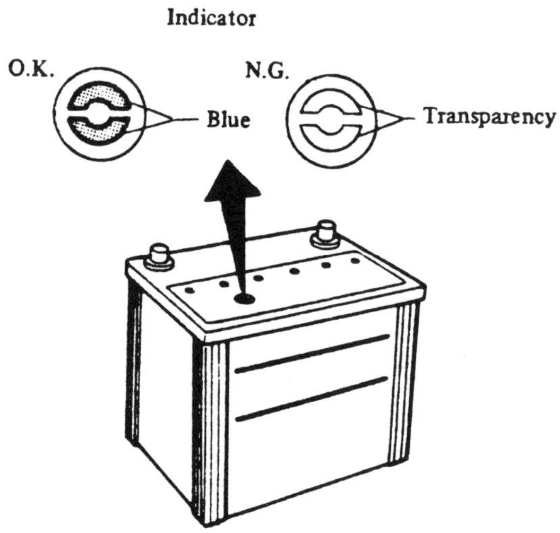

Fig. 10.1 Low maintenance type battery (Sec 2)

Fig. 10.2 Maintenance-free type battery (Sec 2)

Maintenance-free type

6 Later models are fitted with a maintenance-free battery which is similar to the low maintenance type but does not incorporate a charge indicator. However the plug on the top of the battery may be removed in order to check the specific gravity of the electrolyte in the normal way. Note that distilled water must not be added through the plug aperture.

7 Fast charging methods are not recommended, in particular if the specific gravity of the electrolyte is less than 1.100.

Standard type

8 If a replacement battery of conventional type is substituted for the original one, then carry out the following maintenance procedures.

9 Normal weekly battery maintenance consists of checking the electrolyte level of each cell to ensure that the separators are covered by 0.2 in (5 mm) of electrolyte. If the level has fallen, top up the battery using purified (distilled) water only. Do not overfill. If the battery is overfilled or any electrolyte spilled, immediately wipe away the excess as the electrolyte, which is dilute sulphuric acid, attacks and corrodes most metals it comes into contact with very quickly.

10 As well as keeping the terminals clean and covered with a light film of petroleum jelly, the top of the battery, and especially the top of the cells, should be kept clean and dry. This helps prevent corrosion and ensures that the battery does not become partially discharged by leakage through dampness and dirt.

11 Every three months remove the battery and inspect the support tray (photo), the battery clamp and the battery terminals for corrosion. This has the appearance of white fluffy deposits and if it exists it should be cleaned off using warm water to which a little ammonia or baking soda has been added. Treat the battery terminals with petroleum jelly and other metalwork with rust preventative paint.

12 If topping-up the battery becomes excessive and there has been no leakage of electrolyte then it is likely that the battery is being overcharged and it will have to be checked by an auto-electrician. An elderly battery may need more frequent topping-up than a new one because it will take a bigger charge. There is no need to worry about this provided that it gives good service.

13 With the battery on the bench at the three monthly interval check, measure the specific gravity of the electrolyte with a hydrometer to determine the state of charge and condition of the electolyte. There should be very little variation between individual cells and, if a variation in excess of 0.025 exists it will be due to either:

(a) Loss of electrolyte from the battery at some time caused by spillage or a leak, resulting in a drop in the specific gravity of the electrolyte when the deficiency was made up with purified water instead of fresh electrolyte

(b) An internal short circuit caused by buckling of the plates or similar malady pointing to the likelihood of total battery failure in the near future.

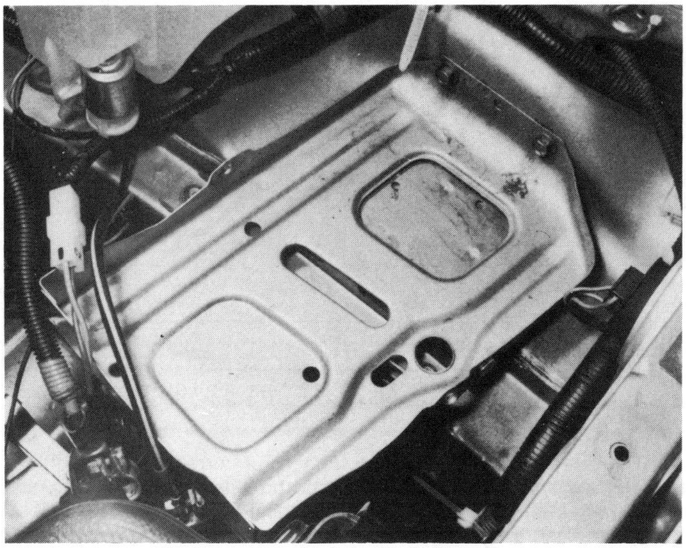

2.11 Battery support tray

14 The specific gravity of the electrolyte for fully charged and fully discharged conditions at different temperatures of the electrolyte is given below:

Fully discharged	Electrolyte temperature	Fully charged
1.098	38°C (100°F)	1.268
1.102	32°C (90°F)	1.272
1.106	27°C (80°F)	1.276
1.110	21°C (70°F)	1.280
1.114	16°C (60°F)	1.284
1.118	10°C (50°F)	1.288
1.122	4°C (40°F)	1.292
1.126	-1.5°C (30°F)	1.296

15 Do not attempt to add acid to a battery. If it is known that electrolyte has been spilled from a cell, leave the mixing of fresh electrolyte and replenishment to your dealer or service station.

16 If the vehicle is to be started from the battery in another vehicle, run one booster cable between the positive terminals of both batteries.

Connect the remaining booster cable between the negative terminal of the rescue vehicle battery and the discharged vehicle's battery negative terminal or bodywork.
17 Disconnect the negative booster cable first.

3 Battery – charging

1 The need for charging a battery from the mains has largely been eliminated with the advent of the alternator.
2 If short daily journeys are made, with much use of the starter and electrical accessories, it is still possible for the battery to become discharged as the alternator is not in use long enough to replace the current being used.
3 A trickle charger can safely be used overnight at a charging rate of 1.5A.
4 Specially rapid 'boost' charges which are claimed to restore the power of the battery in one to two hours should be avoided as they can cause serious damage to the battery plates through overheating.
5 While charging the battery note that the temperature of the electrolyte should never exceed 38°C (100°F) and remember that the gas produced in the cells contains hydrogen which is flammable and explosive, so do not smoke or bring naked lights near the top of the battery.
6 Always disconnect the battery leads before connecting the mains charger to the battery.

4 Battery – removal and refitting

1 Open the bonnet and disconnect the negative and then the positive battery leads.
2 Unscrew the nuts which hold the battery retainer in place (photo). Lift off the crossbar and unhook the clamp rods.
3 Lift out the battery, taking care not to tilt it.
4 Refitting is a reversal of removal, but make sure that it is located on the battery tray correctly with respect to the positive and negative lead connections to the battery.

5 Alternator – description, maintenance and precautions

1 The alternator is mounted on the crankcase at the timing belt end of the engine.
2 The unit is driven by a belt from the crankshaft pulley. A voltage regulator is integral with the brush holder plate.
3 Keep the drivebelt correctly tensioned (see Chapter 2) and the electrical connections tight.
4 Keep the outside of the alternator free from grease and dirt.
5 It is important that the battery leads are always disconnected if the battery is to be charged. Also, if body repairs are to be carried out using electrical welding equipment, the alternator must be disconnected otherwise serious damage can be caused.
6 Do not stop the engine by pulling a lead from the battery.

6 Alternator – removal and refitting

1 Disconnect the battery, and where necessary remove the air cleaner.
2 Disconnect the leads from the rear of the alternator (photo).
3 Release the alternator mounting and adjuster link bolts, push the unit fully in towards the engine and slip the drivebelt off the pulley.

Left-hand drive versions

4 Remove the mounting and adjuster link bolts and lift the alternator from the engine.

Right-hand drive versions

5 Unscrew the nuts which hold the brake master cylinder to the face of the vacuum servo unit.
6 Remove the mounting and adjuster link bolts.
7 Tilt the brake master cylinder just enough to provide clearance for the alternator to be able to pass between the intake manifold and the master cylinder.
8 Remove the alternator from the engine.

Refitting

9 When refitting the alternator, note the position of the spacer washer on the pivot mounting (photo).
10 Tension the drivebelt, as described in Chapter 2.

7 Alternator – overhaul

1 In the event of the charge (ignition) warning lamp not going out after the engine has started or if the battery is being overcharged, indicated by frequent topping-up or by undercharging, the following operations may be carried out to rectify worn brushes or a faulty voltage regulator. If more extensive overhaul is required, or if the alternator has had a long service life, it is recommended that a new or factory-rebuilt unit is obtained.
2 Remove the alternator and clean away the external dirt.
3 Mark the relative position of the rear cover to the front (drive end) cover by scribing a line on them. Unscrew the tie-bolts.
4 Pull off the rear cover with the stator.
5 Unscrew the fixing nuts and separate the rear cover from the stator.
6 If the brush length is less than the specified minimum the brushes must be renewed. To do this, the brush leads must be unsoldered. The new brushes must be attached quickly to prevent the heat leaking away and damaging adjacent components. When locating the new brushes, note that the brushes must extend 11.0 mm (0.43 in) from their holders and the lead must be coiled in the terminal groove.
7 Unless the voltage regulator is faulty ignore the operations described in the next three paragraphs.
8 Remove the regulator securing rivets and unsolder the terminals.
9 Remove the bolts which retain the brush holder/regulator unit.

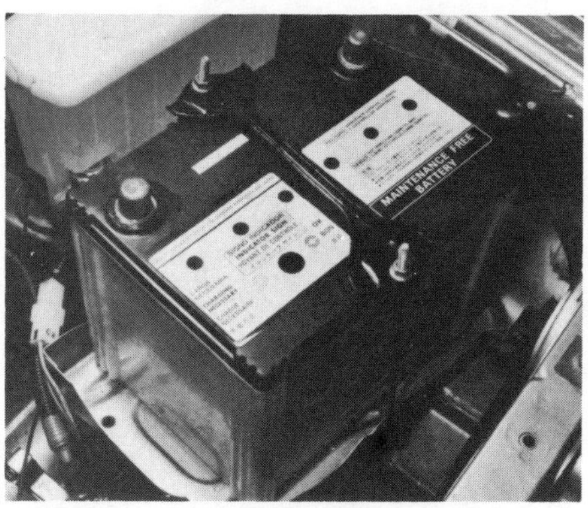

4.2 Showing battery retaining clamp

6.2 Wiring connectors on rear of alternator

6.9 Spacer location on alternator mounting

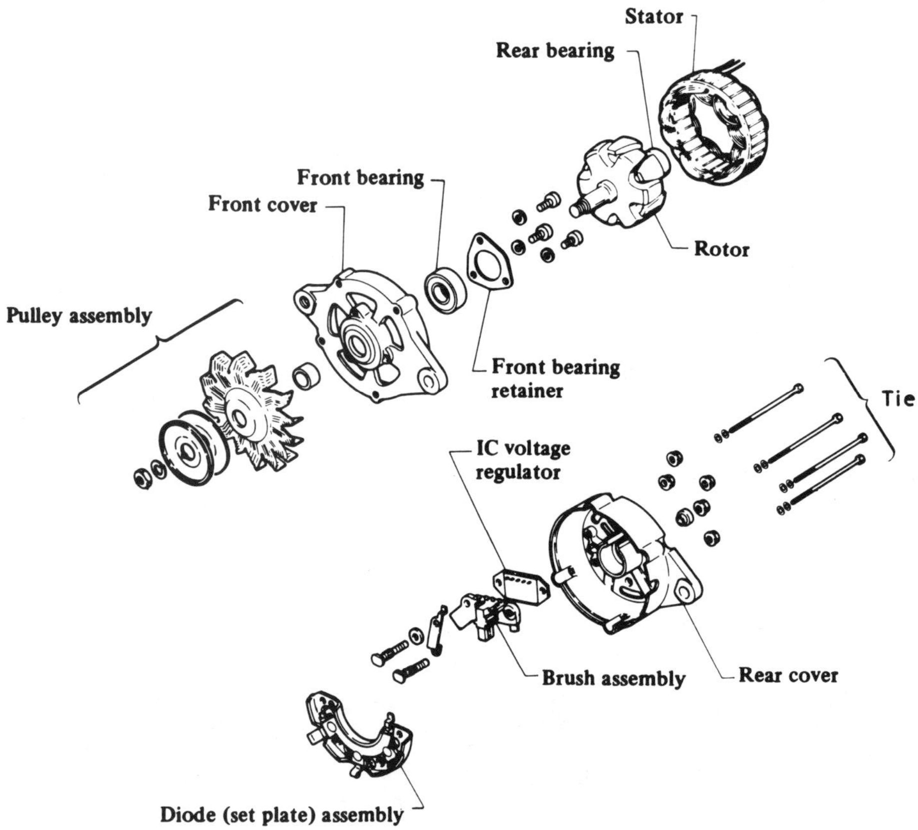

Fig. 10.3 Exploded view of the alternator (Sec 7)

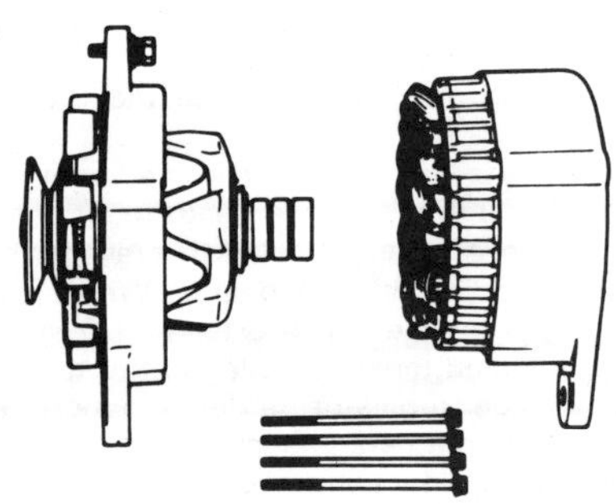

Fig. 10.4 Alternator stator separated from rotor (Sec 7)

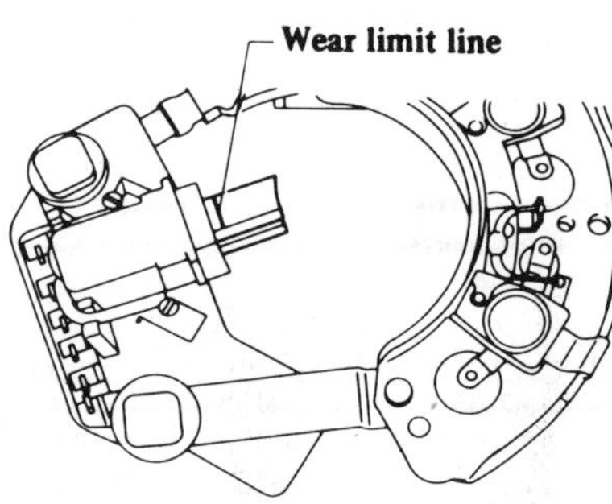

Fig. 10.5 Alternator brush wear limit line (Sec 7)

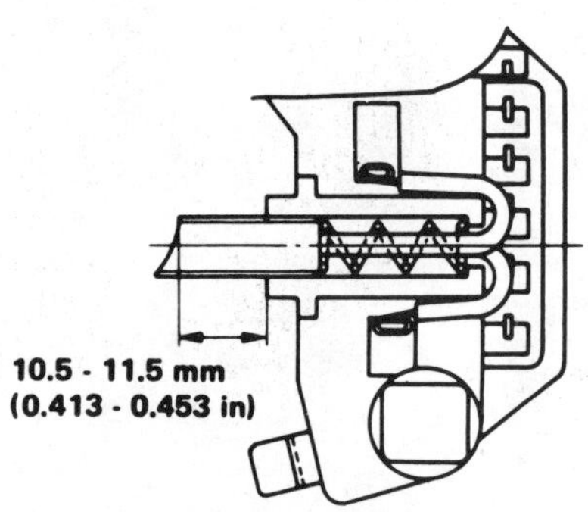

Fig. 10.6 Alternator brush positioning diagram (Sec 7)

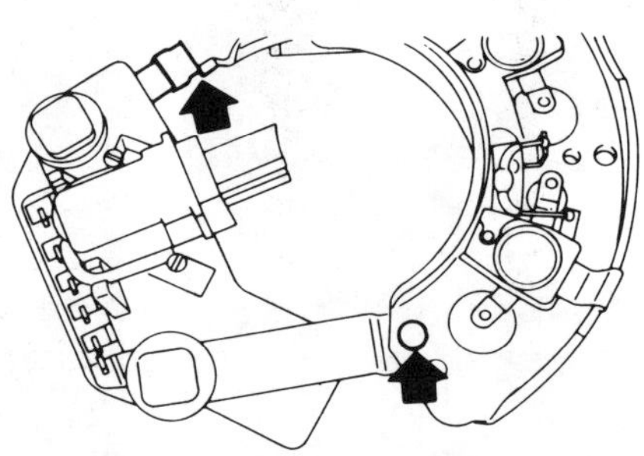

Fig. 10.7 Alternator brush holder/voltage regulator fixing points (Sec 7)

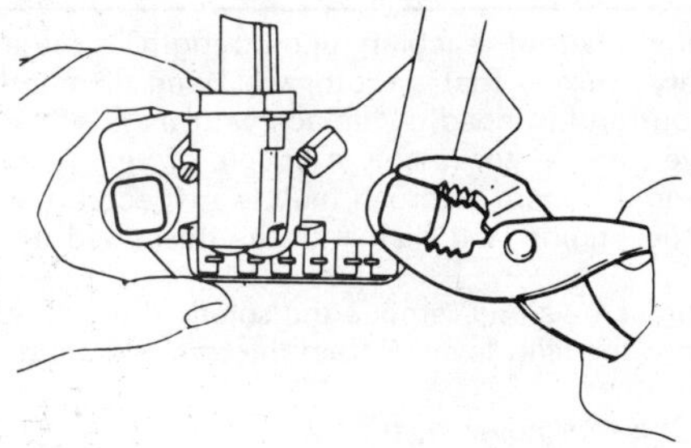

Fig. 10.8 Removing brush holder/voltage regulator fixing bolt (Sec 7)

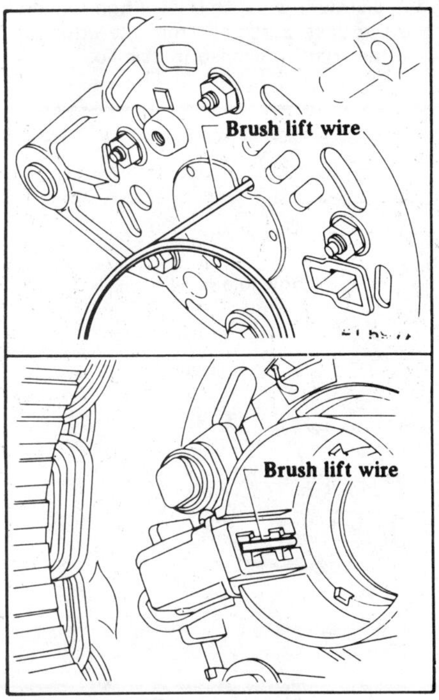

Fig. 10.9 Wire or rod used to retain alternator brushes in the retracted position (Sec 7)

10 Refit the new regulator by reversing the disconnection and removal operations. Peen the rivets securely.
11 Before reassembling the rear cover/brush holder assembly to the motor/front cover, the brushes must be held in the raised position in order to be able to locate them on the slip rings. Do this by passing a thin rod through the hole provided in the alternator rear cover.
12 Align the marks made on the front and rear covers and join the sections of the alternator together.
13 Fit the tie-bolts, remove the temporary brush lift rod.
14 If for any reason the alternator pulley must be removed, the pulley retaining nut can be unscrewed if the rotor shaft is prevented from rotating by inserting a splined socket in the shaft recess or by gripping the rotor in a vice with the alternator dismantled.

8 Starter motor – description

1 The starter motor is of pre-engaged type.
2 When the starter switch is operated, current flows from the battery to the solenoid switch which is mounted on the starter body. The plunger in the solenoid moves inwards, so causing a centrally pivoted lever to push the drive pinion into mesh with the starter ring gear. When the solenoid plunger reaches the end of its travel, it closes an internal contact and full starting current flows to the starter field coils. The armature is then able to rotate the crankshaft, so starting the engine.
3 A special freewheel clutch is fitted to the starter drive pinion so that as soon as the engine fires and starts to operate on its own it does not drive the starter motor.
4 When the starter switch is released, the solenoid is de-energised and a spring moves the plunger back to its rest position. This operates the pivoted lever to withdraw the drive pinion from engagement with the starter ring.
5 On Turbo and automatic transmission models, an idler gear is incorporated at the drive end of the starter motor.

9 Starter motor – testing in situ

1 If the starter motor fails to turn the engine when the switch is operated there are five possible causes:

(a) The battery is faulty
(b) The electrical connections between the switch, solenoid battery and starter motor are somewhere failing to pass the necessary current from the battery through the starter to earth
(c) The solenoid switch is faulty
(d) The starter motor is mechanically or electrically defective
(e) The starter motor pinion and/or flywheel ring gear is badly worn and in need of replacement

2 To check the battery, switch on the headlights. If they dim after a few seconds the battery is in a discharged state. If the lights glow brightly, operate the starter switch and see what happens to the lights. If they dim then you know that power is reaching the starter motor but failing to turn it. If the starter turns slowly when switched on, proceed to the next check.
3 If, when the starter switch is operated, the lights stay bright, then insufficient power is reaching the motor. Remove the battery connections, starter/solenoid power connections and the engine earth strap and thoroughly clean them and refit them. Smear petroleum jelly around the battery connections to prevent corrosion. Corroded connections are the most frequent cause of electric system malfunctions.
4 When the above checks and cleaning tasks have been carried out, but without success, you will possibly have heard a clicking noise each time the starter switch was operated. This was the solenoid switch operating, but it does not necessarily follow that the main contacts were closing properly (if no clicking has been heard from the solenoid, it is certainly defective). The solenoid contact can be checked by putting a voltmeter or bulb across the main cable connection on the starter side of the solenoid and earth. When the switch is operated, there should be a reading or lighted bulb. If there is no reading or lighted bulb, the solenoid unit is faulty and should be renewed.
5 If the starter motor operates but doesn't turn the engine over then it is most probable that the starter pinion and/or flywheel ring gear are badly worn, in which case the starter motor will normally be noisy in operation.
6 Finally, if it is established that the solenoid is not faulty and 12 volts are getting to the starter, then the motor is faulty and should be removed for inspection.

10 Starter motor – removal and refitting

1 Disconnect the battery.
2 Disconnect the leads from the starter motor and solenoid terminals (photo).
3 Unscrew the starter motor fixing bolts and lift the unit from the engine.
4 Refitting is a reversal of removal.

10.2 Wiring connectors on starter motor solenoid

11 Starter motor – overhaul

1 Such is the inherent reliability and strength of the starter motors fitted, it is very unlikely that a motor will need dismantling until it is totally worn out and in need of replacement as a whole.
2 If, however, the motor is only a couple of years old and a pinion carriage, solenoid system or brush fault is suspected then remove the motor from the engine and dismantle as described in the following paragraphs.
3 Extract the screws and remove the solenoid by tilting it to release its plunger from the shift lever. Retain the torsion spring and adjusting plate.
4 Prise off the rear cover dust cap.
5 Prise off the E-ring (Fig. 10.12), and remove the thrust washers.
6 Remove the brush holder screws and the tie-bolts, and withdraw the rear cover.
7 Remove the brush holder. To do this, pull the brush springs upwards and partially withdraw the brushes. If the springs are now released they will apply pressure to the side of the brushes and retain them in the partially withdrawn position.
8 Withdraw the yoke.
9 Withdraw the armature and shift lever from the drive end housing.
10 To remove the drive assembly from the armature shaft, tap the stop ring down the shaft to expose the circlip.

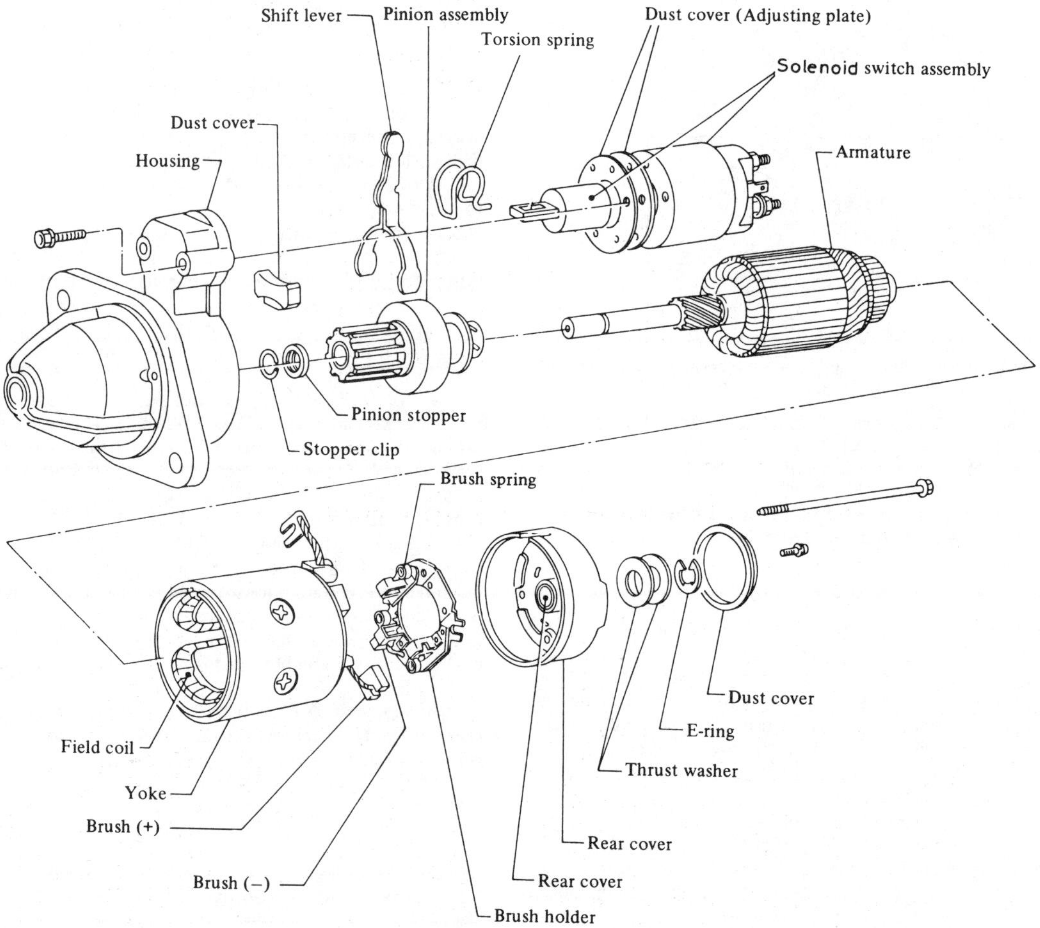

Fig. 10.10 Exploded view of starter motor – manual transmission type (Sec 11)

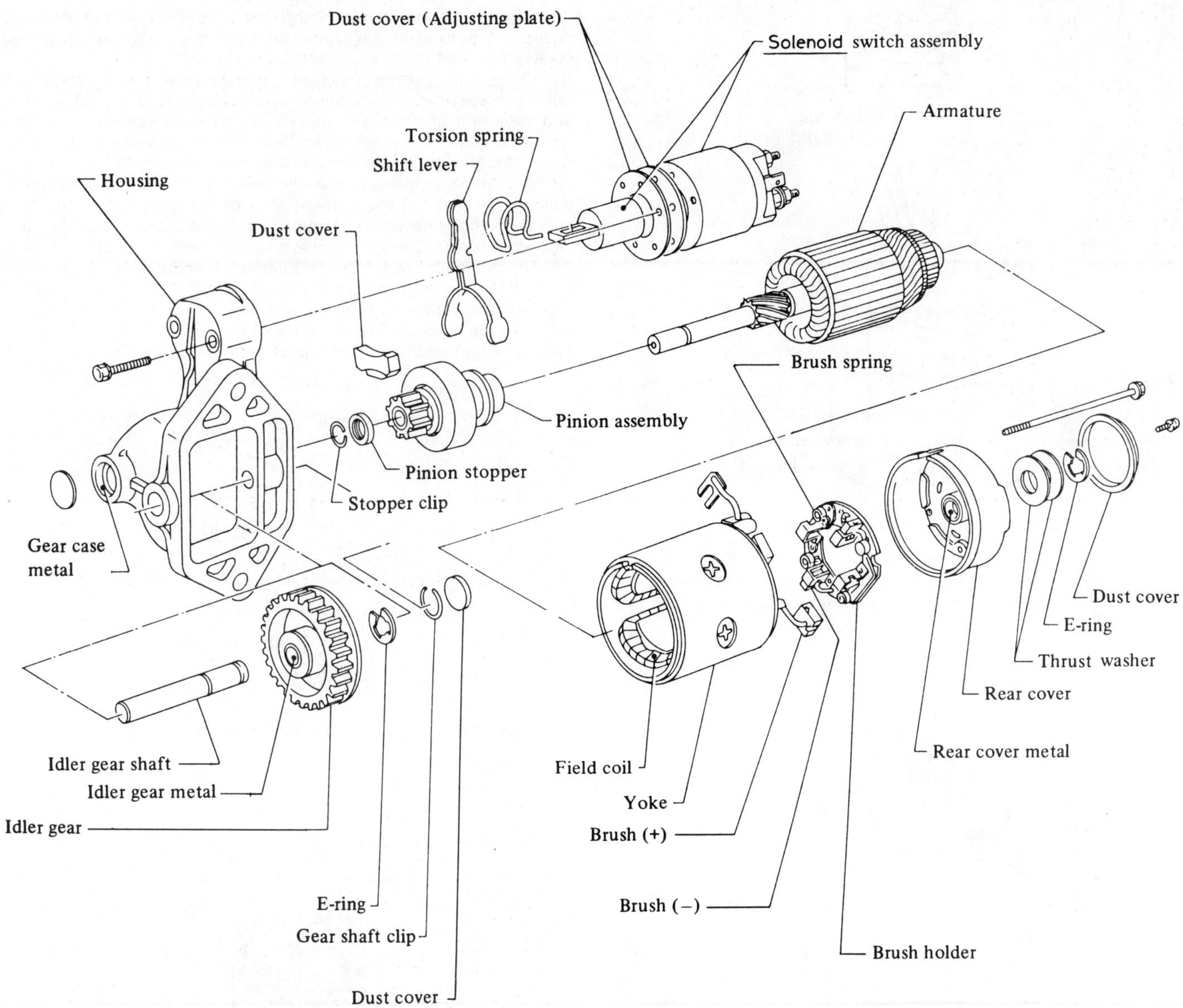

Fig. 10.11 Exploded view of starter motor – Turbo and automatic transmission type (Sec 11)

11 Prise the circlip from its groove, pull the stop ring off the shafts.
12 Remove the drive pinion assembly from the shaft.
13 On Turbo and automatic transmission models remove the idler gear by prising out the E-ring and dust cap and driving out the idler shaft. Note which way round the gear is fitted.
14 With the motor dismantled, inspect all the components for wear.
15 If the commutator appears dirty or burned, clean it with a solvent-soaked rag and, if necessary, burnish it with very fine glass paper.
16 If the segment insulators are flush with the surface of the segments, then the insulators must be undercut, as shown in Fig. 10.13. Use a thin hacksaw blade, or similar, and make sure that the undercut corners are square.
17 If an ohmmeter is available, test the armature for continuity between adjacent segments. The insulation can be tested by placing one probe of the test instrument on the armature shaft and the other on each segment in turn. If continuity is found to exist, the armature must be renewed.
18 Now check for continuity between the field coil positive terminal and the positive brush. If it does not exist, the field coils will have to be renewed.

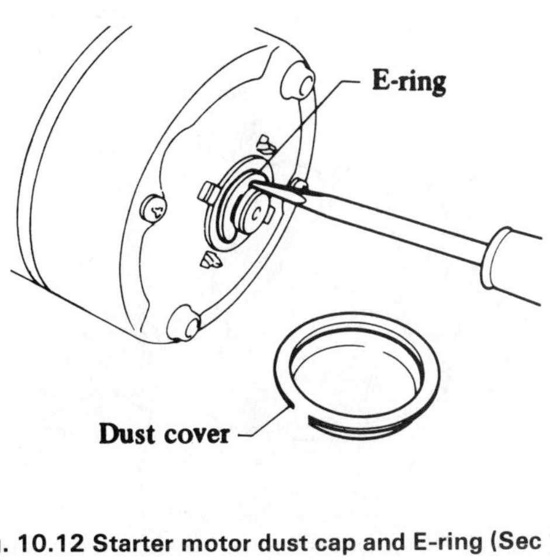

Fig. 10.12 Starter motor dust cap and E-ring (Sec 11)

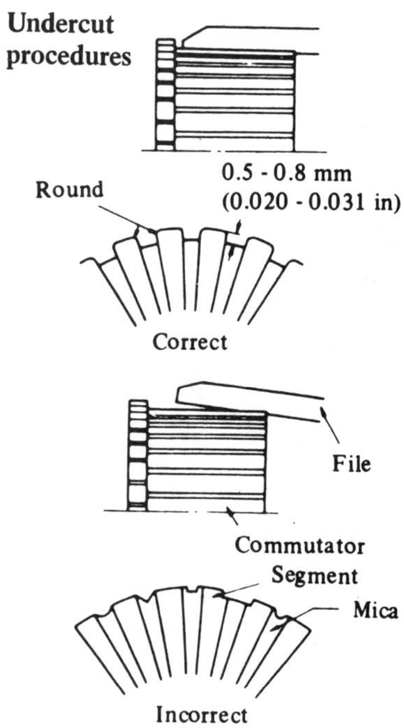

Fig. 10.13 Starter motor commutator insulator undercut (Sec 11)

19 Check the insulation by connecting one probe of the test instrument to the field coil positive terminal and the yoke. If continuity exists, the field coils must be renewed.

20 Renewal of the starter motor field coils is a job best left to your dealer or auto-electrical agent due to the need for a pressure screwdriver and other equipment.

21 Check the brushes for wear. If they have worn down to the minimum specified length, renew them by removing the old brush lead and soldering on the new. Carry out the work quickly to avoid the spread of heat to the field coils, and do not allow the solder to seep down the lead or its flexibility will be impaired.

22 The brush holder can be checked for insulation breakdown by placing one probe of the tester on the positive side of the brush holder and the other one on the negative (baseplate) side. If continuity is indicated, renew the brush holder.

23 The solenoid switch can be checked for continuity by connecting the test instrument between the S terminal and the switch body. If no continuity is indicated, renew the switch.

24 Now place the probes of the tester on the S and M terminals of the switch. If no continuity is indicated, renew the switch.

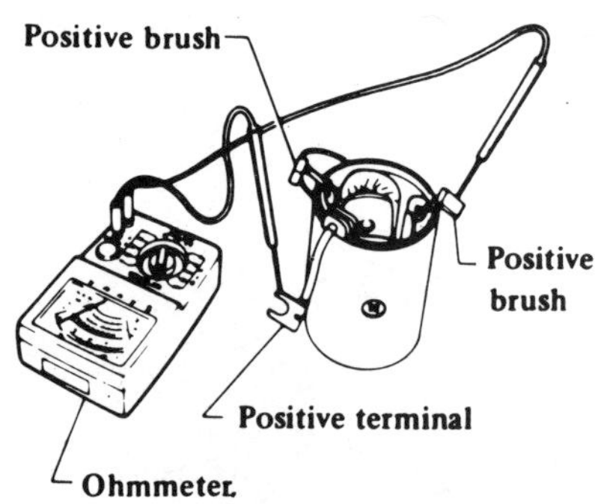

Fig. 10.16 Testing starter field coil for continuity (Sec 11)

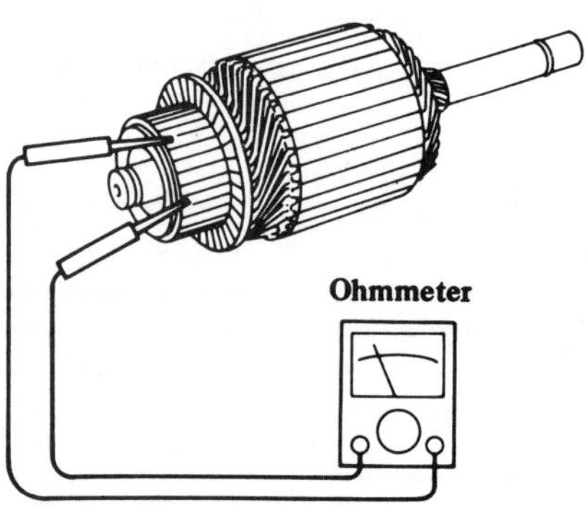

Fig. 10.14 Testing starter armature for continuity (Sec 11)

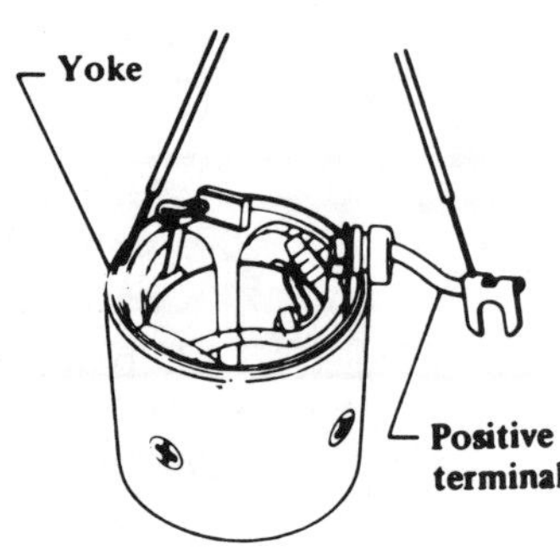

Fig. 10.17 Testing starter field coil insulation (Sec 11)

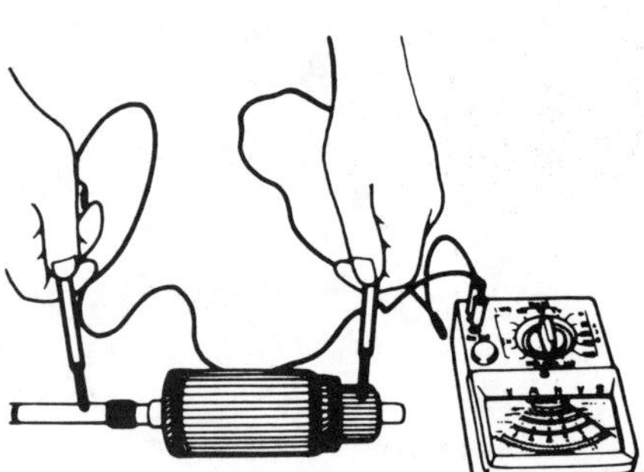

Fig. 10.15 Testing starter armature insulation (Sec 11)

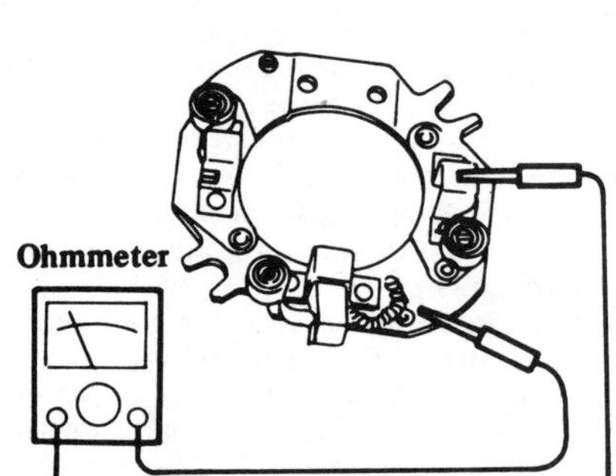

Fig. 10.18 Testing starter brush holder insulation (Sec 11)

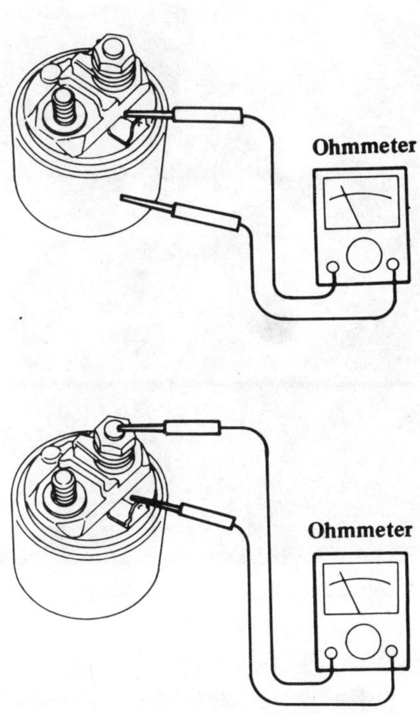

Fig. 10.19 Testing starter solenoid switch for continuity (Sec 11)

25 Finally check the teeth of the drive pinion and idler gear, (where applicable). If they are worn or chipped renew the component. Test the pinion/clutch assembly for correct operation. It should turn smoothly in the drive direction and lock when turned in the reverse direction.
26 Reassemble by reversing the dismantling procedure. Lightly grease the friction surfaces, bushes, bearings and pivots as work proceeds.
27 On Turbo and automatic transmission models, make sure that the idler gear is fitted the right way round with its collar opposite the groove in the pinion.
28 The pinion projection should now be checked by connecting the solenoid to a 12V battery to actuate it.
29 *On manual transmission models,* the clearance between the face of the pinion and the stop plate should be as specified in the Specifications at the beginning of this Chapter. If it is not, change the adjustment plate under the solenoid switch for one of different thickness. The plates are available in thicknesses of 0.5 mm (0.020 in) and 0.8 mm (0.031 in).
30 *On Turbo and automatic transmission models,* with the solenoid actuated, measure the projection of the front face of the idler gear. Now measure again when the idler gear is pulled out by hand. The difference between the two measurements should be within the specified tolerance. If it is not, change the adjusting plate for one of suitable thickness.

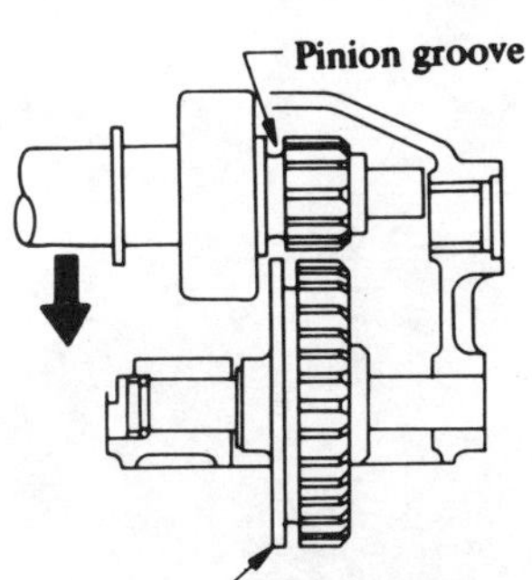

Fig. 10.20 Starter pinion-to-idler gear relationship – Turbo and automatic transmission type (Sec 11)

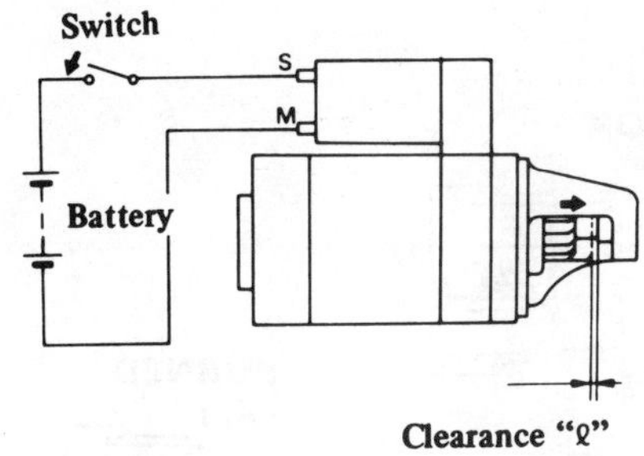

Fig. 10.21 Starter pinion setting diagram – manual transmission type (Sec 11)

Clearance = 0.3 to 2.5 mm (0.012 to 0.098 in)

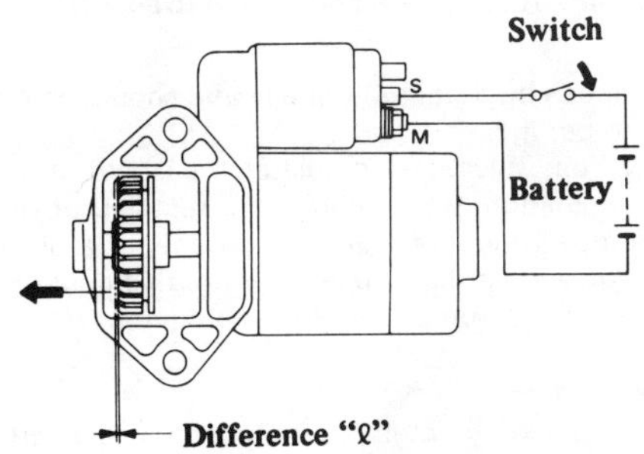

Fig. 10.22 Starter pinion setting diagram – Turbo and automatic transmission type (Sec 11)

Difference = 0.3 to 2.5 mm (0.012 to 0.098 in)

12 Fuses, fusible links and relays – general

Fuses

1 The fusebox (photo) is located inside the vehicle under the right-hand side of the facia panel on RHD models, left-hand side on LHD models.
2 Place the fingers under the fusebox cover and pull it off.
3 The fuses are of 10 to 20A rating according to the circuit which is protected.

12.1 Fusebox

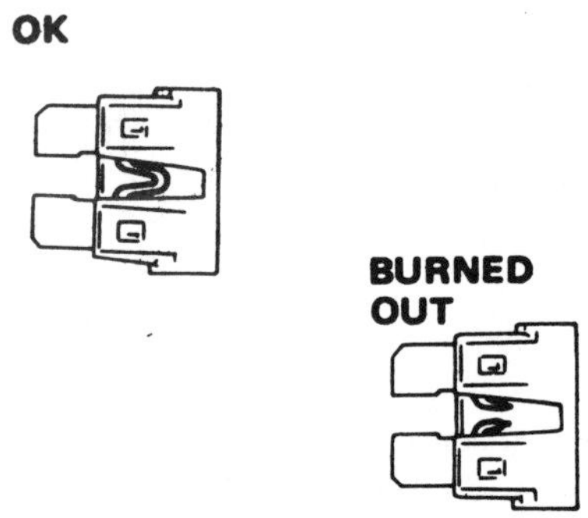

Fig. 10.23 Fuse conditions (Sec 12)

4 A blown fuse can be detected visually and should be renewed with one of similar amperage.
5 If the new fuse blows immediately, suspect a short circuit, probably faulty insulation, which should be rectified at once.
6 Never substitute a fuse of higher amperage, or a piece of wire or foil as a means of preventing a fuse blowing, this could lead to a fire or severely damage the components of the circuit.

Fusible links

7 These are designed to melt in the event of a short in a major current carrying circuit (photo).
8 The links must never be taped up or placed in contact with adjacent wiring, plastic or rubber parts.
9 Before renewing a melted fusible link, rectify the cause or have a thorough check carried out on the vehicle wiring harness.
10 The circuits which the fusible links supply are shown in the wiring diagrams at the end of the manual.

Relays

11 The number and purpose of the relays fitted depends upon the particular model vehicle and its equipment.
12 The flasher unit is located under the facia panel next to the steering column.
13 The ignition and accessory relays are located on top of the fusebox.
14 At the front corners of the engine compartment relays may be located which actuate the following:

Air conditioner
Air conditioner condenser fan
Horn
Headlamp dimmer
Rear foglamp
Automatic choke
Transmission switch

15 On the engine compartment rear bulkhead a relay may be located which actuates the following:

Windscreen wiper (intermittent) (photo)
Boost sensor relay (Turbo models)

13 Steering column combination switch – removal and refitting

1 Remove the steering wheel, as described in Chapter 11. Remove the column shrouds (photo).

Pre 1985 models

2 Slacken the screw at the bottom of the left-hand side of the switch.
3 Withdraw the switch from the column (photo) and disconnect the wiring plugs as it moves up the column (photo).

12.7 Fusible link connection at battery terminal

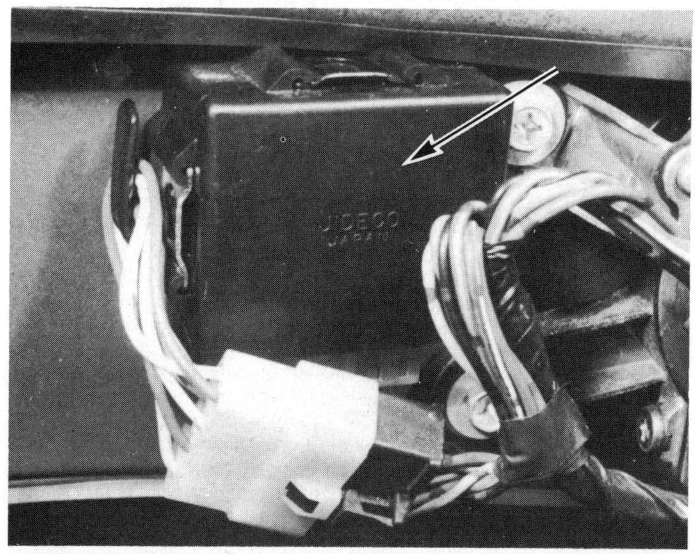

12.15 Windscreen wiper intermittent control relay

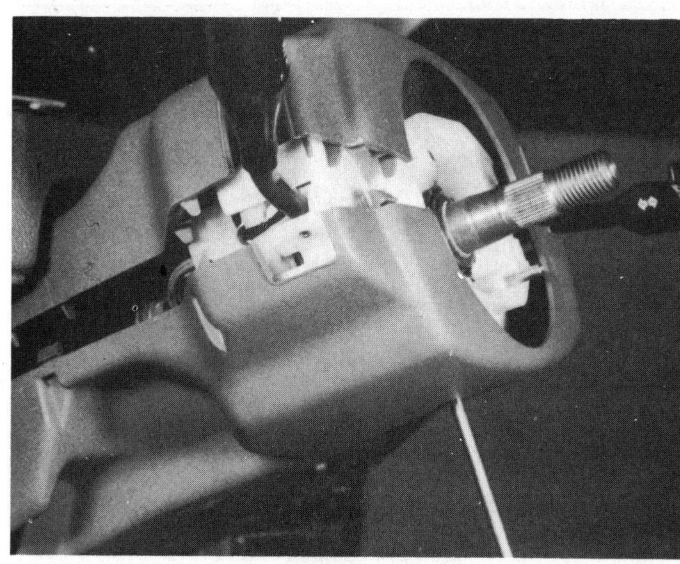

13.1 Removing the steering column shrouds

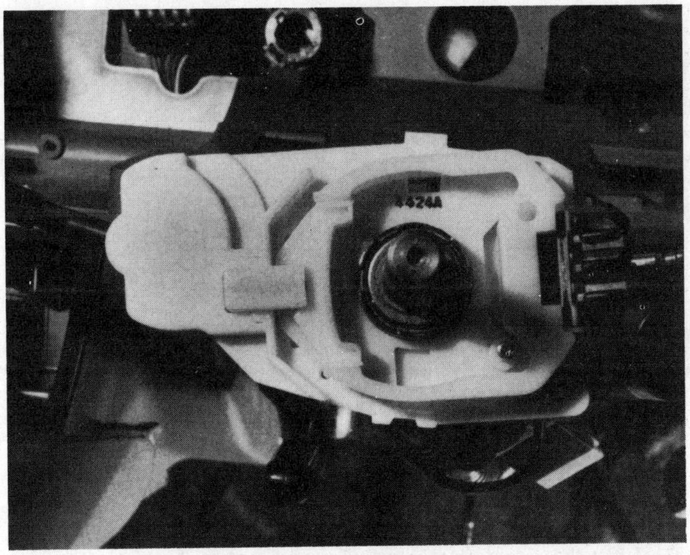

13.3A Steering column combination switch

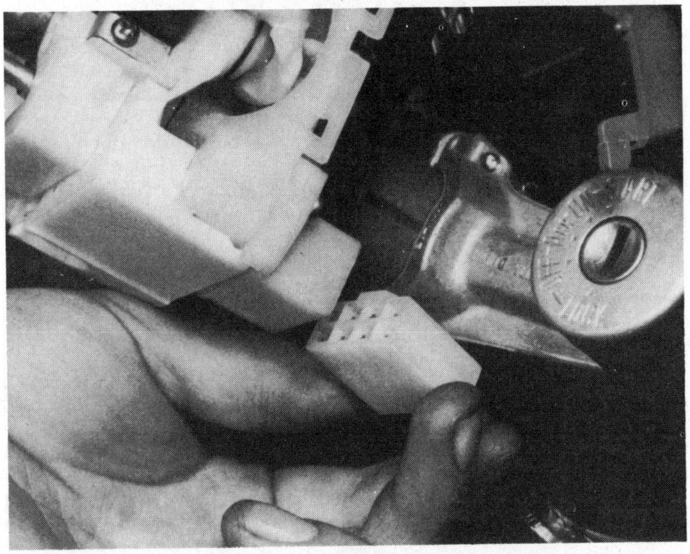

13.3B Disconnecting the combination switch wiring plug

1985 on models

4 The lighting switch and wiper/washer switches are separately attached to the switch base. To remove a switch, disconnect the wiring then remove the retaining screws.

5 To remove the switch base, loosen the clip, then push and turn the switch clockwise.

All models

6 Refitting is a reversal of removal.

14 Courtesy lamp switch – removal and refitting

1 The courtesy lamp switch is located on the front door pillar (photo).

2 To remove the switch, extract the screw and pull the switch with leads from its hole.

3 If the leads are to be disconnected, tape them to the body panel to prevent them from slipping inside the body cavity.

4 Smear the switch contacts and plunger with petroleum jelly before fitting as an aid to preventing corrosion.

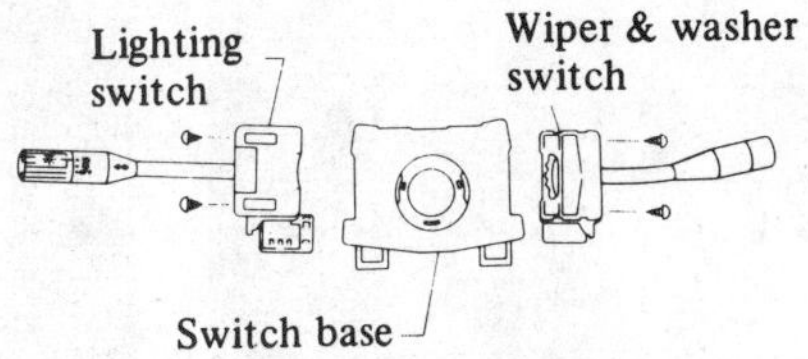

Fig. 10.24 Steering column combination switch components on 1985 on models (Sec 13)

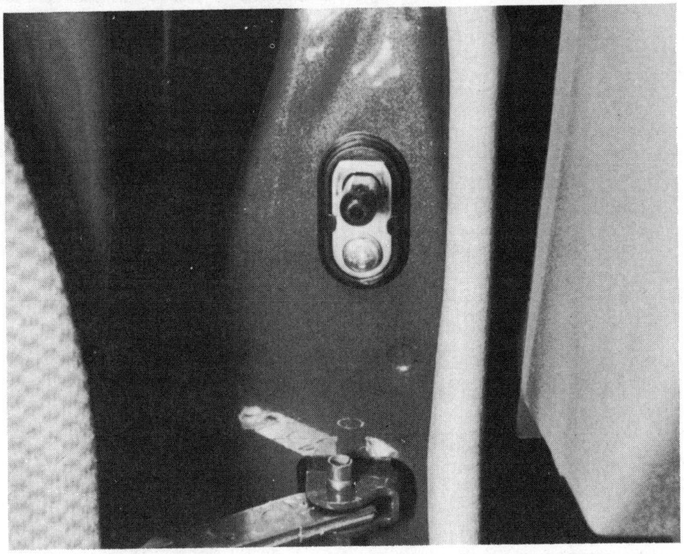

14.1 Courtesy lamp switch

15 Facia panel mounted switches – removal and refitting

1 These switches are held in position by plastic tabs.

2 Reach up behind the switch and compress the tabs. Withdraw the switch far enough to be able to disconnect the wiring plug and remove the switch (photo).

3 Refitting is a reversal of removal.

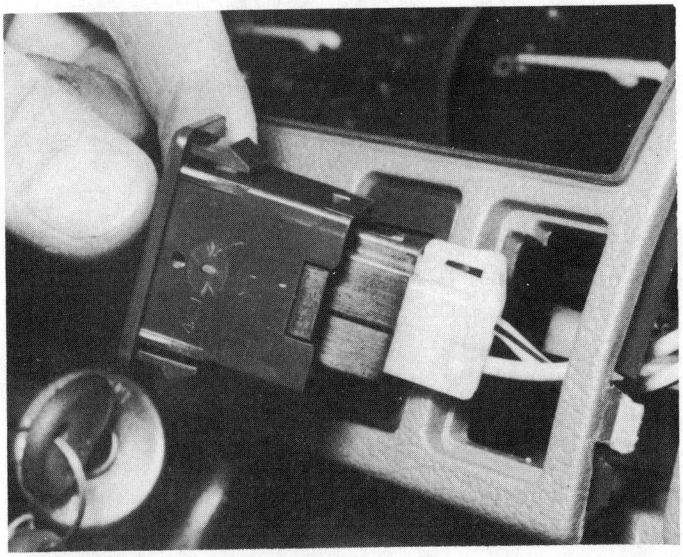

15.2 Removing a facia panel switch

16.1 Pull off the connector ...

16.2 ... remove the dust excluder ...

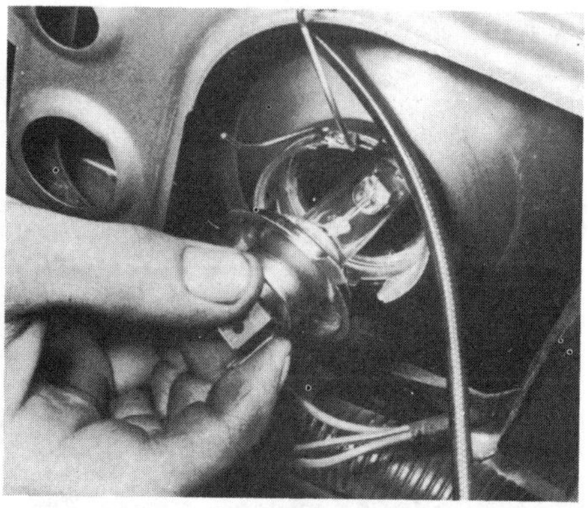

16.3 ... and withdraw the headlamp bulb

16 Headlamp bulb or sealed beam unit – removal and refitting

Bulb type (except Coupe)

1 Open the bonnet, pull the wiring connector from the rear of the headlamp (photo).
2 Peel off the rubber dust excluder (photo).
3 Prise back the spring clips and withdraw the bulb (photo).
4 With halogen type bulbs, avoid touching the glass of the bulb with the fingers, as any residual grease will shorten their life.
5 Refit by reversing the removal operations.

Bulb type (Coupe)

6 Switch on the headlights then, after the headlamp units have opened fully, disconnect the battery negative lead.
7 Remove the side screws and clip, and withdraw the cover.
8 Remove the screws and withdraw the retaining ring, then lift out the headlamp.
9 Remove the rubber dust excluder and pull off the wiring connector.
10 Twist the retainer anti-clockwise and extract the bulb. Note the precaution given in paragraph 4.
11 Refit by reversing the removal operations.

Sealed beam unit type

12 With the bonnet open pull the wiring connector from the rear of the headlamp.
13 Remove the front radiator grille as described in Chapter 12.
14 Release the clip and pull off the side grille.
15 Remove the screws and withdraw the retaining ring, then lift out the unit.
16 Refit by reversing the removal operations. Provided that the adjuster screws are not touched, the beam alignment will not have been altered.

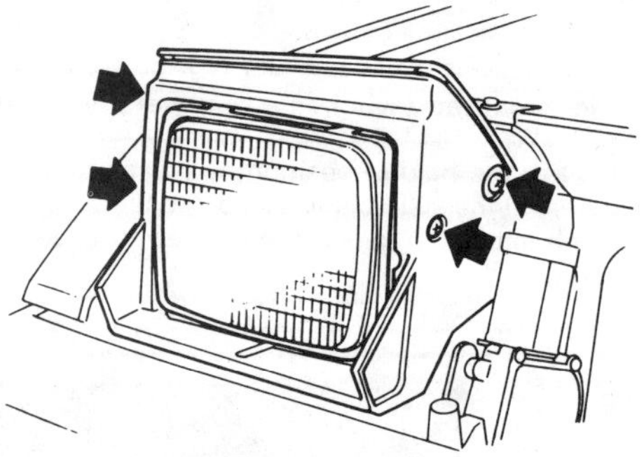

Fig. 10.25 Headlamp side screw locations on Coupe models (Sec 16)

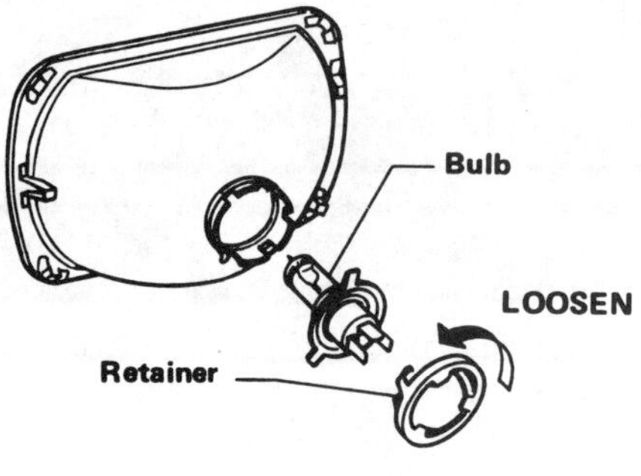

Fig.10.27 Removing the headlamp bulb on Coupe models (Sec 16)

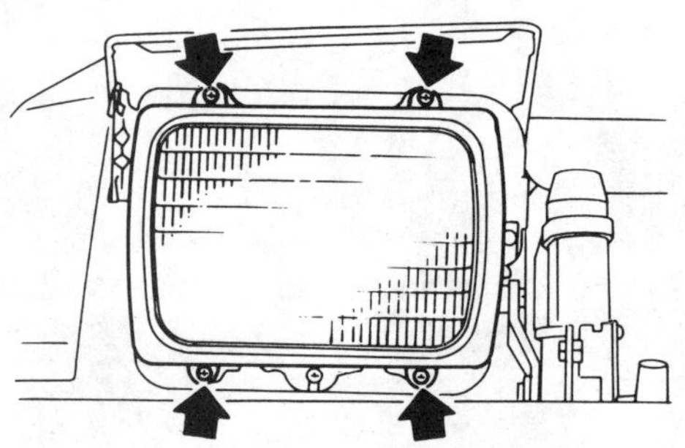

Fig. 10.26 Headlamp retaining ring screw locations on Coupe models (Sec 16)

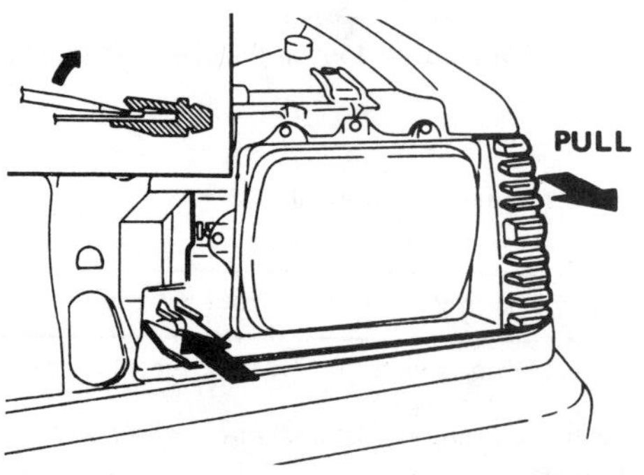

Fig. 10.28 Removing headlamp side grille on North American saloon models (Sec 16)

17 Headlamp beam alignment

1 It is recommended that this work is left to your dealer or a service station with the necessary optical beam setting equipment.
2 In an emergency the beams can be set in the following way.
3 Position the vehicle square to a wall or screen during the hours of darkness. The specified distance for vehicles with sealed beam headlamps is 7.6 m (25.0 ft) or for bulb type headlamps 5.0 m (16.0 ft).
4 Measure the height and separation of the centres of the headlamps and then transpose the measurements onto the wall.
5 Switch the headlamps to dipped beam and adjust the beam so that the upper edge of the brightest spot is level with the lamp centre mark on the wall. On RHD models adjust the beam horizontally so that the right-hand edge of the spot is on the mark. Use the left-hand edge for LHD models.
6 Adjustment screws are provided as shown in Figs. 10.29, 10.30 and 10.31.

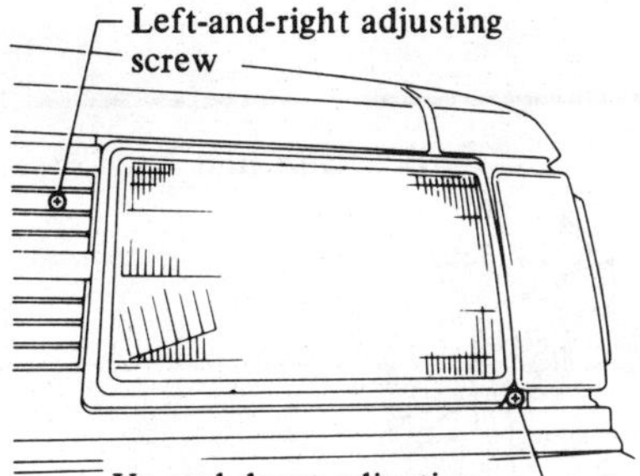

Fig. 10.29 Headlamp beam adjustment screw locations on UK saloon models (Sec 17)

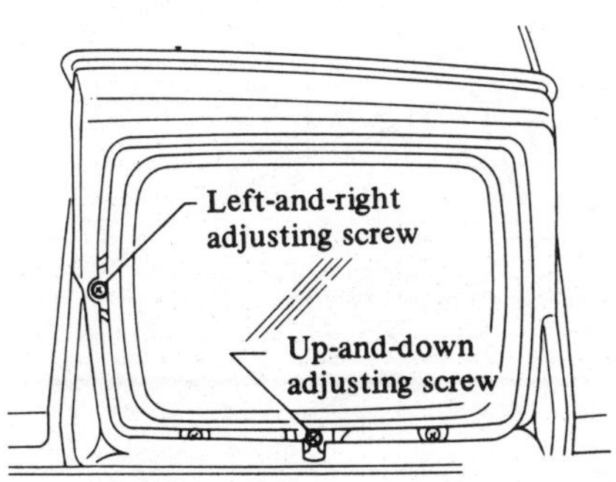

Fig. 10.30 Headlamp beam adjustment screw locations on Coupe models (Sec 17)

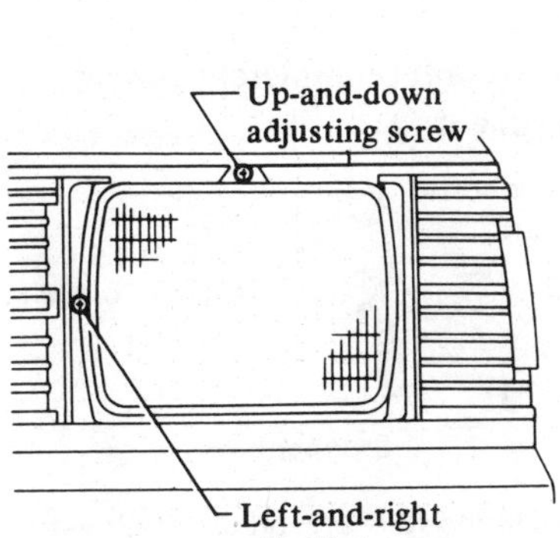

Fig. 10.31 Sealed beam unit adjustment screw locations (Sec 17)

18 Headlamp unit – removal and refitting

UK saloon models

1 Remove the headlamp bulb (Section 16).
2 Remove the radiator grille as described in Chapter 12.
3 Remove the front direction indicator lamp (Section 19).
4 Unscrew the upper mounting nut (photo).
5 Swivel the headlamp and disconnect the anchor brackets from the adjuster screws (photos).
6 Disconnect the tension spring and withdraw the headlamp (photo).
7 Refitting is a reversal of removal, but check the beam alignment (Section 17).

Coupe models

8 Remove the semi-sealed headlamp unit as described in Section 16. Note that the headlamps may be opened manually in an emergency by disconnecting the battery, removing the motor cap, and turning the knob on the end of the shaft.
9 Disconnect the motor wiring connector.
10 Remove the grille panel.

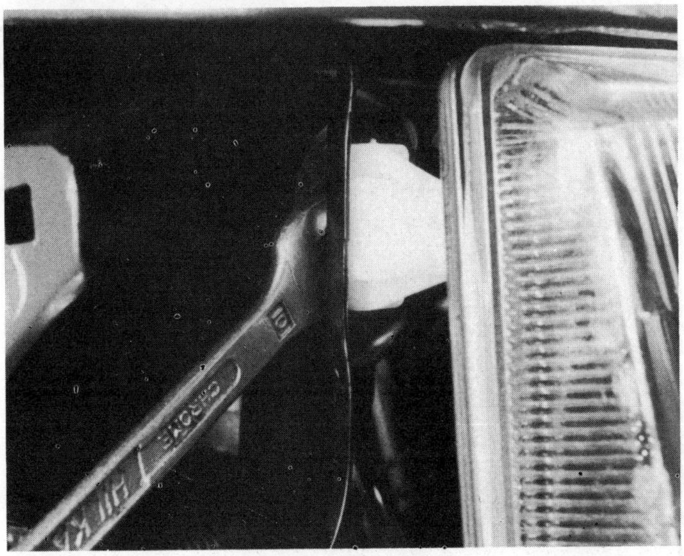

18.4 Unscrewing headlamp unit upper mounting nut (UK Saloon models)

18.5A Headlamp unit lower anchor bracket (UK Saloon models)

18.5B Headlamp unit upper anchor bracket (UK Saloon models)

18.6 Headlamp unit tension spring (UK Saloon models)

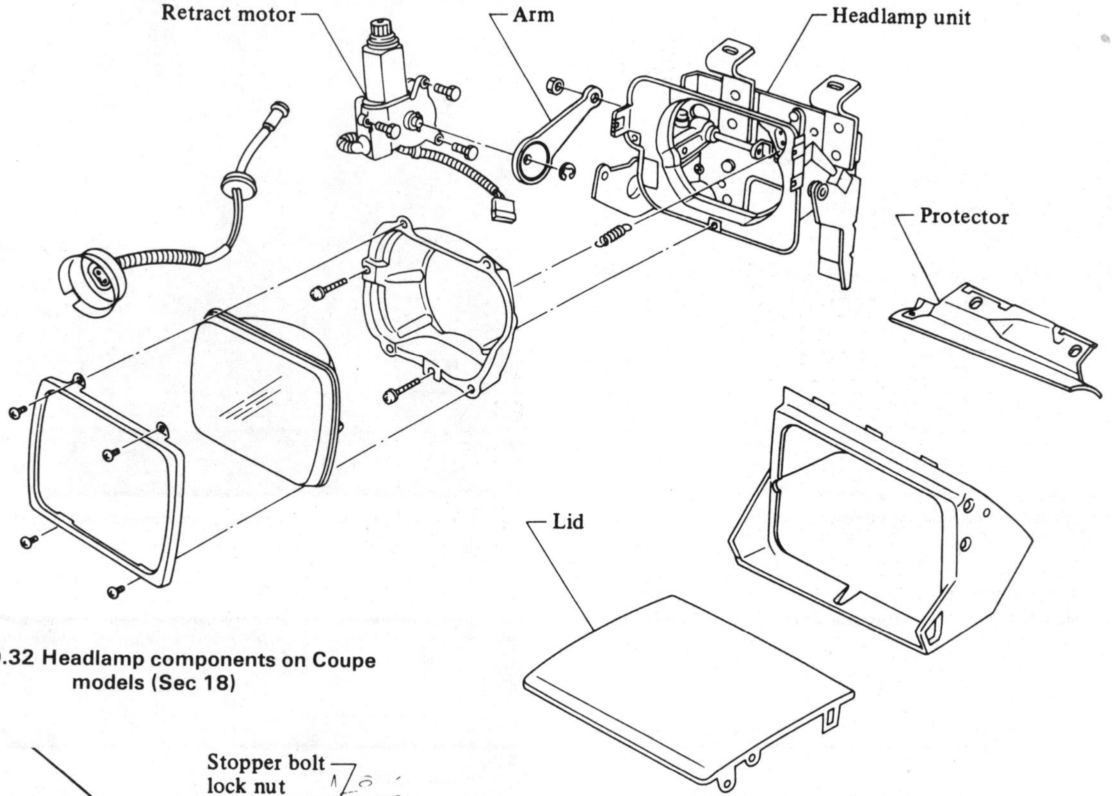

Fig. 10.32 Headlamp components on Coupe models (Sec 18)

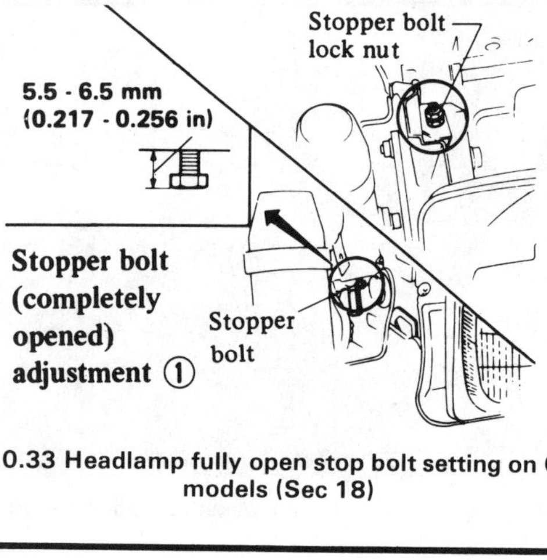

Fig. 10.33 Headlamp fully open stop bolt setting on Coupe models (Sec 18)

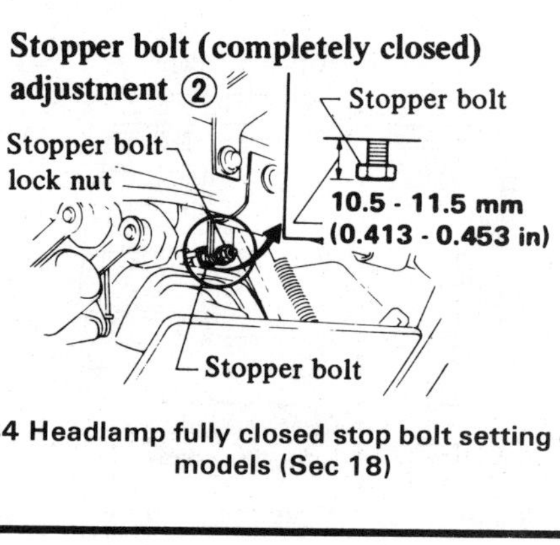

Fig. 10.34 Headlamp fully closed stop bolt setting on Coupe models (Sec 18)

11 Unbolt and remove the headlamp assembly. If necessary separate the motor from the headlamp base.
12 Refitting is a reversal of removal, but align the motor with the fully open mark on the body, and set the open and closed stop bolts to the dimensions shown in Figs. 10.33 and 10.34. Grease all bearing surfaces. Adjust the base and lid so that the headlamp is central within the aperture and flush with the surrounding body panels.

North American saloon models

13 Remove the sealed beam unit as described in Section 16.
14 Unbolt and remove the base.
15 Refitting is a reversal of removal, but check the beam alignment (Section 17).

19 Lamp bulbs – renewal

Front direction indicator lamp (UK saloon models)

1 Open the bonnet, remove the top and bottom mounting screws, and withdraw the lamp sufficient to disconnect the wiring (photo).
2 Remove the four screws noting the location of the top bracket, and separate the lens (photos).

19.2B ... and separate the lens (UK Saloon models)

19.1 Removing the front direction indicator lamp (UK Saloon models)

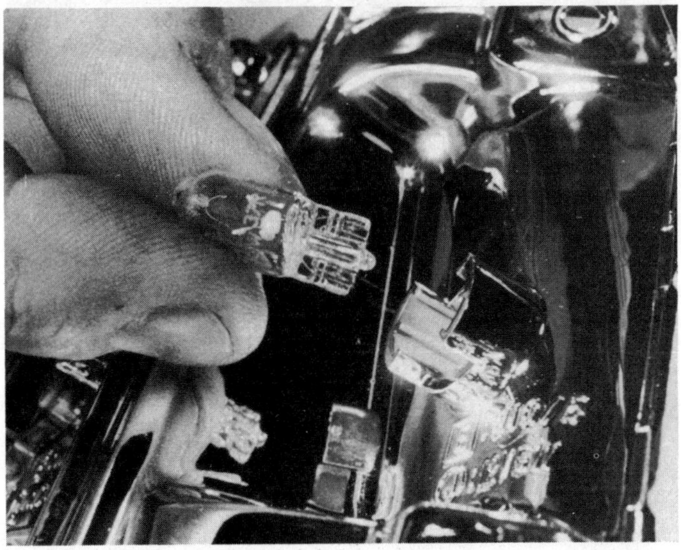

19.3 Removing the front direction indicator bulb (UK Saloon models)

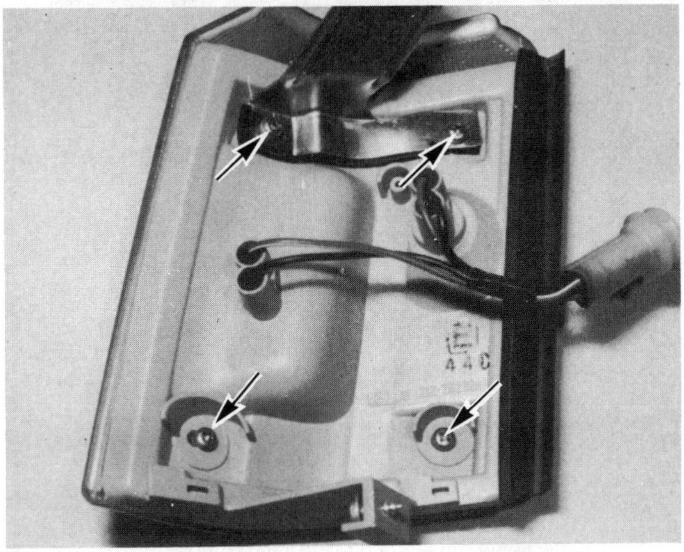

19.2A Remove the screws ...

3 Pull out the wedge-type bulb (photo).
4 Refitting is a reversal of removal.

Front combination lamp (Coupe models)

5 Remove the screws and withdraw the lens.
6 Depress and twist the bulb to remove it.
7 Refitting is a reversal of removal.

Front direction indicator lamp (North American Saloon models)

8 Remove the screws and withdraw the lens.
9 Pull out the wedge-type bulb.
10 Refitting is a reversal of removal.

Front side marker lamp (Coupe models)

11 Remove the plastic under wing shield.
12 Withdraw the bulbholder then pull out the wedge-type bulb.
13 Refitting is a reversal of removal.

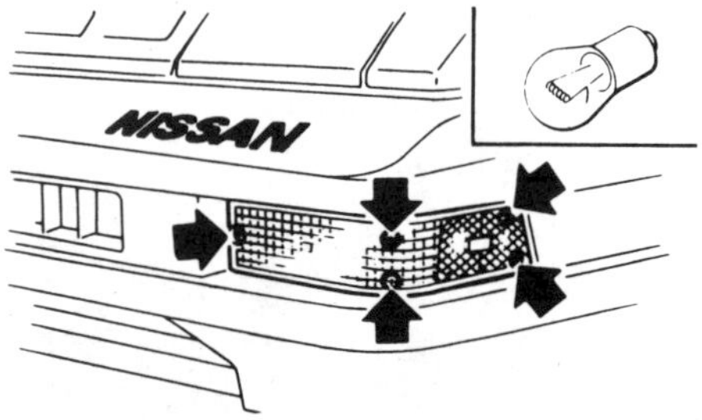

Fig. 10.35 Front combination lamp lens retaining screws on Coupe models (Sec 19)

Front sidelamp (Saloon models)
14 Remove the screws and withdraw the lens (photo).
15 Depress and twist the bulb to remove it.
16 Refitting is a reversal of removal.

Rear lamp cluster (Saloon models)
17 Pull off the cover located in the rear compartment.
18 Twist and remove the bulbholder (photo).

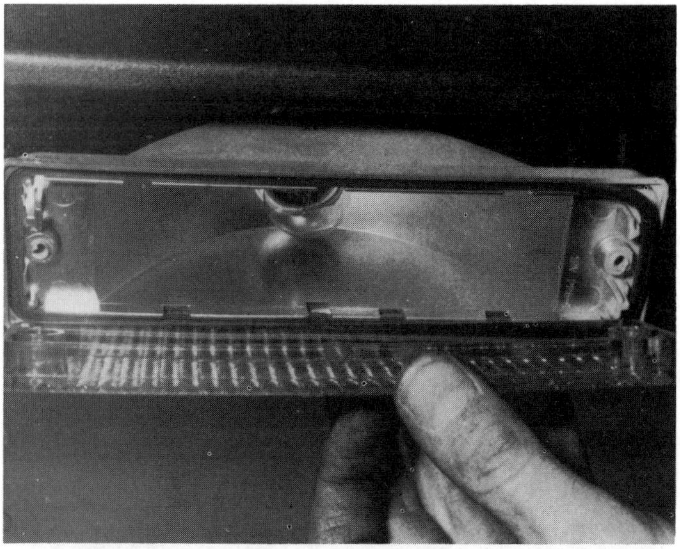

19.14 Removing the front sidelamp lens (Saloon models)

19 Depress and twist the bulb to remove it.
20 Refitting is a reversal of removal.

Rear lamp cluster (Coupe models)
21 Twist and remove the bulbholder located in the rear compartment after removing the cover (if fitted).
22 Depress and twist the bulb to remove it.
23 Refitting is a reversal of removal.

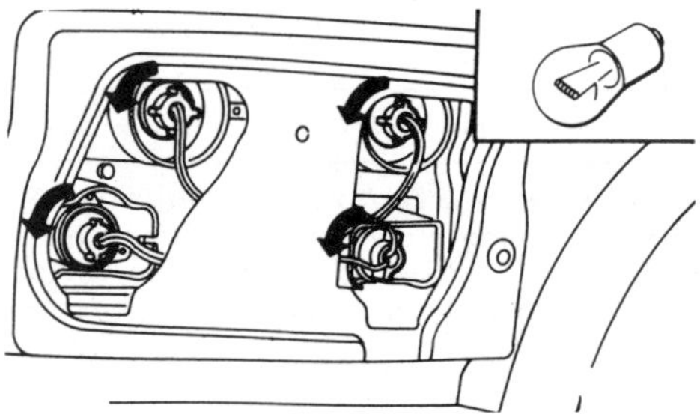

Fig. 10.36 Rear lamp bulb removal on Coupe models (Sec 19)

Rear side marker lamp
24 Remove the screws and withdraw the lens.
25 Pull out the wedge-type bulb.
26 Refitting is a reversal of removal.

Luggage compartment lamp
27 Twist the lens anti-clockwise and remove it.
28 Pull out the wedge type bulb (photo).
29 Refitting is a reversal of removal.

Number plate lamp (UK models)
30 Prise the lamp from the rear bumper.
31 Remove the screws and take off the lens (photo).
32 Depress and twist the bulb to remove it.
33 Refitting is a reversal of removal.

Number plate lamp (North American models)
34 Twist and remove the bulbholder located in the rear compartment.
35 Depress and twist the bulb to remove it.
36 Refitting is a reversal of removal.

19.18 Removing a rear lamp cluster bulb (Saloon models)

19.28 Luggage compartment lamp with lens removed

19.31 Number plate lamp and bulb (UK models)

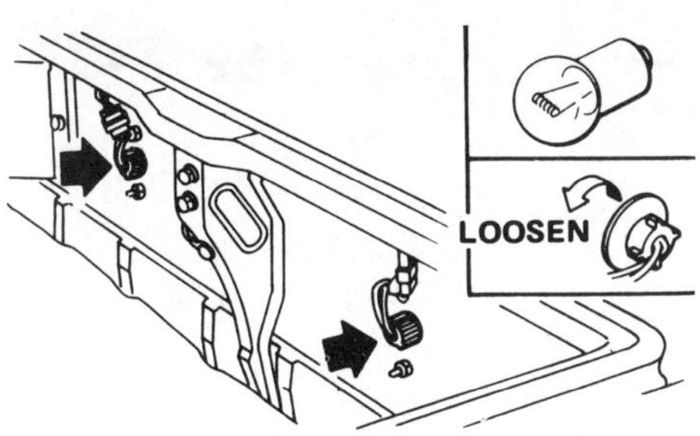

Fig. 10.37 Number plate lamp bulb removal on North American models (Sec 19)

Interior lamp

37 Twist the lamp lens anti-clockwise and remove it (photo).
38 Prise the festoon type bulb from the spring terminals.
39 Refitting is a reversal of removal, but if necessary tension the terminals to hold the bulb firmly.

19.37 Removing the interior lamp lens

Instrument panel warning lamps

40 Remove the instrument panel as described in Section 20.
41 Twist and remove the bulbholder from the rear of the panel (photos).
42 Pull out the wedge type bulb.
43 Refitting is a reversal of removal.

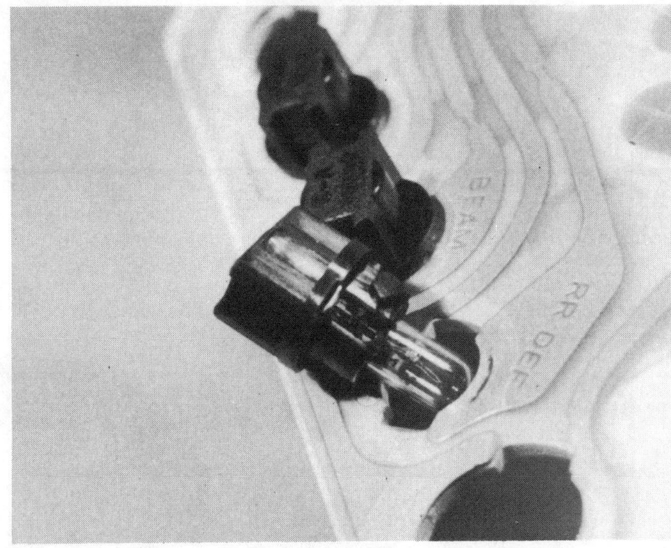

19.41A Instrument panel warning lamp – Type A

19.41B Instrument panel warning lamp – Type B

20 Instrument panel – removal and refitting

1 Remove the steering wheel and steering column shrouds with reference to Chapter 11.
2 Remove the bottom mounting screws and withdraw the surround (photo). Note the tabs and slots to attach the surround to the facia. Disconnect the switches and withdraw the surround.
3 Remove the mounting screws and withdraw the instrument panel sufficient to disconnect the wiring plugs and speedometer cable (photos).
4 Withdraw the instrument panel (photos).
5 Refitting is a reversal of removal.

20.2 Removing the instrument panel surround bottom mounting screws

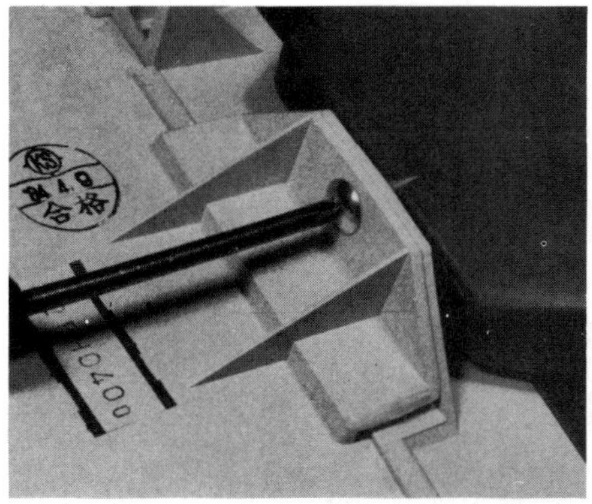

20.3A Removing the instrument panel top mounting screw

20.3B Removing the instrument panel bottom mounting screws

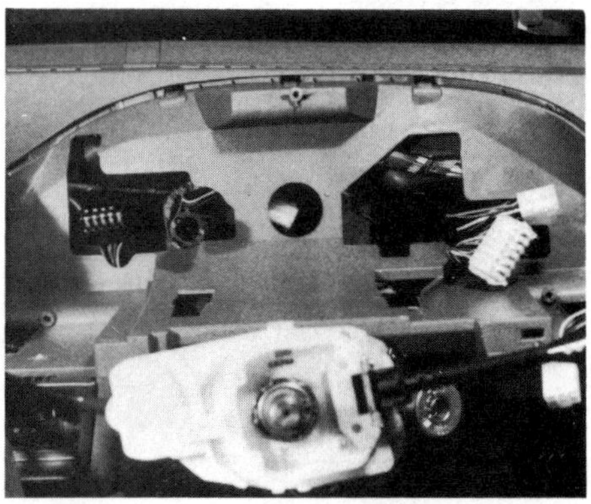

20.3C Wiring connectors and speedometer cable end with instrument panel removed

20.4A Front view of instrument panel

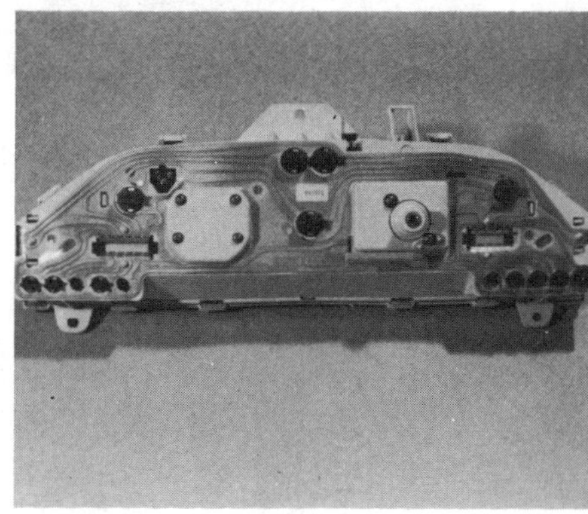

20.4B Rear view of instrument panel

21 Windscreen wiper blades and arms – removal and refitting

1 The wiper blades should be renewed as soon as they cease to wipe the glass cleanly.
2 The complete blade assembly or just the rubber insert are available as replacements.
3 Pull the wiper arm from the glass until it locks.
4 Depress the small tab and slide the blade off the arm (photo). On some North American models the blade is secured to the arm by screws.
5 Refitting is a reversal of removal.
6 Before removing a wiper arm, it is worthwhile sticking a strip of masking tape on the glass against the edge of the wiper blade as a guide to wiper arm setting when refitting.
7 Lift up the cap to expose the nut which holds the wiper arm to the driving spindle (photo).

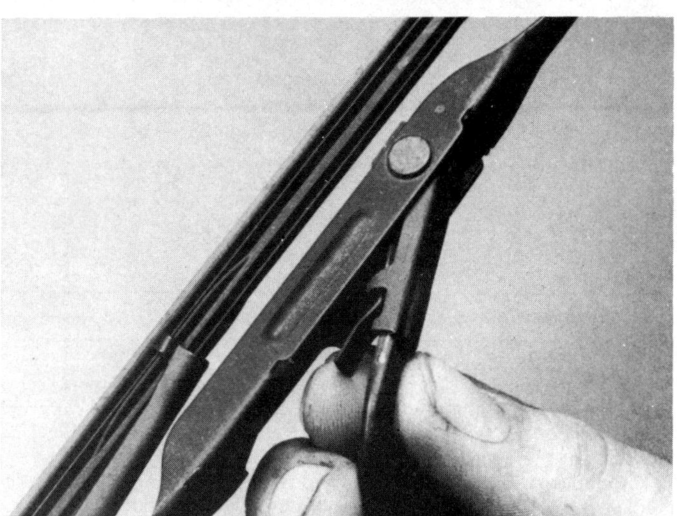

21.4 Disconnecting wiper blade from arm

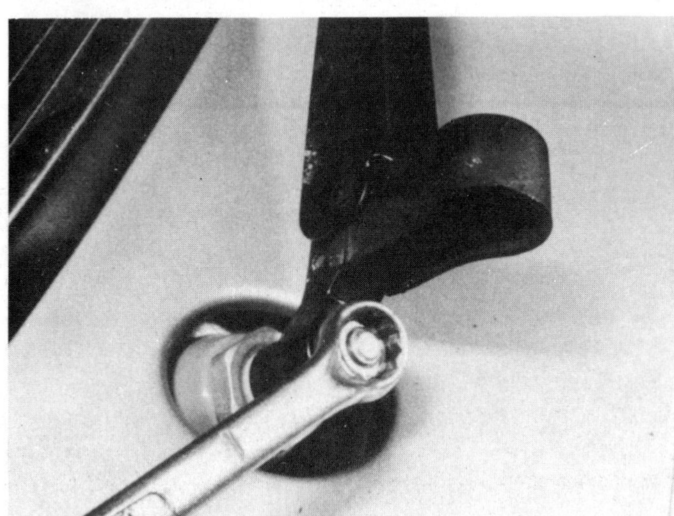

21.7 Unscrewing the wiper arm nut

8 Unscrew the nut and pull the arm/blade assembly from the spindle splines.
9 Refit by reversing the removal operations.
10 Wet the glass and operate the wipers to check their arc of travel. If it is incorrect, remove the arm and move it a spline or two in the required direction.

22 Windscreen wiper motor/linkage – removal and refitting

1 Remove the wiper arms, as described in the preceding Section, then disconnect the battery.
2 Open the bonnet and disconnect the wiper motor wiring. Disconnect the wiper motor crankarm by unscrewing the nut (photos).
3 Unscrew the mounting bolts and withdraw the wiper motor.
4 Unscrew the nuts from the wiper drive spindle units.
5 Remove the linkage.
6 Refitting is a reversal of removal.

23 Tailgate wiper motor – removal and refitting

1 The tailgate wiper motor is of direct drive type, without linkages.
2 The wiper arm/blade is removed as described for the windscreen wiper in Section 21.
3 Open the tailgate and remove the trim panel by carefully prising out the clips. Disconnect the battery.
4 Disconnect the wiper motor wiring (photo).
5 Prise off the cap and unscrew the drive spindle nut.
6 Unbolt and remove the wiper motor.
7 Refitting is a reversal of removal.

24 Washer system – general

1 All models are equipped with a windscreen washer.
2 The washer reservoir is located within the engine compartment.

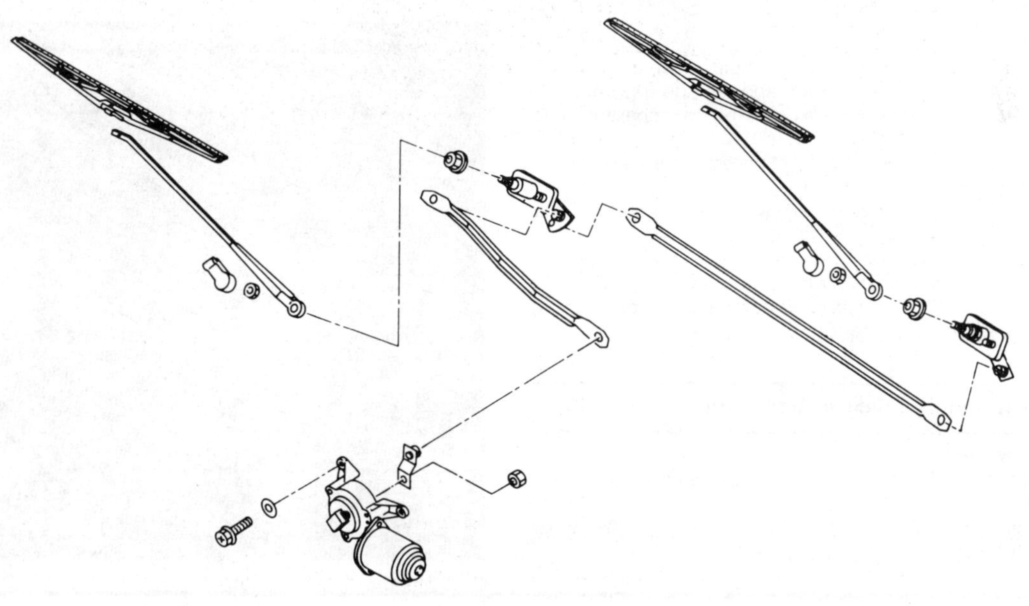

Fig. 10.38 Windscreen wiper components (Sec 22)

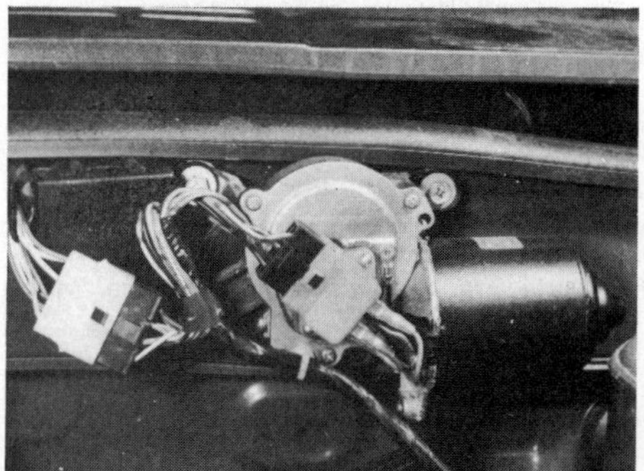

22.2A Windscreen wiper motor

22.2B Windscreen wiper motor crankarm and linkage

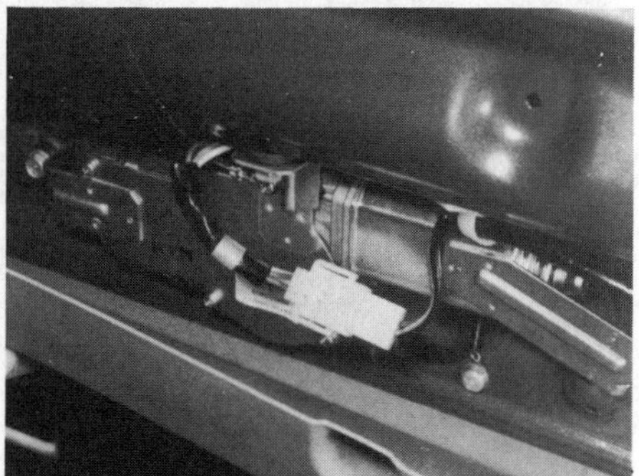

23.4 Tailgate wiper motor

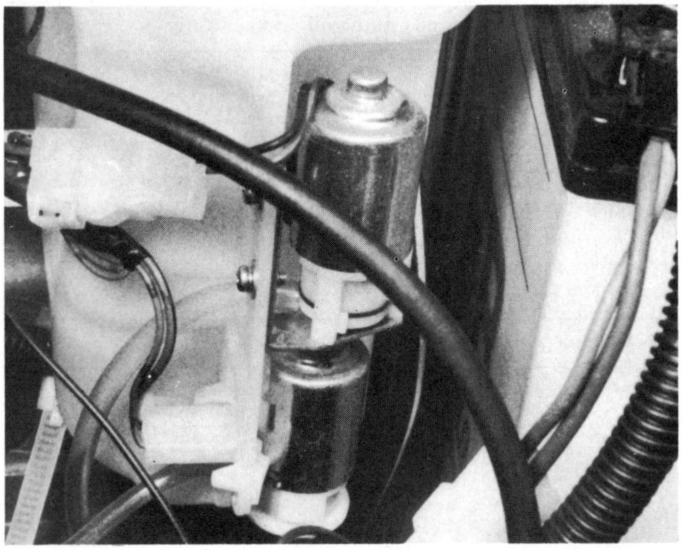

24.3 Washer reservoir showing windscreen and tailgate pumps

3 On vehicles with a tailgate washer, the fluid reservoir with its electric pumps serves both the windscreen and tailgate (photo).
4 Where a headlamp washer system is fitted, a separate fluid reservoir is used.
5 The windscreen wiper/washer is controlled by a stalk switch on the steering column.
6 The tailgate wiper/washer is controlled by a double action rocker switch.
7 The headlamp washer switch is of single action rocker type.
8 The washer jets which are located within the slats of the air intake grille are adjusted by inserting a pin in their nozzles and repositioning them to give a satisfactory spray pattern.

25 Headlamp wiper motor – removal and refitting

1 Disconnect the battery.
2 Remove the screws and separate the wiper blades from the arms. Also disconnect the washer tube.
3 Prise off the caps, unscrew the nuts, and remove the wiper arms from the spindles (photos).

25.3A Headlamp wiper arm cap

25.3B Headlamp wiper arm nut

25.5 Headlamp wiper motor

4 Remove the headlamp unit as described in Section 18.
5 Disconnect the wiring and unbolt the motor (photo).
6 Refitting is a reversal of removal, but adjust the wiper arm positions as required.

26 Horn – removal and refitting

1 Remove the headlamp unit as described in Section 18.
2 Unbolt the horn from the bracket and disconnect the wiring (photo).
3 If the horn emits a weak sound it may be possible to improve it by adjusting the screw on the rear of the unit.
4 Refitting is a reversal of removal.

27 Speedometer cable – removal and refitting

1 The speedometer drive cable should be renewed as an assembly if it breaks.

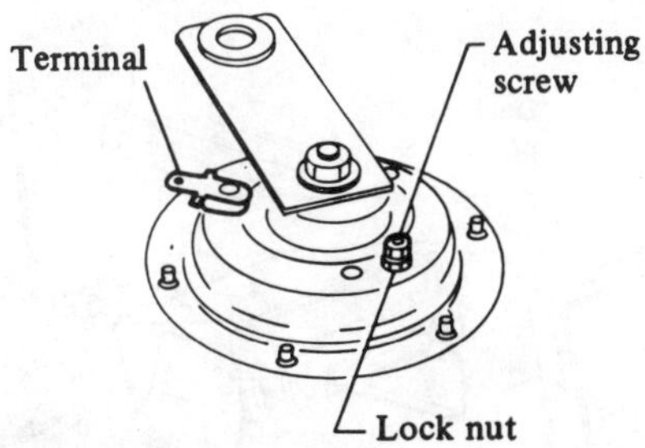

Fig. 10.39 Horn showing adjusting screw (Sec 26)

26.2 Horn and mounting bolt

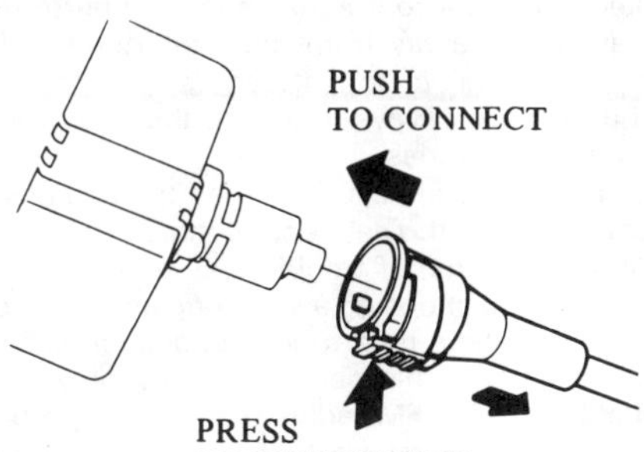

Fig. 10.40 Speedometer cable fixing to instrument panel (Sec 27)

2 Disconnect the cable from the speedometeer head after partially withdrawing the instrument panel, as described in Section 20 (photo).
3 Disconnect the cable from the transmission by unscrewing its knurled retaining ring (photo).
4 Slip the cable out of its clip on the engine compartment rear bulkhead and then withdraw it, with its grommet, through the bulkhead.
5 Refit the new cable by reversing the removal operations.

27.2 Speedometer cable end at the instrument panel

27.3 Speedometer cable end at the transmission

28 Heated rear window – general

1 Care should be taken to avoid damage to the element for the heated rear window or tailgate.
2 Avoid scratching with rings on the fingers when cleaning, and do not allow luggage to rub against the glass.
3 Do not stick labels over the element on the inside of the glass.
4 If the element grids do become damaged, a special conductive paint is available from most motor factors to repair it.
5 Do not leave the heated rear window switched on unnecessarily as it draws a high current from the electrical system.

29 Seat belt warning system – general

1 On certain models the seat belt on the driver's side incorporates a visual and audible warning as a reminder that the belt is not connected when starting the vehicle.
2 The warnings are given for a six second period if the ignition key is turned to ON without the seat belt having been fastened.

30 Radio – removal and refitting

1 Disconnect the battery.
2 Remove the ashtray then remove the bracket (one screw).
3 Remove the two screws from the underside of the radio.
4 Pull off the radio control knobs on pre 1985 models.
5 Remove the two upper screws and withdraw the surround.
6 Disconnect the wiring harness and the aerial plug, and withdraw the radio.
7 Refitting is a reversal of removal, but, if a new receiver has been fitted, the aerial should be trimmed in the following way.
8 Extend the aerial fully, switch on the radio and turn to maximum volume.
9 Tune to a station which is barely audible and is found at around 1400 kHz.
10 Turn the trim screw in the front of the receiver until the signal is at its strongest. The screw should not be turned more than one half turn in either direction.

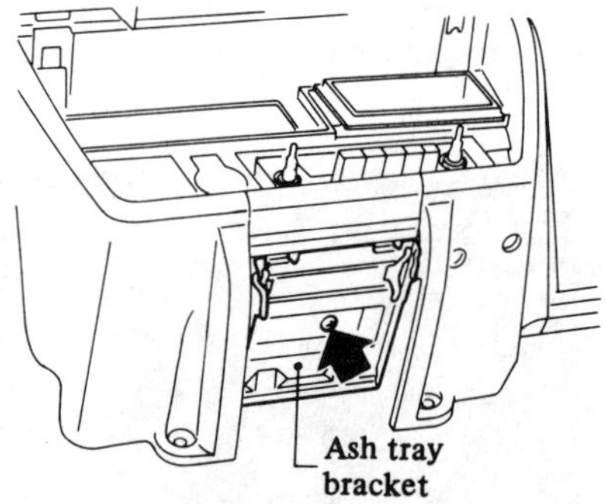

Fig. 10.41 Ashtray bracket retaining screw location (Sec 30)

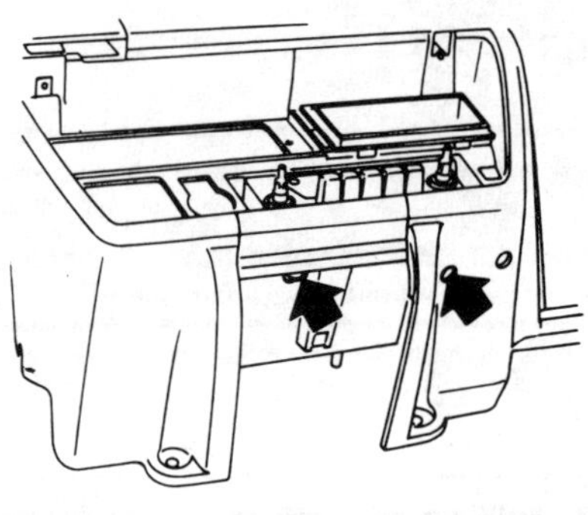

Fig. 10.42 Radio mounting screw locations (Sec 30)

31 Cassette player – removal and refitting

1 Remove the radio as described in Section 30.
2 Unscrew the cassette mounting bolts.
3 Withdraw the unit and disconnect the wiring harness.
4 Refitting is a reversal of removal.

32 Mobile radio equipment – interference-free installation

Aerials – selection and fitting

The choice of aerials is now very wide. It should be realised that the quality has a profound effect on radio performance, and a poor, inefficient aerial can make suppression difficult.

A wing-mounted aerial is regarded as probably the most efficient

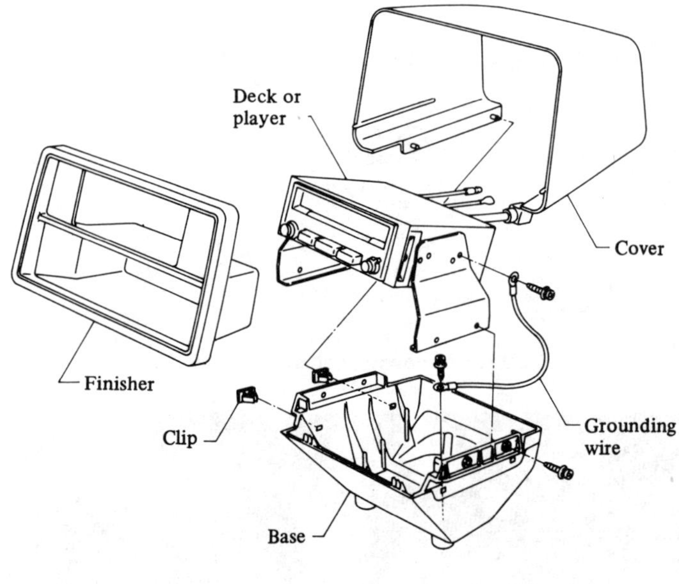

Fig. 10.43 Cassette player components (Sec 31)

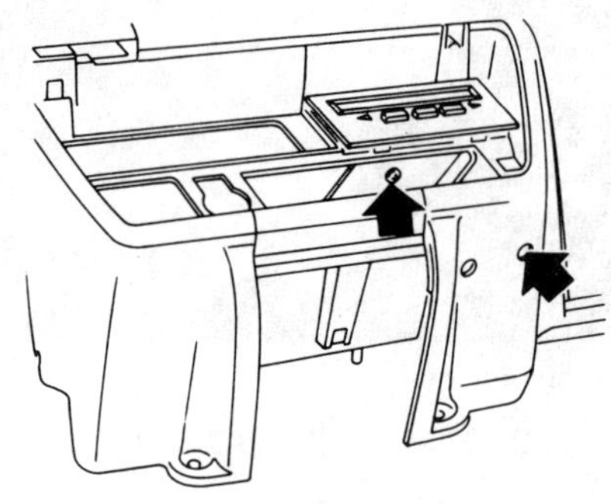

Fig. 10.44 Cassette mounting bolt locations (Sec 31)

for signal collection, but a roof aerial is usually better for suppression purposes because it is away from most interference fields. Stick-on wire aerials are available for attachment to the inside of the windscreen, but are not always free from the interference field of the engine and some accessories.

Motorised automatic aerials rise when the equipment is switched on and retract at switch-off. They require more fitting space and supply leads, and can be a source of trouble.

There is no merit in choosing a very long aerial as, for example, the type about three metres in length which hooks or clips on to the rear of the car, since part of this aerial will inevitably be located in an interference field. For VHF/FM radios the best length of aerial is about one metre. Active aerials have a transistor amplifier mounted at the base and this serves to boost the received signal. The aerial rod is sometimes rather shorter than normal passive types.

A large loss of signal can occur in the aerial feeder cable, especially over the Very High Frequency (VHF) bands. The design of feeder cable is invariably in the co-axial form, ie a centre conductor surrounded by a flexible copper braid forming the outer (earth) conductor. Between the inner and outer conductors is an insulator material which can be in solid or stranded form. Apart from insulation, its purpose is to maintain the correct spacing and concentricity. Loss of signal occurs in this insulator, the loss usually being greater in a poor quality cable. The quality of cable used is reflected in the price of the aerial with the attached feeder cable.

The capacitance of the feeder should be within the range 65 to 75 picofarads (pF) approximately (95 to 100 pF for Japanese and American equipment), otherwise the adjustment of the car radio aerial

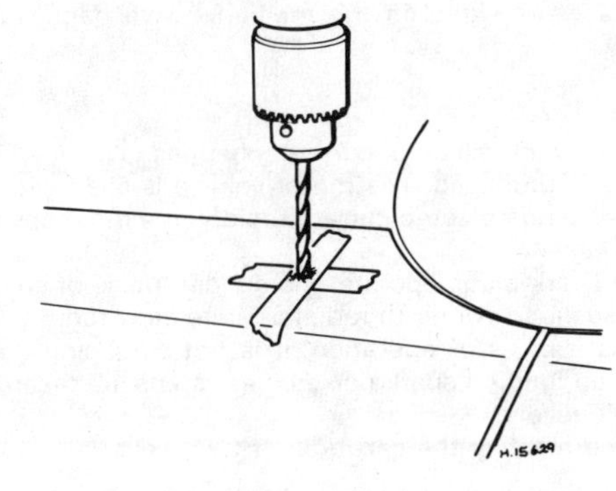

Fig. 10.45 Drilling the bodywork for aerial mounting (Sec 32)

trimmer may not be possible. An extension cable is necessary for a long run between aerial and receiver. If this adds capacitance in excess of the above limits, a connector containing a series capacitor will be required, or an extension which is labelled as 'capacity-compensated'.

Fitting the aerial will normally involve making a $\frac{7}{8}$ in (22 mm) diameter hole in the bodywork, but read the instructions that come with the aerial kit. Once the hole position has been selected, use a centre punch to guide the drill. Use sticky masking tape around the area for this helps with marking out and drill location, and gives protection to the paintwork should the drill slip. Three methods of making the hole are in use:

(a) Use a hole saw in the electric drill. This is, in effect, a circular hacksaw blade wrapped round a former with a centre pilot drill.
(b) Use a tank cutter which also has cutting teeth, but is made to shear the metal by tightening with an Allen key.
(c) The hard way of drilling out the circle is using a small drill, say $\frac{1}{8}$ in (3 mm), so that the holes overlap. The centre metal drops out and the hole is finished with round and half-round files.

Whichever method is used, the burr is removed from the body metal and paint removed from the underside. The aerial is fitted tightly ensuring that the earth fixing, usually a serrated washer, ring or clamp, is making a solid connection. *This earth connection is important in reducing interference.* Cover any bare metal with primer paint and topcoat, and follow by underseal if desired.

Aerial feeder cable routing should avoid the engine compartment and areas where stress might occur, eg under the carpet where feet will be located. Roof aerials require that the headlining be pulled back and that a path is available down the door pillar. It is wise to check with the vehicle dealer whether roof aerial fitting is recommended.

Loudspeakers

Speakers should be matched to the output stage of the equipment, particularly as regards the recommended impedance. Power transistors used for driving speakers are sensitive to the loading placed on them.

Before choosing a mounting position for speakers, check whether the vehicle manufacturer has provided a location for them. Generally door-mounted speakers give good stereophonic reproduction, but not all doors are able to accept them. The next best position is the rear parcel shelf, and in this case speaker apertures can be cut into the shelf, or pod units may be mounted.

For door mounting, first remove the trim, which is often held on by 'poppers' or press studs, and then select a suitable gap in the inside door assembly. Check that the speaker would not obstruct glass or winder mechanism by winding the window up and down. A template is often provided for marking out the trim panel hole, and then the four fixing holes must be drilled through. Mark out with chalk and cut cleanly with a sharp knife or keyhole saw. Speaker leads are then threaded through the door and door pillar, if necessary drilling 10 mm diameter holes. Fit grommets in the holes and connect to the radio or tape unit correctly. Do not omit a waterproofing cover, usually supplied with door speakers. If the speaker has to be fixed into the metal of the door itself, use self-tapping screws, and if the fixing is to the door trim use self-tapping screws and flat spire nuts.

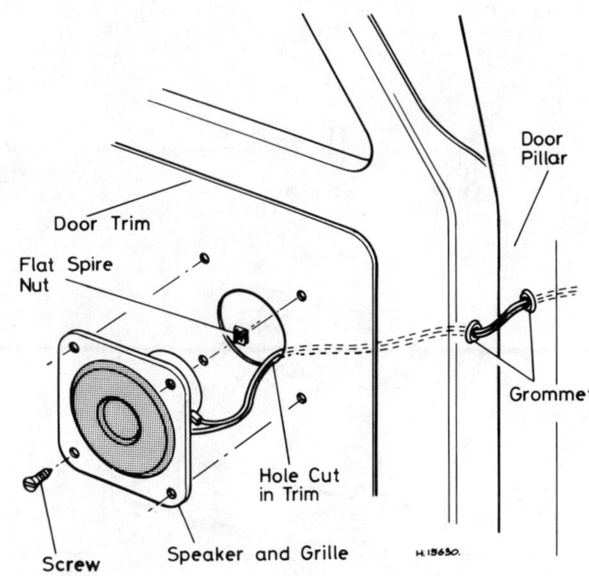

Fig. 10.46 Door-mounted speaker installation (Sec 32)

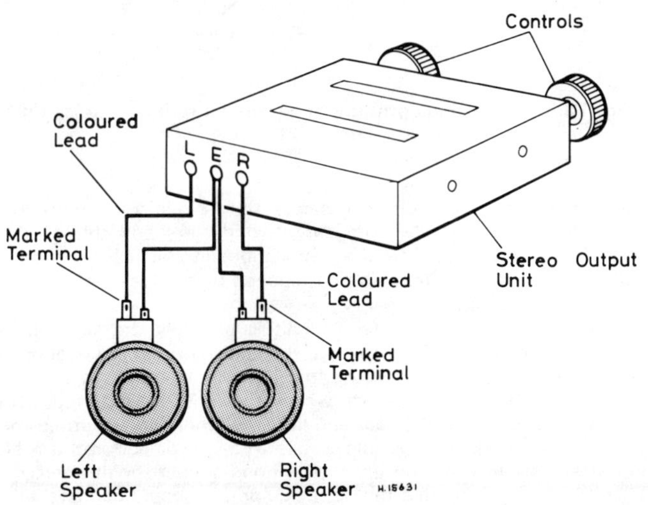

Fig. 10.47 Speaker connections must be correctly made as shown (Sec 32)

Rear shelf mounting is somewhat simpler but it is necessary to find gaps in the metalwork underneath the parcel shelf. However, remember that the speakers should be as far apart as possible to give a good stereo effect. Pod-mounted speakers can be screwed into position through the parcel shelf material, but it is worth testing for the best position. Sometimes good results are found by reflecting sound off the rear window.

Unit installation

Many vehicles have a dash panel aperture to take a radio/audio unit, a recognised international standard being 189.5 mm x 60 mm. Alternatively a console may be a feature of the car interior design and this, mounted below the dashboard, gives more room. If neither facility is available a unit may be mounted on the underside of the parcel shelf; these are frequently non-metallic and an earth wire from the

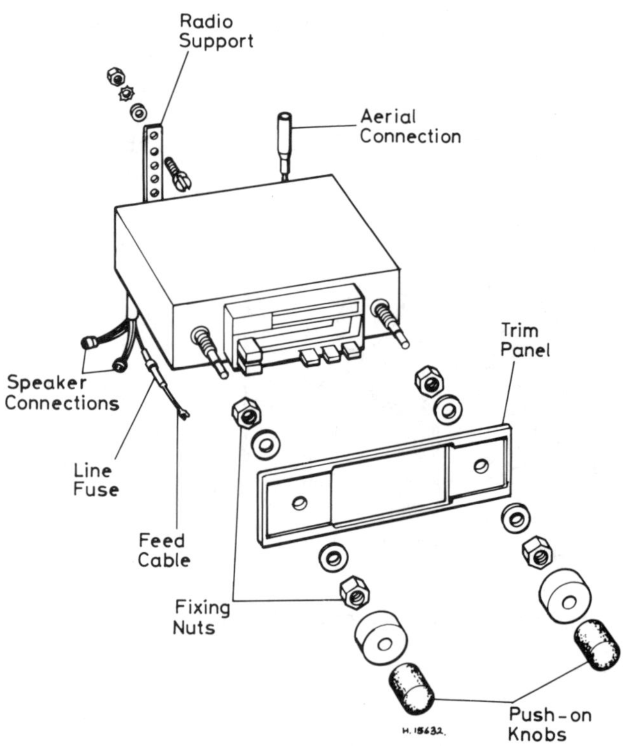

Fig. 10.48 Mounting component details for radio/cassette unit (Sec 32)

case to a good earth point is necessary. A three-sided cover in the form of a cradle is obtainable from car radio dealers and this gives a professional appearance to the installation; in this case choose a position where the controls can be reached by a driver with his seat belt on.

Installation of the radio/audio unit is basically the same in all cases, and consists of offering it into the aperture after removal of the knobs *(not* push buttons) and the trim plate. In some cases a special mounting plate is required to which the unit is attached. It is worthwhile supporting the rear end in cases where sag or strain may occur, and it is usually possible to use a length of perforated metal strip attached between the unit and a good support point nearby. In general it is recommended that tape equipment should be installed at or nearly horizontal.

Connections to the aerial socket are simply by the standard plug terminating the aerial downlead or its extension cable. Speakers for a stereo system must be matched and correctly connected, as outlined previously.

Note: *While all work is carried out on the power side, it is wise to disconnect the battery earth lead.* Before connection is made to the vehicle electrical system, check that the polarity of the unit is correct. Most vehicles use a negative earth system, but radio/audio units often have a reversible plug to convert the set to either + or – earth. *Incorrect connection may cause serious damage.*

The power lead is often permanently connected inside the unit and terminates with one half of an in-line fuse carrier. The other half is fitted with a suitable fuse (3 or 5 amperes) and a wire which should go to a power point in the electrical system. This may be the accessory terminal on the ignition switch, giving the advantage of power feed with ignition or with the ignition key at the 'accessory' position. Power to the unit stops when the ignition key is removed. Alternatively, the lead may be taken to a live point at the fusebox with the consequence of having to remember to switch off at the unit before leaving the vehicle.

Before switching on for initial test, be sure that the speaker connections have been made, for running without load can damage the output transistors. Switch on next and tune through the bands to ensure that all sections are working, and check the tape unit if applicable. The aerial trimmer should be adjusted to give the strongest reception on a weak signal in the medium wave band, at say 200 metres.

Interference

In general, when electric current changes abruptly, unwanted electrical noise is produced. The motor vehicle is filled with electrical devices which change electric current rapidly, the most obvious being the contact breaker.

When the spark plugs operate, the sudden pulse of spark current causes the associated wiring to radiate. Since early radio transmitters used sparks as a basis of operation, it is not surprising that the car radio will pick up ignition spark noise unless steps are taken to reduce it to acceptable levels.

Interference reaches the car radio in two ways:

(a) by conduction through the wiring.
(b) by radiation to the receiving aerial.

Initial checks presuppose that the bonnet is down and fastened, the radio unit has a good earth connection *(not* through the aerial downlead outer), no fluorescent tubes are working near the car, the aerial trimmer has been adjusted, and the vehicle is in a position to receive radio signals, ie not in a metal-clad building.

Switch on the radio and tune it to the middle of the medium wave (MW) band off-station with the volume (gain) control set fairly high. Switch on the ignition (but do not start the engine) and wait to see if irregular clicks or hash noise occurs. Tapping the facia panel may also produce the effects. If so, this will be due to the voltage stabiliser, which is an on-off thermal switch to control instrument voltage. It is located usually on the back of the instrument panel, often attached to the speedometer. Correction is by attachment of a capacitor and, if still troublesome, chokes in the supply wires.

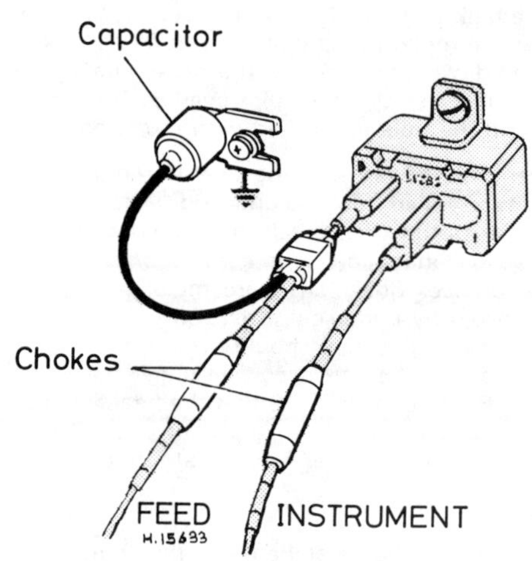

Fig. 10.49 Voltage stabiliser interference suppression (Sec 32)

Switch on the engine and listen for interference on the MW band. Depending on the type of interference, the indications are as follows.

A harsh crackle that drops out abruptly at low engine speed or when the headlights are switched on is probably due to a voltage regulator.

A whine varying with engine speed is due to the dynamo or alternator. Try temporarily taking off the fan belt – if the noise goes this is confirmation.

Regular ticking or crackle that varies in rate with the engine speed is due to the ignition system. With this trouble in particular and others in general, check to see if the noise is entering the receiver from the wiring or by radiation. To do this, pull out the aerial plug, (preferably shorting out the input socket or connecting a 62 pF capacitor across it). If the noise disappears it is coming in through the aerial and is *radiation noise.* If the noise persists it is reaching the receiver through the wiring and is said to be *line-borne.*

Interference from wipers, washers, heater blowers, turn-indicators, stop lamps, etc is usually taken to the receiver by wiring, and simple treatment using capacitors and possibly chokes will solve the problem. Switch on each one in turn (wet the screen first for running wipers!) and listen for possible interference with the aerial plug in place and again when removed.

Electric petrol pumps are now finding application again and give rise to an irregular clicking, often giving a burst of clicks when the ignition is on but the engine has not yet been started. It is also possible to receive whining or crackling from the pump.

Note that if most of the vehicle accessories are found to be creating interference all together, the probability is that poor aerial earthing is to blame.

Component terminal markings

Throughout the following sub-sections reference will be found to various terminal markings. These will vary depending on the manufacturer of the relevant component. If terminal markings differ from those mentioned, reference should be made to the following table, where the most commonly encountered variations are listed.

Alternator	*Alternator terminal (thick lead)*	*Exciting winding terminal*
DIN/Bosch	B+	DF
Delco Remy	+	EXC
Ducellier	+	EXC
Ford (US)	+	DF
Lucas	+	F
Marelli	+B	F

Ignition coil	*Ignition switch terminal*	*Contact breaker terminal*
DIN/Bosch	15	1
Delco Remy	+	–
Ducellier	BAT	RUP
Ford (US)	B/+	CB/–
Lucas	SW/+	–
Marelli	BAT/+B	D

Voltage regulator	*Voltage input terminal*	*Exciting winding terminal*
DIN/Bosch	B+/D+	DF
Delco Remy	BAT/+	EXC
Ducellier	BOB/BAT	EXC
Ford (US)	BAT	DF
Lucas	+/A	F
Marelli		F

Suppression methods – ignition

Suppressed HT cables are supplied as original equipment by manufacturers and will meet regulations as far as interference to neighbouring equipment is concerned. It is illegal to remove such suppression unless an alternative is provided, and this may take the form of resistive spark plug caps in conjunction with plain copper HT cable. For VHF purposes, these and 'in-line' resistors may not be effective, and resistive HT cable is preferred. Check that suppressed cables are actually fitted by observing cable identity lettering, or measuring with an ohmmeter – the value of each plug lead should be 5000 to 10 000 ohms.

A 1 microfarad capacitor connected from the LT supply side of the ignition coil to a good nearby earth point will complete basic ignition interference treatment. *NEVER fit a capacitor to the coil terminal to the contact breaker – the result would be burnt out points in a short time.*

If ignition noise persists despite the treatment above, the following sequence should be followed:

(a) Check the earthing of the ignition coil; remove paint from fixing clamp.
(b) If this does not work, lift the bonnet. Should there be no change in interference level, this may indicate that the bonnet is not electrically connected to the car body. Use a proprietary braided strap across a bonnet hinge ensuring a first class electrical connection. If, however, lifting the bonnet increases the interference, then fit resistive HT cables of a higher ohms-per-metre value.
(c) If all these measures fail, it is probable that re-radiation from metallic components is taking place. Using a braided strap between metallic points, go round the vehicle systematically – try the following: engine to body, exhaust system to body, front suspension to engine and to body, steering column to body (especially French and Italian cars), gear lever to engine and to body (again especially French and Italian cars), Bowden cable to body, metal parcel shelf to body. When an offending component is located it should be bonded with the strap permanently.
(d) As a next step, the fitting of distributor suppressors to each lead at the distributor end may help.
(e) Beyond this point is involved the possible screening of the distributor and fitting resistive spark plugs, but such advanced treatment is not usually required for vehicles with entertainment equipment.

Electronic ignition systems have built-in suppression components, but this does not relieve the need for using suppressed HT leads. In some cases it is permitted to connect a capacitor on the low tension supply side of the ignition coil, but not in every case. Makers' instructions should be followed carefully, otherwise damage to the ignition semiconductors may result.

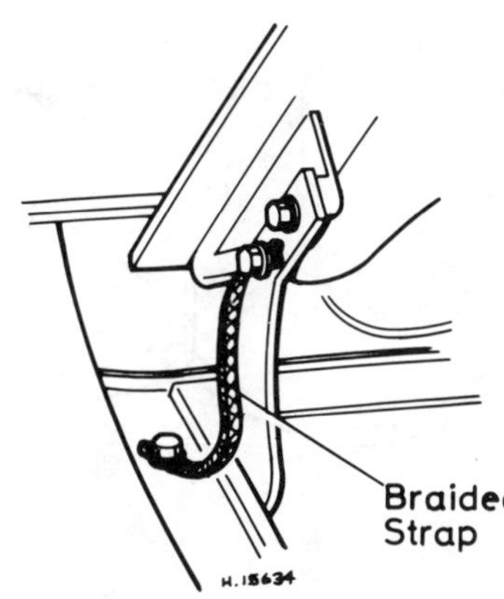

Fig. 10.50 Braided earth strap between bonnet and body (Sec 32)

Suppression methods – generators

For older vehicles with dynamos a 1 microfarad capacitor from the D (larger) terminal to earth will usually cure dynamo whine. Alternators should be fitted with a 3 microfarad capacitor from the B+ main output terminal (thick cable) to earth. Additional suppression may be obtained by the use of a filter in the supply line to the radio receiver.

It is most important that:

(a) Capacitors are never connected to the field terminals of either a dynamo or alternator.
(b) Alternators must not be run without connection to the battery.

Suppression methods – voltage regulators

Voltage regulators used with DC dynamos should be suppressed by connecting a 1 microfarad capacitor from the control box D terminal to earth.

Alternator regulators come in three types:

(a) Vibrating contact regulators separate from the alternator. Used extensively on continental vehicles.
(b) Electronic regulators separate from the alternator.
(c) Electronic regulators built-in to the alternator.

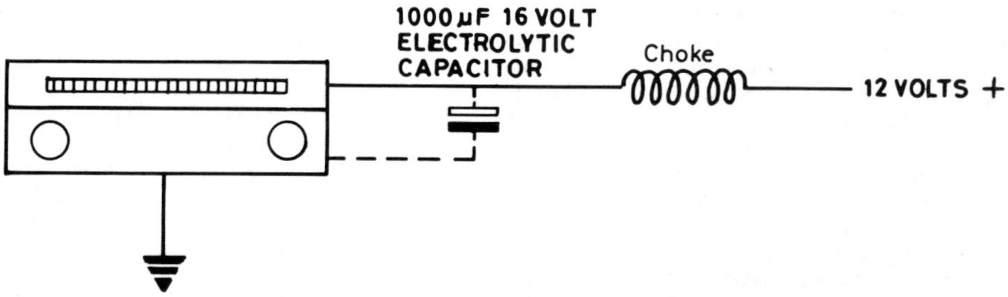

Fig. 10.51 Line-borne interference suppression (Sec 32)

In case (a) interference may be generated on the AM and FM (VHF) bands. For some cars a replacement suppressed regulator is available. Filter boxes may be used with non-suppressed regulators. But if not available, then for AM equipment a 2 microfarad or 3 microfarad capacitor may be mounted at the voltage terminal marked D+ or B+ of the regulator. FM bands may be treated by a feed-through capacitor of 2 or 3 microfarad.

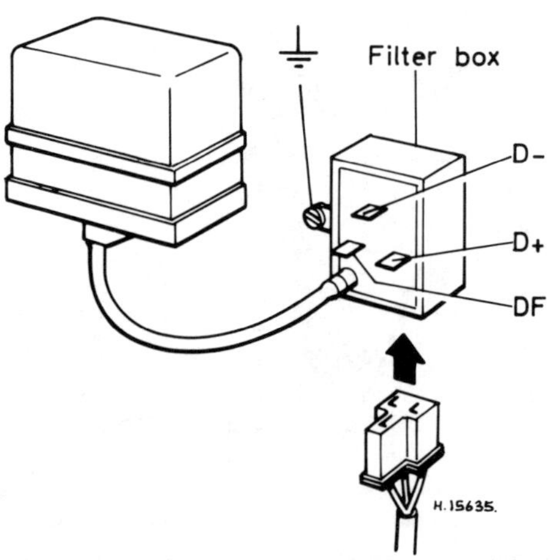

Fig. 10.52 Typical filter box for vibrating contact voltage regulator (alternator equipment) (Sec 32)

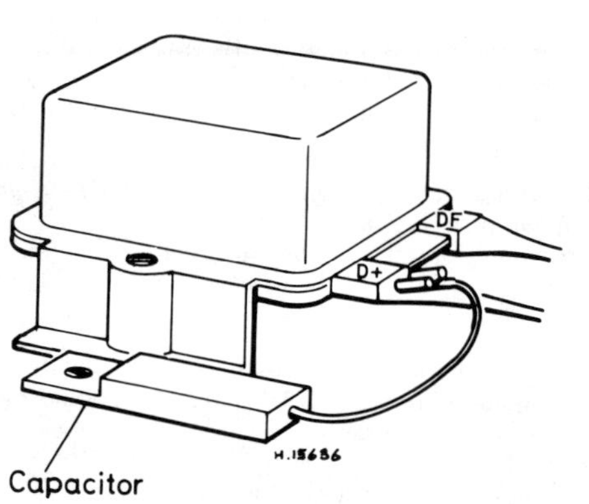

Fig. 10.53 Suppression of AM interference by vibrating contact voltage regulator (alternator equipment) (Sec 32)

Electronic voltage regulators are not always troublesome, but where necessary, a 1 microfarad capacitor from the regulator + terminal will help.

Integral electronic voltage regulators do not normally generate much interference, but when encountered this is in combination with alternator noise. A 1 microfarad or 2 microfarad capacitor from the warning lamp (IND) terminal to earth for Lucas ACR alternators and Femsa, Delco and Bosch equivalents should cure the problem.

Suppression methods – other equipment

Wiper motors – Connect the wiper body to earth with a bonding strap. For all motors use a 7 ampere choke assembly inserted in the leads to the motor.

Heater motors – Fit 7 ampere line chokes in both leads, assisted if necessary by a 1 microfarad capacitor to earth from both leads.

Electronic tachometer – The tachometer is a possible source of ignition noise – check by disconnecting at the ignition coil CB terminal. It usually feeds from ignition coil LT pulses at the contact breaker terminal. A 3 ampere line choke should be fitted in the tachometer lead at the coil CB terminal.

Horn – A capacitor and choke combination is effective if the horn is directly connected to the 12 volt supply. The use of a relay is an alternative remedy, as this will reduce the length of the interference-carrying leads.

Electrostatic noise – Characteristics are erratic crackling at the receiver, with disappearance of symptoms in wet weather. Often shocks may be given when touching bodywork. Part of the problem is the build-up of static electricity in non-driven wheels and the acquisition of charge on the body shell. It is possible to fit spring-loaded contacts at the wheels to give good conduction between the rotary wheel parts and the vehicle frame. Changing a tyre sometimes helps – because of tyres' varying resistances. In difficult cases a trailing flex which touches the ground will cure the problem. If this is not acceptable it is worth trying conductive paint on the tyre walls.

Fuel pump – Suppression requires a 1 microfarad capacitor between the supply wire to the pump and a nearby earth point. If this is insufficient a 7 ampere line choke connected in the supply wire near the pump is required.

Fluorescent tubes – Vehicles used for camping/caravanning frequently have fluorescent tube lighting. These tubes require a relatively high voltage for operation and this is provided by an inverter (a form of oscillator) which steps up the vehicle supply voltage. This can give rise to serious interference to radio reception, and the tubes themselves can contribute to this interference by the pulsating nature of the lamp discharge. In such situations it is important to mount the aerial as far away from a fluorescent tube as possible. The interference problem may be alleviated by screening the tube with fine wire turns spaced an inch (25 mm) apart and earthed to the chassis. Suitable chokes should be fitted in both supply wires close to the inverter.

Radio/cassette case breakthrough

Magnetic radiation from dashboard wiring may be sufficiently intense to break through the metal case of the radio/cassette player. Often this is due to a particular cable routed too close and shows up as ignition interference on AM and cassette play and/or alternator whine on cassette play.

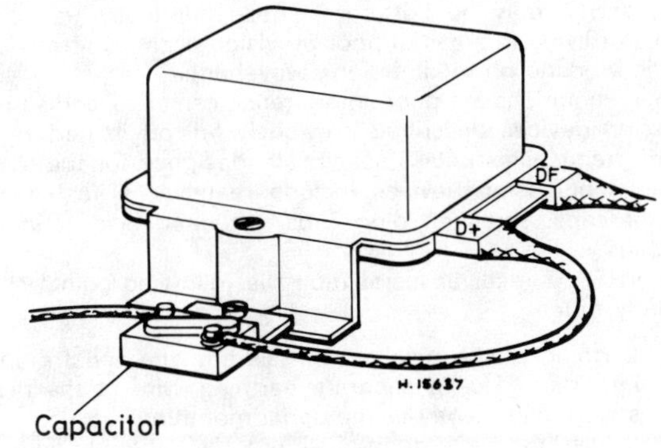

Fig. 10.54 Suppression of FM interference by vibrating contact voltage regulator (alternator equipment) (Sec 32)

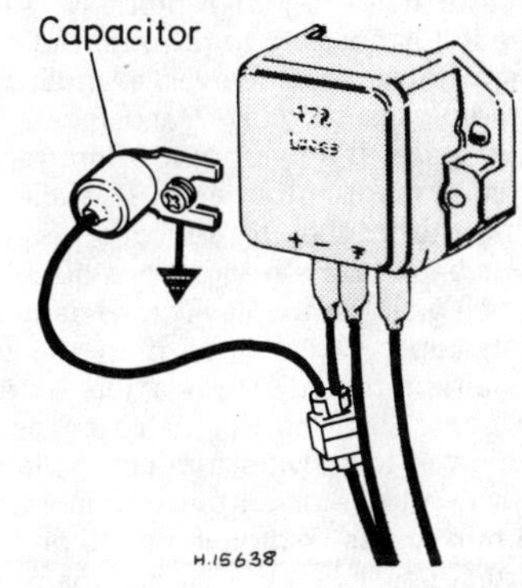

Fig. 10.55 Electronic voltage regulator suppression (Sec 32)

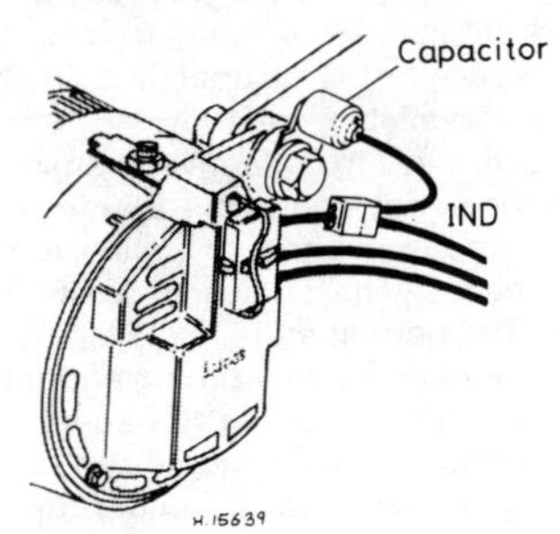

Fig. 10.56 Suppression of interference from electronic voltage regulator when integral with alternator (Sec 32)

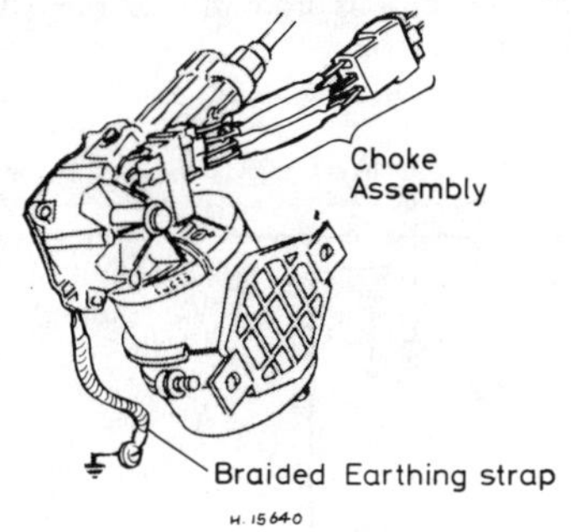

Fig. 10.57 Wiper motor suppression (Sec 32)

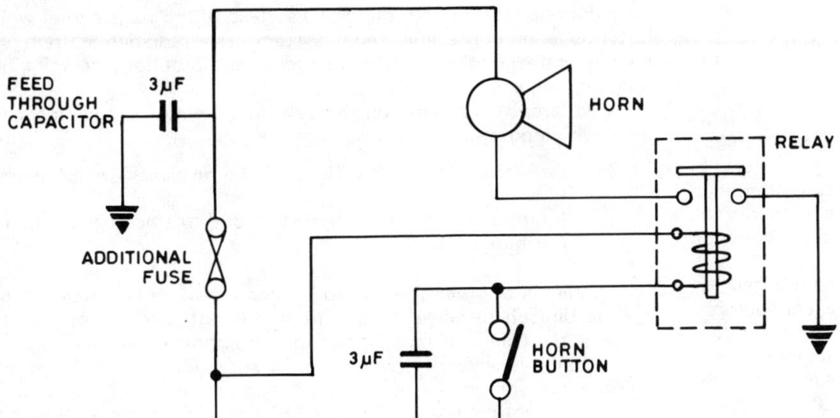

Fig. 10.58 Use of relay to reduce horn interference (Sec 32)

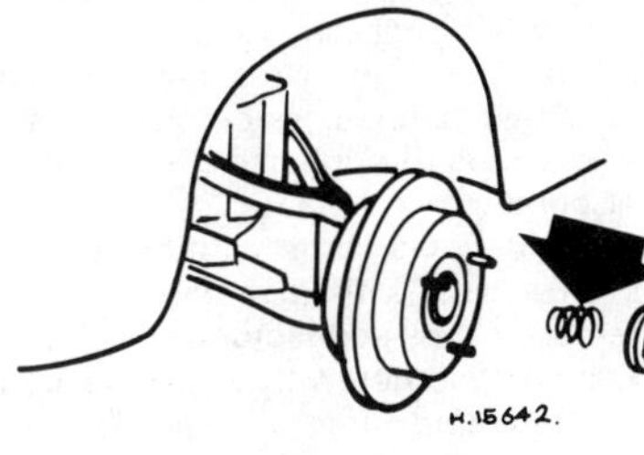

Fig. 10.59 Use of spring contacts at wheels (Sec 32)

The first point to check is that the clips and/or screws are fixing all parts of the radio/cassette case together properly. Assuming good earthing of the case, see if it is possible to re-route the offending cable – the chances of this are not good, however, in most cars.

Next release the radio/cassette player and locate it in different positions with temporary leads. If a point of low interference is found, then if possible fix the equipment in that area. This also confirms that local radiation is causing the trouble. If re-location is not feasible, fit the radio/cassette player back in the original position.

Alternator interference on cassette play is now caused by radiation from the main charging cable which goes from the battery to the output terminal of the alternator, usually via the + terminal of the starter motor relay. In some vehicles this cable is routed under the dashboard, so the solution is to provide a direct cable route. Detach the original cable from the alternator output terminal and make up a new cable of at least 6 mm^2 cross-sectional area to go from alternator to battery with the shortest possible route. *Remember – do not run the engine with the alternator disconnected from the battery.*

Ignition breakthrough on AM and/or cassette play can be a difficult problem. It is worth wrapping earthed foil round the offending cable run near the equipment, or making up a deflector plate well screwed down to a good earth. Another possibility is the use of a suitable relay to switch on the ignition coil. The relay should be mounted close to the ignition coil; with this arrangement the ignition coil primary current is not taken into the dashboard area and does not flow through the ignition switch. A suitable diode should be used since it is possible that at ignition switch-off the output from the warning lamp alternator terminal could hold the relay on.

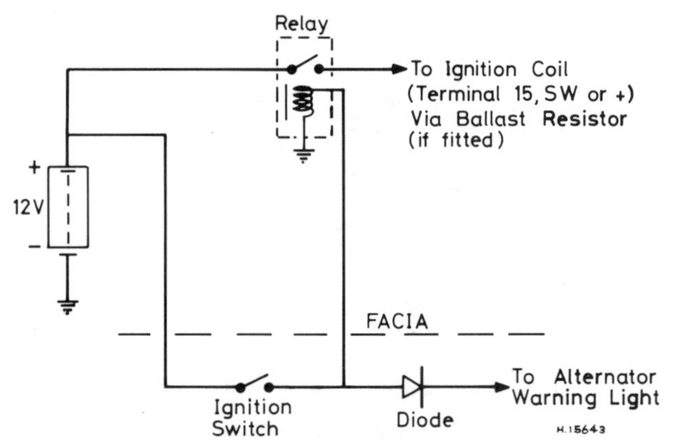

Fig. 10.60 Use of ignition coil relay to suppress case breakthrough (Sec 32)

Connectors for suppression components

Capacitors are usually supplied with tags on the end of the lead, while the capacitor body has a flange with a slot or hole to fit under a nut or screw with washer.

Connections to feed wires are best achieved by self-stripping connectors. These connectors employ a blade which, when squeezed down by pliers, cuts through cable insulation and makes connection to the copper conductors beneath.

Chokes sometimes come with bullet snap-in connectors fitted to the wires, and also with just bare copper wire. With connectors, suitable female cable connectors may be purchased from an auto-accessory shop together with any extra connectors required for the cable ends after being cut for the choke insertion. For chokes with bare wires, similar connectors may be employed together with insulation sleeving as required.

VHF/FM broadcasts

Reception of VHF/FM in an automobile is more prone to problems than the medium and long wavebands. Medium/long wave transmitters are capable of covering considerable distances, but VHF transmitters are restricted to line of sight, meaning ranges of 10 to 50 miles, depending upon the terrain, the effects of buildings and the transmitter power.

Because of the limited range it is necessary to retune on a long journey, and it may be better for those habitually travelling long distances or living in areas of poor provision of transmitters to use an AM radio working on medium/long wavebands.

When conditions are poor, interference can arise, and some of the suppression devices described previously fall off in performance at very high frequencies unless specifically designed for the VHF band. Available suppression devices include reactive HT cable, resistive distributor caps, screened plug caps, screened leads and resistive spark plugs.

For VHF/FM receiver installation the following points should be particularly noted:

(a) Earthing of the receiver chassis and the aerial mounting is important. Use a separate earthing wire at the radio, and scrape paint away at the aerial mounting.
(b) If possible, use a good quality roof aerial to obtain maximum height and distance from interference generating devices on the vehicle.
(c) Use of a high quality aerial downlead is important, since losses in cheap cable can be significant.
(d) The polarisation of FM transmissions may be horizontal, vertical, circular or slanted. Because of this the optimum mounting angle is at 45° to the vehicle roof.

Citizens' Band radio (CB)

In the UK, CB transmitter/receivers work within the 27 MHz and 934 MHz bands, using the FM mode. At present interest is concentrated on 27 MHz where the design and manufacture of equipment is less difficult. Maximum transmitted power is 4 watts, and 40 channels spaced 10 kHz apart within the range 27.60125 to 27.99125 MHz are available.

Aerials are the key to effective transmission and reception. Regulations limit the aerial length to 1.65 metres including the loading coil and any associated circuitry, so tuning the aerial is necessary to obtain optimum results. The choice of a CB aerial is dependent on whether it is to be permanently installed or removable, and the performance will hinge on correct tuning and the location point on the vehicle. Common practice is to clip the aerial to the roof gutter or to employ wing mounting where the aerial can be rapidly unscrewed. An alternative is to use the boot rim to render the aerial theftproof, but a popular solution is to use the 'magmount' – a type of mounting having a strong magnetic base clamping to the vehicle at any point, usually the roof.

Aerial location determines the signal distribution for both transmission and reception, but it is wise to choose a point away from the engine compartment to minimise interference from vehicle electrical equipment.

The aerial is subject to considerable wind and acceleration forces. Cheaper units will whip backwards and forwards and in so doing will alter the relationship with the metal surface of the vehicle with which it forms a ground plane aerial system. The radiation pattern will change correspondingly, giving rise to break-up of both incoming and outgoing signals.

Interference problems on the vehicle carrying CB equipment fall into two categories:

(a) Interference to nearby TV and radio receivers when transmitting.
(b) Interference to CB set reception due to electrical equipment on the vehicle.

Problems of break-through to TV and radio are not frequent, but can be difficult to solve. Mostly trouble is not detected or reported because the vehicle is moving and the symptoms rapidly disappear at the TV/radio receiver, but when the CB set is used as a base station any trouble with nearby receivers will soon result in a complaint.

It must not be assumed by the CB operator that his equipment is faultless, for much depends upon the design. Harmonics (that is, multiples) of 27 MHz may be transmitted unknowingly and these can fall into other user's bands. Where trouble of this nature occurs, low pass filters in the aerial or supply leads can help, and should be fitted in base station aerials as a matter of course. In stubborn cases it may be necessary to call for assistance from the licensing authority, or, if possible, to have the equipment checked by the manufacturers.

Interference received on the CB set from the vehicle equipment is, fortunately, not usually a severe problem. The precautions outlined previously for radio/cassette units apply, but there are some extra points worth noting.

It is common practice to use a slide-mount on CB equipment enabling the set to be easily removed for use as a base station, for example. Care must be taken that the slide mount fittings are properly earthed and that first class connection occurs between the set and slide-mount.

Vehicle manufacturers in the UK are required to provide suppression of electrical equipment to cover 40 to 250 MHz to protect TV and VHF radio bands. Such suppression appears to be adequately effective at 27 MHz, but suppression of individual items such as alternators, clocks, stabilisers, flashers, wiper motors, etc, may still be necessary. The suppression capacitors and chokes available from auto-electrical suppliers for entertainment receivers will usually give the required results with CB equipment.

Other vehicle radio transmitters

Besides CB radio already mentioned, a considerable increase in the use of transceivers (ie combined transmitter and receiver units) has taken place in the last decade. Previously this type of equipment was fitted mainly to military, fire, ambulance and police vehicles, but a large business radio and radio telephone usage has developed.

Generally the suppression techniques described previously will suffice, with only a few difficult cases arising. Suppression is carried out to satisfy the 'receive mode', but care must be taken to use heavy duty chokes in the equipment supply cables since the loading on 'transmit' is relatively high.

Glass-fibre bodied vehicles

Such vehicles do not have the advantage of a metal box surrounding the engine as is the case, in effect, of conventional vehicles. It is usually necessary to line the bonnet, bulkhead and wing valances with metal foil, which could well be the aluminium foil available from builders merchants. Nonding of sheets one to another and the whole down to the chassis is essential.

Wiring harness may have to be wrapped in metal foil which again should be earthed to the vehicle chassis. The aerial base and radio chassis must be taken to the vehicle chassis by heavy metal braid. VHF radio suppression in glass-fibre cars may not be a feasible operation.

In addition to all the above, normal suppression components should be employed, but special attention paid to earth bonding. A screen enclosing the entire ignition system usually gives good improvement, and fabrication from fine mesh perforated metal is convenient. Good bonding of the screening boxes to several chassis points is essential.

33 Fault diagnosis – electrical system

Symptom	Reason(s)
No voltage at starter motor	Battery discharged Battery defective internally Battery terminals loose or earth lead not securely attached to body Loose or broken connections in starter motor circuit Starter motor switch or solenoid faulty
Voltage at starter motor – faulty motor	Starter brushes badly worn, sticking, or brush wires loose Commutator dirty, worn or burnt Starter motor armature faulty Field coils earthed
Electrical defects	Battery in discharged condition Starter brushes badly worn, sticking, or brush wires loose Loose wires in starter motor circuit Starter motor pinion sticking on the screwed sleeve Dirt or oil on drivegear
Starter motor noisy or rough in engagement	Pinion or flywheel gear teeth broken or worn Starter drive main spring broken Starter motor retaining bolts loose
Alternator not charging*	Drivebelt loose and slipping, or broken Brushes worn, sticking, broken or dirty Brush springs weak or broken
If all appears to be well but the alternator is still not charging, take the car to an automobile electrician for checking of the alternator	
Battery will not hold charge for more than a few days	Battery defective internally Electrolyte level too low or electrolyte too weak due to leakage Plate separators no longer fully effective Battery plates severely sulphated Drivebelt slipping Battery terminal connections loose or corroded Alternator not charging properly Short in lighting circuit causing continual battery drain
Ignition light fails to go out, battery runs flat in a few days	Drivebelt loose and slipping, or broken Alternator faulty
Failure of individual electrical equipment to function correctly is dealt with alphabetically below	
Fuel gauge gives no reading	Fuel tank empty Electric cable between tank sender unit and gauge earthed or loose Fuel gauge case not earthed Fuel gauge supply cable interrupted Fuel gauge unit broken

33 Fault diagnosis – electrical system

Symptom	Reason(s)
Fuel gauge registers full all the time	Electric cable between tank unit and gauge broken or disconnected
Horn operates all the time	Horn push either earthed or stuck down Horn cable to horn push earthed
Horn fails to operate	Blown fuse Cable or cable connection loose, broken or disconnected Horn has an internal fault
Horn emits intermittent or unsatisfactory noise	Cable connections loose Horn incorrectly adjusted
Lights do not come on	If engine not running, battery discharged Light bulb filament burnt out or bulbs broken Wire connections loose, disconnected or broken Light switch shorting or otherwise faulty
Lights come on but fade	If engine not running, battery discharged
Lights give very poor illumination	Lamp glasses dirty Reflector tarnished or dirty Lamps badly out of adjustment Incorrect bulb with too low wattage fitted Existing bulbs old and badly discoloured Electrical wiring too thin not allowing full current to pass
Lights work erratically, flashing on and off, especially over bumps	Battey terminals or earth connections loose Lights not earthing properly Contacts in light switch faulty
Wiper motor fails to work	Blown fuse Brushes badly worn Wire connections loose, disconnected or broken Armature worn or faulty Field coils faulty
Wiper motor works very slowly and takes excessive current	Commutator dirty, greasy or burnt Drive to spindles bent or unlubricated Drive spindle binding or damaged Armature bearings dry or misaligned Armature badly worn or faulty
Wiper motor works slowly and takes little current	Brushes badly worn Commutator dirty, greasy or burnt Armature badly worn or faulty
Wiper motor works but wiper blades remain static	Linkage disengaged or faulty Drive spindle damaged or worn Wiper motor gearbox parts badly worn

Chapter 11 Suspension and steering

Contents

Specifications

Front suspension

Type .... MacPherson strut with anti-roll bar

Rear suspension

Type .... Trailing arm, telescopic shock absorbers and coil springs

Steering

Type .... Rack and pinion with a universally-jointed column. A tilt column and power-assisted steering are optional

Steering angles (unladen)

UK models:	
Camber – except Turbo	–0°25′ to +1°05′
Camber – Turbo	–0°35′ to +1°05′
Castor	+0°45′ to +2°15′
King pin inclination	+12°10′ to +13°40′
Toe-in	0 to 2.0 mm (0 to 0.08 in)
North American models:	
Camber – 1983	+0°35′ to + 1°05′
Camber – 1984, 1985	–0°25′ to +1°05′
Castor	+0°45′ to +2°15′
King pin inclination	+12°10′ to +13°40′
Toe-in – 1983	0 to 2.0 mm (0 to 0.08 in)
Toe-in – 1984, 1985	3.0 to 5.0 mm (0.12 to 0.20 in)

Steering ratio

Manual steering	21.55 to 1
Power assisted steering	18.45 to 1, or 17.04 to 1

Number of turns lock to lock

Manual steering	3.7 or 3.9
Power-assisted steering	3.33

Power-assisted steering system fluid capacity

Power-assisted steering system fluid capacity	1.76 Imp pt, 1.06 US qt, 1.0 litre

Roadwheels

Type	Pressed steel or light alloy
Size	
Pressed steel	4J x 12 or $4\frac{1}{2}$J x 13
Light alloy	5J x 13

Tyres

Sizes	6.00 x 12, 6.15 x 13, 145SR13, 155SR12, 155SR13, 165/70SR13, 175/70SR13

Note: *In some territories the spare wheel is fitted with a size P155/80D 13 tyre*

Pressures (cold)

Crossply	24 lbf/in^2 (1.7 bar) front and rear
Radial	26 lbf/in^2 (1.8 bar) front and rear
Spare (P155/80D13)	34 lbf/in^2 (2.4 bar)

Note: *Refer also to vehicle decal*

Torque wrench settings

	lbf ft	Nm
Front suspension		
Strut piston rod self-locking nut	52	70
Strut upper mounting nuts	17	23
Strut to stub axle carrier	59	80
Lower balljoint to track control arm	45	61
Lower balljoint to stub axle carrier	35	48
Tie-rod balljoint taper pin nut	35	48
Track control pivot arm bolts	70	95
Track control arm support plate	70	95
Track control arm pivot nuts	81	110
Roadwheel nuts	70	95
Anti-roll bar clamp	9	12
Anti-roll bar end links	9	12
Rear suspension		
Shock absorber upper mounting	9	12
Shock absorber lower mounting	33	45
Trailing arm pivot bolt	48	65
Steering		
Steering wheel nut	37	50
Steering shaft lower joint pinch-bolt:		
M10	28	38
M8	21	28
Tilt lever bolt	8	11
Column mounting bracket	10	14
Steering gear mounting clamps	55	75
Fluid pump to bracket	18	25
Pump pulley nut	55	75
Pressure hose to gear housing	21	28
Pressure hose to pump	35	48

1 General description

The front suspension is of MacPherson strut type, having coil springs, a lower track control arm and an anti-roll bar.

The rear suspension is of trailing arm type with coil springs and telescopic shock absorbers.

The steering gear is of rack-and-pinion type with a universally-jointed steering column.

Power-assisted steering is available as a factory-fitted option.

2 Routine maintenance

1 Every 6000 miles (10 000 km) or 6 months whichever comes first on UK models, or every 15 000 miles (24 000 km) or 12 months whichever comes first on North American models check the power steering fluid level in the reservoir and top up as necessary. Also check the power steering hoses for condition and security.

2 Every 12 000 miles (20 000 km) or 12 months whichever comes first on UK models or every 30 000 miles (48 000 km) or 24 months whichever comes first on North American models check the power steering pump drivebelt for condition and tension.

3 Every 12 000 miles (20 000 km) or 12 months whichever comes first on UK models or every 15 000 miles (24 000 km) or 12 months whichever comes first on North American models check the steering and suspension components including all balljoints for excessive wear, adjust the rear wheel bearings, and check the front wheel bearings for wear and leakage of grease. Check and if necessary adjust the front wheel alignment. Change the roadwheel positions and have them balanced.

4 If the car is operated under extreme conditions the maintenance intervals should be more frequent, and the mileages reduced accordingly.

Fig. 11.1 Front suspension components (Sec 1)

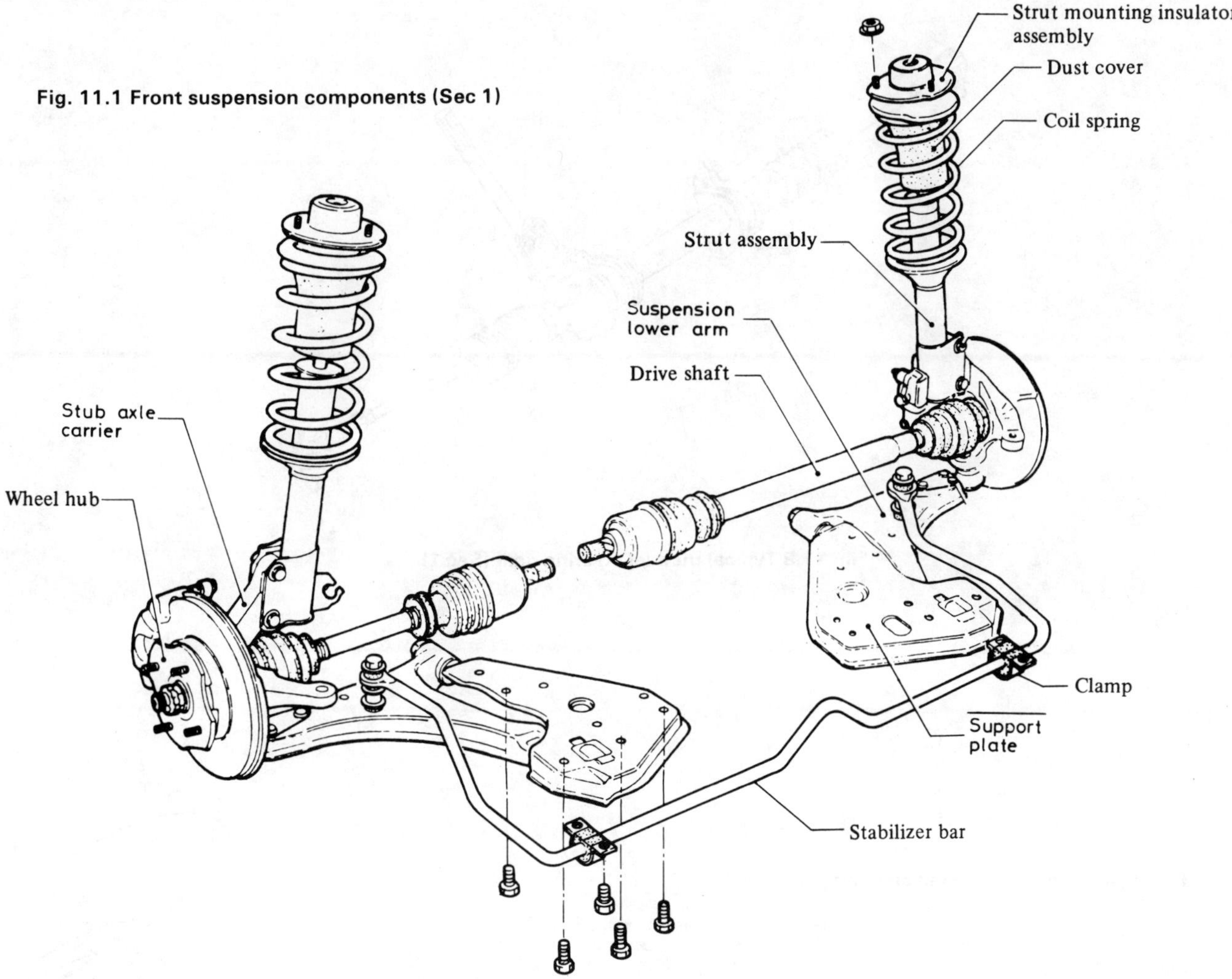

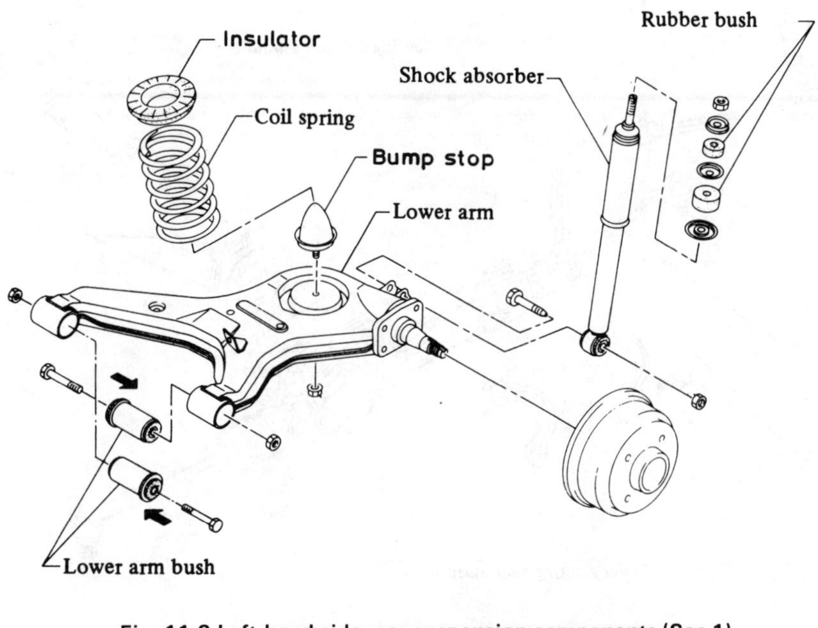

Fig. 11.2 Left-hand side rear suspension components (Sec 1)

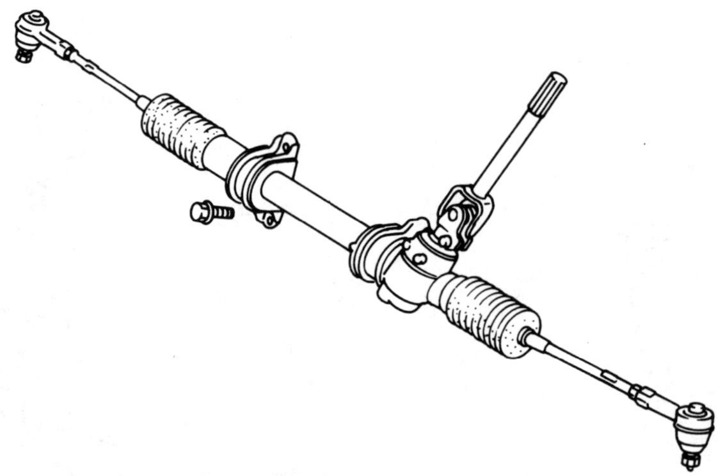

Fig. 11.3 Typical manual steering gear (Sec 1)

Fig. 11.4 Typical power-assisted steering gear (Sec 1)

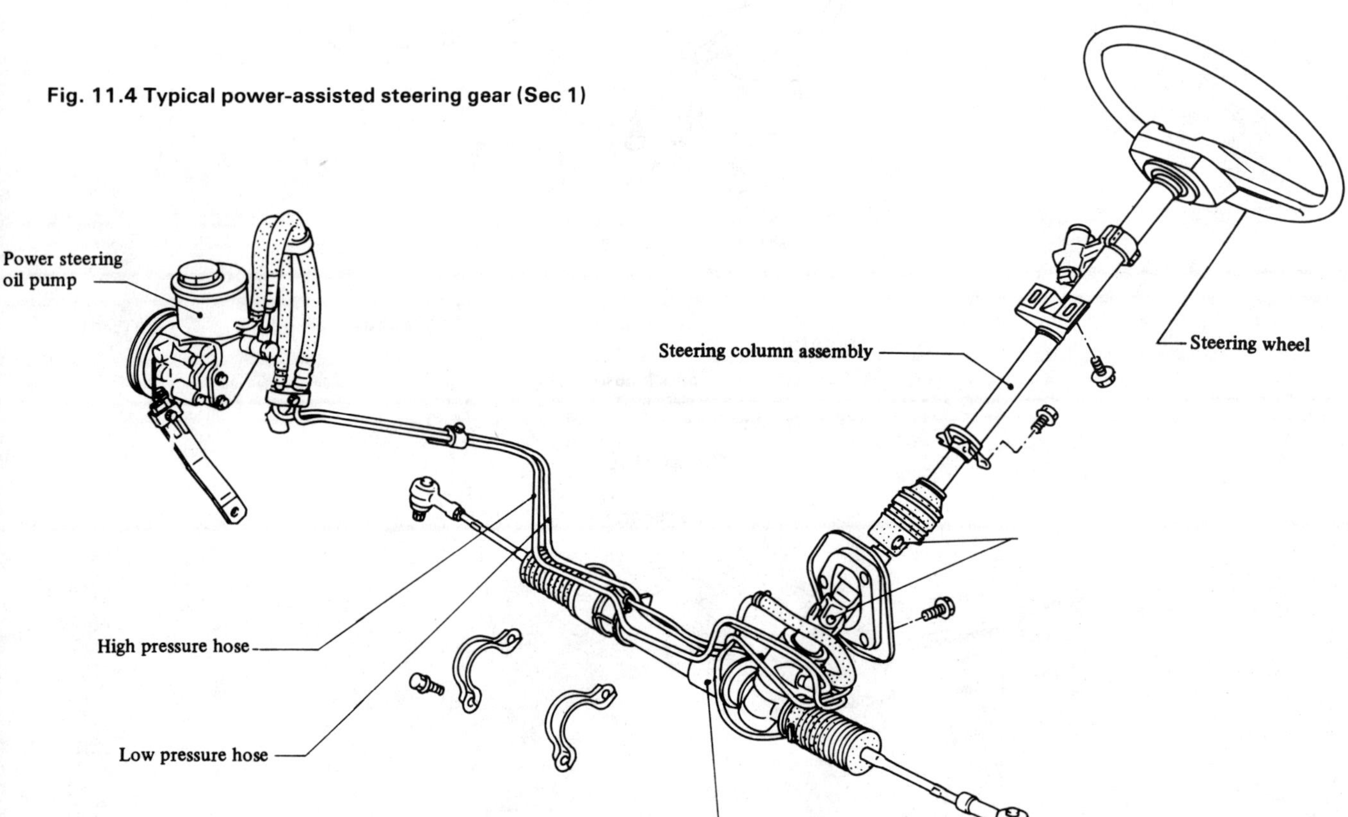

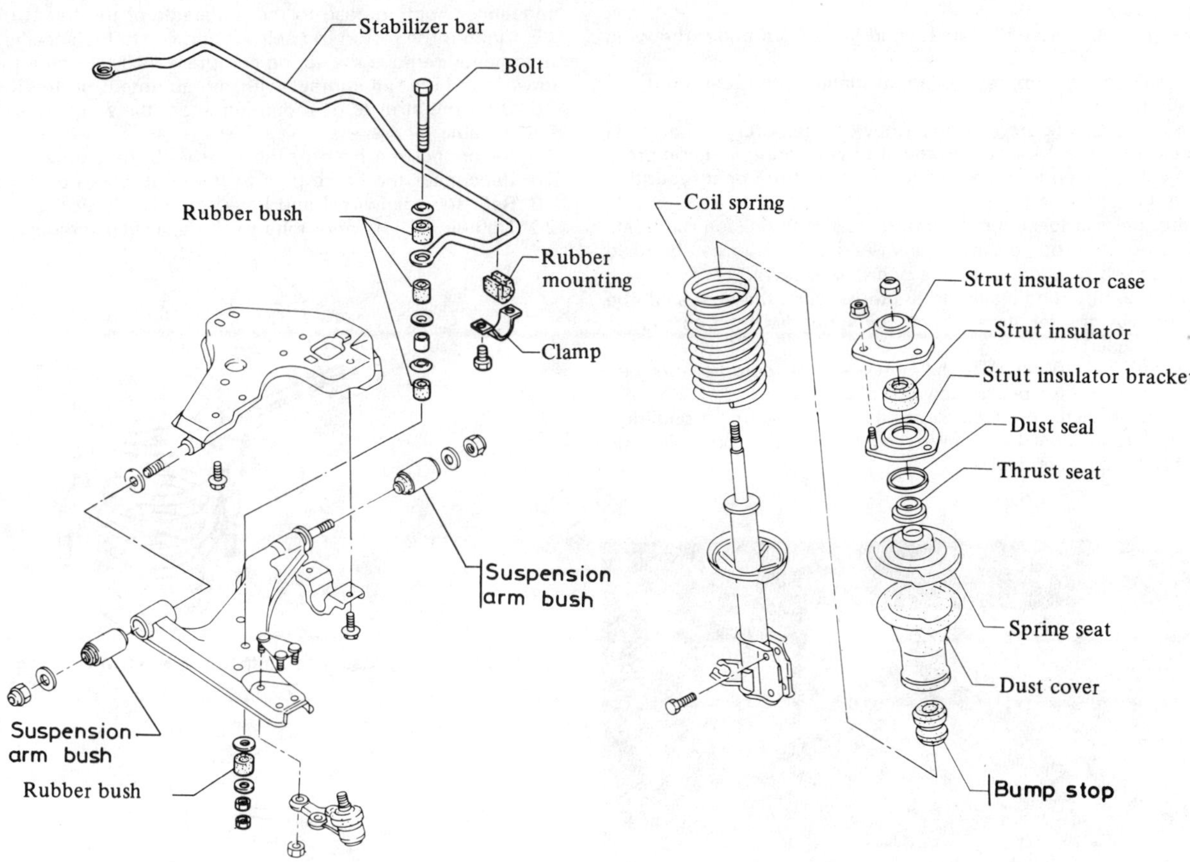

Fig. 11.5 Exploded view of the front suspension (Secs 3, 4, 5 and 6)

3 Front anti-roll bar – removal and refitting

1 Disconnect the exhaust pipe front mounting bracket and the flexible mounting nearest the front expansion box.
2 Unscrew the anti-roll bar clamp bolts (photo) and the end bolt nuts (photos).
3 Withdraw the anti-roll bar.
4 Refitting is a reversal of removal, but if the vehicle was jacked up to remove the bar do not fully tighten the anti-roll bar nuts and bolts until the weight of the vehicle has been lowered onto the roadwheels.

4 Front suspension strut – removal, overhaul and refitting

1 Raise the front of the vehicle and support it securely on axle stands placed under the side-members.
2 Remove the roadwheel.
3 Disconnect the hydraulic brake pipe from the suspension strut where applicable.
4 Support the suspension lower arm on a jack.
5 Unscrew the two pinch-bolts which hold the stub axle carrier to the base of the suspension strut (photo).

3.2A Anti-roll bar mounting clamp

3.2B Anti-roll bar end mounting

4.5 Front suspension strut to stub axle carrier fixing bolts

6 Working at the top of the inner wing within the engine compartment, unscrew and remove the three nuts which secure the strut top mounting (photo).
7 Support the strut assembly and withdraw it from under the wing (photo).
8 Unless coil spring compressors are available do not carry out any further dismantling.
9 Where compressors are available (they can be purchased at most motor accessory stores) fit them to the strut coil spring and compress the spring just sufficiently to be able to turn the strut upper mounting insulator by hand.
10 Unscrew the self-locking nut from the top of the piston rod. Flats are machined on the rod so that an open-ended spanner can be used to prevent the rod rotating while the unit is unscrewed.
11 Take off the mounting insulator, the thrust seat, the dust seal, the spring upper seat, the rebound rubber, coil spring (with compressors) and the dust excluder.
12 Unless the coil spring is to be renewed, the compressors can remain on the spring for reassembly.
13 If the strut is distorted, leaking or has lost its damping qualities, then the strut tube must be renewed, no repair being possible. The renewal of both struts is advised.

4.6 Front suspension strut top mounting

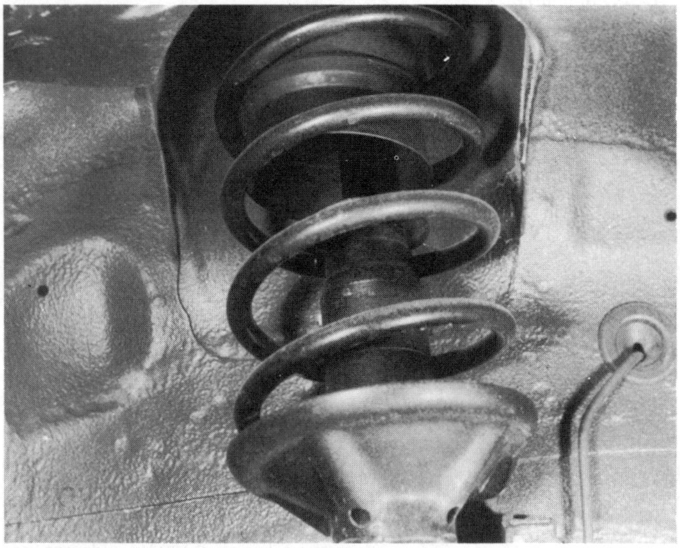

4.7 Removing the front suspension strut from under the wing

14 Reassemble the strut by fitting the spring in its compressed state followed by the upper mounting components in their originally fitted sequence. Apply grease to the underside of the thrust plate.
15 Tighten the piston rod self-locking nut to the specified torque and then gently remove the spring compressors. Make sure that the spring lower end is in full contact with the abutment on the lower seat.
16 Offer the strut to its mounting under the wing. Screw on the nuts finger tight.
17 Reconnect the base of the strut with the stub axle carrier.
18 Reconnect the brake pipe to the strut where applicable.
19 Refit the roadwheel and lower the vehicle to the ground.
20 Tighten all nuts and bolts to the specified torques.

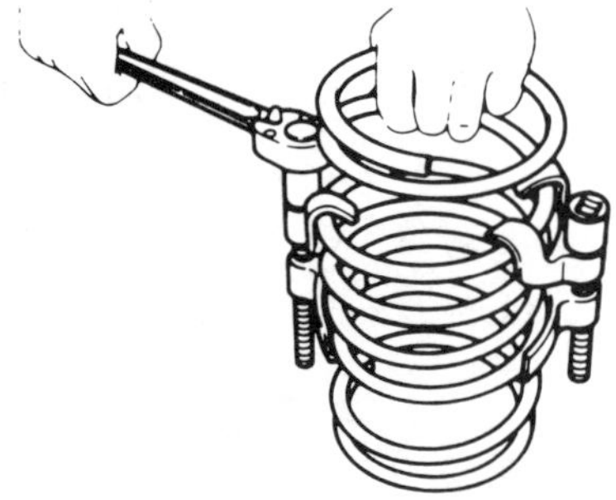

Fig. 11.6 Compressor tools fitted to the front coil spring (Sec 4)

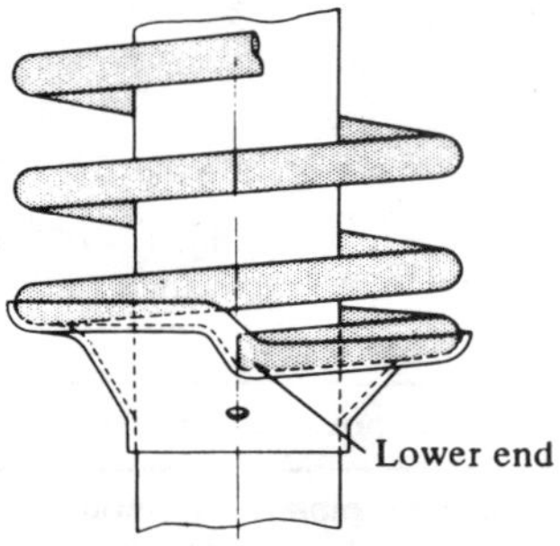

Fig. 11.7 Coil spring lower seat abutment (Sec 4)

5 Front suspension lower track control arm – removal and refitting

1 Raise the vehicle and support on axle stands placed under the side-members.
2 Remove the roadwheel.
3 Remove the anti-roll bar (see Section 3).
4 Disconnect the suspension lower balljoint from the track control arm by unscrewing the three securing nuts. It is recommended that these nuts are renewed at reassembly (photos).
5 Working at the inboard end of the control arm, unbolt the pivot rod clamp and pivot end nut (photos). Remove the control arm.
6 Unbolt and remove the support plate, if required, by unscrewing the fixing bolts (photo).
7 The flexible bushes may be renewed. Do this by pressing them out or by drawing them out using a bolt, nut, washers and tubular spacers. Smear the new bushes with hydraulic fluid to make fitting easier.
8 Refit the support plate and the control arm by reversing the removal operations.
9 Do not fully tighten the nuts and bolts to the specified torques until the weight of the vehicle is on the roadwheels.

5.4A Front suspension track control arm

5.4B Showing lower balljoint to track control arm nuts

5.5A Track control arm rear mounting

5.5B Track control arm front mounting

5.6 Showing the track control arm support plate

6 Front suspension lower balljoint – removal and refitting

1 In order to obtain access to the balljoint fixing nut the driveshaft must first be removed, as described in Chapter 8.
2 Unscrew the balljoint nut and then, using a suitable balljoint extractor tool, disconnect the balljoint from the stub axle carrier.
3 Unscrew the three nuts which hold the balljoint to the track control arm and remove the balljoint. It is recommended that new nuts are used at reassembly.
4 Refitting is a reversal of removal, tighten all nuts and bolts to the specified torque. Do not apply any grease to the balljoint taper pin or eye before connecting. It is sometimes found that when tightening the taper pin nut the ball turns in the joint socket rather than the nut tightening. Should this happen, raise a jack under the balljoint in order to seat the taper pin more firmly in the eye of the stub axle carrier.

7 Front suspension stub axle carrier and hub bearings – removal and refitting

1 Remove the driveshaft as described in Chapter 8.
2 If necessary separate the lower balljoint from the stub axle carrier by unscrewing the nut and using an extractor tool.
3 The hub should now be separated from the stub axle carrier using a slide hammer attached to the roadwheel studs.
4 Unscrew the nuts which hold the brake disc and rotor together, and remove the disc.
5 Remove the outer bearing from the hub either by using a suitable bearing extractor or by supporting the bearing and pressing the hub from it. Remove and discard the oil seal.
6 Prise the oil seal from the inner end of the stub axle carrier and remove the inner bearing.
7 Drive both bearing outer tracks from the stub axle carrier using a soft metal drift or alternatively use a puller and adaptor (see Fig. 11.9).
8 The hub bearings should be renewed in pairs (inner and outer), not singly.
9 Clean the recesses in the stub axle carrier and drive in the new bearing outer tracks – using a piece of brass or copper tubing as a drift.
10 Pack each bearing with grease then locate the inner bearing in its outer track and press a new oil seal into position with the spring tensioned lip facing outwards. Smear some grease onto the lip.
11 Press a new oil seal onto the hub shoulder (Fig. 11.10) then drive on the new bearing and inner track using metal tubing on the inner track only. Smear grease onto the oil seal lip.
12 Locate the spacer. If the original one is damaged, renew it with one which carries the same markings as the old one. The spacer thickness need only be changed if the stub axle carrier is renewed. To calculate the thickness of the new spacer, measure the distance (L) between the two hub bearing outer tracks (see Fig. 11.11). From this figure subtract 0.16 mm (0.0063 in) for models with the E15 engine, or 1.19 mm (0.0469 in) for all other models, and the remainder will be the thickness of the spacer required to obtain a satisfactory preload. Spacers are available in eighteen increments, and your dealer's parts department will be able to supply the correct spacer once they know the dimension which resulted from your calculation.
13 Fit the disc and hub together and tighten the connecting bolts to the specified torque.
14 Connect the hub with the stub axle carrier. Do this using a long bolt or length of studding with large washers and nuts to draw the hub fully into the carrier, but take great care not to damage the oils seals.
15 Refit the lower balljoint to the stub axle carrier and tighten the nut.
16 Refit the driveshaft with reference to Chapter 8 (photo).

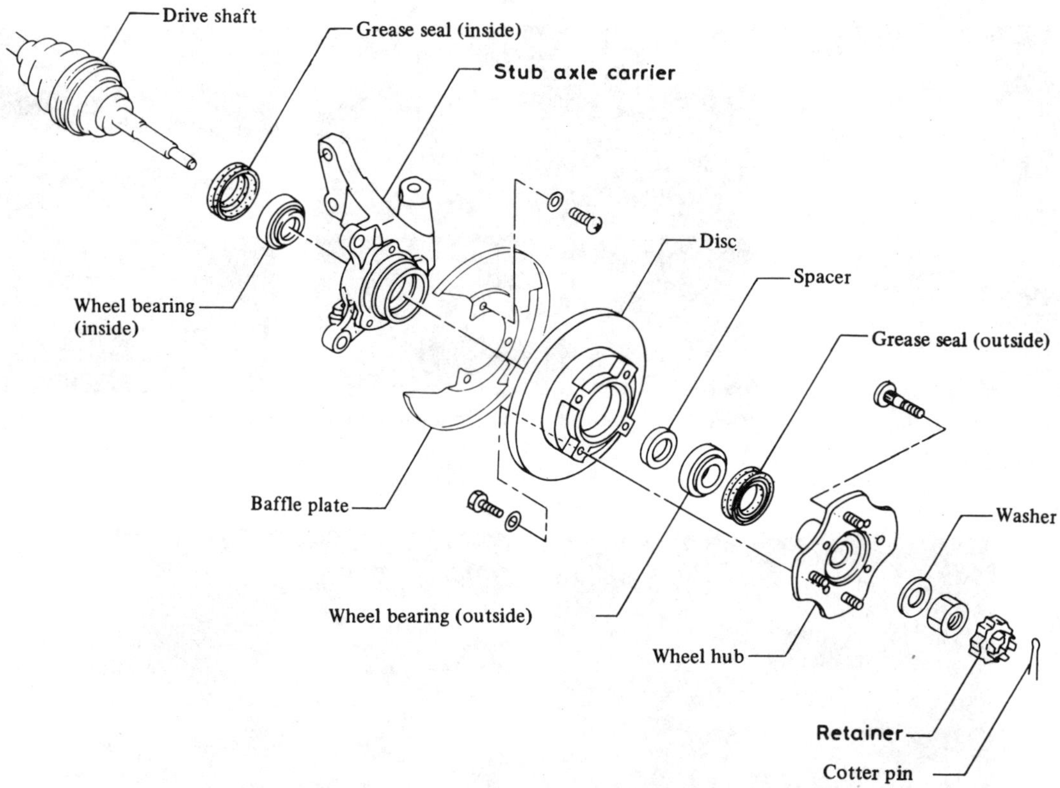

Fig. 11.8 Exploded view of the front hub (Sec 7)

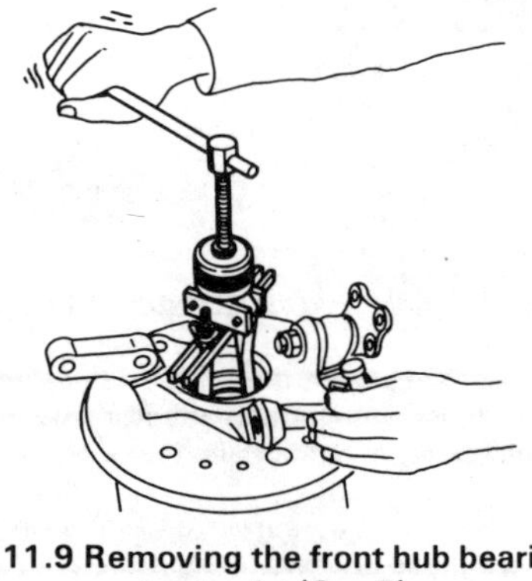

Fig. 11.9 Removing the front hub bearing outer tracks (Sec 7)

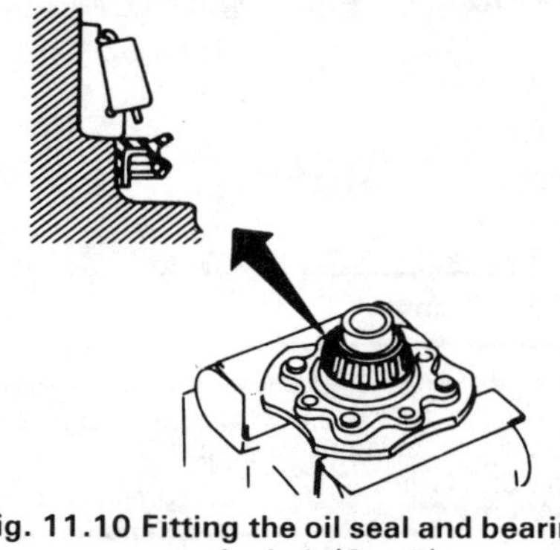

Fig. 11.10 Fitting the oil seal and bearing onto the hub (Sec 7)

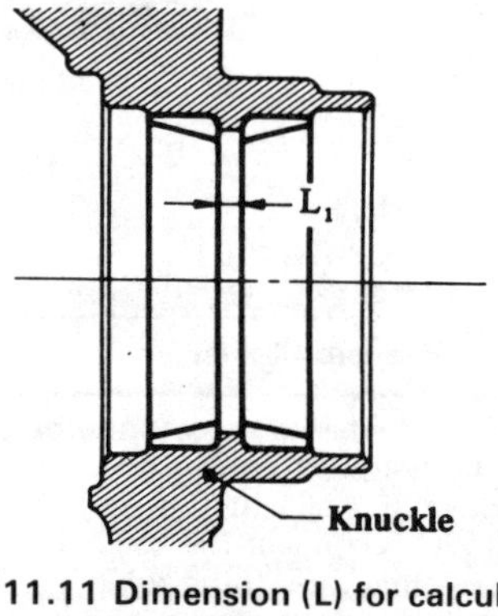

Fig. 11.11 Dimension (L) for calculating front hub bearing spacer thickness (Sec 7)

7.16 Front hub and driveshaft showing correct split pin fitting

8.3 Rear shock absorber upper mounting

8.4 Rear shock absorber lower mounting

8 Rear shock absorber – removal and refitting

1 Jack up the rear of the car and support it on axle stands. Chock the front wheels.
2 Using a trolley jack slightly raise the suspension trailing arm.
3 Working inside the rear compartment, prise off the plastic cover then unscrew the nut and take the seats and rubber cushion from the top of the shock absorber (photo).
4 From under the car unscrew and remove the lower mounting bolt then withdraw the shock absorber (photo).
5 Refitting is a reversal of removal, but make sure that all the rubber bushes and cushions are in good condition and make sure that the upper mounting components are located in their correct sequence.
6 Tighten the mounting nuts and bolts to the specified torque.

9 Rear coil spring – removal and refitting

1 Raise the rear of the vehicle and support it securely. Remove the roadwheel. Chock the front wheels.
2 Support the suspension trailing arm on a jack.
3 Disconnect the shock absorber lower mounting.
4 Lower the jack under the suspension arm until the coil spring is free and can be removed (photo).
5 Refitting is a reversal of removal, but make sure that the rubber insulator is in good condition and that the flattened end of the spring engages correctly with the insulator.

10 Rear suspension trailing arm – removal and refitting

1 Remove the coil spring, as described in Section 9.
2 Unscrew the union and disconnect the rigid brake pipe from the flexible hose. Cap the end of the rigid pipe with a bleed screw dust cap, or similar, to reduce fluid loss.
3 Bend back the clamp and release the handbrake cable from the suspension arm.
4 Refer to Chapter 9 and remove the brake drum, shoes, wheel cylinder and brake backplate.
5 Unbolt and remove the suspension arm pivot bolts (photos). Withdraw the suspension arm from the vehicle.
6 If the suspension arm flexible bushes are worn they can be renewed using a press or a bolt, nut, washers and tubular distance pieces. Before the bushes can be removed, however, the metal flange will first have to be ground off.
7 When fitting the new bushes, align their cut-outs as shown in the diagram (Fig. 11.13).
8 Refit the suspension arm by reversing the removal operations but do not tighten the nuts and bolts to the specified torque settings until the weight of the vehicle is on the roadwheels.
9 Bleed the brake circuit, as described in Chapter 9.

11 Rear hub bearings – adjustment

1 Jack up the rear of the car and support on axle stands. Chock the front wheels.
2 With the roadwheel removed, prise off the grease cap, remove the split pin and nut retainer (photos).
3 Unscrew the nut and remove the thrust washer (photo).
4 Pull the brake drum towards you and catch the outboard bearing which will be displaced. Remove the drum (photos).
5 If the bearings appear dry, work some multi-purpose grease into them.
6 Refit the brake drum, the outboard bearing, and thrust washer and screw on the nut finger tight.
7 Using a torque wrench tighten the nut to 39.0 to 44.0 Nm (29.0

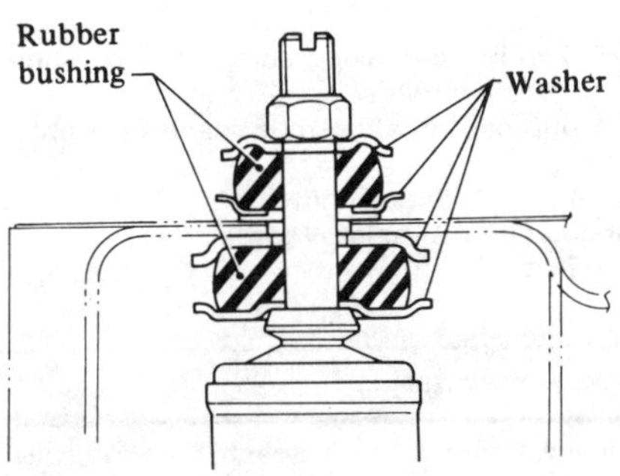

Fig. 11.12 Rear shock absorber upper mounting components (Sec 8)

9.4 Rear coil spring

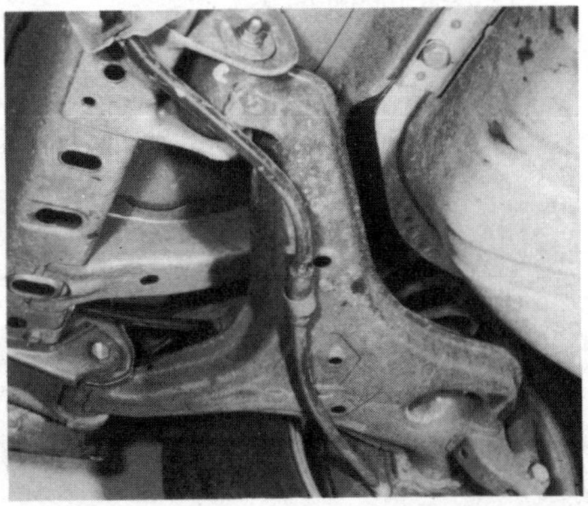

10.5A Rear suspension trailing arm

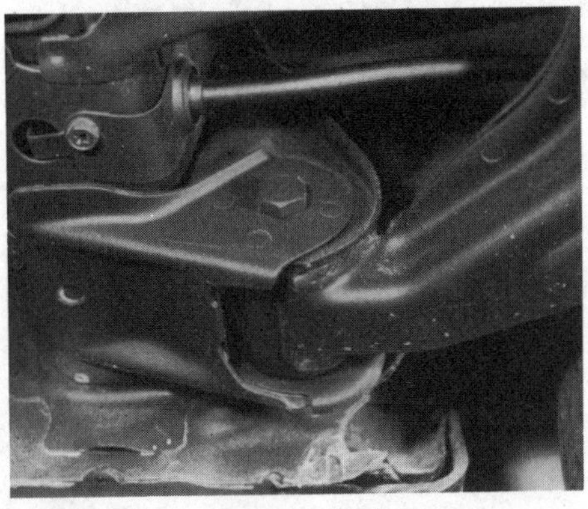

10.5B Trailing arm outer pivot

10.5C Trailing arm inner pivot

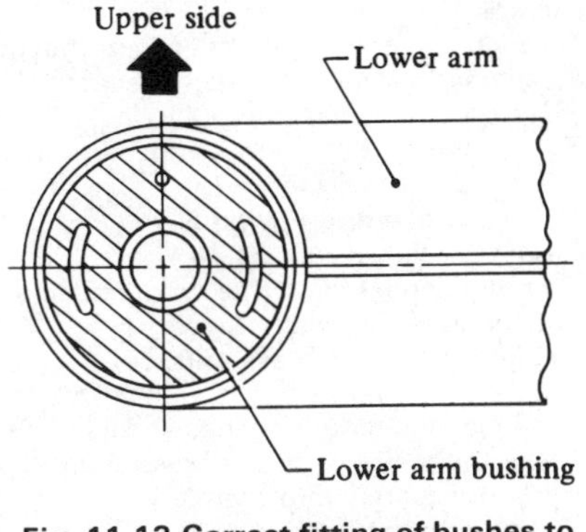

Fig. 11.13 Correct fitting of bushes to rear trailing arm (Sec 10)

11.2A Rear hub grease cap

11.2B Showing rear hub nut split pin

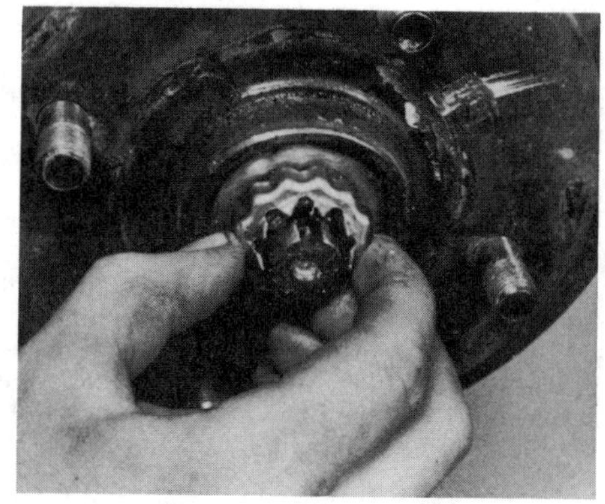
11.2C Removing rear hub nut retainer

11.3 Remove the thrust washer ...

11.4A ... followed by the outer bearing ...

11.4B ... and brake drum

to 33.0 lbf ft), then turn the drum several times in both directions to settle the bearings. Re-tighten the nut to the same torque.
8 Unscrew the nut through 90° which is equivalent to one and a half flats.
9 Fit the nut retainer so that its slots align with the split pin hole and fit a new split pin.
10 Check that the drum/hub turns smoothly without drag. There must not be any endfloat. If there is, and the adjustment has been carried out correctly, suspect worn bearings (see the next Section).
11 Bend the ends of the split pin around the nut retainer.
12 Fill the grease cap only one third full with multi-purpose grease, check that the O-ring is in good condition and tap the cap squarely into position.

12 Rear hub bearings – renewal

1 Worn rear hub bearings may be detected by noise, especially when cornering, or by failure to eliminate bearing endfloat during adjustment.
2 Raise the rear of the car, chock the front wheels, release the handbrake and remove the roadwheel.
3 Prise off the grease cap.
4 Extract the split pin, remove the nut retainer, unscrew the nut and remove the thrust washer.
5 Pull the brake drum towards you and catch the outboard bearing which will be displaced.
6 From the inner face of the drum prise out and discard the oil seal. Lift out the inner bearing.
7 Using a copper or brass drift, drive the bearing outer tracks from the drum.
8 The inner and outer bearings should be renewed at the same time.
9 Clean the recesses in the brake drum and drive the new bearing outer tracks squarely into them.
10 Fit the inboard bearing race and a new oil seal, having worked multi-purpose grease into the bearing rollers and the oil seal lips.
11 Offer the drum onto the stub axle, taking care not to damage the lips of the oil seal with the shaft threads.
12 Work grease into the outboard bearing rollers and fit it onto the stub axle.
13 Fit the thrust washer and nut, finger tight.
14 Adjust the bearing preload and complete reassembly, as described in Section 11 paragraphs 7 to 12.

13 Steering rack bellows – renewal

1 The steering rack bellows should be inspected periodically for splits. Have an assistant turn the steering to full lock while doing this, otherwise the split will not be immediately apparent.
2 Unscrew the nut from the balljoint taper pin and, using an extractor, separate the balljoint from the eye of the steering arm (photo).
3 Release the locknut and then unscrew the balljoint from the tie-rod, counting the number of turns required to remove it.
4 Release the bellows securing bands and pull the bellows from the rack housing and off the tie-rod.
5 If the bellows have been split for some time and dirt has entered, wipe away all the old lubricant and smear the rack (extended) and the rack end balljoint with a suitable grease.
6 Slide on the new bellows and fit the securing bands.
7 Screw the balljoint onto the tie-rod by the same number of turns as was recorded at removal and tighten the locknut. Reconnect the balljoint to the steering arm of the stub axle carrier, use a new split pin (photo).
8 Check the front wheel alignment by referring to Section 26.

14 Tie-rod end balljoint – renewal

1 The removal and refitting of a balljoint is covered in the preceding Section.

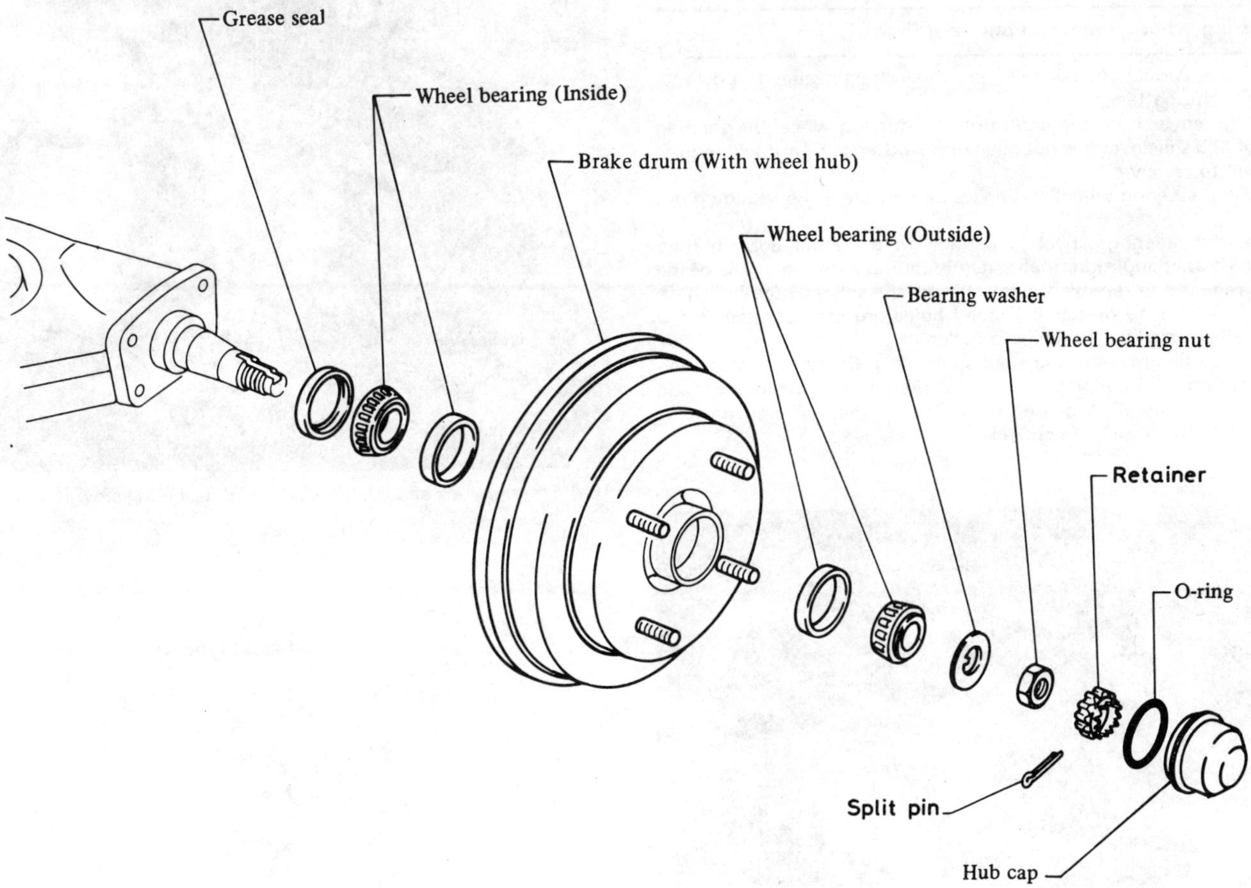

Fig. 11.14 Exploded view of the rear hub (Sec 12)

13.2 Using a separator tool to release the track rod end from the steering arm

13.7 Fitting a new split pin to the track rod end

2 Always check the front wheel alignment after having fitted a new balljoint – refer to Section 26.

15 Steering wheel – removal and refitting

1 Set the front roadwheels in the straight-ahead position, disconnect the battery.
2 Pull the cover from the centre of the steering wheel by gripping the top of it. If the cover has not been removed since new it will require a firm pull to remove it.
3 Hold the steering wheel stationary and unscrew the retaining nut (photo).
4 Mark the steering wheel in relation to the inner column then remove it. If anything more than a gentle thump with the palms of the hands is required to remove the wheel from the splined steering shaft, a puller will have to be used. Tapped holes are incorporated in the steering wheel hub for attaching a puller.
5 Apply a little grease to the shaft splines and the horn contact ring before refitting the steering wheel. Make sure that the previously made marks are correctly aligned. Tighten the nut to the specified torque.
6 Refit the cover and re-connect the battery lead.

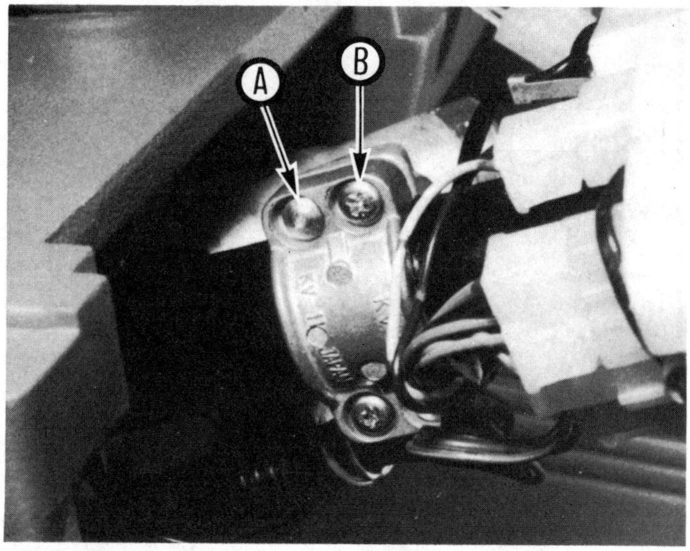

16.4 Steering lock shear screw (A) and crosshead screw (B)

15.3 Steering wheel retaining nut with cover removed

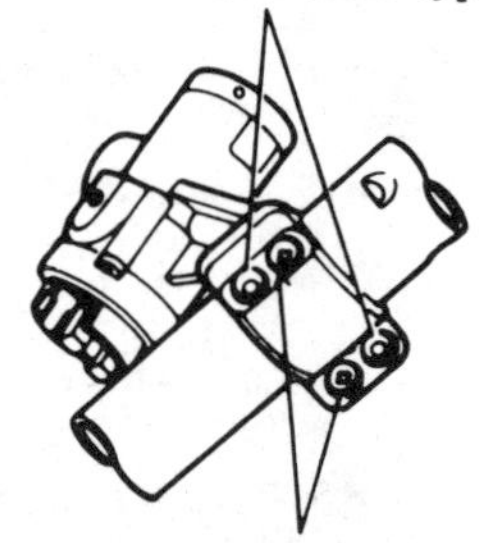

Fig. 11.15 Correct location of steering lock retaining screws (Sec 16)

16 Steering lock – removal and refitting

1 Remove the steering wheel as described in Section 15.
2 Remove the column shrouds (one screw).
3 Disconnect the wiring from the ignition switch.
4 The steering lock is secured with two cross-head screws and two shear type screws (photo). Unscrew the cross-head screws and drill out the shear type screws. Remove the steering lock.
5 To refit the lock align it with the hole in the outer column, locate the clamp plate, and insert the screws finger tight (Fig. 11.15).
6 Check that the lock operates correctly then tighten the cross-head screws and tighten the shear type screws until the heads break off.
7 Re-connect the wiring then refit the shrouds and steering wheel.

17 Steering shaft lower joint – removal and refitting

1 To provide better access remove the steering column aperture dust excluding cover from the bulkhead. Set the steering in the straight-ahead position.
2 Unscrew the pinch-bolts from both universally-jointed couplings, prise the jaws of the couplings open just enough to be able to withdraw the lower joint from the upper coupling and then from the steering gear pinion shaft (photo).

17.2 Steering shaft lower joint

3 When refitting the joint, connect it first to the upper coupling and then to the pinion. Note that the pinch-bolt cut-out must be aligned with the upper coupling clamp jaws and that the jaws of the lower coupling must be aligned with the mark on the steering pinion spacer.
4 Tighten the pinch-bolts to the specified torque.

18 Steering column – removal and refitting

Non-tilt type

1 Remove the steering wheel (Section 15).
2 Remove the shaft lower joint (Section 17).
3 Remove the shrouds from the upper steering column. Disconnect the battery.
4 Extract the screws, disconnect the wiring harness plugs and remove the steering column switch.
5 Remove the heater ducts.
6 Remove the nuts from the column dust excluder on the bulkhead.
7 Unscrew the upper and lower column retaining bolts and withdraw the column from under the facia panel. Recover the spacer plates noting their location.
8 Check the column length A (Fig. 11.17). If it does not conform, renew the assembly.

Tilt type

9 The operations are very similar to the non-tilt type, except for the tilt lever components at the upper end of the column which are shown in Figs. 11.18 and 11.19. Check the column length A (Fig. 11.20). If it does not conform, renew the assembly.

Refitting – all types

10 Offer the column into position making sure that the sliding plates are correctly located. Fit the bolts finger tight at this stage.
11 Fit the lower joint (Section 17), again leaving the pinch-bolts finger tight.
12 Check that none of the brackets or couplings are under stress and then tighten the column bolts to the specified torque.

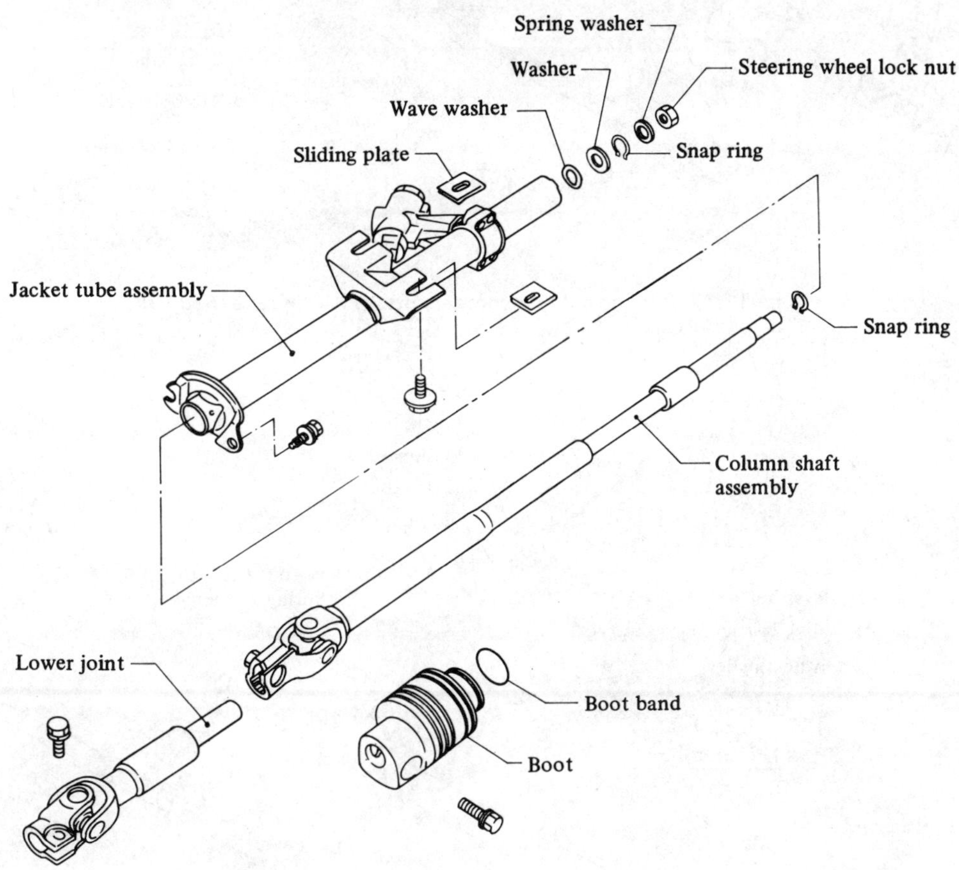

Fig. 11.16 Steering column components – non-tilt type (Sec 18)

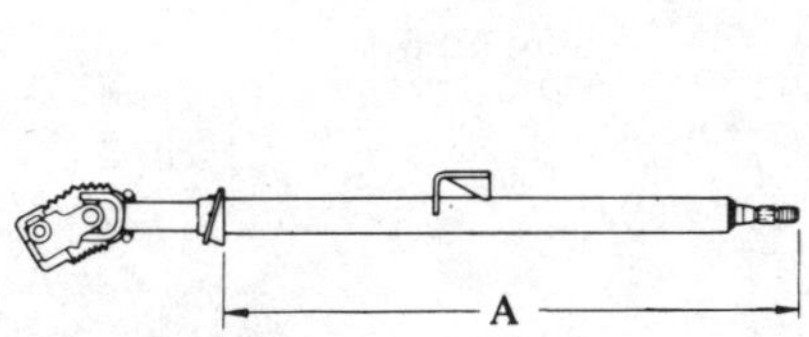

Fig. 11.17 Steering column length – non-tilt type (Sec 18)

A = 481.3 to 484.3 mm (18.95 to 19.07 in)

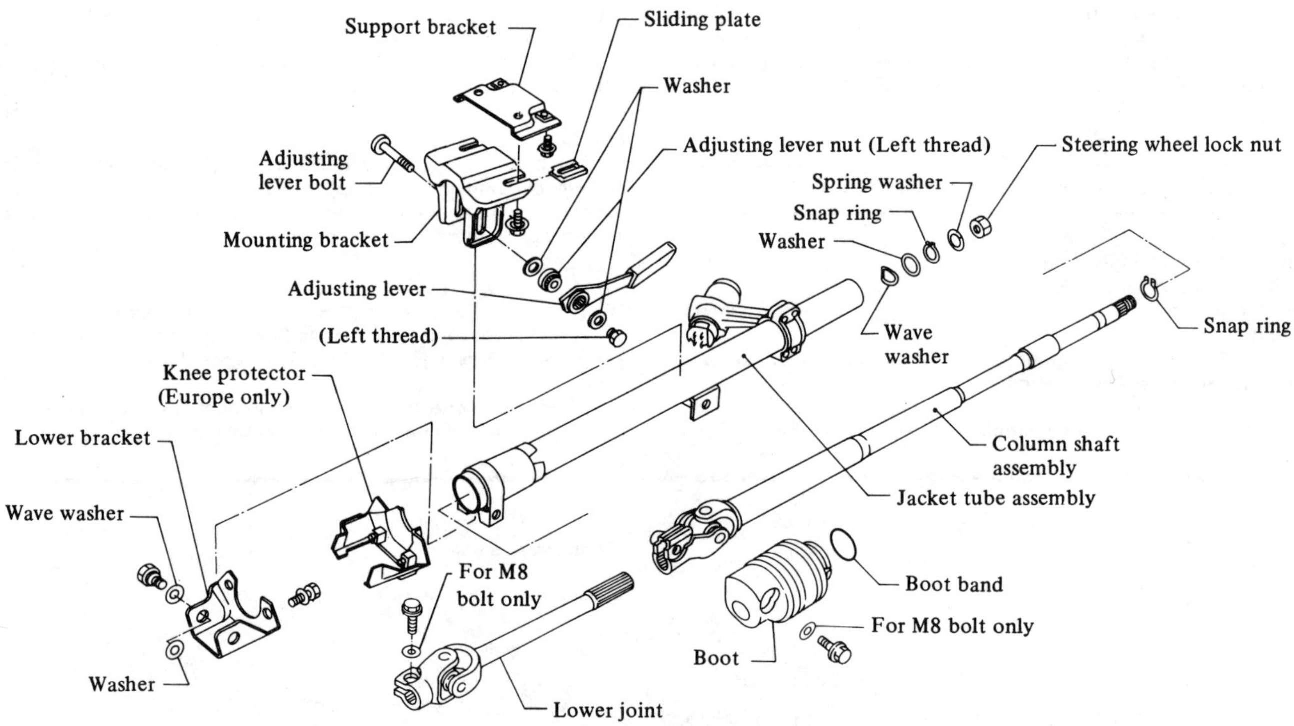

Fig. 11.18 Steering column components – FKK tilt type (Sec 18)

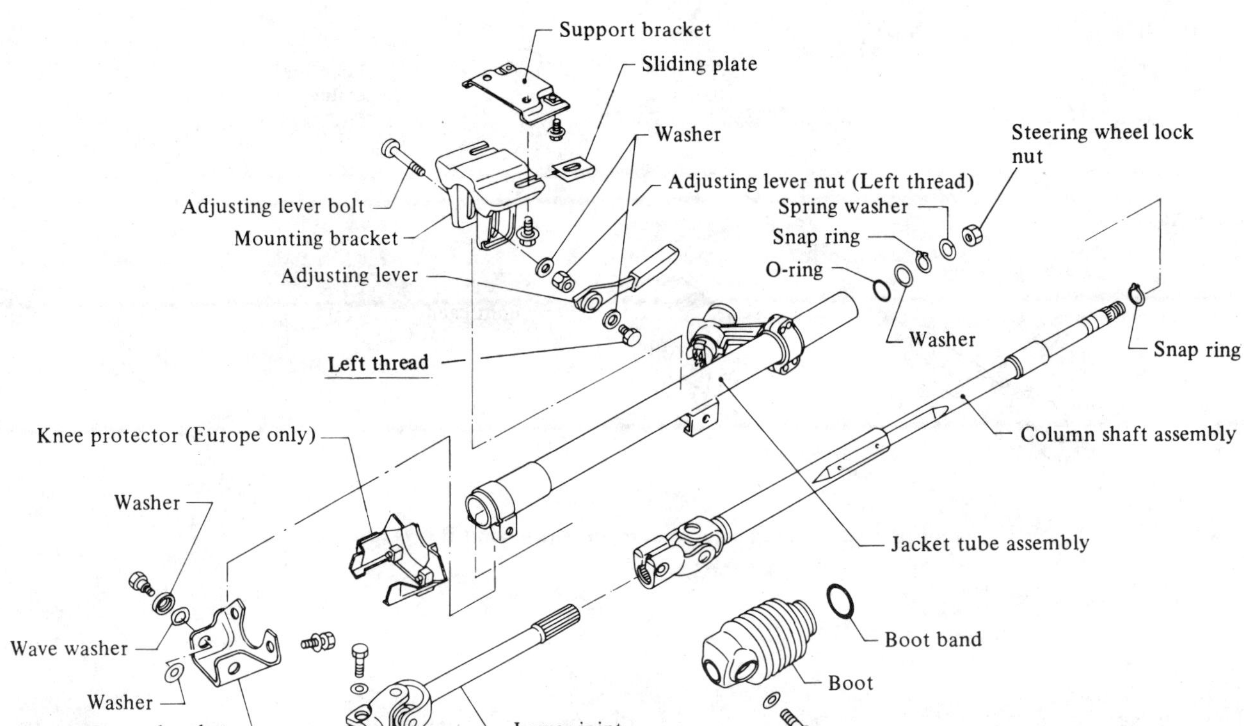

Fig. 11.19 Steering column components – NSK tilt type (Sec 18)

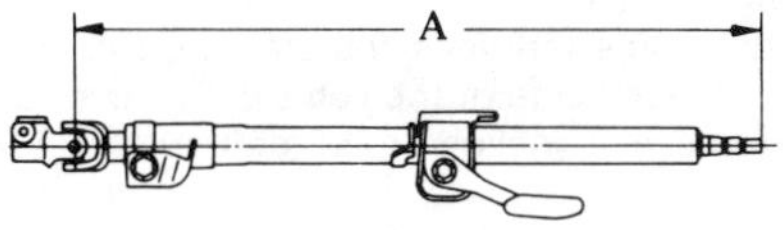

Fig. 11.20 Steering column length – tilt type (Sec 18)

A = 593.6 to 595.2 mm (23.37 to 23.43 in)

19 Steering column (non-tilt type) – overhaul

1 The steering column shaft can be removed from the column jacket tube if the bearings require greasing. If the bearings are worn a new jacket tube, complete with bearings, will be required.
2 Remove the steering column, as described in the preceding Section, and unlock the steering lock using the ignition key.
3 With a pair of circlip pliers, extract the circlip from the upper end of the column.
4 Remove the plain and wave washers and then pull the shaft out of the jacket tube.
5 Grease the bearings and bushes, push the shaft into the jacket tube and locate the wave washer followed by the plain washer.
6 Make sure that the circlip which is located below the upper bearing, and the one which is to be fitted above the plain washer both have their rounded edges towards the bearing. New circlips should always be used at reassembly.

20 Steering column (tilt type) – overhaul

1 The operations are similar to those described in the preceding Section, but there are slight differences in detail between the two different makes of column. At the upper end of the FKK column a wave washer is used, as on non-tilt types, while on NSK columns an O-ring is fitted.
2 Make sure that the column lever mounting bracket is correctly assembled according to column make.
3 The sequence of fitting the components of the tilt mechanism is important and a little grease should be applied to the sliding surfaces. The adjusting lever nut and bolt have *left-hand threads* and the nut must only be tightened while the tilt lever is held at between 20° and 30° to the centre-line of the column, as shown in Fig. 11.21.

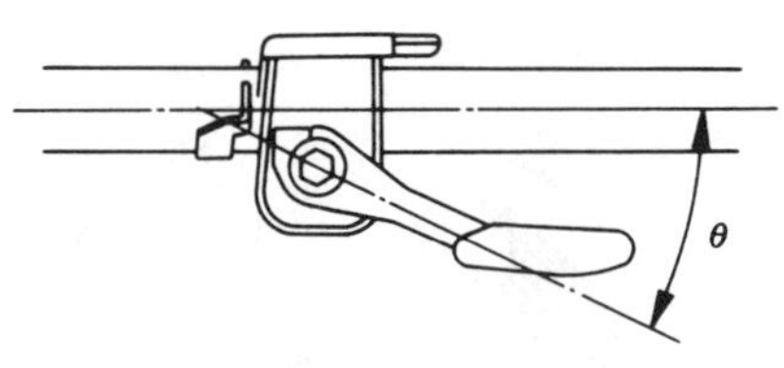

Fig. 11.21 Tilt lever setting angle (Sec 20)

$\theta = 20° \text{ to } 30°$

21 Manual steering gear – removal and refitting

1 Raise the front of the vehicle and support it securely using axle stands under the side-members. Apply the handbrake and remove the front roadwheels.
2 Disconnect both tie-rod balljoints from the steering arms on the stub axle carriers.
3 Remove the steering shaft lower joint, as described in Section 17.
4 Unscrew the rack housing mounting nuts and lift the steering gear from the bulkhead (photo).

21.4 Showing the manual steering gear and a mounting bolt

5 Refitting is a reversal of removal, but make sure that the arrows on the clamp housings point upwards in order to give the rack housing an inclination of 13.5° rearwards.
6 Tighten all nuts and bolts to the specified torques, and finally check the front wheel alignment as described in Section 26.

22 Power-assisted steering gear – removal and refitting

1 Raise the front of the vehicle and support it on axle stands by the side-members. Apply the handbrake and remove the front roadwheels.
2 Disconnect the hydraulic hose clamp at the pinion housing on the steering gear to provide access to the union nut on the pipeline. Unscrew the union nut and allow the fluid to drain into a suitable container.
3 Extract the split pins and unscrew the castellated nuts from the tie-rod end balljoints. With a suitable tool, disconnect the balljoints from the steering arms on the stub axle carriers.
4 Support the transmission on a jack with a block of wood as an insulator.
5 Disconnect the exhaust downpipe from the manifold and also disconnect the exhaust support bracket just ahead of the steering gear.
6 Unbolt and remove the engine rear mounting.
7 Unscrew and remove the bolts from the steering gear mounting clamps.
8 Remove the steering shaft lower joint, as described in Section 17.
9 Withdraw the steering gear from under the front wing, moving it in the direction of the arrows as shown in the diagram (Fig. 11.22).

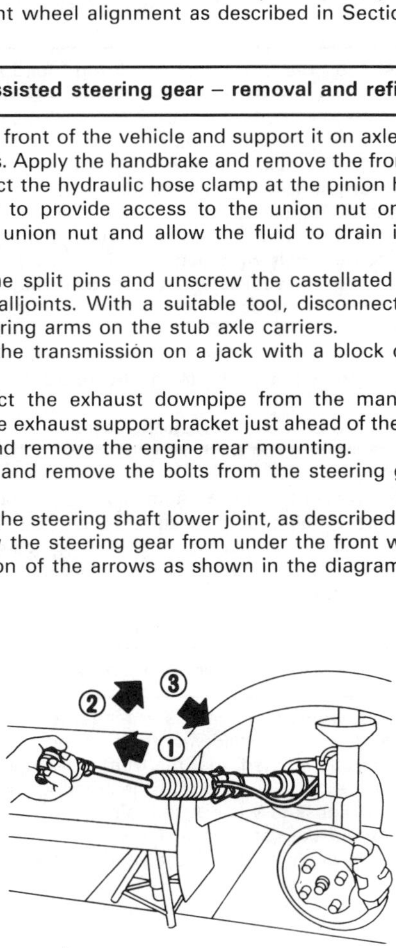

Fig. 11.22 Method of removing the power assisted steering gear (Sec 22)

10 Refitting is a reversal of removal, but make sure that the arrows on the mounting clamps are pointing upwards in order to give the rack housing an inclination of 19.3°.
11 Fill and bleed the system, as described in Section 25 and finally check the front wheel alignment as described in Section 26.

23 Power steering pump – removal and refitting

1 Release the pump drivebelt adjuster link lockbolt and turn the adjuster bolt to release the tension on the belt. Slip the belt from the pulleys.
2 Disconnect the pressure hose from the pump by unscrewing the banjo union bolt. Allow the fluid to drain into a suitable container.
3 Take off the return hose clamp.
4 Unbolt and remove the pump.
5 If the connecting lines are to be removed, unscrew the union nuts and hose clips.
6 Refitting and reconnection are reversals of disconnection and removal.
7 Tension the pump drivebelt, as described in Chapter 2.
8 Fill and bleed the system, as described in Section 25.

24 Steering gear – overhaul

1 It is not recommended that the manual or power-assisted steering gear or the power steering pump are overhauled.
2 Due to the precise nature of the assembly work and the need for special tools for measuring turning torque, it is preferable to purchase a new or factory-reconditioned unit when the original one becomes worn or develops a fault.

25 Power-assisted steering – fluid level and bleeding

1 At the intervals specified in Section 2 unscrew the power steering pump filler cap when the engine and pump are cold and observe the level of fluid on the dipstick. Add fluid of the correct type to bring the fluid level between the low and high marks.
2 If the system pipelines have been disconnected or new components fitted then, after reassembly, the system must be bled.
3 Fill the pump reservoir with fluid.
4 Raise the front of the vehicle until the roadwheels are off the floor.
5 Turn the steering from lock to lock ten times and then top up the fluid in the reservoir so that it is at the correct level on the dipstick.
6 Start the engine and turn the steering wheel left and right lock until the fluid becomes hot to the touch (60 to 80°C – 140 to 176° F).
7 Switch off the engine and top up the reservoir, if necessary.
8 Start the engine and run for five seconds. Switch off and top up the fluid, if necessary.
9 If air is still present in the system, which will be indicated by the steering wheel being stiff to turn, repeat the operations as previously described. When turning the steering from lock to lock during bleeding, do not hold it at full lock for more than fifteen seconds while the engine is running.

26 Steering angles and front wheel alignment – general

1 Accurate front wheel alignment is essential to good steering and for even tyre wear. Before considering the steering angles, check that the tyres are correctly inflated, that the front wheels are not buckled, the hub bearings are not worn and that the steering linkage is in good order, without slackness or wear at the joints.
2 Wheel alignment consists of four factors:

Camber, is the angle at which the roadwheels are set from the vertical when viewed from the front or rear of the vehicle. Positive camber is the angle (in degrees) that the wheels are tilted outwards at the top from the vertical.

Castor, is the angle between the steering axis and a vertical line when viewed from each side of the vehicle. Positive castor is indicated when the steering axis is inclined towards the rear of the vehicle at its upper end.

Steering axis inclination, is the angle, when viewed from the front or rear of the vehicle, between the vertical and an imaginary line drawn between the upper and lower front suspension strut mountings.

Toe, is the amount by which the distance between the front inside edges of the roadwheel rim differs from that between the rear inside edges. If the distance between the front edges is less than that at the rear, the wheels are said to toe-in. If the distance between the front inside edges is greater than that at the rear, the wheels toe-out.

3 Owing to the need for precision gauges to measure the small angles of the steering and suspension settings, it is preferable that checking of camber and castor is left to a service station having the necessary equipment. Camber and castor are set during production of the vehicle, and any deviation from the specified angle will be due to accident damage or gross wear in the suspension mountings.
4 To check the front wheel alignment, first make sure that the lengths of both tie-rods are equal when the steering is in the straight-ahead position. The tie-rod lengths can be adjusted for length if necessary by releasing the locknuts from the balljoint ends and rotating the rods. Flats are provided on the rods in order to hold them still with an open-ended spanner when the locknut is undone.
5 Obtain a tracking gauge. These are available in various forms from accessory stores, or one can be fabricated from a length of steel tubing suitably cranked to clear the sump and bellhousing and having a setscrew and locknut at one end.
6 With the gauge, measure the distances between the two wheel inner rims (at hub height) at the rear of the wheel. Push the vehicle forward to rotate the wheel through 180° (half a turn) and measure the distance between the wheel inner rims, again at hub height, at the front of the wheel. This last measurement should differ from the first by the appropriate toe-in which is given in the Specifications. The vehicle must be on level ground.
7 Where the toe-in is found to be incorrect, release the tie-rod balljoint locknut and turn the tie-rods equally. Only turn them a quarter of a turn at a time before re-checking the alignment. Do not grip the threaded part of the tie-rod during adjustment, but use an open-ended spanner on the flats provided. It is important not to allow the tie-rods to become unequal in length during adjustment, otherwise the alignment of the steering wheel will become incorrect and tyre scrubbing will occur on turns.
8 On completion, tighten the locknuts without disturbing the setting. Check that the balljoint is at the centre of its arc of travel.

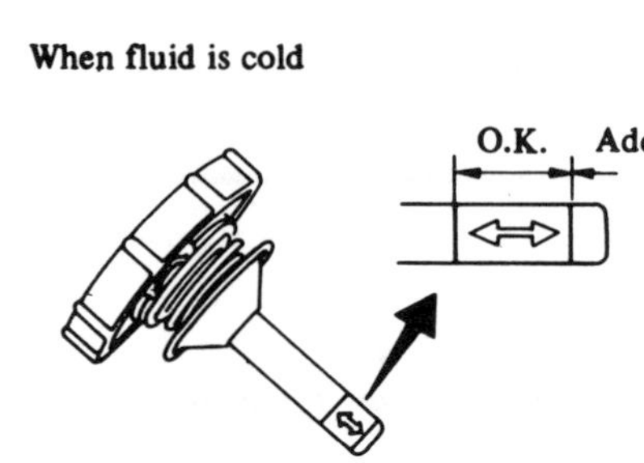

Fig. 11.23 Power-assisted steering pump filler cap and dipstick markings (Sec 25)

27 Fault diagnosis – suspension and steering

Symptom	Reason(s)
Front suspension	
Vehicle wanders	Incorrect wheel alignment Worn front control arm balljoints
Heavy or stiff steering	Incorrect front wheel alignment Incorrect tyre pressures
Wheel wobble or vibration	Roadwheels out of balance Roadwheel buckled Incorrect front wheel alignment Faulty strut Weak coil spring
Excessive pitching or rolling on corners or during braking	Faulty strut Weak or broken coil spring
Tyre squeal when cornering	Incorrect front wheel alignment Incorrect tyre pressures
Abnormal tyre wear	Incorrect tyre pressures Incorrect front wheel alignment Worn hub bearing
Rear suspension	
Poor roadholding and wander	Faulty shock absorber Weak coil spring Worn or incorrectly adjusted hub bearing Worn trailing arm bush
Manual steering gear	
Stiff action	Lack of rack lubrication Seized tie-rod end balljoint Seized suspension lower balljoint
Free movement at steering wheel	Wear in tie-rod balljoint Wear in rack teeth
Knocking when traversing uneven surface	Incorrectly adjusted rack slipper
Power-assisted steering gear	
The symptoms and reasons applicable to manual steering gear will apply, plus the following:	
Stiff action or no return action	Slipping pump drivebelt Air in fluid Steering column out of alignment Castor angle incorrect due to damage or gross wear in bushes and mountings
Steering effort on both locks unequal	Leaking seal in steering gear Clogged fluid passage within gear assembly
Noisy pump	Loose pulley Kinked hose Clogged filter in fluid reservoir Low fluid level

Chapter 12 Bodywork and fittings

Contents

1 General description

The bodywork on all versions is of welded steel, unitary construction. In the interests of economical repair, the front wings are readily detachable, but other body panels are not.

The vehicles are protected against corrosion by dipping and also by the provision of stone guards under the front wings.

All models in the range are well-equipped and certain factory-fitted options are also available.

2 Maintenance – bodywork and underframe

1 The general condition of a vehicle's bodywork is the one thing that significantly affects its value. Maintenance is easy but needs to be regular. Neglect, particularly after minor damage, can lead quickly to further deterioration and costly repair bills. It is important also to keep watch on those parts of the vehicle not immediately visible, for instance the underside, inside all the wheel arches and the lower part of the engine compartment.

2 The basic maintenance routine for the bodywork is washing – preferably with a lot of water, from a hose. This will remove all the loose solids which may have stuck to the vehicle. It is important to flush these off in such a way as to prevent grit from scratching the finish. The wheel arches and underframe need washing in the same way to remove any accumulated mud which will retain moisture and tend to encourage rust. Paradoxically enough, the best time to clean the underframe and wheel arches is in wet weather when the mud is thoroughly wet and soft. In very wet weather the underframe is usually cleaned of large accumulations automatically and this is a good time for inspection.

3 Periodically, it is a good idea to have the whole of the underframe of the vehicle steam cleaned, engine compartment included, so that a thorough inspection can be carried out to see what minor repairs and renovations are necessary. Steam cleaning is available at many garages and is necessary for removal of the accumulation of oily grime which sometimes is allowed to become thick in certain areas. If steam cleaning facilities are not available, there are one or two excellent grease solvents available which can be brush applied. The dirt can then be simply hosed off.

4 After washing paintwork, wipe off with a chamois leather to give an unspotted clear finish. A coat of clear protective wax polish will give added protection against chemical pollutants in the air. If the

2.4A Door drain hole

2.4B Clearing a sill drain hole

paintwork sheen has dulled or oxidised, use a cleaner/polisher combination to restore the brilliance of the shine. This requires a little effort, but such dulling is usually caused because regular washing has been neglected. Always check that the door and ventilator opening drain holes and pipes are completely clear so that water can be drained out (photos). Bright work should be treated in the same way as paintwork. Windscreens and windows can be kept clear of the smeary film which often appears, by adding a little ammonia to the water. If they are scratched, a good rub with a proprietary metal polish will often clear them. Never use any form of wax or other body or chromium polish on glass.

3 Maintenance – upholstery and carpets

Mats and carpets should be brushed or vacuum cleaned regularly to keep them free of grit. If they are badly stained remove them from the vehicle for scrubbing or sponging and make quite sure they are dry before refitting. Seats and interior trim panels can be kept clean by wiping with a damp cloth. If they do become stained (which can be more apparent on light coloured upholstery) use a little liquid detergent and a soft nail brush to scour the grime out of the grain of the material. Do not forget to keep the headlining clean in the same way as the upholstery. When using liquid cleaners inside the vehicle do not over-wet the surfaces being cleaned. Excessive damp could get into the seams and padded interior causing stains, offensive odours or even rot. If the inside of the vehicle gets wet accidentally it is worthwhile taking some trouble to dry it out properly, particularly where carpets are involved. *Do not leave oil or electric heaters inside the vehicle for this purpose.*

4 Minor body damage – repair

The photographic sequences on pages 234 and 235 illustrate the operations detailed in the following sub-sections.

Repair of minor scratches in bodywork

If the scratch is very superficial, and does not penetrate to the metal of the bodywork, repair is very simple. Lightly rub the area of the scratch with a paintwork renovator, or a very fine cutting paste, to remove loose paint from the scratch and to clear the surrounding bodywork of wax polish. Rinse the area with clean water.

Apply touch-up paint to the scratch using a fine paint brush; continue to apply fine layers of paint until the surface of the paint in the scratch is level with the surrounding paintwork. Allow the new paint at least two weeks to harden: then blend it into the surrounding paintwork by rubbing the scratch area with a paintwork renovator or a very fine cutting paste. Finally, apply wax polish.

Where the scratch has penetrated right through to the metal of the bodywork, causing the metal to rust, a different repair technique is required. Remove any loose rust from the bottom of the scratch with a penknife, then apply rust inhibiting paint to prevent the formation of rust in the future. Using a rubber or nylon applicator fill the scratch with bodystopper paste. If required, this paste can be mixed with cellulose thinners to provide a very thin paste which is ideal for filling narrow scratches. Before the stopper-paste in the scratch hardens, wrap a piece of smooth cotton rag around the top of a finger. Dip the finger in cellulose thinners and then quickly sweep it across the surface of the stopper-paste in the scratch; this will ensure that the surface of the stopper-paste is slightly hollowed. The scratch can now be painted over as described earlier in this Section.

Repair of dents in bodywork

When deep denting of the vehicle's bodywork has taken place, the first task is to pull the dent out, until the affected bodywork almost attains its original shape. There is little point in trying to restore the original shape completely, as the metal in the damaged area will have stretched on impact and cannot be reshaped fully to its original contour. It is better to bring the level of the dent up to a point which is about $\frac{1}{8}$ in (3 mm) below the level of the surrounding bodywork. In cases where the dent is very shallow anyway, it is not worth trying to pull it out at all. If the underside of the dent is accessible, it can be hammered out gently from behind, using a mallet with a wooden or plastic head. Whilst doing this, hold a suitable block of wood firmly against the outside of the panel to absorb the impact from the hammer blows and thus prevent a large area of the bodywork from being 'belled-out'.

Should the dent be in a section of the bodywork which has a double skin or some other factor making it inaccessible from behind, a different technique is called for. Drill several small holes through the metal inside the area – particularly in the deeper section. Then screw long self-tapping screws into the holes just sufficiently for them to gain a good purchase in the metal. Now the dent can be pulled out by pulling on the protruding heads of the screws with a pair of pliers.

The next stage of the repair is the removal of the paint from the damaged area, and from an inch or so of the surrounding 'sound' bodywork. This is accomplished most easily by using a wire brush or abrasive pad on a power drill, although it can be done just as effectively by hand using sheets of abrasive paper. To complete the preparation for filling, score the surface of the bare metal with a screwdriver or the tang of a file, or alternatively, drill small holes in the affected area. This will provide a really good 'key' for the filler paste.

To complete the repair see the Section on filling and re-spraying.

Repair of rust holes or gashes in bodywork

Remove all paint from the affected area and from an inch or so of the surrounding 'sound' bodywork, using an abrasive pad or a wire brush on a power drill. If these are not available a few sheets of abrasive paper will do the job just as effectively. With the paint removed you will be able to gauge the severity of the corrosion and therefore decide whether to renew the whole panel (if this is possible) or to repair the affected area. New body panels are not as expensive as most people think and it is often quicker and more satisfactory to fit a new panel than to attempt to repair large areas of corrosion.

Remove all fittings from the affected area except those which will act as a guide to the original shape of the damaged bodywork (eg headlamp shells etc). Then, using tin snips or a hacksaw blade, remove all loose metal and any other metal badly affected by corrosion. Hammer the edges of the hole inwards in order to create a slight depression for the filler paste.

Wire brush the affected area to remove the powdery rust from the surface of the remaining metal. Paint the affected area with rust inhibiting paint; if the back of the rusted area is accessible treat this also.

Before filling can take place it will be necessary to block the hole in some way. This can be achieved by the use of aluminium or plastic mesh, or aluminium tape.

Aluminium or plastic mesh is probably the best material to use for a large hole. Cut a piece to the approximate size and shape of the hole to be filled, then position it in the hole so that its edges are below the level of the surrounding bodywork. It can be retained in position by several blobs of filler paste around its periphery.

This sequence of photographs deals with the repair of the dent and paintwork damage shown in this photo. The procedure will be similar for the repair of a hole. It should be noted that the procedures given here are simplified – more explicit instructions will be found in the text

In the case of a dent the first job – after removing surrounding trim – is to hammer out the dent where access is possible. This will minimise filling. Here, the large dent having been hammered out, the damaged area is being made slightly concave

Now all paint must be removed from the damaged area, by rubbing with coarse abrasive paper. Alternatively, a wire brush or abrasive pad can be used in a power drill. Where the repair area meets good paintwork, the edge of the paintwork should be 'feathered', using a finer grade of abrasive paper

In the case of a hole caused by rusting, all damaged sheet-metal should be cut away before proceeding to this stage. Here, the damaged area is being treated with rust remover and inhibitor before being filled

Mix the body filler according to its manufacturer's instructions. In the case of corrosion damage, it will be necessary to block off any large holes before filling – this can be done with aluminium or plastic mesh, or aluminium tape. Make sure the area is absolutely clean before ...

... applying the filler. Filler should be applied with a flexible applicator, as shown, for best results; the wooden spatula being used for confined areas. Apply thin layers of filler at 20-minute intervals, until the surface of the filler is slightly proud of the surrounding bodywork

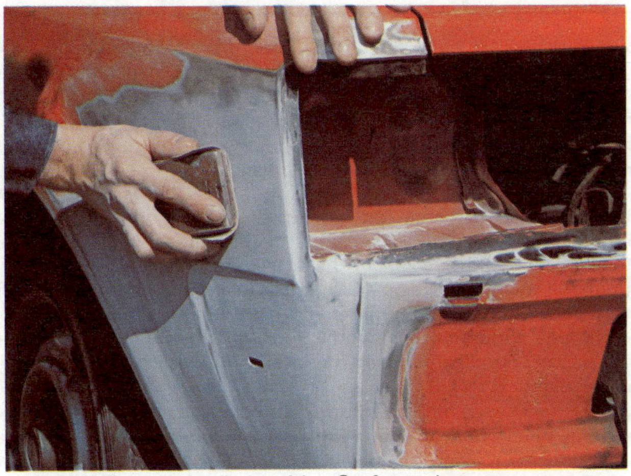

Initial shaping can be done with a Surform plane or Dreadnought file. Then, using progressively finer grades of wet-and-dry paper, wrapped around a sanding block, and copious amounts of clean water, rub down the filler until really smooth and flat. Again, feather the edges of adjoining paintwork

The whole repair area can now be sprayed or brush-painted with primer. If spraying, ensure adjoining areas are protected from over-spray. Note that at least one inch of the surrounding sound paintwork should be coated with primer. Primer has a 'thick' consistency, so will find small imperfections

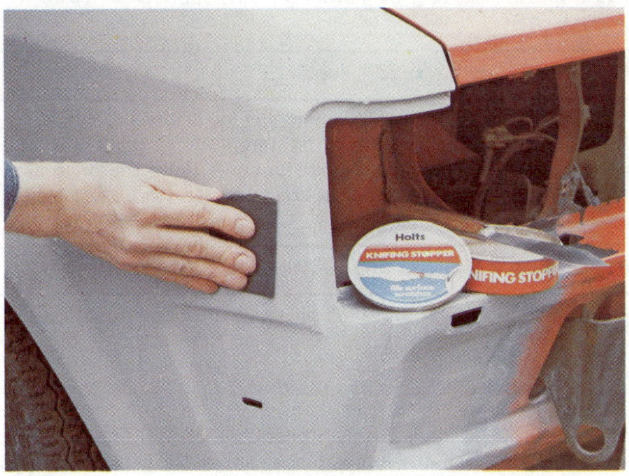

Again, using plenty of water, rub down the primer with a fine grade wet-and-dry paper (400 grade is probably best) until it is really smooth and well blended into the surrounding paintwork. Any remaining imperfections can now be filled by carefully applied knifing stopper paste

When the stopper has hardened, rub down the repair area again before applying the final coat of primer. Before rubbing down this last coat of primer, ensure the repair area is blemish-free – use more stopper if necessary. To ensure that the surface of the primer is really smooth use some finishing compound

The top coat can now be applied. When working out of doors, pick a dry, warm and wind-free day. Ensure surrounding areas are protected from over-spray. Agitate the aerosol thoroughly, then spray the centre of the repair area, working outwards with a circular motion. Apply the paint as several thin coats

After a period of about two weeks, which the paint needs to harden fully, the surface of the repaired area can be 'cut' with a mild cutting compound prior to wax polishing. When carrying out bodywork repairs, remember that the quality of the finished job is proportional to the time and effort expended

Aluminium tape should be used for small or very narrow holes. Pull a piece off the roll and trim it to the approximate size and shape required, then pull off the backing paper (if used) and stick the tape over the hole; it can be overlapped if the thickness of one piece is insufficient. Burnish down the edges of the tape with the handle of a screwdriver or similar, to ensure that the tape is securely attached to the metal underneath.

Bodywork repairs – filling and re-spraying

Before using this Section, see the Sections on dent, deep scratch, rust holes and gash repairs.

Many types of bodyfiller are available, but generally speaking those proprietary kits which contain a tin of filler paste and a tube of resin hardener are best for this type of repair. A wide, flexible plastic or nylon applicator will be found invaluable for imparting a smooth and well contoured finish to the surface of the filler.

Mix up a little filler on a clean piece of card or board – measure the hardener carefully (follow the maker's instructions on the pack) otherwise the filler will set too rapidly or too slowly.

Using the applicator apply the filler paste to the prepared area; draw the applicator across the surface of the filler to achieve the correct contour and to level the filler surface. As soon as a contour that approximates to the correct one is achieved, stop working the paste – if you carry on too long the paste will become sticky and begin to 'pick up' on the applicator. Continue to add thin layers of filler paste at twenty-minute intervals until the level of the filler is just proud of the surrounding bodywork.

Once the filler has hardened, excess can be removed using a metal plane or file. From then on, progressively finer grades of abrasive paper should be used, starting with a 40 grade production paper and finishing with 400 grade wet-and-dry paper. Always wrap the abrasive paper around a flat rubber, cork, or wooden block – otherwise the surface of the filler will not be completely flat. During the smoothing of the filler surface the wet-and-dry paper should be periodically rinsed in water. This will ensure that a very smooth finish is imparted to the filler at the final stage.

At this stage the 'dent' should be surrounded by a ring of bare metal, which in turn should be encircled by the finely 'feathered' edge of the good paintwork. Rinse the repair area with clean water, until all of the dust produced by the rubbing-down operation has gone.

Spray the whole repair area with a light coat of primer – this will show up any imperfections in the surface of the filler. Repair these imperfections with fresh filler paste or bodystopper, and once more smooth the surface with abrasive paper. If bodystopper is used, it can be mixed with cellulose thinners to form a really thin paste which is ideal for filling small holes. Repeat this spray and repair procedure until you are satisfied that the surface of the filler, and the feathered edge of the paintwork are perfect. Clean the repair area with clean water and allow to dry fully.

The repair area is now ready for final spraying. Paint spraying must be carried out in a warm, dry, windless and dust free atmosphere. This condition can be created artificially if you have access to a large indoor working area, but if you are forced to work in the open, you will have to pick your day very carefully. If you are working indoors, dousing the floor in the work area with water will help to settle the dust which would otherwise be in the atmosphere. If the repair area is confined to one body panel, mask off the surrounding panels; this will help to minimise the effects of a slight mis-match in paint colours. Bodywork fittings (eg chrome strips, door handles etc) will also need to be masked off. Use genuine masking tape and several thicknesses of newspaper for the masking operations.

Before commencing to spray, agitate the aerosol can thoroughly, then spray a test area (an old tin, or similar) until the technique is mastered. Cover the repair area with a thick coat of primer; the thickness should be built up using several thin layers of paint rather than one thick one. Using 400 grade wet-and-dry paper, rub down the surface of the primer until it is really smooth. While doing this, the work area should be thoroughly doused with water, and the wet-and-dry paper periodically rinsed in water. Allow to dry before spraying on more paint.

Spray on the top coat, again building up the thickness by using several thin layers of paint. Start spraying in the centre of the repair area and then, using a circular motion, work outwards until the whole repair area and about 2 inches of the surrounding original paintwork is covered. Remove all masking material 10 to 15 minutes after spraying on the final coat of paint.

Allow the new paint at least two weeks to harden, then, using a paintwork renovator or a very fine cutting paste, blend the edges of the paint into the existing paintwork. Finally, apply wax polish.

5 Major body damage – repair

This should be left to your dealer or a specialist body repairer. Special jigs and gauges will be required to check for body and underframe distortion. This must be corrected if the original steering and roadholding characteristics are to be retained.

6 Bonnet – removal and refitting

1 Open the bonnet and have an assistant support its weight.
2 Mark the position of the hinges on the underside of the bonnet with a soft pencil.
3 Unscrew the hinge bolts (photo) and then lift the bonnet from the vehicle.
4 Refitting is a reversal of removal, but before fully tightening the bolts, gently close the bonnet and check its alignment. Adjust as necessary before fully tightening the bolts.
5 Now close the bonnet. If it does not shut smoothly and positively adjust the bonnet lock and striker, as described in Section 8.

7 Radiator grille – removal and refitting

1 Open the bonnet then use a screwdriver to prise open the plastic clips along the top and bottom of the grille (photo). Where necessary, remove the headlamp wiper blades and arms.
2 Withdraw the grille.
3 Refitting is a reversal of removal.

6.3 Bonnet hinge

7.1 Prising open a radiator grille clip

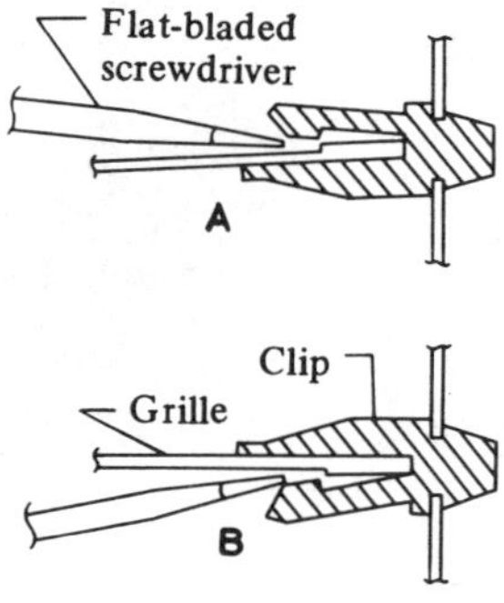

Fig. 12.1 Radiator grille clips (Sec 7)

A Upper *B Lower*

8 Bonnet lock and release assembly – removal, refitting and adjustment

1 Open the bonnet and support it on its strut.
2 Remove the radiator grille, as described in the preceding Section.
3 Unscrew the three lock retaining bolts (photo), pull the lock from its mounting bracket and unhook the release cable from it.
4 The control lever and cable may be removed after unscrewing the lever fixing screws under the facia panel and withdrawing the cable through its bulkhead grommet.
5 Refitting is a reversal of removal, but set the lock bolts finger tight so that when the bonnet is closed the striker enters the lock slot centrally.
6 Close the bonnet and check that there is no up and down movement when the front of the bonnet is depressed with the hand. If there is, move the lock downwards to eliminate any free play. Fully tighten the lock fixing bolts. If necessary, adjust the rubber corner stops so that the bonnet is supported firmly when shut. Apply a little grease to the lock lever end and pivot.

8.3 Bonnet lock retaining bolt locations

9 Front wing – removal and refitting

1 Raise the front end of the vehicle and support it securely, placing axle stands under the side-members. Remove the roadwheel.
2 Remove the radiator grille (Section 7).
3 Remove the headlamp (Chapter 10).
4 Open the bonnet and remove the row of fixing bolts from the top edge of the wing.
5 Working under the wing, disconnect the wing stay where fitted.
6 Remove the bumper and screws (Section 10).
7 Remove the screws from the lower edge of the wing.
8 Open the front door fully and, working between the hinges, extract the wing fixing screws.
9 Extract the retaining screws and clips and remove the plastic under-wing protective shield.
10 Disconnect the electrical leads from the side indicator lamp where applicable.
11 Using a sharp knife, cut along the mastic seal all round the wing sealing edge.
12 Remove the wing. If necessary, remove the indicator assembly.
13 Before fitting the new wing, clean the mating flange on the body and apply a new bead of mastic.
14 Offer the wing into position and fit the retaining screws.
15 Seal the joint between the front of the wing and the front apron.
16 Apply underseal to the underneath of the wing and refinish the outside to match the vehicle paintwork.
17 Fit the plastic undershield.

10 Bumpers – removal and refitting

1 The bumpers are shown in Figs. 12.3 to 12.6. They are mounted on brackets or hydraulic shock absorbers with additional side mounting bolts.

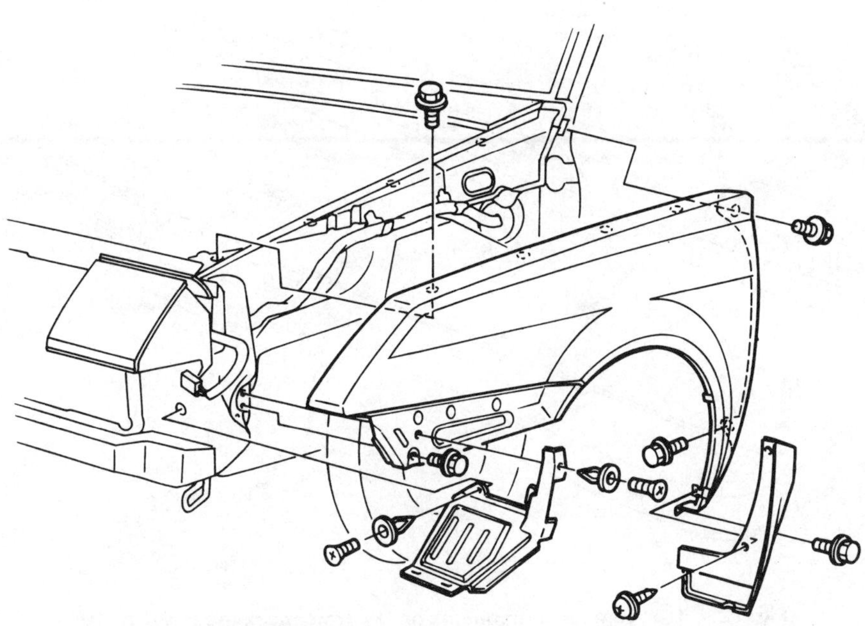

Fig. 12.2 Front wing components on Coupe models (Sec 9)

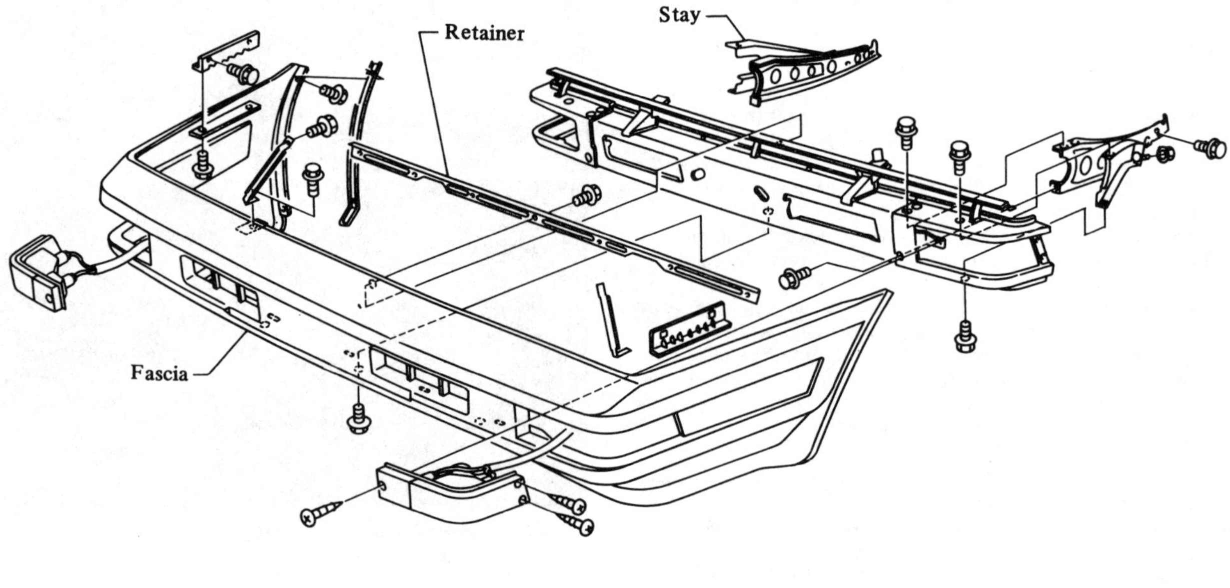

Fig. 12.3 Front bumper components on UK Coupe models (Sec 10)

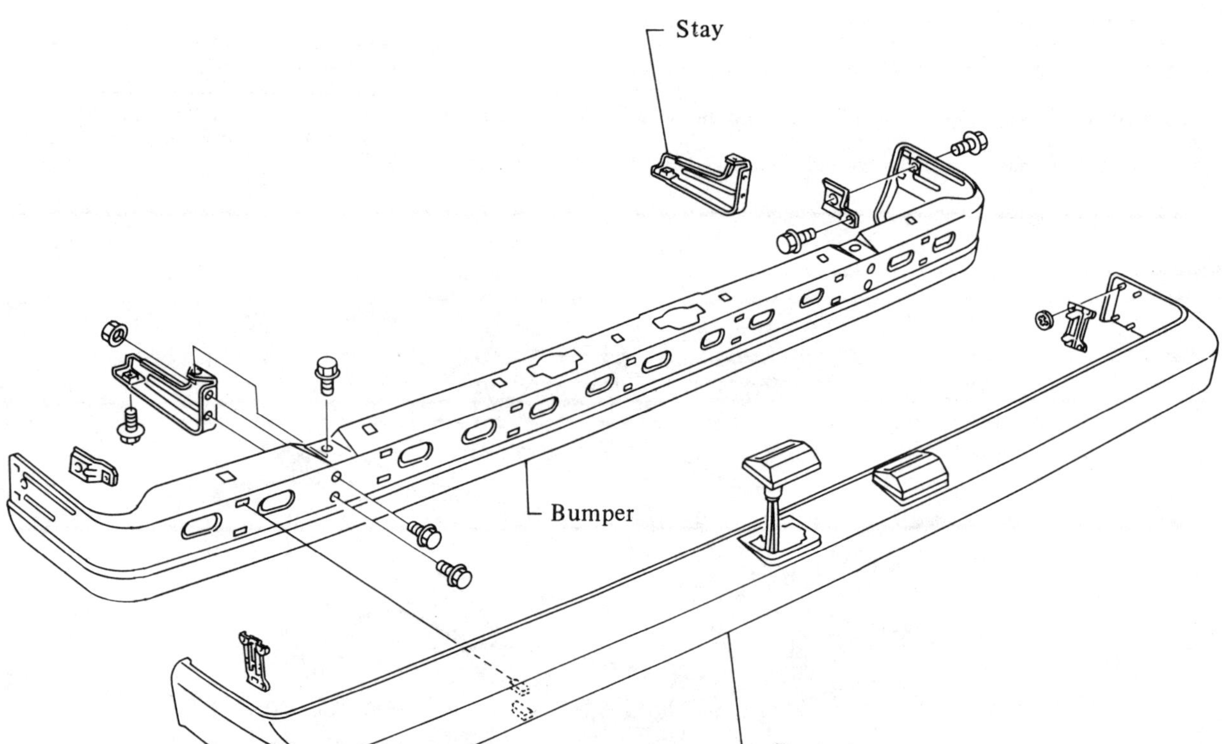

Fig. 12.4 Rear bumper components on UK Hatchback models (Sec 10)

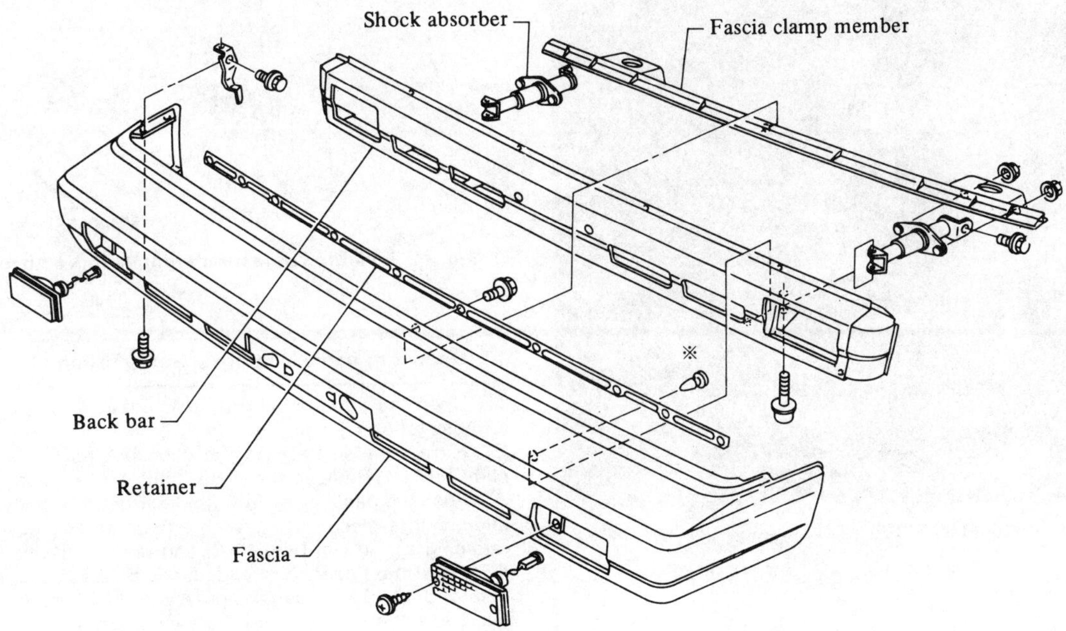

Fig. 12.5 Front bumper component on North American Hatchback models (Sec 10)

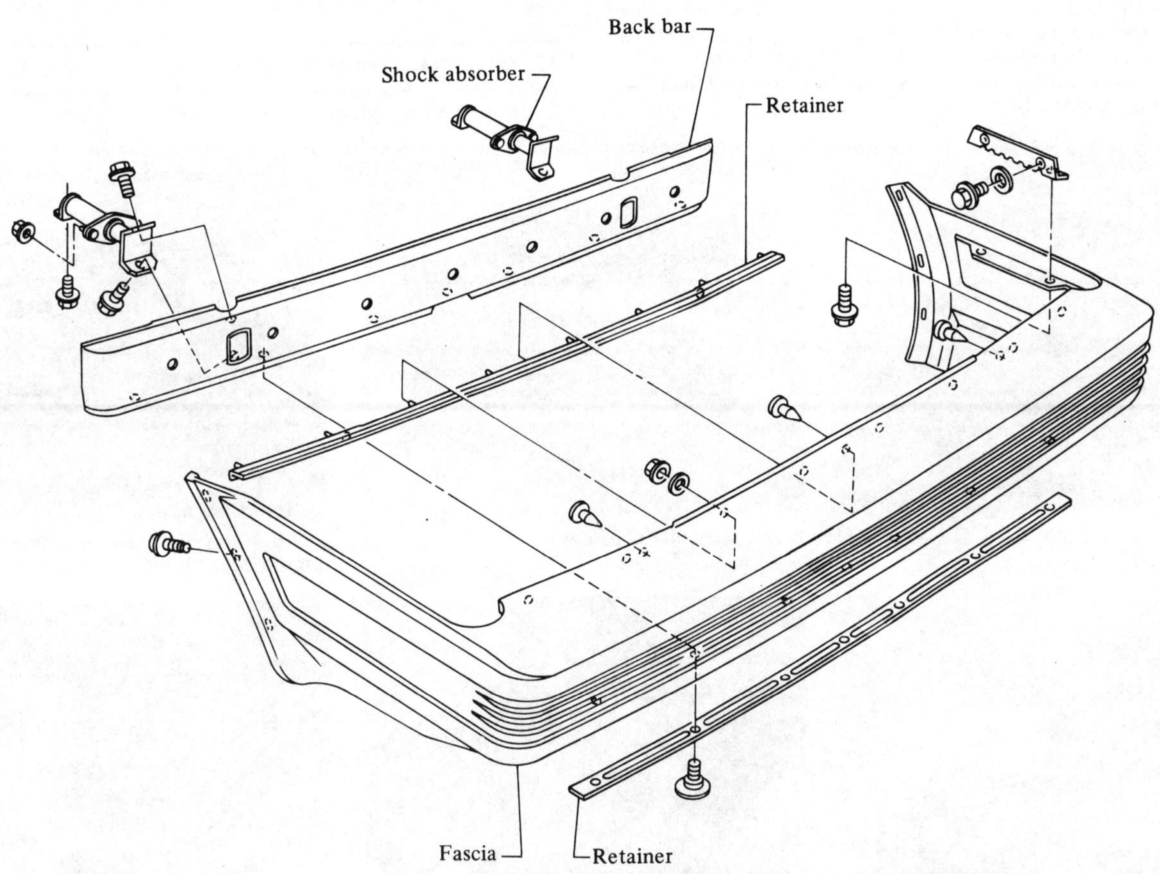

Fig. 12.6 Rear bumper components on North American Coupe models (Sec 10)

10.2 Bumper side mounting bolt location

2 To remove a bumper raise the front or rear of the car, and support with axle stands. Unscrew the mounting bolts and where applicable disconnect the wiring from the lighting or direction indicator lamps, then withdraw the bumper (photo).

3 The efficiency of the bumper shock absorbers may be checked by placing the vehicle square to a wall and then, using a jack placed between the wall and the bumper, compressing the bumper by at least 10.0 mm (0.39 in). Retract the jack and check that the bumper returns to its original position. If it does not, or any fluid leaks are observed from the shock absorber body, renew the unit.

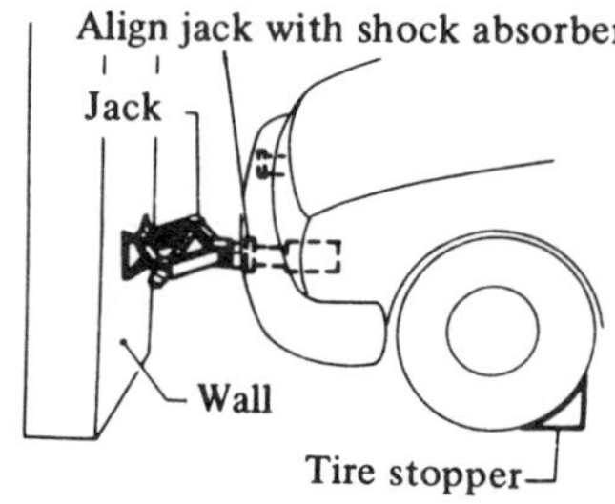

Fig. 12.7 Method of testing bumper shock absorbers (Sec 10)

11 Door trim panel – removal and refitting

1 Open the door, extract the two retaining screws and remove the armrest (photo).

2 Extract the screw from the door lock remote control escutcheon plate (photo). Remove the plate (photo).

3 Push the panel away from the rear of the window regulator handle and, using a piece of wire with a hook at its end, extract the handle retaining spring clip (Fig. 12.8) and remove the handle (photo).

4 Insert the fingers or a wide blade between the trim panel and the door and release the panel clips. A jerking movement will more readily overcome the resistance of the clips (photo).

5 Remove the trim panel and carefully peel away the waterproof sheet (photo).

6 Refitting is a reversal of removal, but with the window wound fully up, set the regulator handle 30° from the horizontal towards the front of the car (Fig. 12.9). Simply press the handle onto the shaft until the spring clip engages the groove.

12 Door lock – removal and refitting

1 Remove the trim panel, as described in the preceding Section.

11.1 Removing the armrest

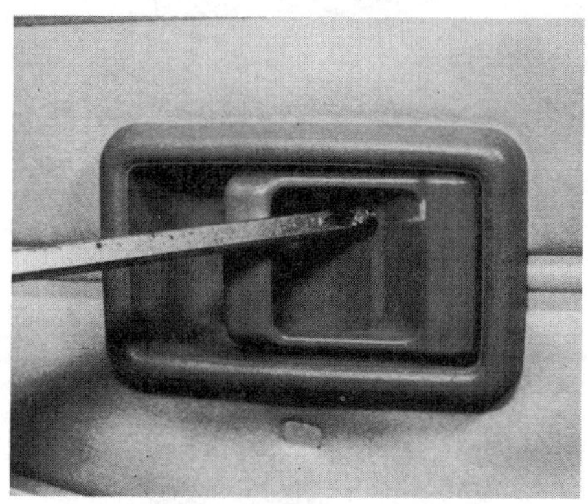

11.2A Remove the screw ...

11.2B ... and withdraw the plate from the remote door handle

11.3 Removing the window regulator handle

11.4 Door trim panel clip

11.5 Door waterproof sheet and sealant

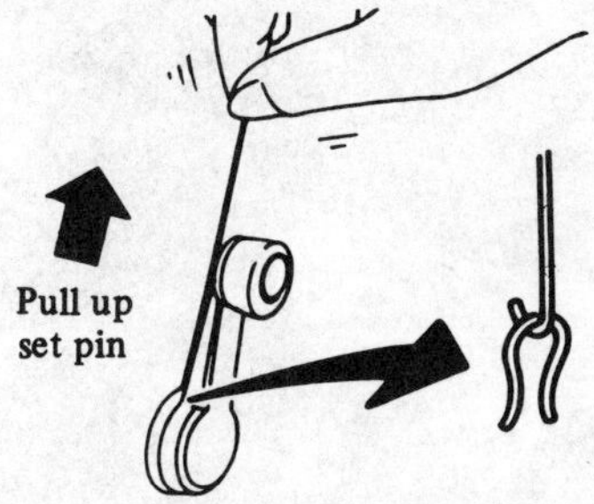

Fig. 12.8 Method of removing the window regulator handle (Sec 11)

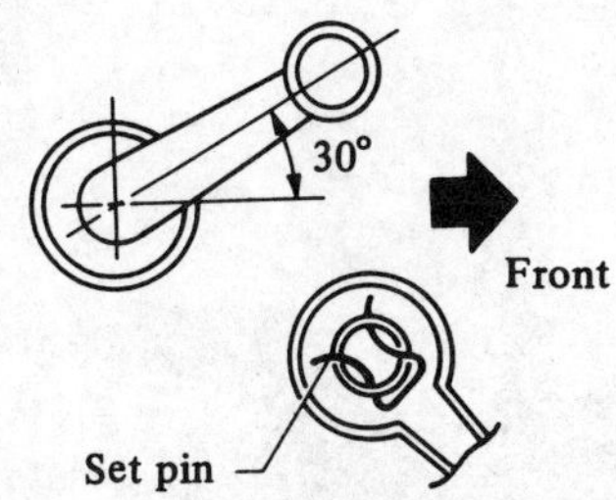

Fig. 12.9 Window regulator handle fitting angle (Sec 11)

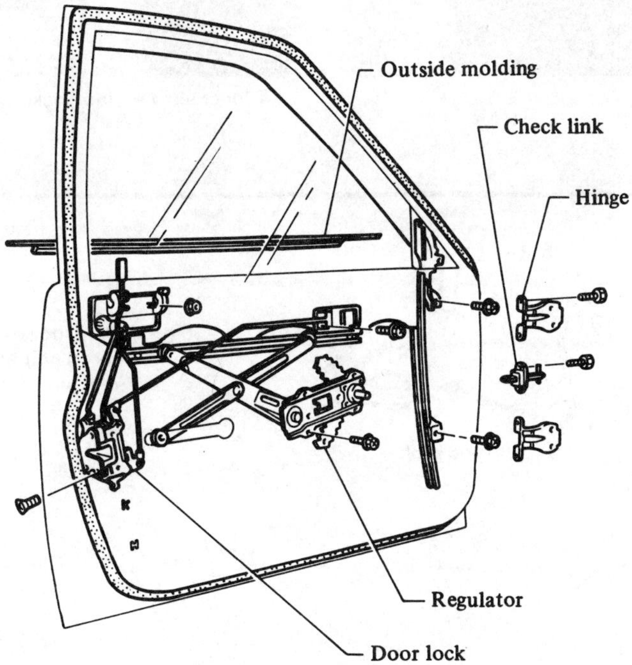

Fig. 12.10 Front door components (Secs 11 to 13)

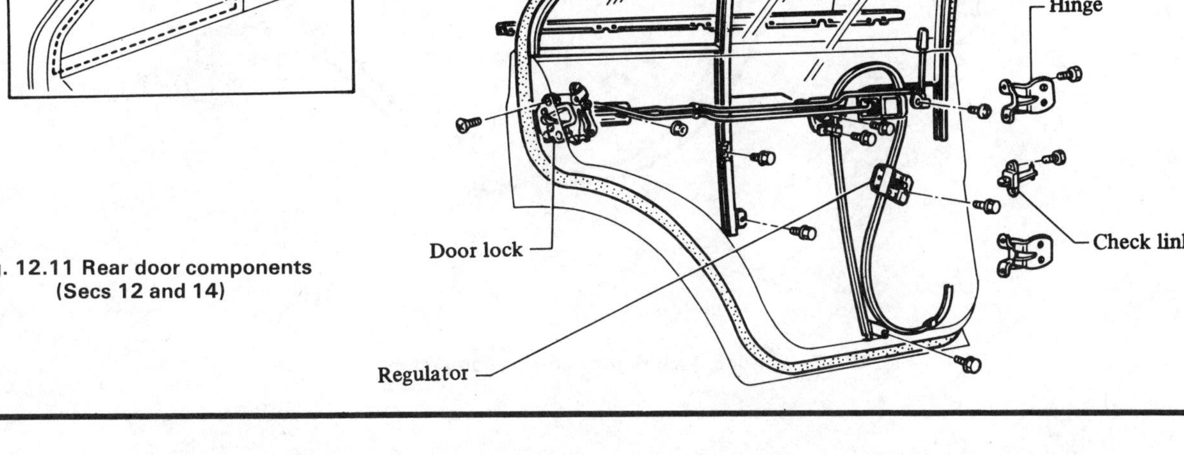

Fig. 12.11 Rear door components (Secs 12 and 14)

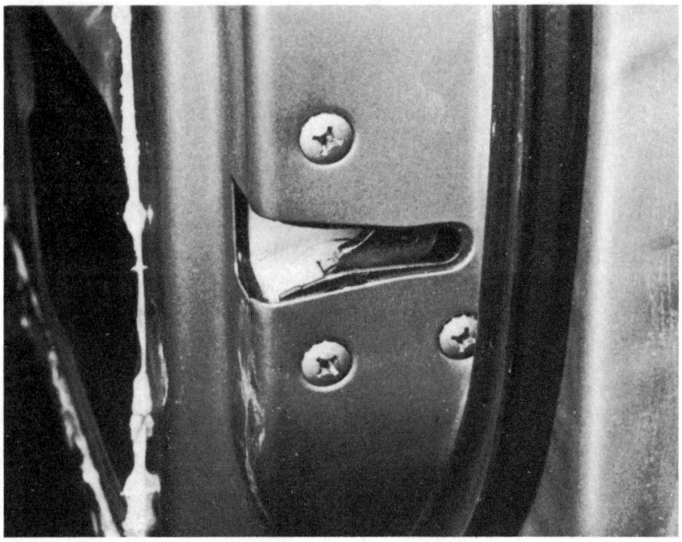

12.2 Door lock retaining screws

12.4 Inner view of door lock cylinder and retaining clip

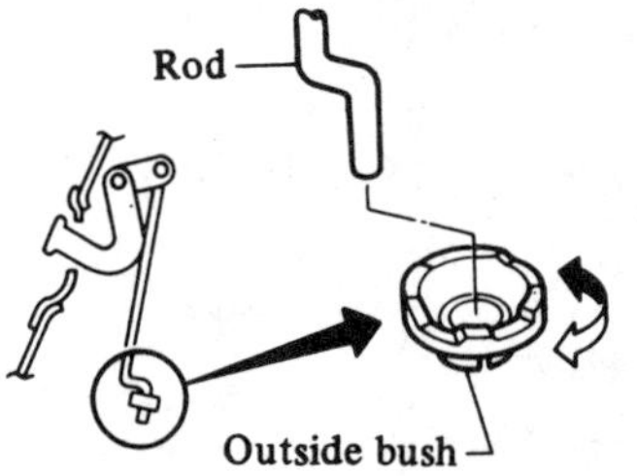

Fig. 12.12 Door exterior handle control rod adjuster (Sec 12)

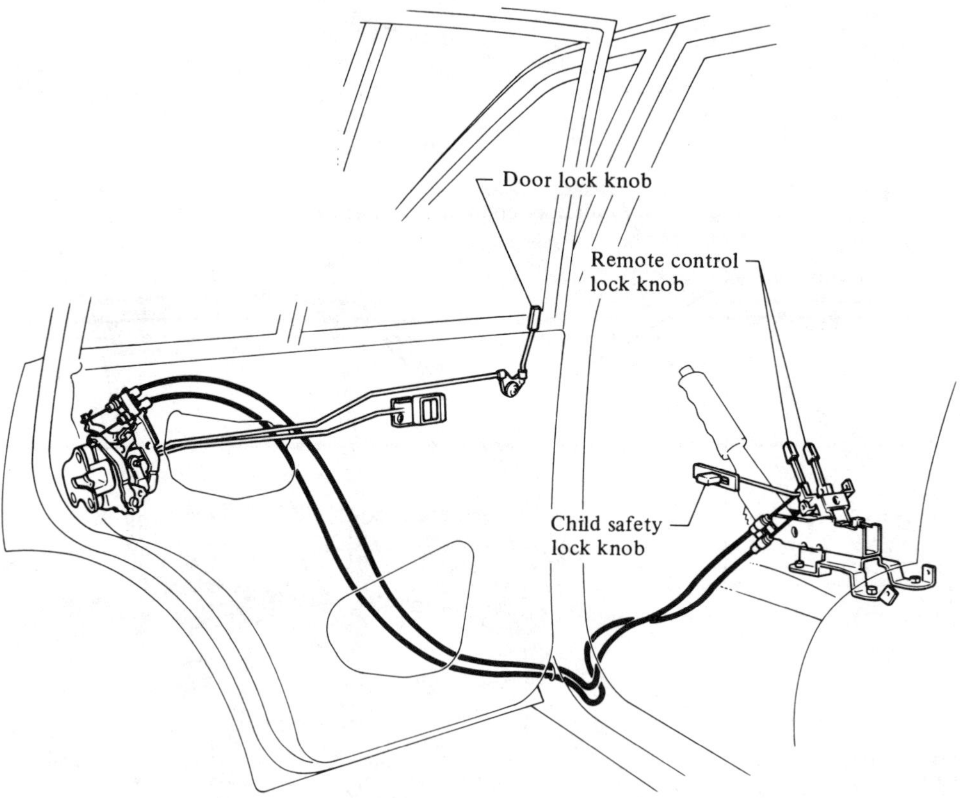

Fig. 12.13 Rear door lock remote control components (Sec 12)

2 Unscrew the lock plunger knob and the screws which hold the lock assembly to the door edge (photo).
3 Disconnect the control rods from the lock and withdraw the lock through the aperture in the door inner panel.
4 The door exterior handle can be removed by unscrewing its two fixing nuts by passing a tool through the hole in the upper part of the door inner panel. The lock cylinder can be removed after prising out its retaining clip (photo).
5 Refitting is a reversal of removal, but adjust the door exterior handle by turning the stepped bush on the control rod as required, and also adjust the door striker if necessary.
6 On certain versions a remote cable-operated rear door lock and child safety lock are fitted.

13 Front door window – removal and refitting

1 Remove the door trim panel, as previously described (Section 11).
2 Unscrew the window regulator fixing screws. Lower and swivel the regulator and release its lifting arms from the channel at the base of the glass.
3 Release the screw from the glass guide channel.
4 Lower the glass fully and remove the two weatherseal strips from the glass slot. These are held by spring clips.
5 Pull the glass upwards, tilt it and remove it from the door.
6 If a new window is being fitted, tap the base channel onto it using a wooden or plastic-faced hammer.
7 Refit by reversing the removal operations. Adjust the position of the glass guide channel before tightening its screw so that the window moves up and down smoothly.

14 Rear door window – removal and refitting

1 The procedure is identical to that for the front door window described in the preceding Section, but additionally the rear quarter light must be removed by pulling it forwards (Fig. 12.11).

15 Door – removal and refitting

1 The door hinges are of bolt-on type (photo).
2 Open the door wide and support its lower edge on jacks or blocks with pads of rag to prevent damage to the paintwork.
3 Disconnect the check link either by removing the fixing screws or by driving out the roll pin (photo).
4 Mark the position of the hinges on the door edge with a soft pencil and then support the weight of the door and remove the hinge bolts.
5 Lift the door from the vehicle.
6 Refitting is a reversal of removal. Provided the hinges are positioned within their original marked areas the door should close satisfactorily. Adjustment may be carried out by releasing the hinge bolts and moving the door.
7 The striker on the door pillar may also be adjusted to ensure smooth positive closure (photo).

15.1 Door hinge

15.3 Door check link roll pin

15.7 Door striker

16 Tailgate – removal and refitting

1 Disconnect the battery negative lead.
2 Fully open the tailgate then disconnect the wiring and pull the harness from the rear pillars.
3 With an assistant supporting the tailgate mark the positions of the struts and unscrew the bolts (photos).
4 Unscrew the hinge bolts after also marking the hinge positions, then withdraw the tailgate from the car (photo).
5 Refitting is a reversal of removal, but check that the tailgate is positioned centrally in the body aperture. Adjustment is made by loosening the hinge nuts on the body and moving the tailgate as required. Adjustment of the striker will also be necessary to ensure smooth operation of the lock (photo) and the lock can also be adjusted within the elongated mounting holes.

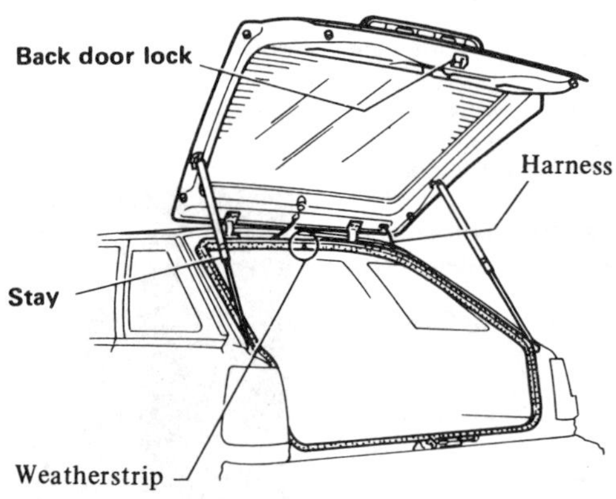

Fig. 12.14 Tailgate components (Sec 16)

16.3A Tailgate strut mounting on tailgate

16.3B Tailgate strut mounting on body

16.4 Tailgate hinge

16.5 Striker location on the tailgate

17 Tailgate or boot lid lock/fuel filler lid remote control – removal, refitting and adjustment

1 To remove the tailgate lock, unscrew the mounting nuts and the cable clamp, withdraw the lock and disconnect the cable (photo).
2 To remove the operating lever unscrew the mounting screws and disconnect the cable.
3 Removal of the cables is straightforward. The double action lever incorporates a cable clamp, and at the fuel filler lid end the cable is simply released from the lock.
4 Refitting is a reversal of removal, but adjust the cable clamps to give 0.5 to 1.5 mm (0.020 to 0.059 in) free play. If necessary the operating lever may be adjusted within the mounting slots.

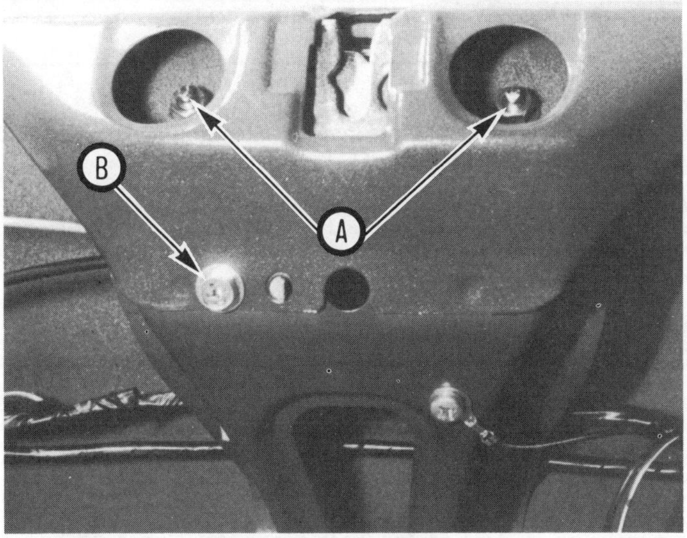

17.1 Tailgate lock mounting nuts (A) and cable clamp screw (B)

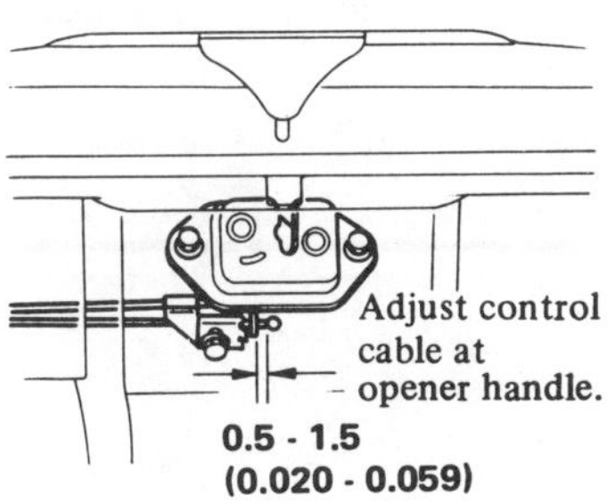

Fig. 12.15 Tailgate lock adjustment – mm/in (Sec 17)

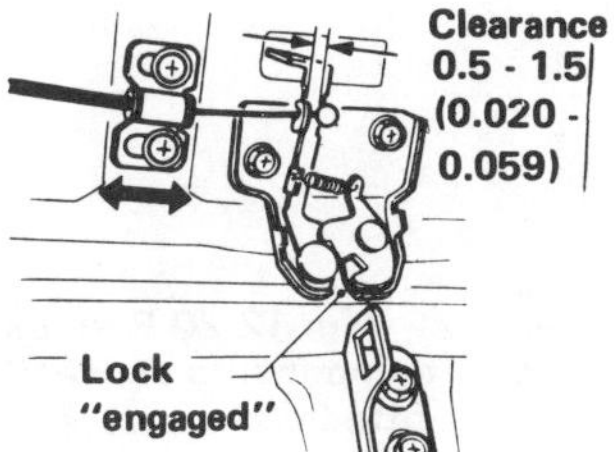

Fig. 12.16 Boot lid lock adjustment – mm/in (Sec 17)

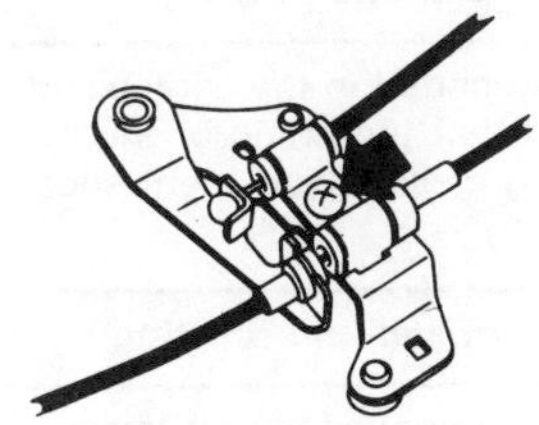

Fig. 12.17 Rear remote control double action lever showing clamp screw (Sec 17)

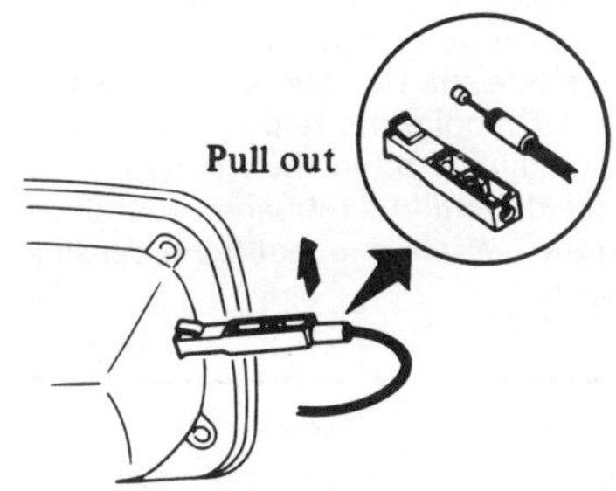

Fig. 12.18 Remote cable removal at the fuel filler lid (Sec 17)

18 Boot lid – removal and refitting

1 Open the boot lid and unbolt the lock and cable (photo).
2 Mark the position of the hinges on the underside of the lid.
3 With the help of an assistant, support the lid and unscrew the bolts from the hinges.
4 Remove the bootlid from the car.
5 Torsion rods are used to counterbalance the boot lid and they should be released gently with a suitable lever before attempting to unbolt the hinges from the body.
6 Refitting is a reversal of removal. If adjustment is required, release the hinge bolts and move the boot lid to align it. Adjust the position of the striker to ensure smooth positive closure. Adjust the lock cable with reference to Section 17.

18.1 Boot lid lock and cable

19 Windscreen – removal and refitting

1 The windscreen is bonded to the body using a primer and sealant kit. It is recommended that the fitting of this type of screen is left to a Nissan dealer or auto glass replacement specialist.

20 Tailgate glass – removal and refitting

1 If the glass is to be removed and/or replaced for any reason, it is a job which is better left to an auto glass replacement specialist. They will do the job in half the time and most important, ensure that it is correctly fitted with no leakages around the surround rubber. However, if you wish to do it yourself proceed as follows.
2 If the glass has shattered first remove all the broken glass from the surround and rear compartment.
3 If the glass is to be removed intact, first release the rubber surround from the tailgate by running a blunt, small screwdriver around and under the rubber weatherstrip both inside and outside. Take care not to damage the paintwork or catch the rubber surround with the screwdriver. Remove the rear wiper arm and blade.
4 Remove the moulding if fitted then carefully push one corner of the glass from the inside until the rubber surround is released from the flange. Move along the glass and repeat the process until the complete glass can be removed.
5 Scrape the old sealant from the edge of the aperture and wipe clean.
6 Fit a new rubber surround to the glass.
7 Cut a piece of strong cord greater in length than the periphery of the glass and insert it into the flange locating channel of the rubber surround. Apply some soapy water to the rubber surround with a brush.
8 Locate the glass on the tailgate with the cord ends loose on the inside bottom centre.
9 While an assistant presses the glass pull one of the cord ends so that the surround lip engages the flange. Repeat with the other cord end until the glass is fully engaged.
10 Apply sealant beneath both outer lips of the rubber surround (Fig. 12.19) then refit the moulding where applicable.
11 Refit the rear wiper arm and blade.

21 Rear side windows – removal and refitting

1 Open the side window then remove the screw securing the toggle catch or cable end fitting to the glass (photo).
2 Remove the screws or nuts from the hinge end of the glass, and withdraw the glass from the car.
3 Refitting is a reversal of removal, but position the glass centrally over the rubber surround before tightening the retaining screws or nuts. Some models are fitted with cable-operated remote control

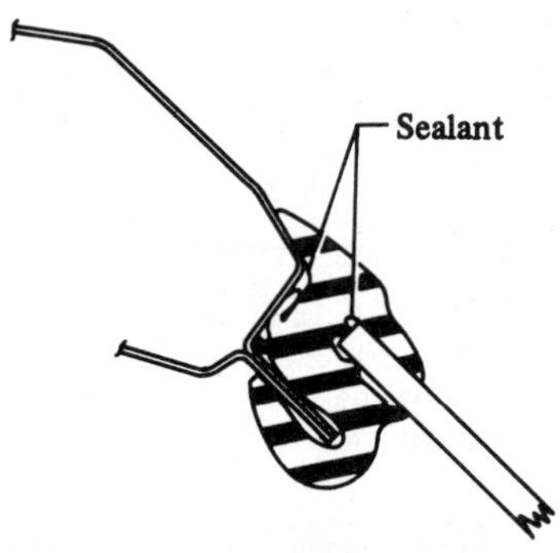

Fig. 12.19 Apply sealant as indicated when fitting the tailgate glass (Sec 20)

21.1 Cable end fitting on remote control rear side window

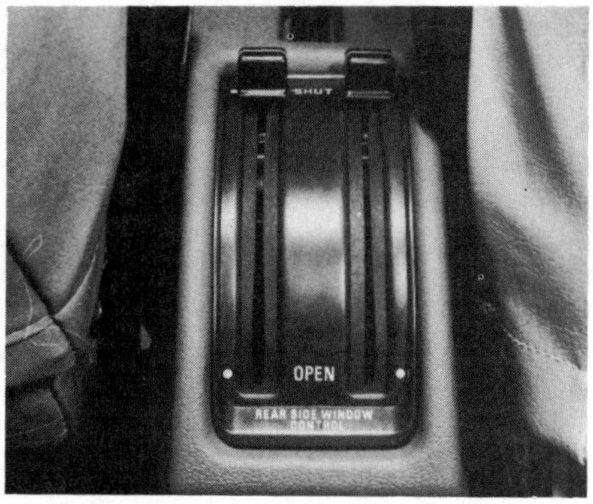

21.3 Rear side window remote control levers

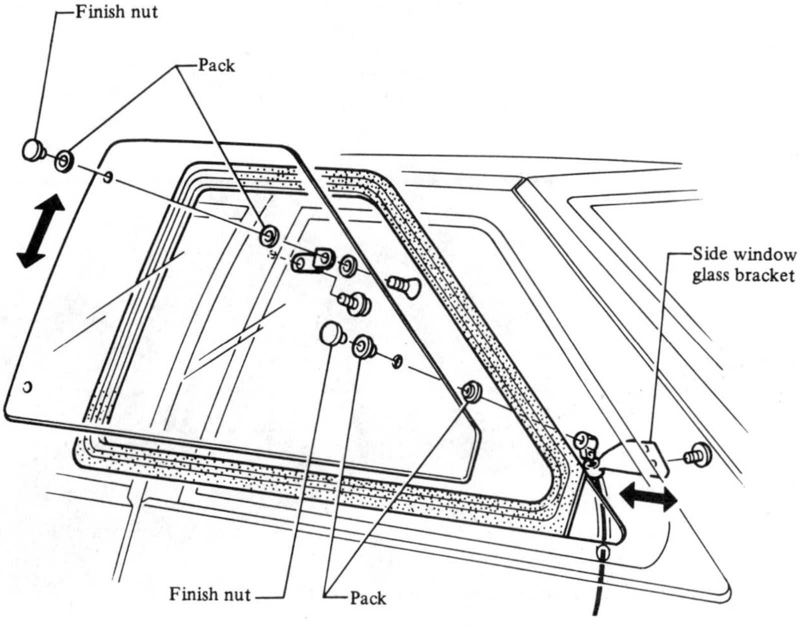

Fig. 12.20 Rear side window components on Hatchback models fitted with remote control system (Sec 21)

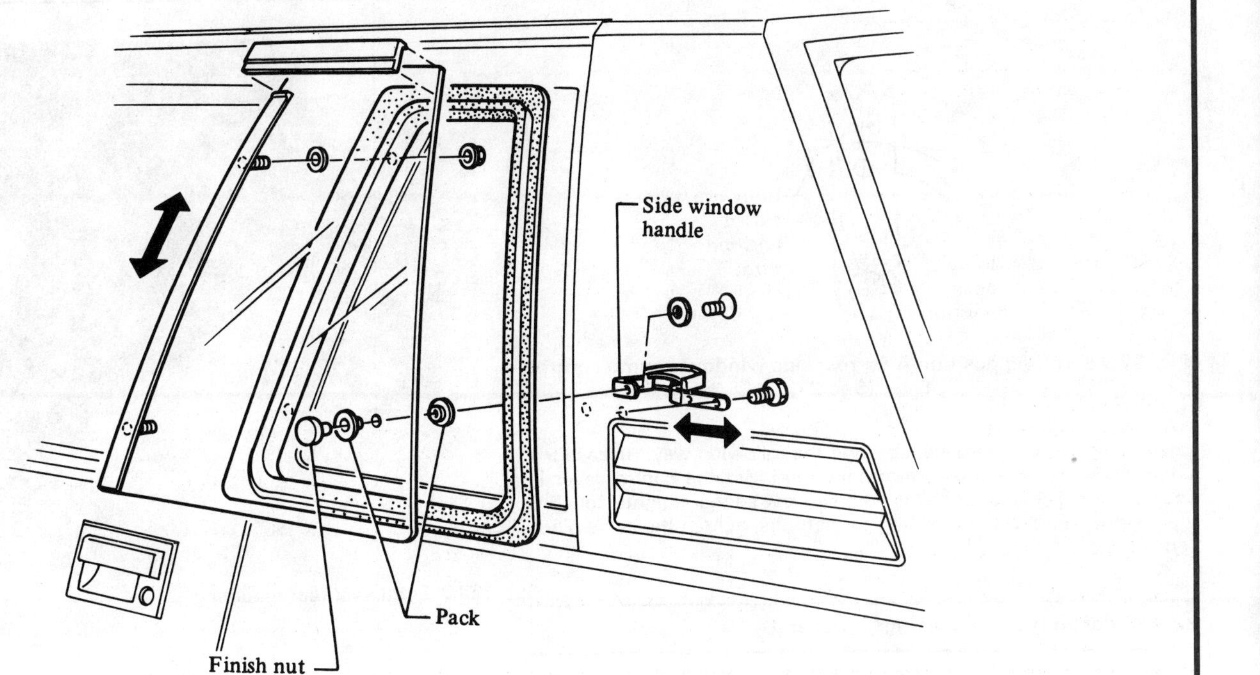

Fig. 12.21 Rear side window components on Coupe models (Sec 21)

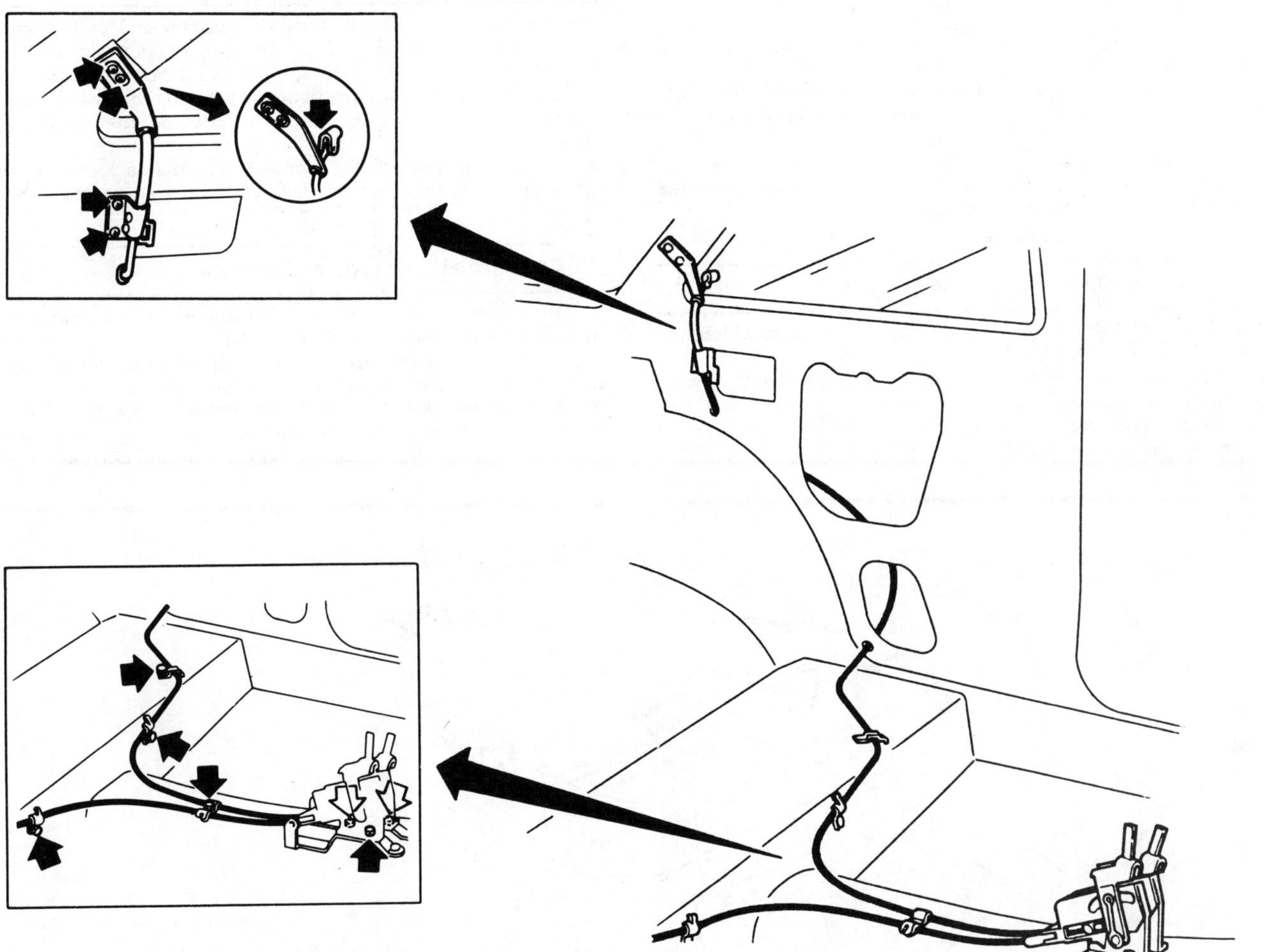

Fig. 12.22 Rear side window remote control components (Sec 21)

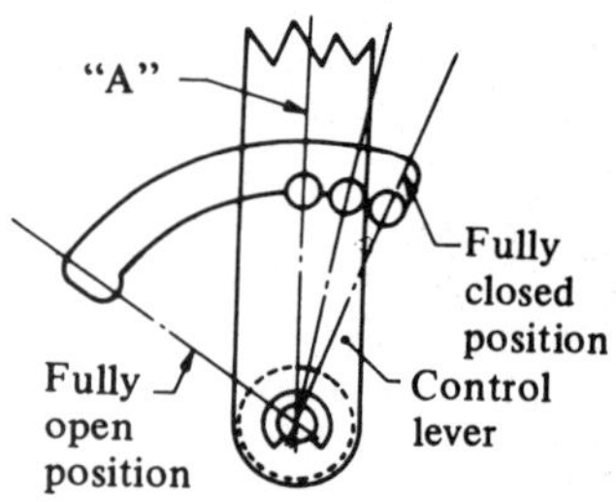

Fig. 12.23 Setting position A for rear side window remote control lever (Sec 21)

catches (photo). If necessary, adjust in the following way. Release the cable pinch screw at the control lever end. Set the control lever in position A (Fig. 12.13) and close the window so that it lightly touches the weatherseal. Without moving the settings, tighten the cable pinch screw.

22 Interior trim and mouldings – general

1 Most of the interior trim and mouldings are of plastic construction and care should be exercised when removing or refitting it.
2 Clips are used extensively to fix the trim, with self-tapping screws in certain positions.
3 Removal of the headlining is best left to an expert.

23 Exterior trim and mouldings – general

1 The exterior body trim is either clipped in position or held by double-sided tape.
2 The side and rear guard mouldings are secured by double-sided tape. To remove a moulding, the use of an electrically-operated heat gun will be found to be the most effective.

24 Front seat – removal and refitting

1 Push the seat fully to the rear and unscrew the bolts which hold the seat runners to the floor.
2 Now push the seat fully forward and unscrew the bolts from the rear of the runners (photo).
3 Remove the seat from the vehicle.
4 Refitting is a reversal of removal.

24.2 Front seat rear mounting

25 Rear seat – removal and refitting

1 Grip the front edge of the seat cushion and pull it upwards to release the tongues from the retaining recesses (photos).
2 Pull the cushion forward and remove it from the car.
3 Either a single or double backrest is fitted. To remove the backrest release it from the upper clip, fold it forwards, then unbolt the pivot brackets.
4 Refitting is a reversal of removal. On hatchback models the upper retaining bracket may be adjusted within the elongated screw holes.

26 Facia panel – removal and refitting

1 Remove the instrument panel, speedometer cable, clock, radio and switches with reference to Chapter 10.
2 Remove the choke cable (Chapter 2), steering wheel (Chapter 11) and the heater control and air ducts (Section 32).
3 Pull out the glovebox pivot pins and withdraw the glovebox.
4 Referring to Fig. 12.25 remove the facia fixing screws noting that the upper screws are concealed beneath plastic covers.

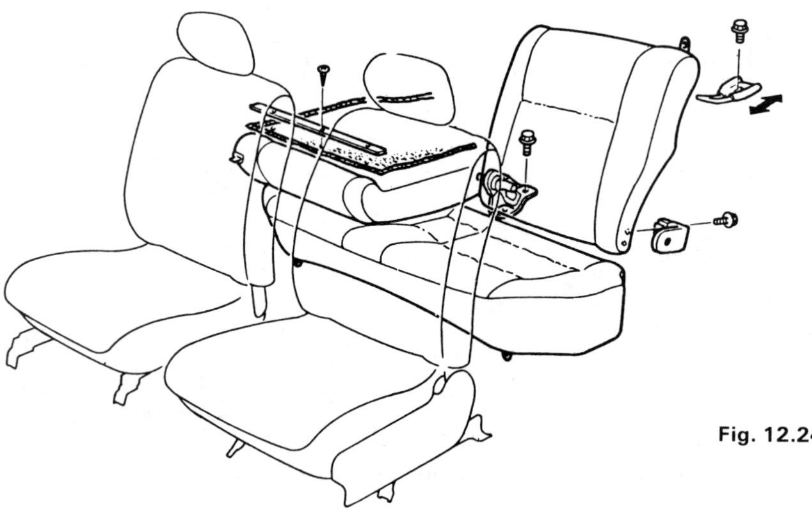

Fig. 12.24 Rear seat components – split backrest (Sec 25)

25.1A Rear seat cushion locating tongue

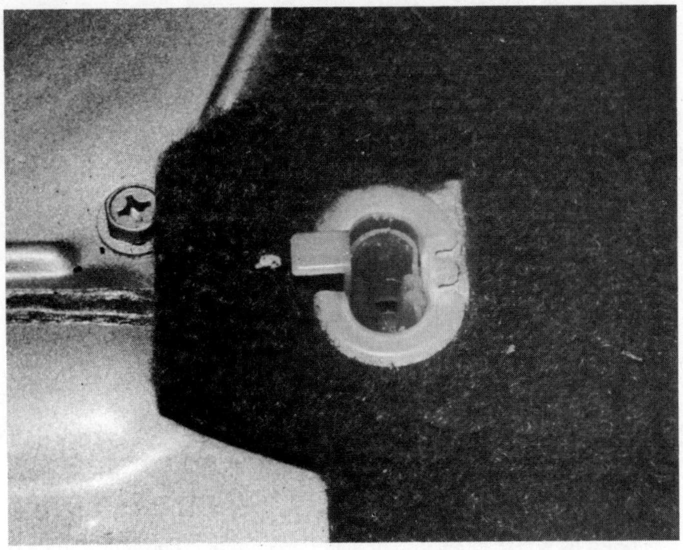

25.1B Rear seat cushion tongue recess

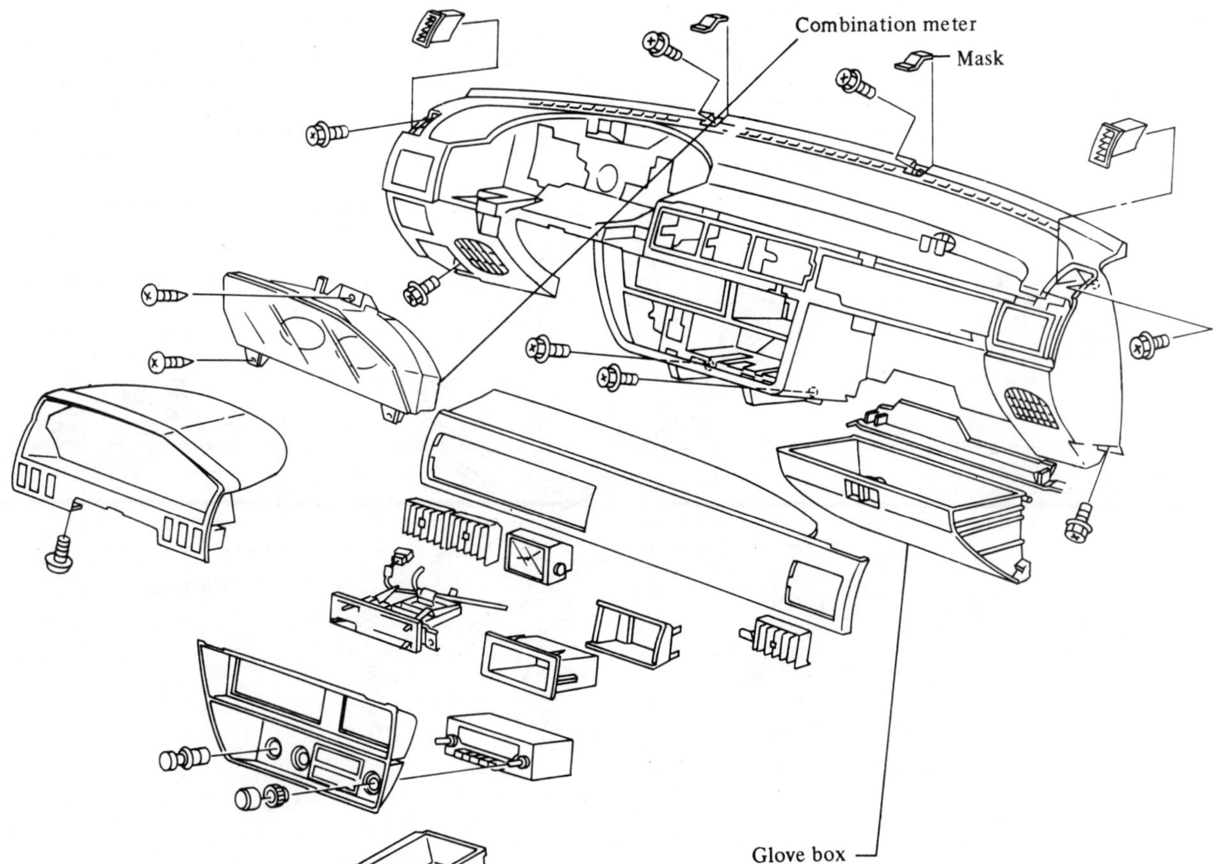

Fig. 12.25 Facia panel components – LHD model shown (Sec 26)

5 Withdraw the facia panel making sure that all wires and ducts are disconnected.
6 Refitting is a reversal of removal.

27 Sunroof – removal and refitting

1 The sunroof consists of a glass roof panel and a sunshade plate attached under the glass.
2 To remove the sunshade plate, hold the plate while unscrewing the retaining bolts. Pull the hooks of the plate out of their holders.
3 To remove the sunroof (always having first removed the sunshade plate), tilt the roof and push in the two buttons on the handle while pushing the sunroof upwards.
4 Raise the roof vertical then slide it to the left to disengage it from the hinges.
5 Lift the two air deflectors at the front corners of the aperture if the car is to be used with the sunroof removed. Store the sunroof in the special bag in the rear compartment.
6 Refitting is a reversal of removal. The sunroof is correctly engaged with the front hinges when the red marks are no longer visible.

28 Seat belts – general

1 Regularly check the condition of the seat belts. If they are frayed or cut, they must be renewed.
2 Clean the webbing by wiping it with warm water and a mild detergent only. Leave the belts unretracted until quite dry.
3 Never alter the attachment points of the belts and, if removed, make quite sure that the original sequence of fitting of the anchor bolt, spacers, washers and connecting plate is retained.
4 On North American vehicles, a warning chime sounds for a six second period if the driver's seat belt is not fastened when the ignition is switched on.

29 Rear view mirrors – removal and refitting

Interior mirror

1 One of two types of interior mirror may be used, depending upon vehicle model. One type of mirror has a sprung base, while the other has a base designed to collapse on impact.
2 Both types of mirror are screwed to the roof rail.

Door mirror

3 To remove the mirror, extract the screw and take off the mirror positioning control knob (photo).
4 Prise off the cover plate to expose the mirror mounting screw (photo).
5 Remove the screws and the mirror (photos).
6 Refitting is a reversal of removal, but apply a bead of RTV type sealant to the cover plate when fitting it.

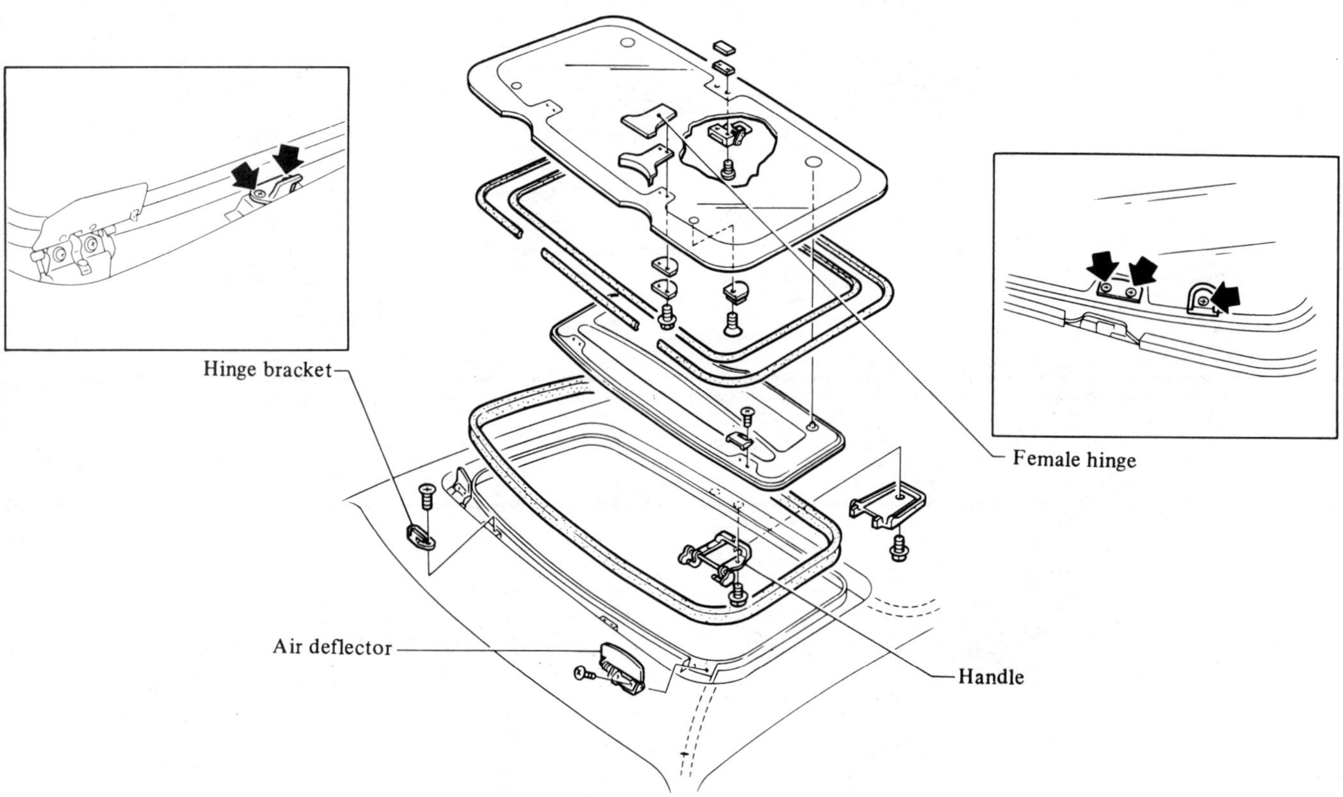

Fig. 12.26 Sunroof components (Sec 27)

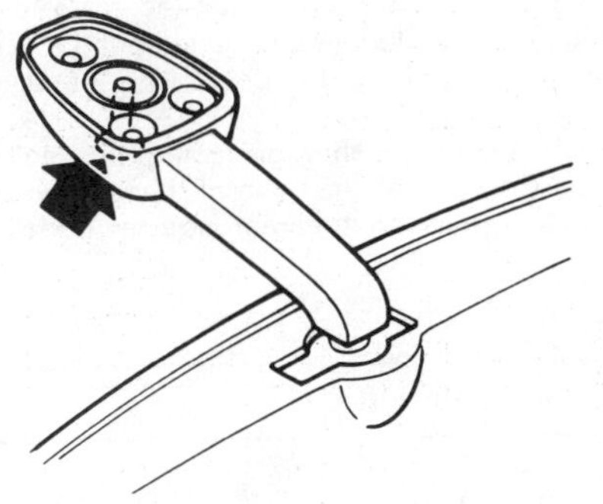

Fig. 12.27 Rigid type interior mirror base (Sec 29)

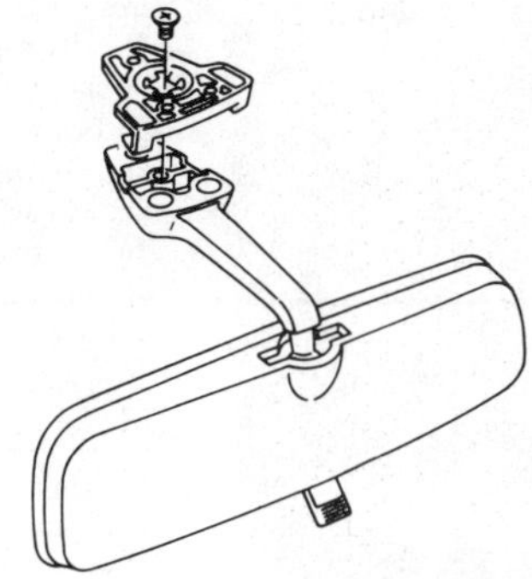

Fig. 12.28 Interior mirror with sprung type base (Sec 29)

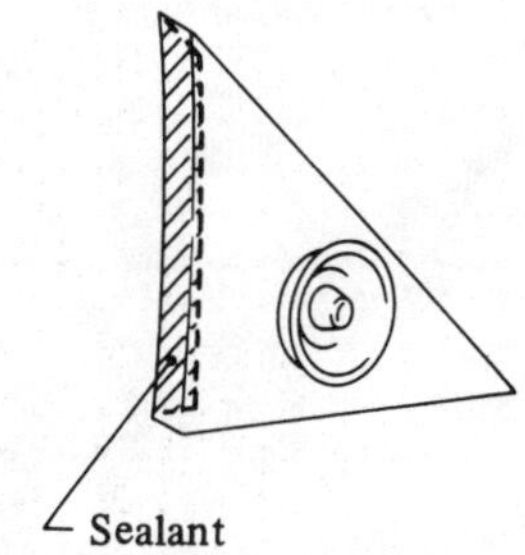

Fig. 12.29 Door mirror cover plate showing area for sealant (Sec 29)

29.3 Removing the door mirror control knob

29.4 Removing the door mirror inner cover plate

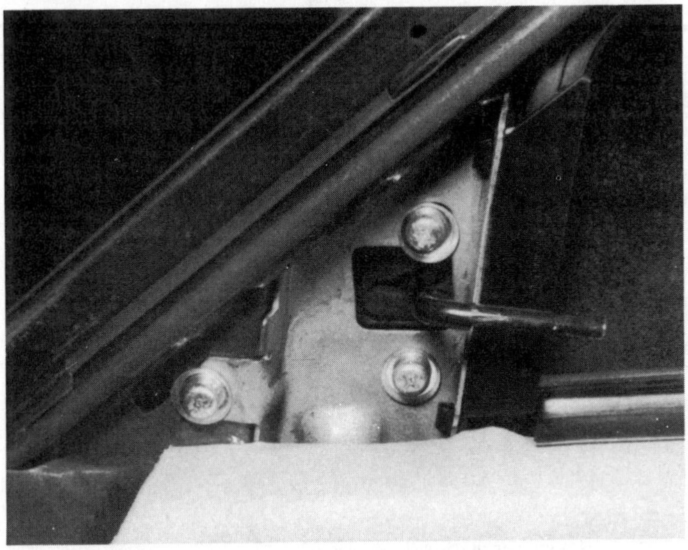

29.5A Door mirror retaining screw locations

29.5B Removing the door mirror

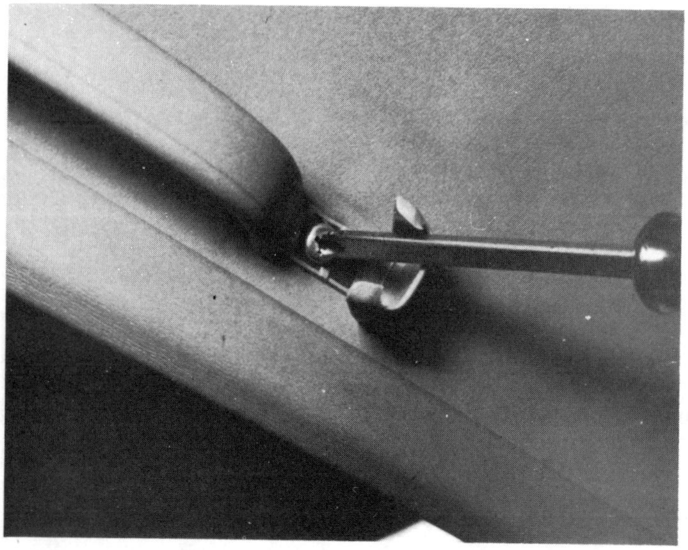

30.2 Removing the interior grab handle

30 Interior grab handle – removal and refitting

1 Using a small screwdriver prise back the screw covers.
2 Extract the screws (photo) and remove the handle.
3 Refitting is a reversal of removal. The screw cover snaps into place.

31 Heating and ventilation system – description

1 The heater incorporates a matrix which is fed with hot coolant from the engine cooling system.
2 A booster (blower) is used to force air through the matrix.
3 Temperature levels and airflow are controlled from levers on a facia-mounted control panel. Heated air can be supplied to the vehicle interior through centre or side ducts, or to the windscreen for demisting or defrosting purposes.
4 Fresh air ventilation is supplied through facia-mounted grilles, air being drawn in through the grille at the base of the windscreen. Stale air is exhausted through grilles on the rear quarter panels or door pillars (photo).

32 Heater – removal and refitting

Control panel

1 Release the control cable clamps and slip the ends of the cables from the pivot pins on the airflow flaps. It is worthwhile marking the position of the clamps in relation to the cables for ease of refitting.
2 Disconnect the wiring harness plug.
3 Remove the screws from the top of the control panel surround, withdraw the surround then unbolt and remove the control panel.
4 Refitting is a reversal of removal, but adjust the control cables in the following way.
5 Set the air intake lever to RECIRC (MAX A/C if an air conditioner is fitted). Set the flap at the recirculation position. Engage the control cable on its pin and fit the cable clamp.
6 Set the temperature control lever to MAX COLD. Move the air mix flap lever in the direction of the arrow (Fig. 12.33) and connect the control rod to the link lever. Hold these positions, engage the control cable with its pin and then clamp the outer cable.
7 Pull the control rod of the coolant valve in the direction of the arrow (Fig. 12.34) until there is a clearance of 2.0 mm (0.08 in) between the end of the rod and the link lever. Connect the rod to the flap lever.
8 Set the air control lever in the DEF position, then move the link lever in the direction of the arrow (Fig. 12.35). Connect the end of the control cable to the link lever and then clamp the outer cable.

Blower motor – UK models

9 Remove the lower cover from the facia panel.
10 Disconnect the blower motor wiring harness.

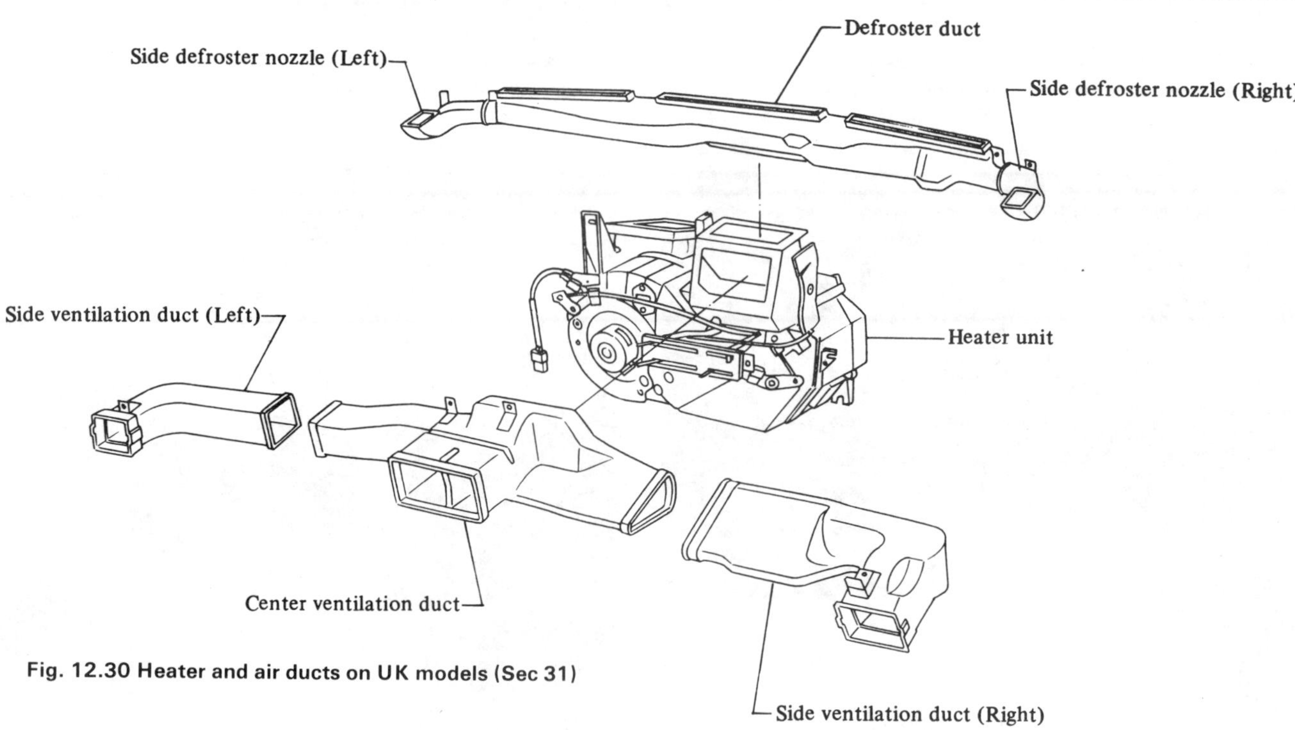

Fig. 12.30 Heater and air ducts on UK models (Sec 31)

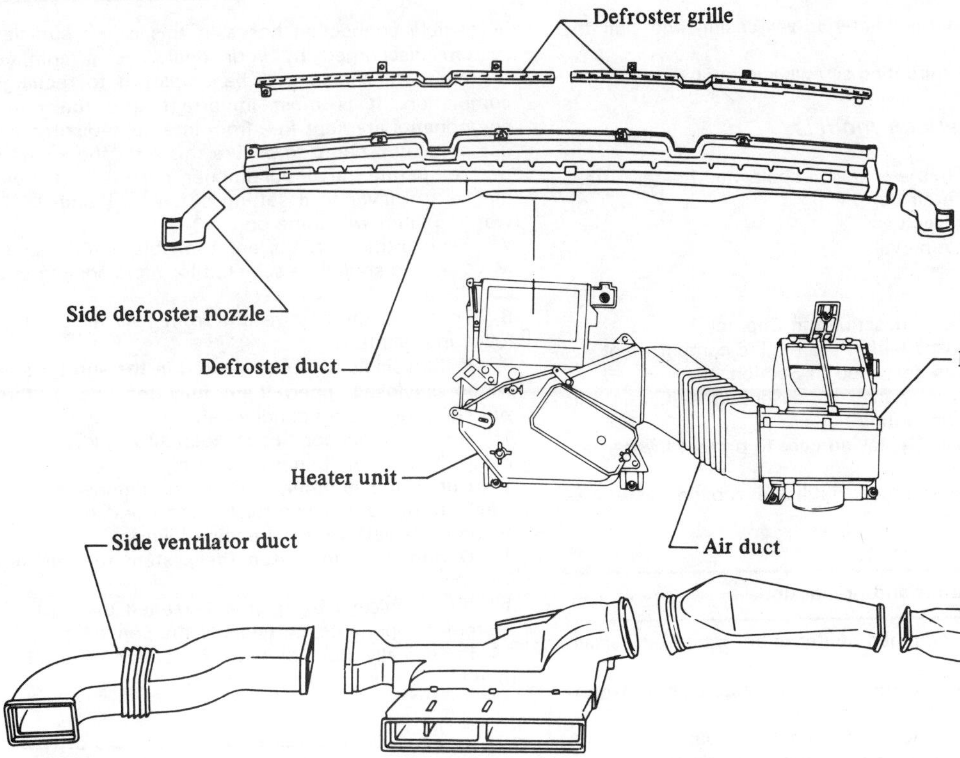

Fig. 12.31 Heater and air ducts on North American models (Sec 31)

31.4 Door pillar ventilation grille

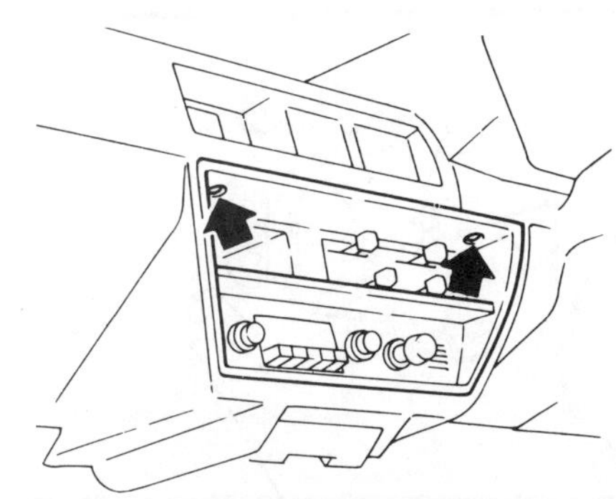

Fig. 12.32 Location of heater control panel surround retaining screws (Sec 32)

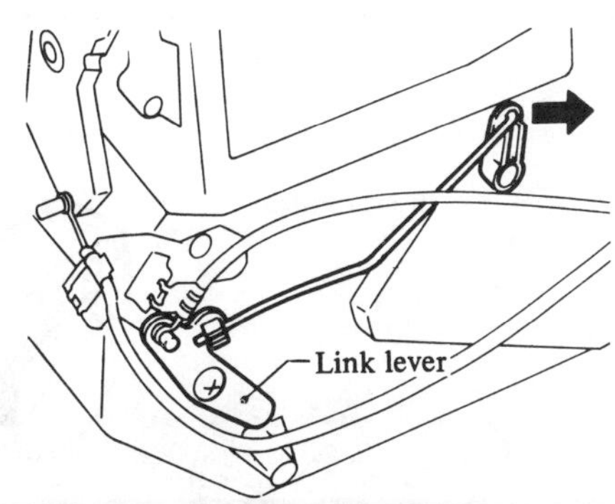

Fig. 12.33 Temperature control cable adjustment (Sec 32)

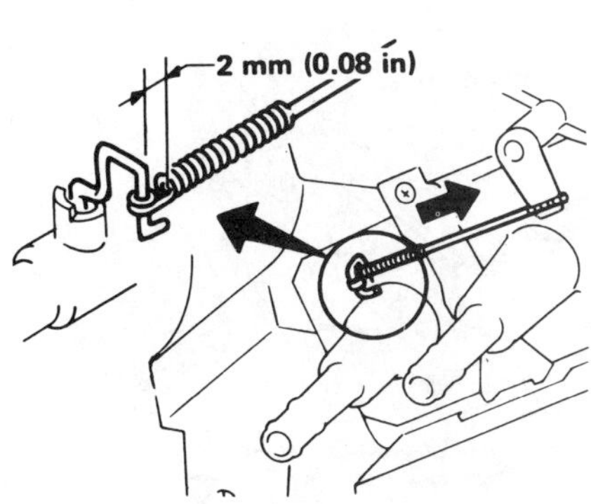

Fig. 12.34 Coolant valve control rod adjustment (Sec 32)

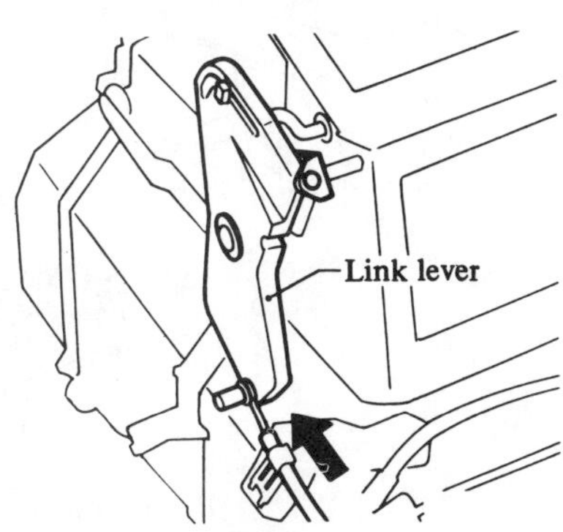

Fig. 12.35 Air control cable adjustment (Sec 32)

Fig. 12.36 Blower motor removal (Sec 32)

11 Extract the screws from the heater lower casing and pull the casing downwards.
12 Extract the blower motor mounting screws and remove the motor.

Blower motor – N. American models

13 Remove the glovebox.
14 Disconnect the wiring harness and unbolt the heater intake assembly. Release it from the air duct.
15 Unclip the blower motor section.
16 Refitting is a reversal of removal.

Heater (complete unit)

17 Drain the cooling system, as described in Chapter 2.
18 Disconnect the heater hoses within the engine compartment.
19 Remove the facia panel as described in Section 26.
20 Remove the heater control panel, as described earlier in this Section. Unscrew the heater mounting bolts.
21 Withdraw the heater assembly, taking care to protect the carpets against coolant spillage.
22 Refitting is a reversal of removal. Refill the cooling system as described in Chapter 2.

33 Air conditioner – operation and precautions

1 An air conditioner with combined heater is an option on certain models.
2 The refrigerant fluid used in the system is odourless and non-poisonous. It is not corrosive nor inflammable.
3 Leaks are not dangerous unless the fluid or vapour comes into contact with a naked flame when a poison gas is created.
4 Refrigerant fluid is dangerous to the eyes and skin and contact should be avoided.
5 If overhaul operations require the removal of any part of the system, try and move the obstructing component within the limits of its flexible connecting hoses. If this is not sufficient then have the system discharged by your dealer or a competent refrigeration engineer. He should also be employed to recharge the system on completion. It is most important that the refrigerant lines and components are kept free from internal moisture and your dealer will use a vacuum pump to ensure this after the circuit has been broken.
6 To operate the air conditioner, push the A/C switch to ON and the fan control lever to a setting between 1 and 4. The air conditioner warning lamp will come on.
7 Two further settings are available – MAX and ECONOMY. The MAX setting should be selected for rapid cooling in very hot or humid conditions.
8 To obtain the best results from your air conditioner, observe the following points.
9 If the vehicle has been parked in the sun for a long time with the windows closed, open them fully for two or three minutes after switching on the air conditioner.
10 Under normal conditions, keep all windows and ventilators closed when the air conditioner is working.
11 During the summer, if the air conditioner has not been used for a week, turn the fan control lever on and off several times at three second intervals with the engine idling.
12 During the winter, run the system for ten minutes at monthly intervals.
13 If the vehicle is being steam cleaned, avoid using the steam on the system components, particularly the condenser.
14 Periodically brush or hose flies and dirt from the condenser which might otherwise restrict its airflow.

34 Air conditioner components – removal and refitting

1 As already explained in the preceding Section, have the system discharged by your dealer before carrying out any of the following operations.

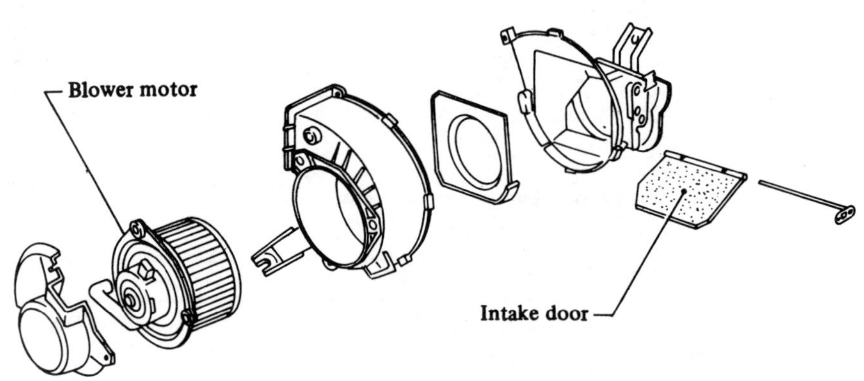

Fig. 12.37 Blower motor components for North American models (Sec 32)

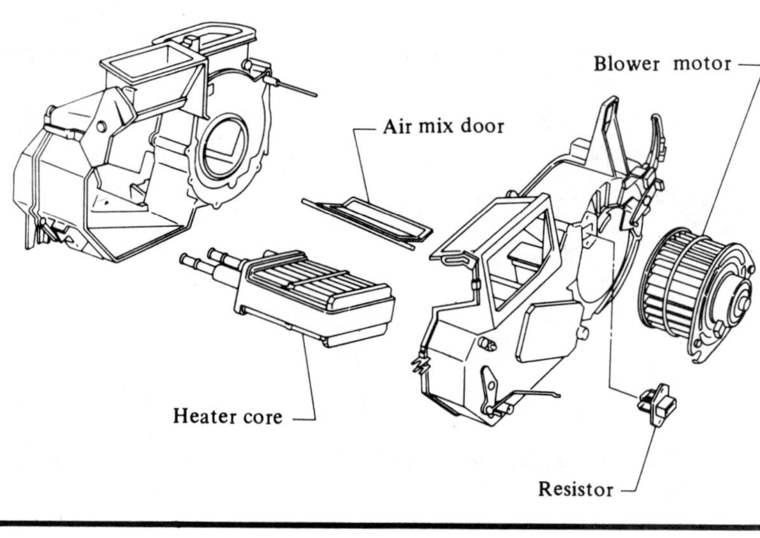

Fig. 12.38 Heater unit components for UK models (Sec 32)

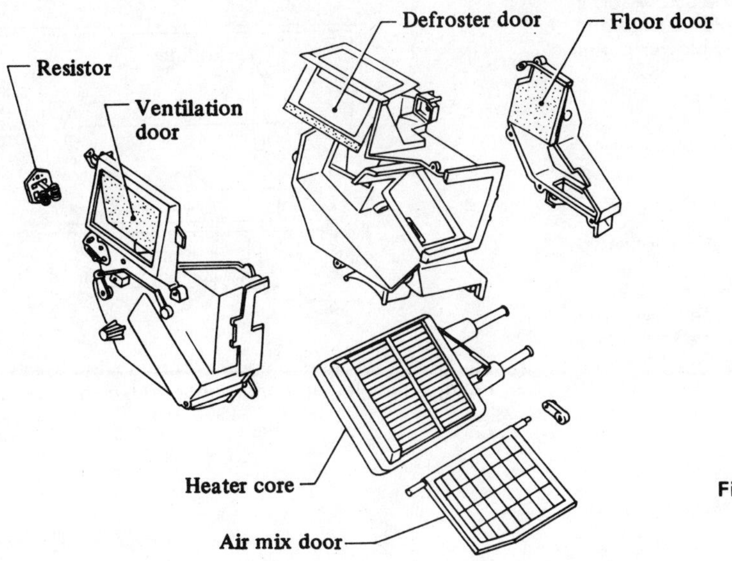

Fig. 12.39 Heater unit components for North American models (Sec 32)

Condenser

2 Disconnect the battery.
3 Remove the radiator grille (Section 7).
4 Disconnect the refrigerant pipelines from the condenser and cap them.
5 Remove the condenser cooling fan.
6 Disconnect the radiator mountings and support the radiator, not allowing it to hang on its hoses.
7 Disconnect the condenser mountings.
8 Push the radiator towards the engine and lift out the condenser.

Compressor

9 Disconnect the battery.
10 Raise the front of the vehicle and support it on axle stands.
11 Remove the compressor drivebelt (refer to Chapter 2).
12 Disconnect the compressor clutch harness.
13 Disconnect the flexible hoses from the compressor and cap them.
14 Remove the compressor after disconnecting its mounting and adjuster bolts.
15 Keep the compressor in its 'in car' attitude, otherwise oil could enter the low pressure chambers. Should this happen, the compressor pulley must be turned through several revolutions when refitted to the vehicle in order to expel the oil. Failure to do this could result in internal damage once the air conditioning system is operated.

Evaporator

16 Disconnect the battery.
17 Disconnect the refrigerant lines from the evaporator and cap the pipes.

UK models

18 Remove the instrument panel as described in Chapter 10.
19 Remove the mounting screws and withdraw the evaporator.

North American models

20 From the passenger side remove the facia lower cover and glovebox.

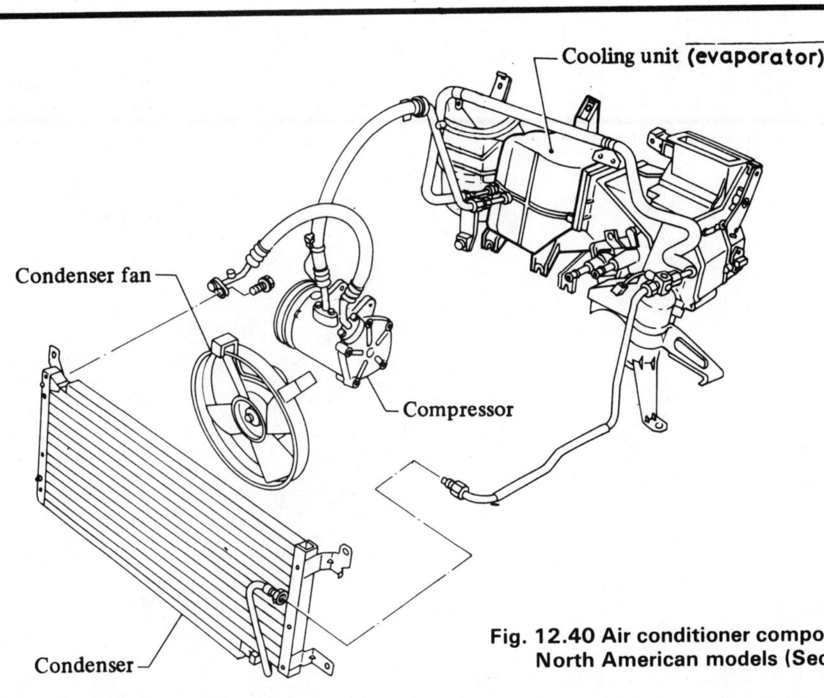

Fig. 12.40 Air conditioner components for North American models (Sec 34)

21 Remove the mounting plate by either removing the screws or, if the air conditioner is factory-installed, by cutting a hole as shown in Fig. 12.41. If the latter method is necessary, cover the blower motor ducts to prevent entry of swarf.
22 Remove the blower motor followed by the evaporator.

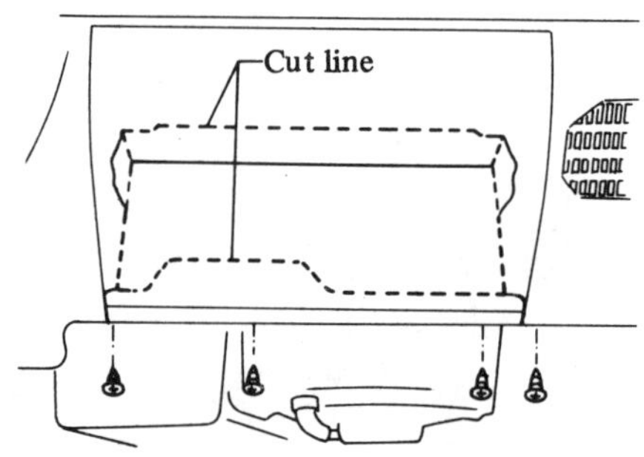

Fig. 12.41 'Cut line' diagram for removing the evaporator on factory-installed air conditioner on North American models (Sec 34)

Refitting

23 Refitting of all components is a reversal of removal, but observe the following points.
24 Use new O-ring seals when reconnecting pipelines and hoses.
25 Once the compressor is refitted with the drivebelt correctly tensioned, release the bolts which secure the flexible mounting knuckles to the engine front left and the rear mounting brackets. Retighten them again with just the normal engine/transmission weight on them.

Index

C

D

E

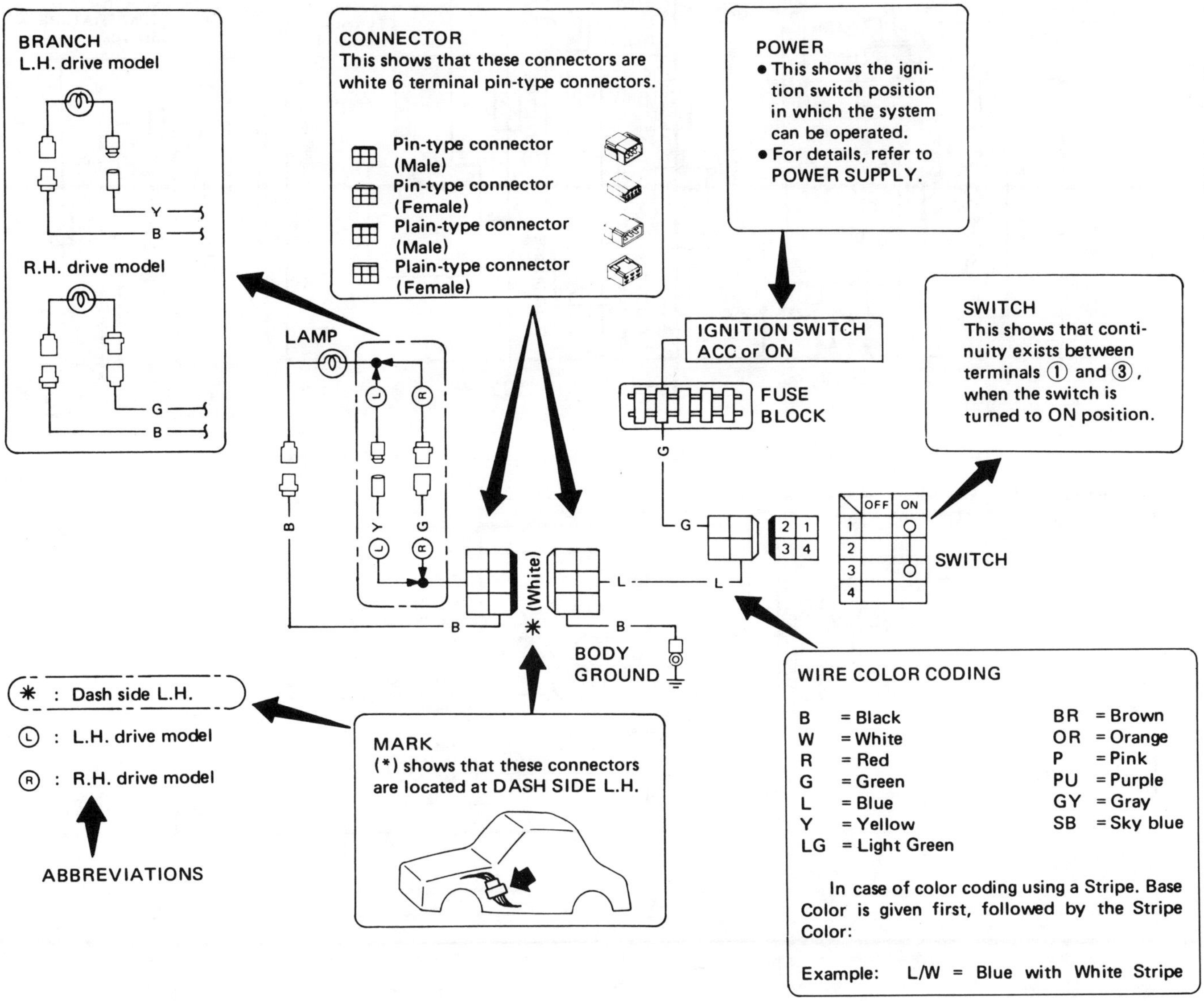

How to use the wiring diagrams for UK models

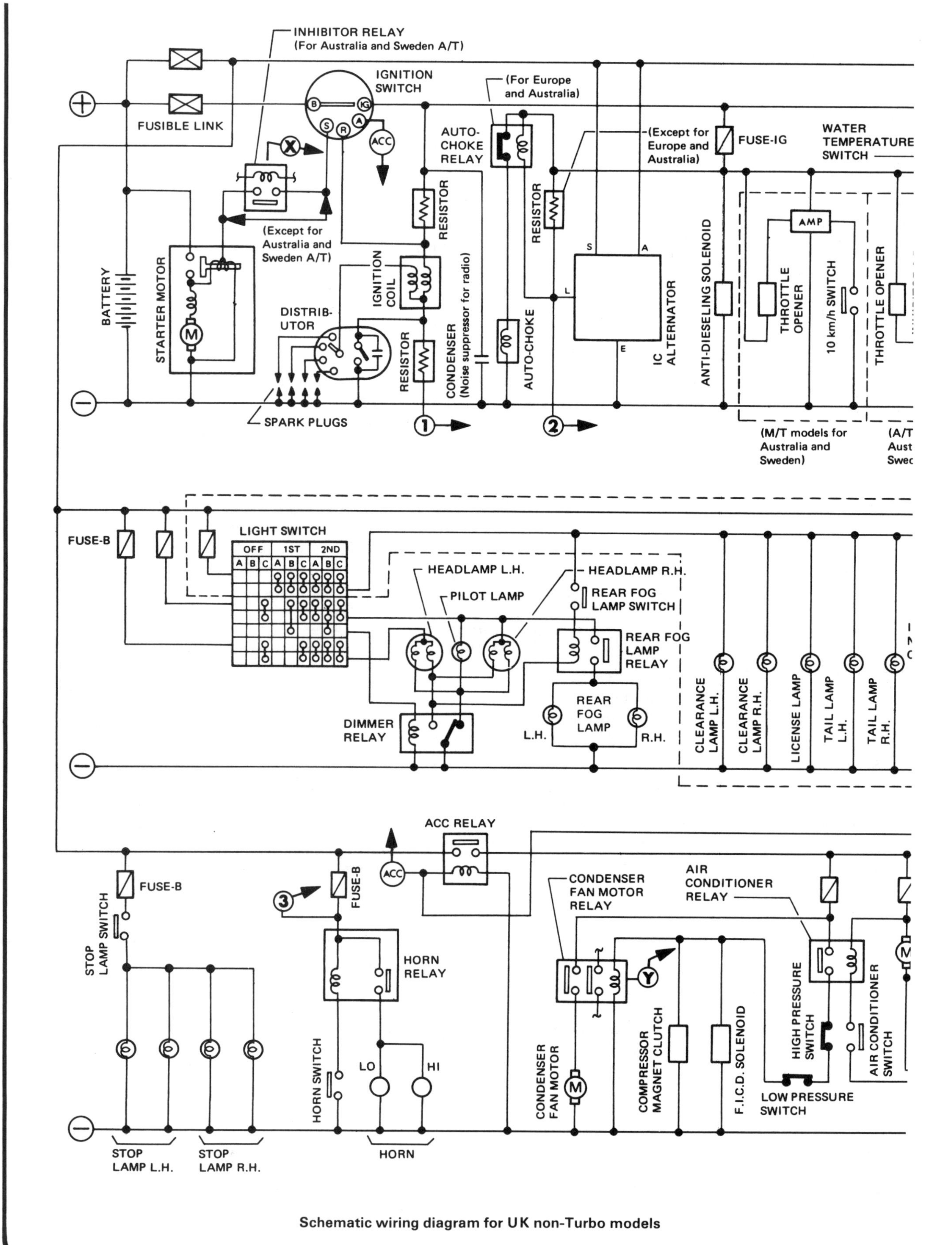

Schematic wiring diagram for UK non-Turbo models

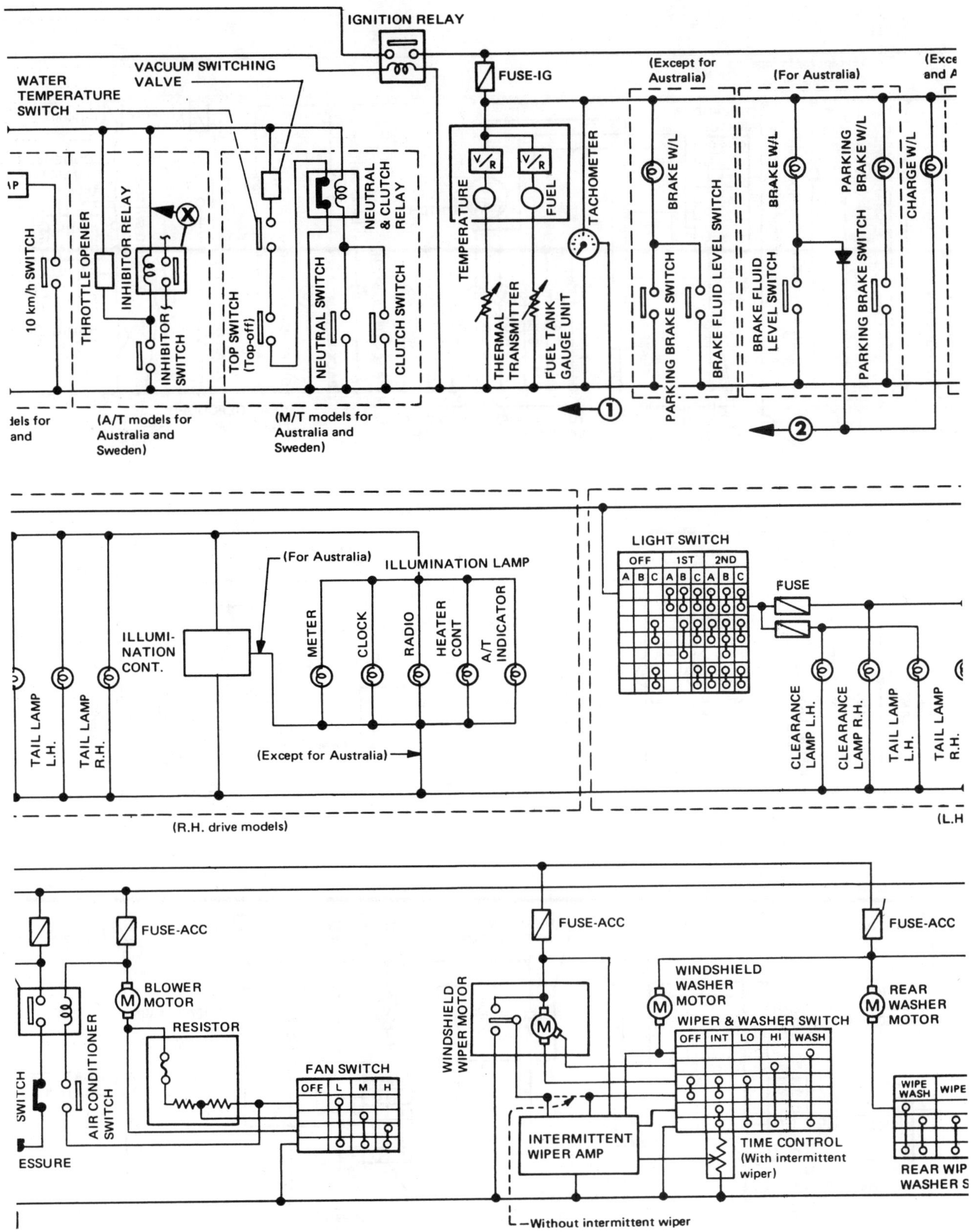

Schematic wiring diagram for UK non-Turbo models (continued)

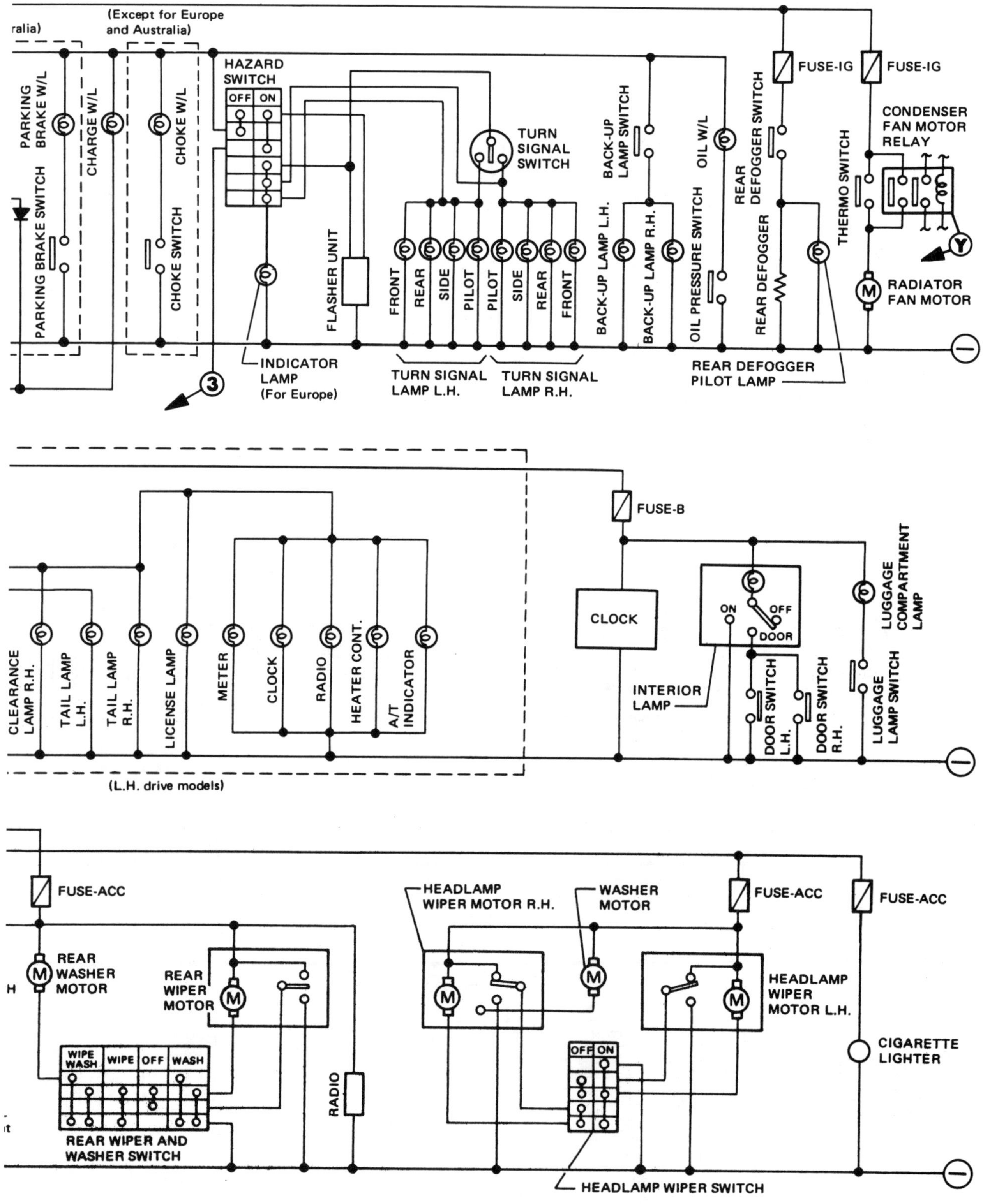

Schematic wiring diagram for UK non-Turbo models (continued)

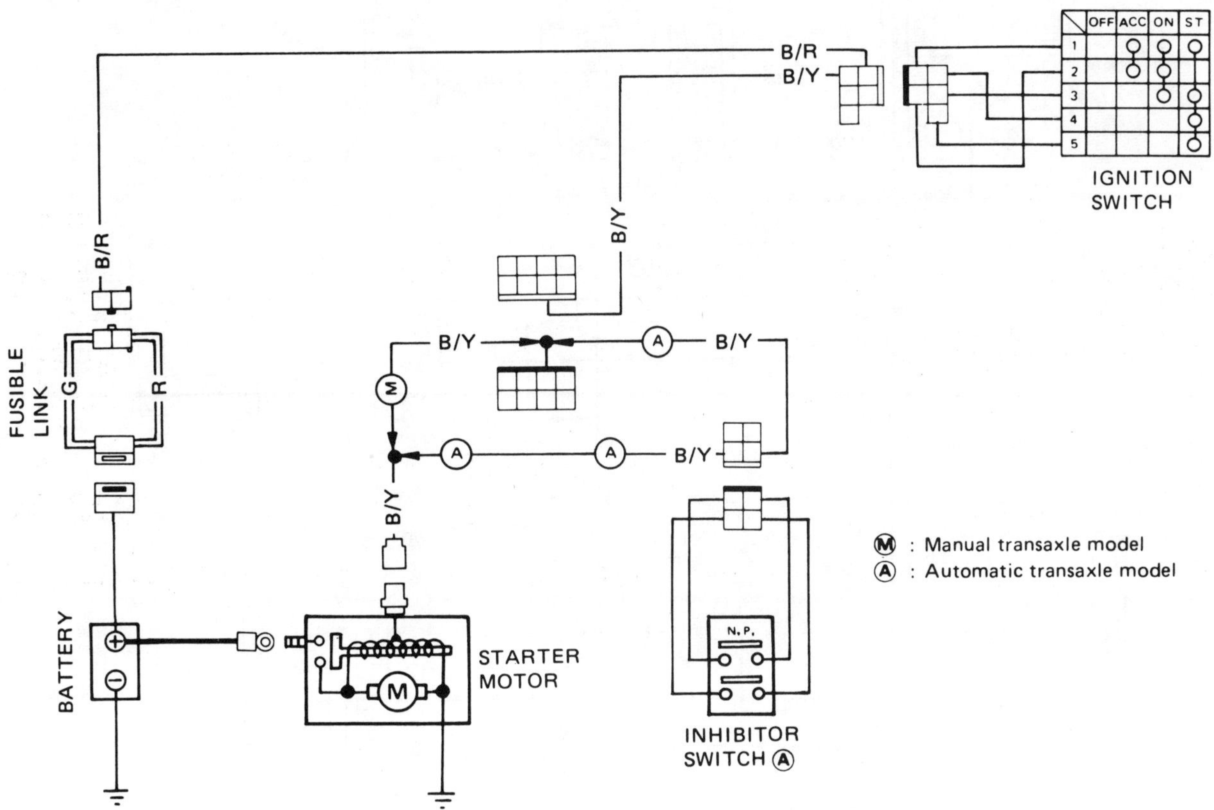

Starter motor circuit for UK non-Turbo models

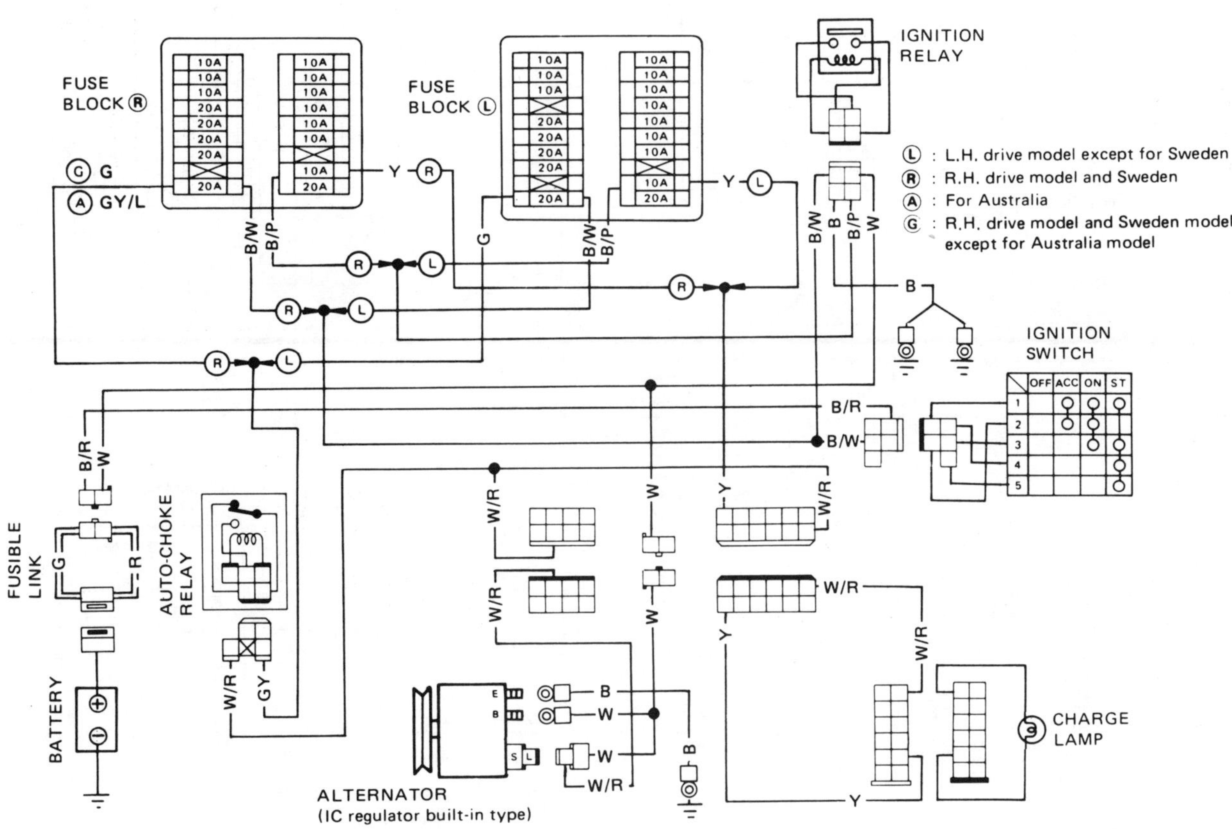

Charging circuit for UK non-Turbo auto-choke models

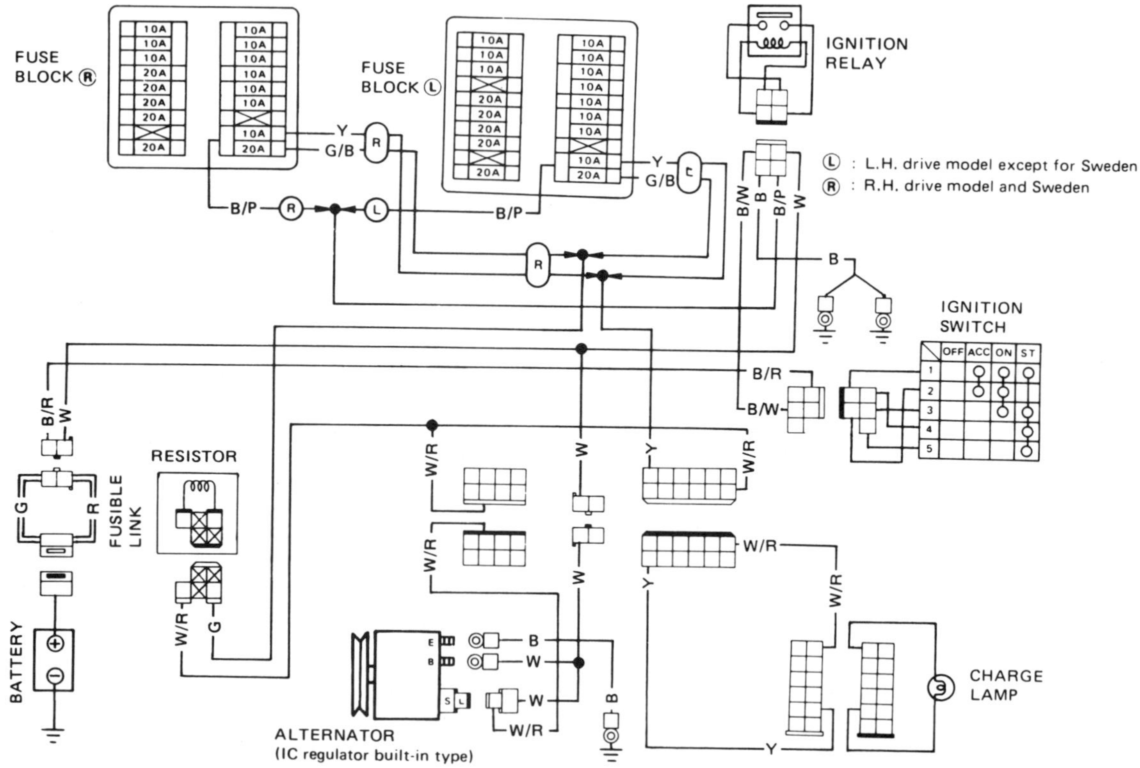

Charging circuit for UK non-Turbo manual choke models

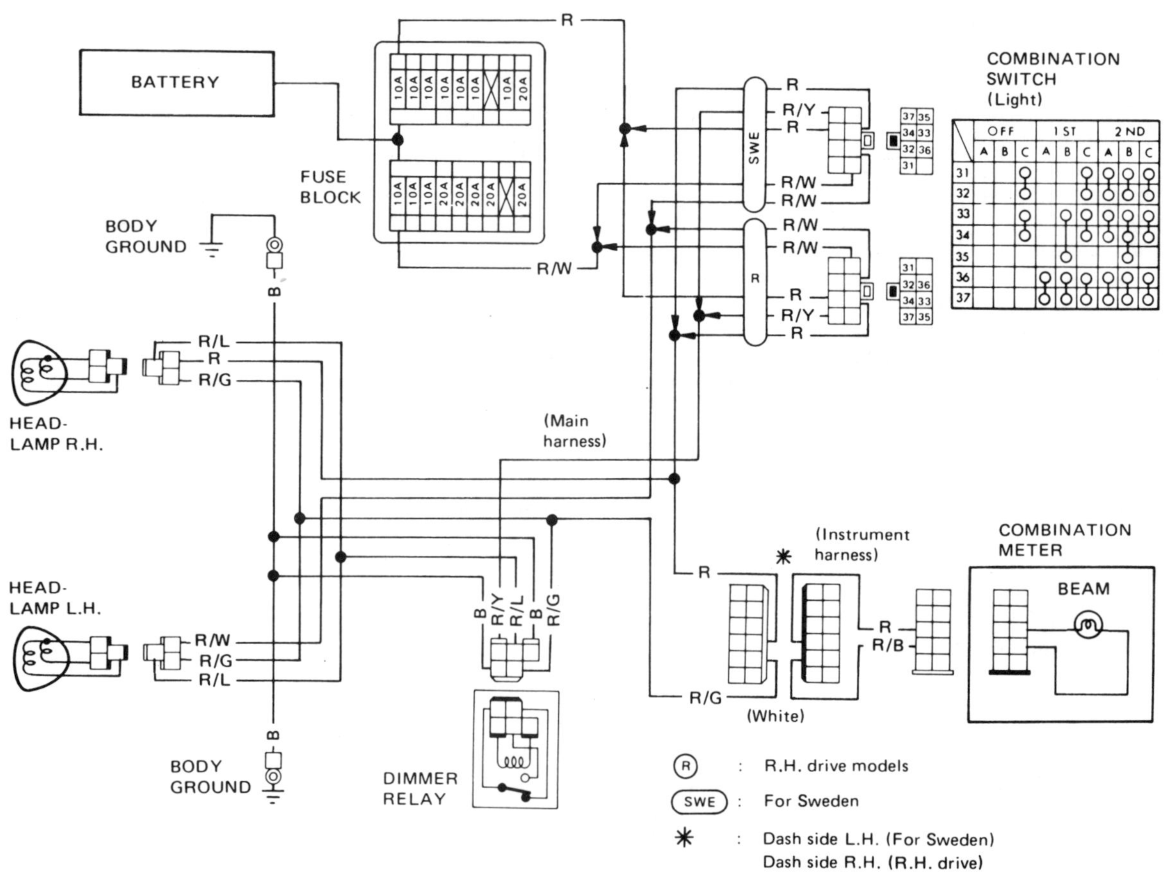

Headlamp circuit for UK non-Turbo models

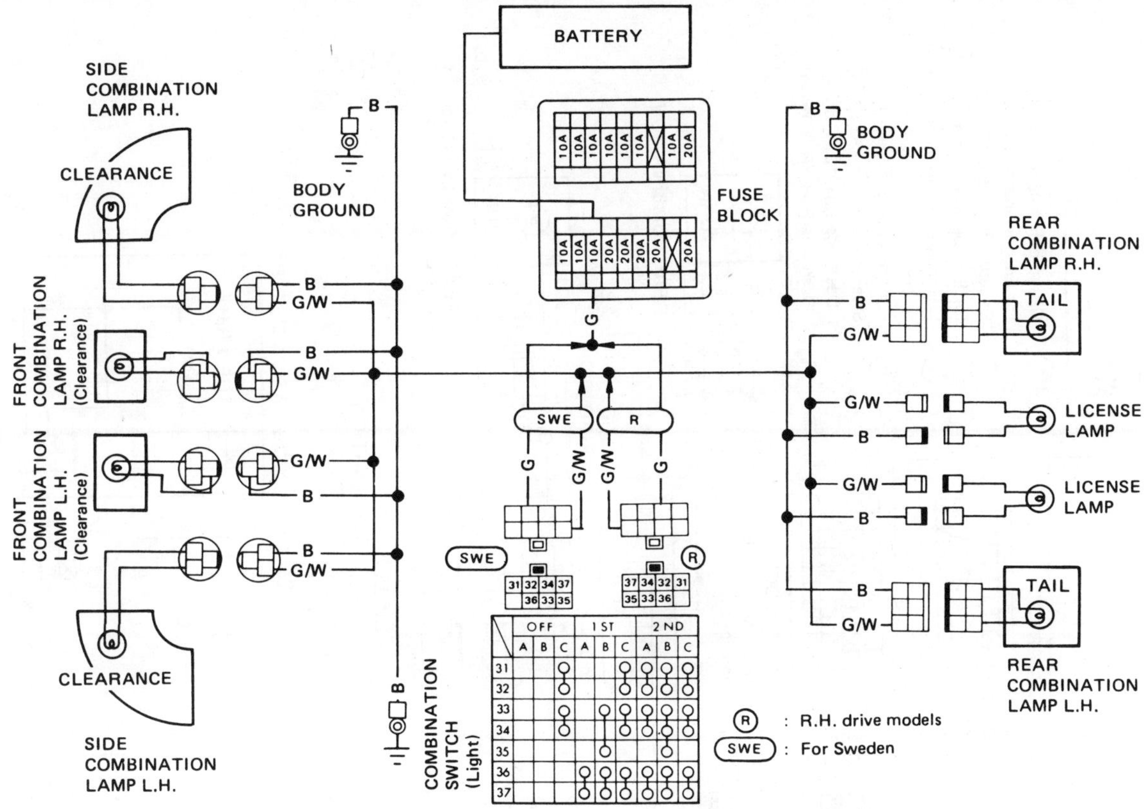

Side, number plate and tail lamp circuit for UK non-Turbo models

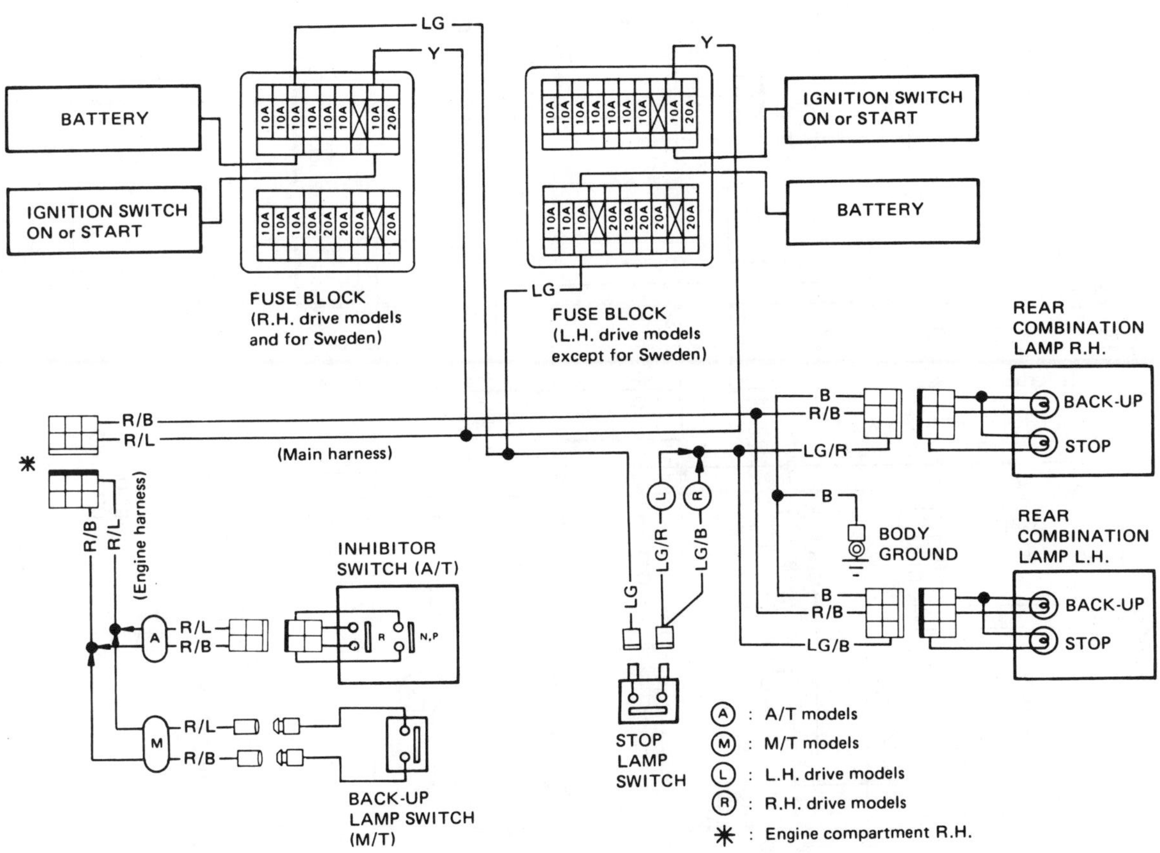

Stop and reversing lamp circuit for UK non-Turbo models

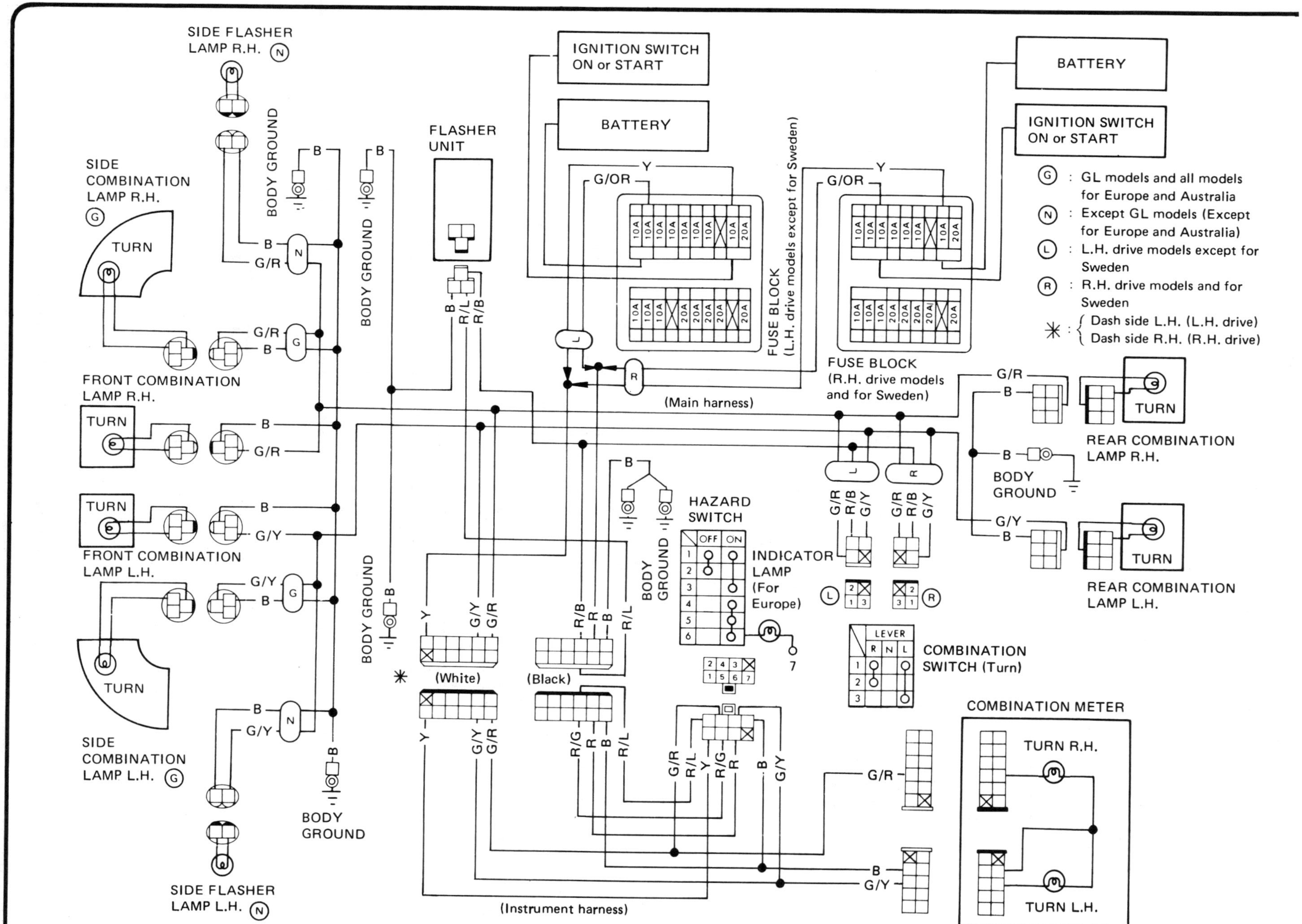

Direction indicator and hazard light circuit for UK non-Turbo models

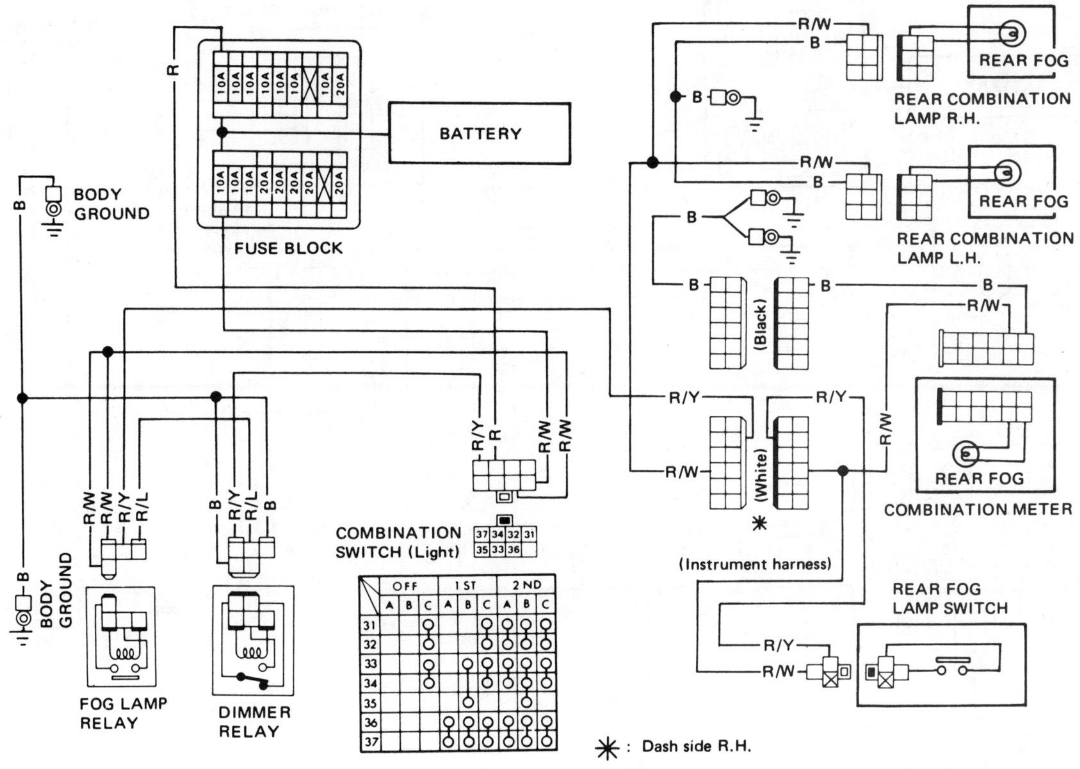

Rear foglamp circuit for UK non-Turbo models

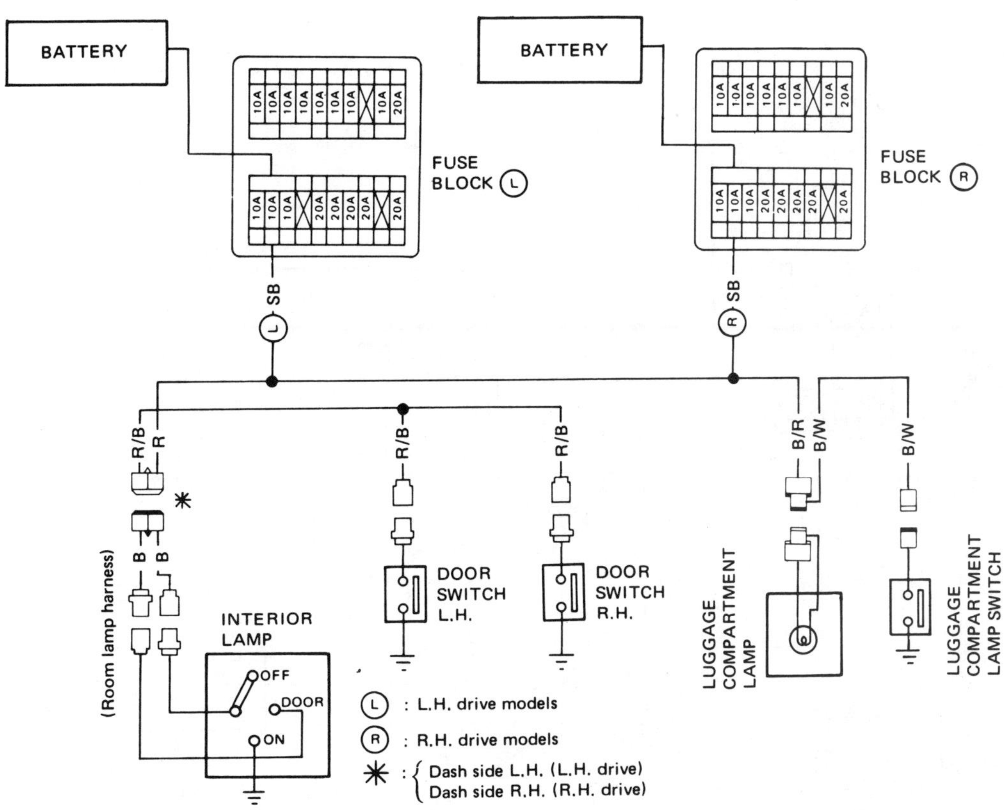

Interior lamps circuit for UK non-Turbo models

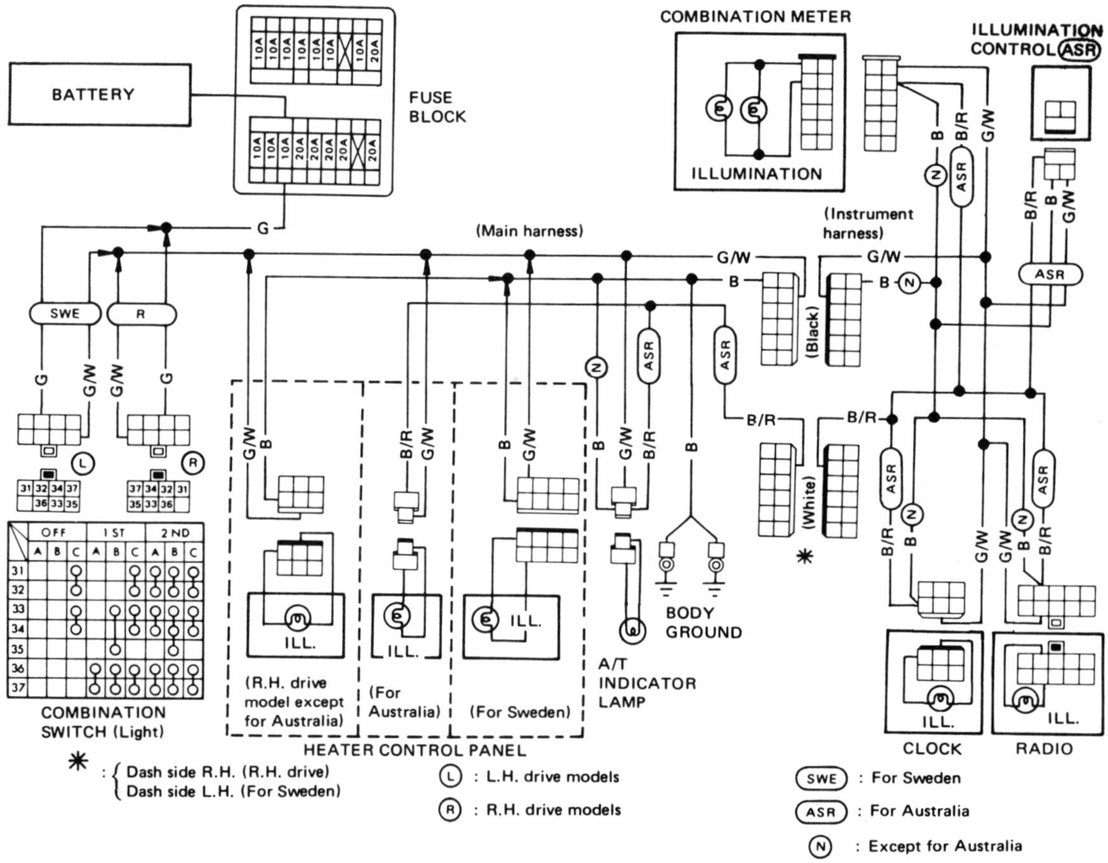

Illumination lamp circuit for UK non-Turbo models

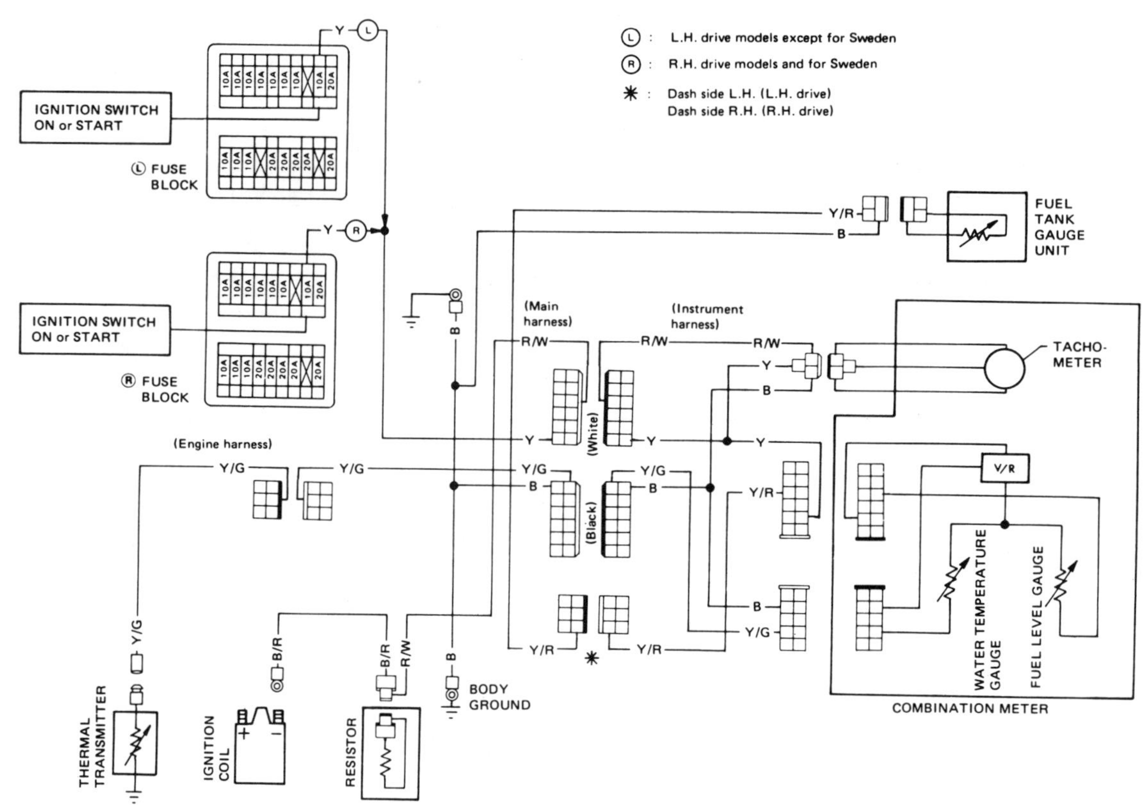

Gauge and tachometer circuit for UK non-Turbo models

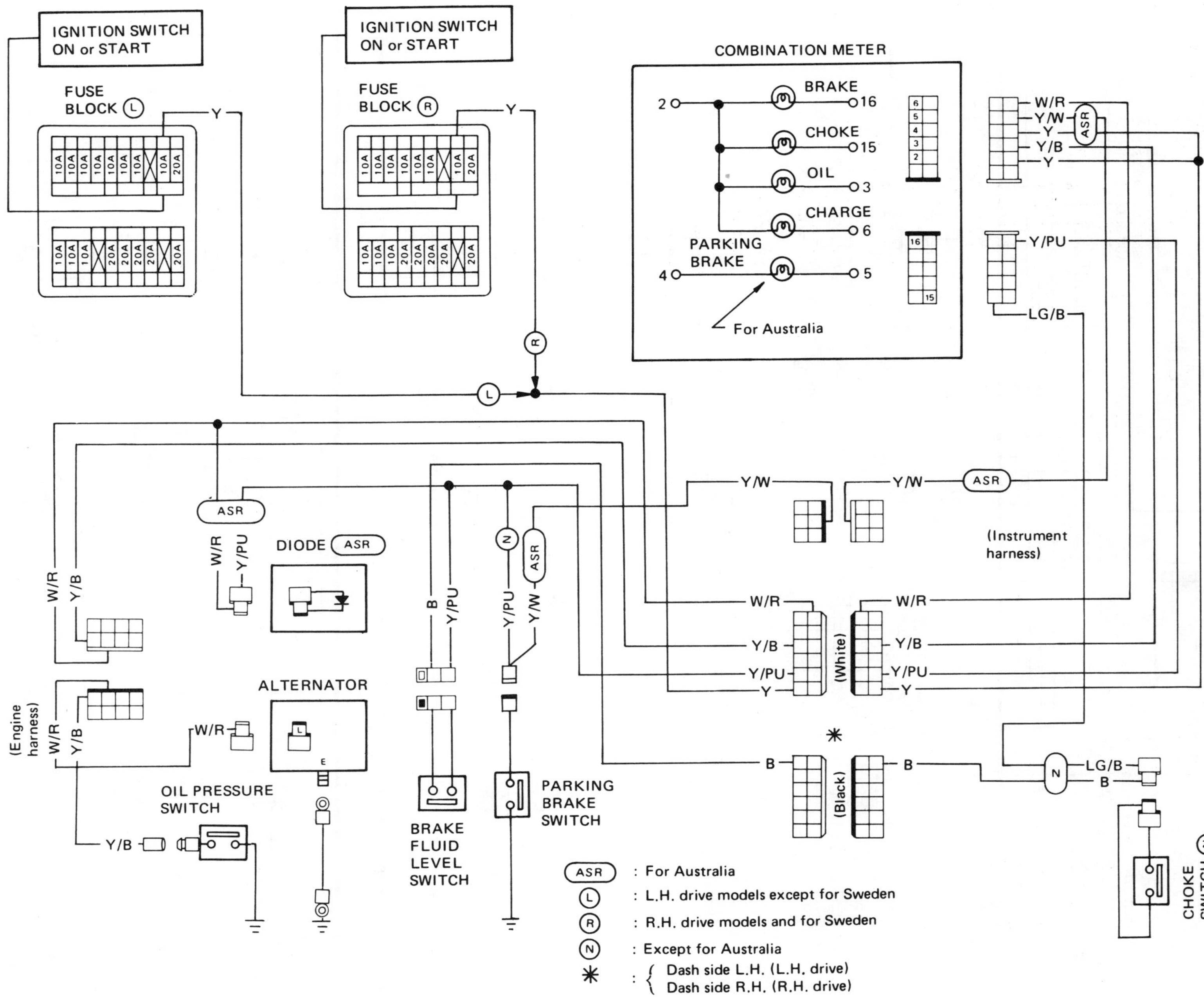

Warning lamp circuit for UK non-Turbo models

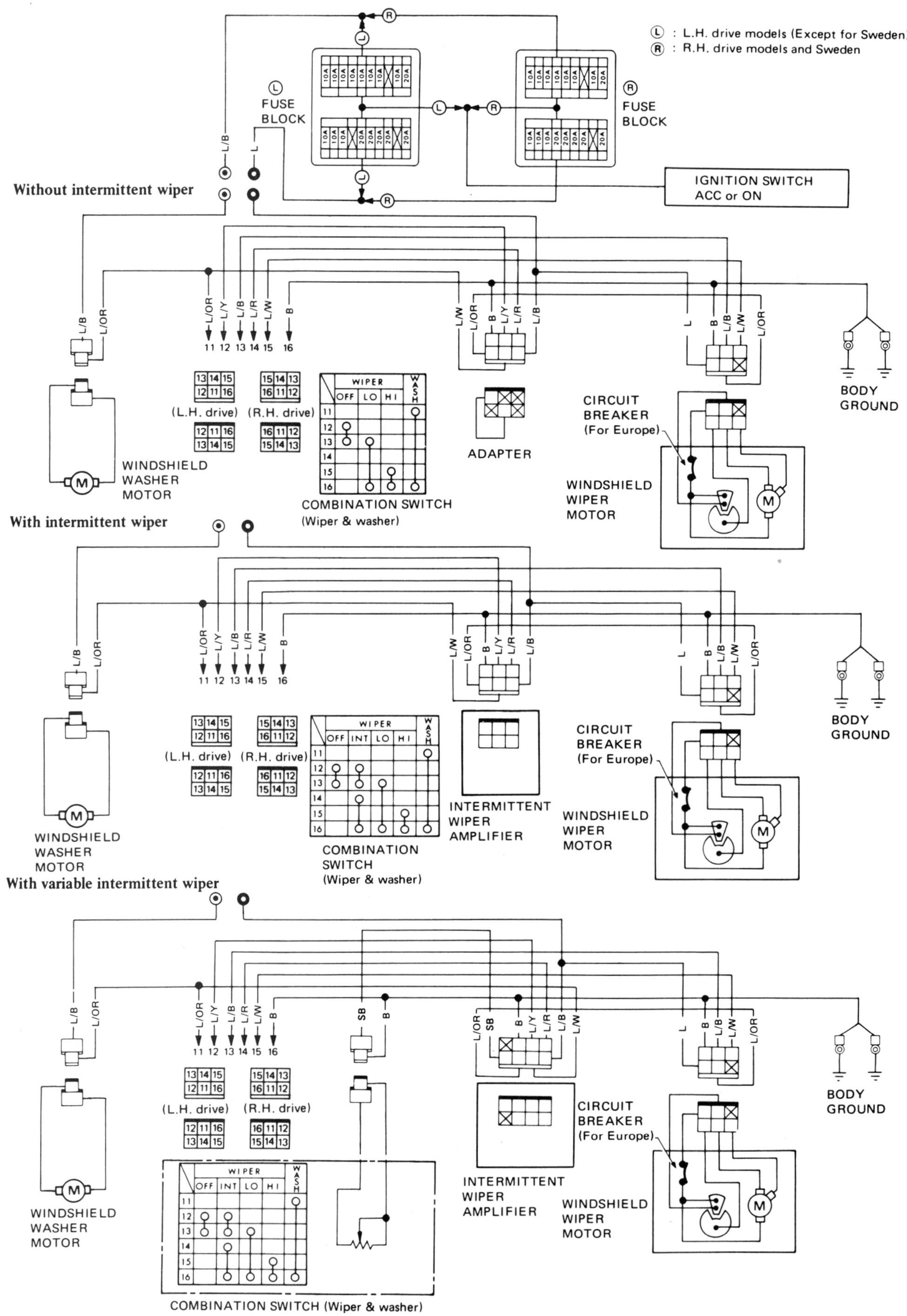

Windscreen wiper and washer circuit for UK non-Turbo models

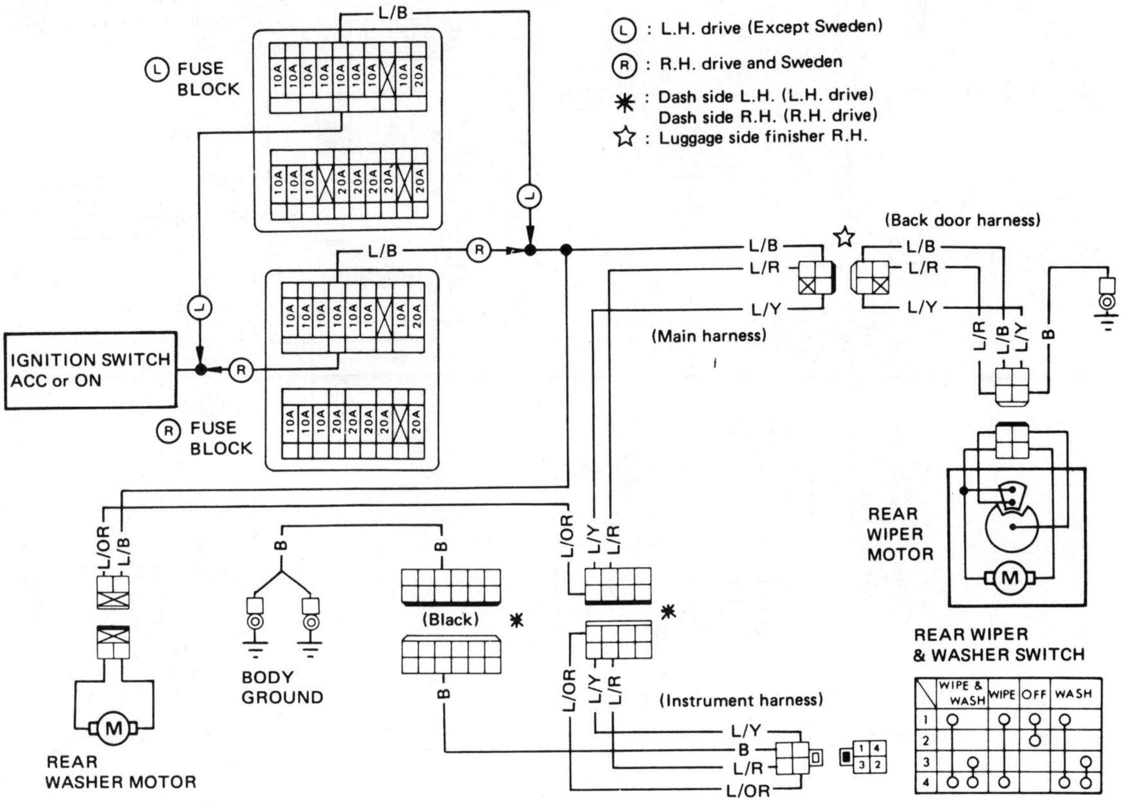

Tailgate wiper and washer circuit for UK non-Turbo models

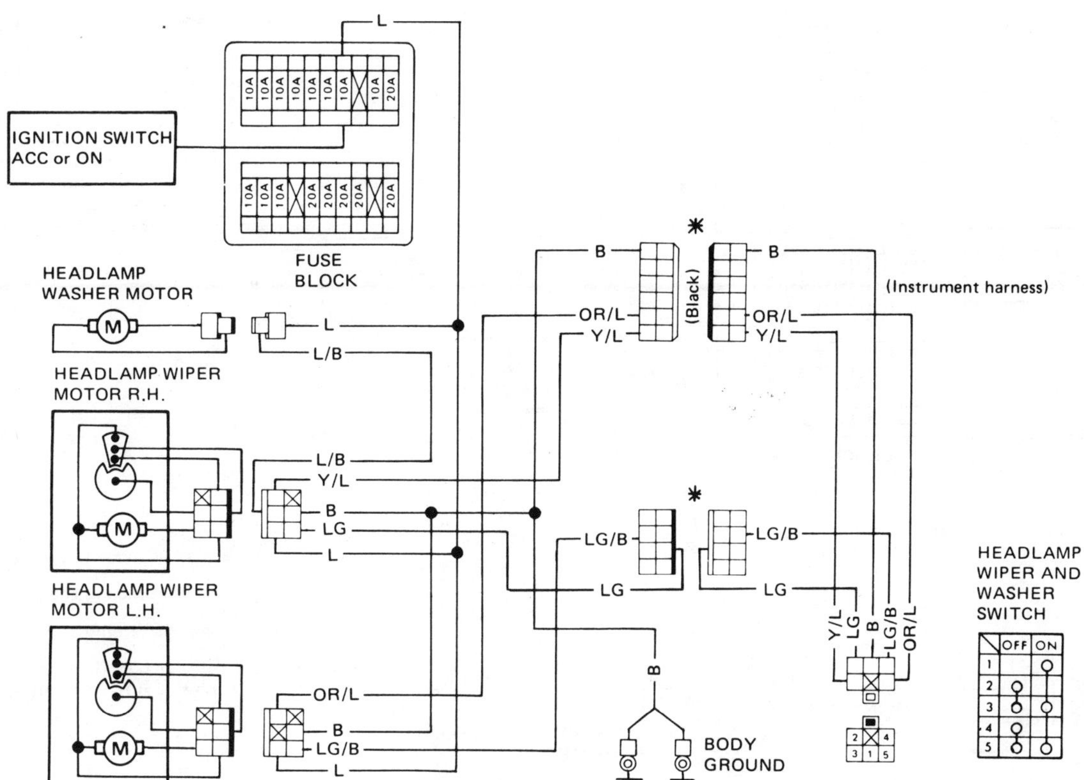

Headlamp wiper and washer circuit for UK non-Turbo models

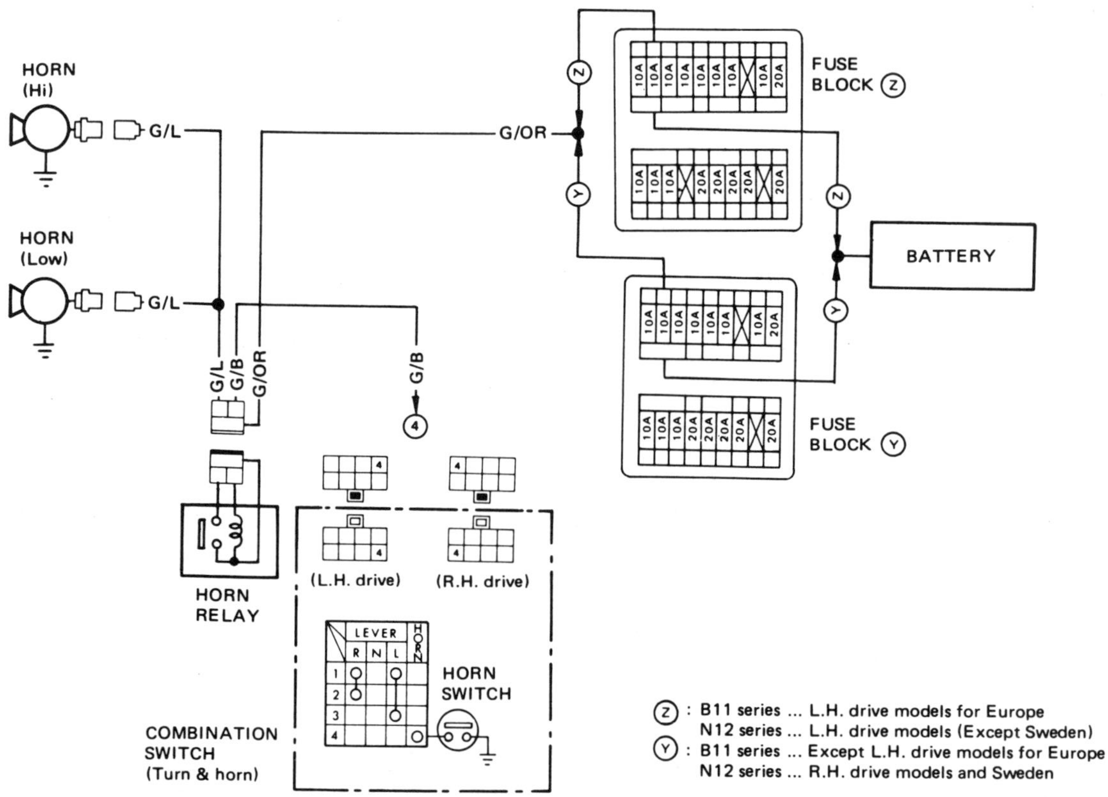

Horn circuit for UK non-Turbo models

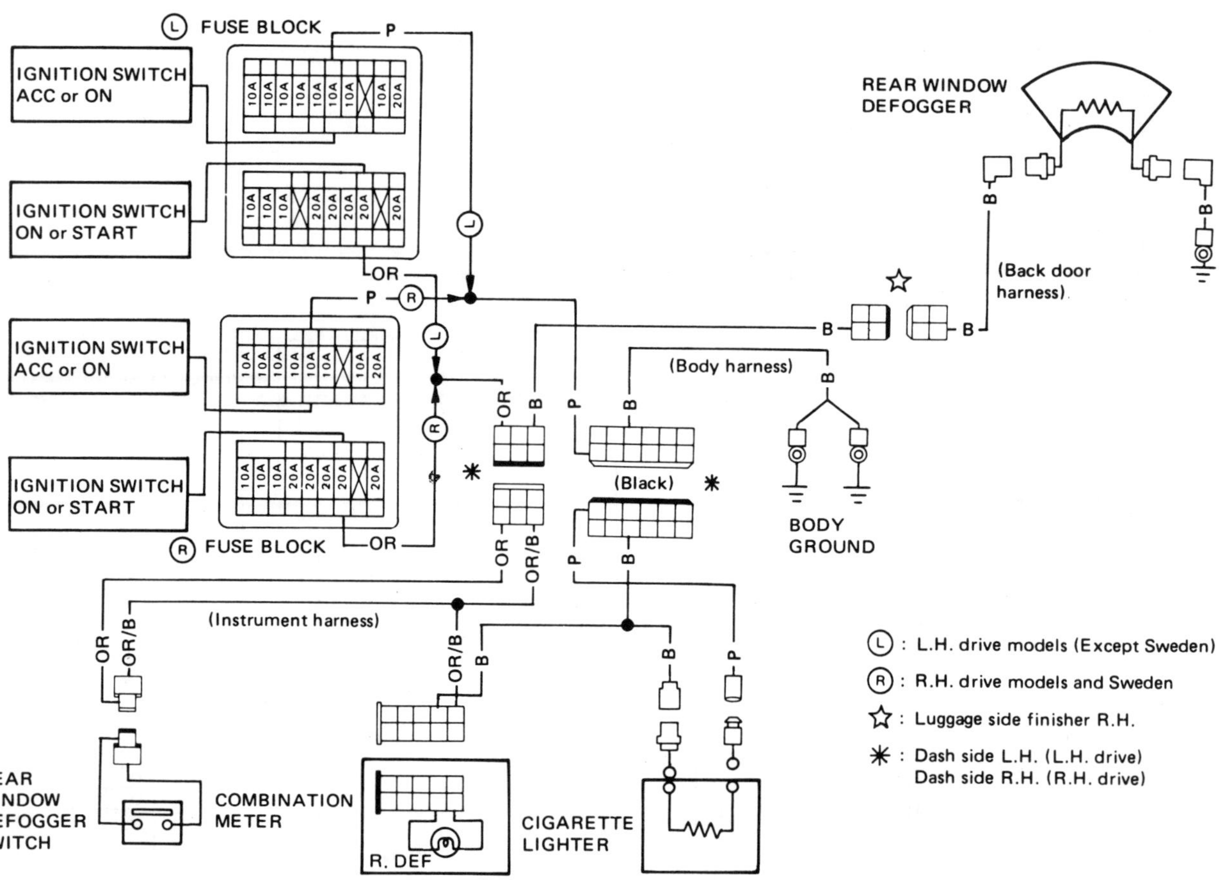

Heated rear window and cigarette lighter circuit for UK non-Turbo models

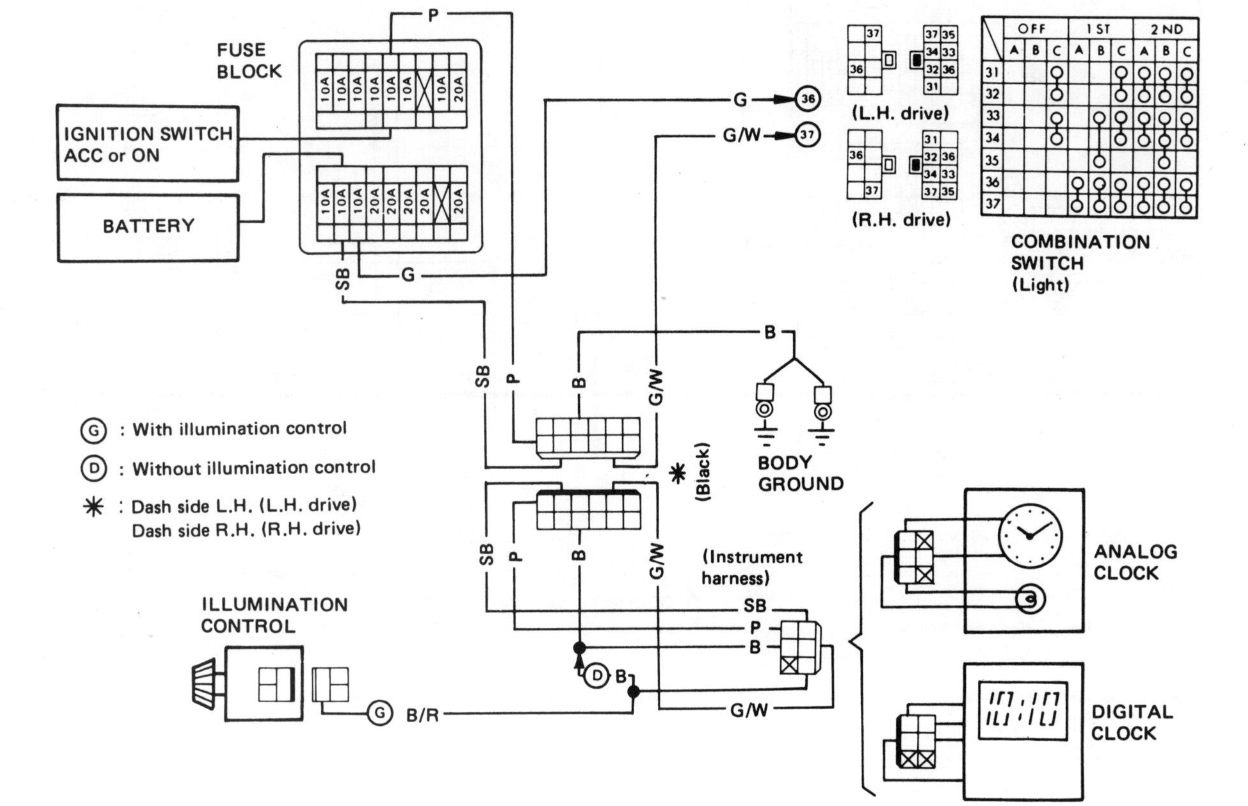

Clock circuit for UK non-Turbo models

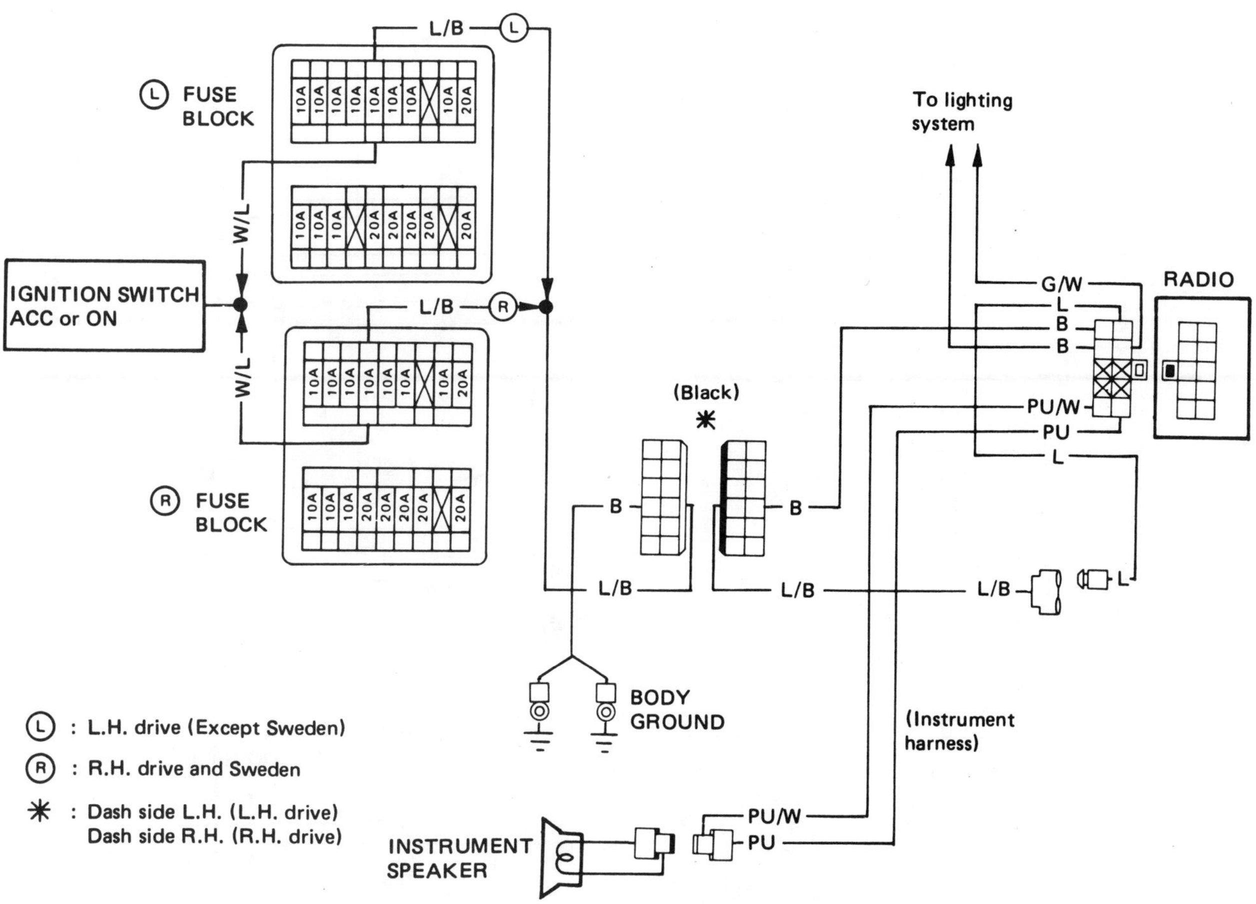

One speaker audio circuit for UK non-Turbo models

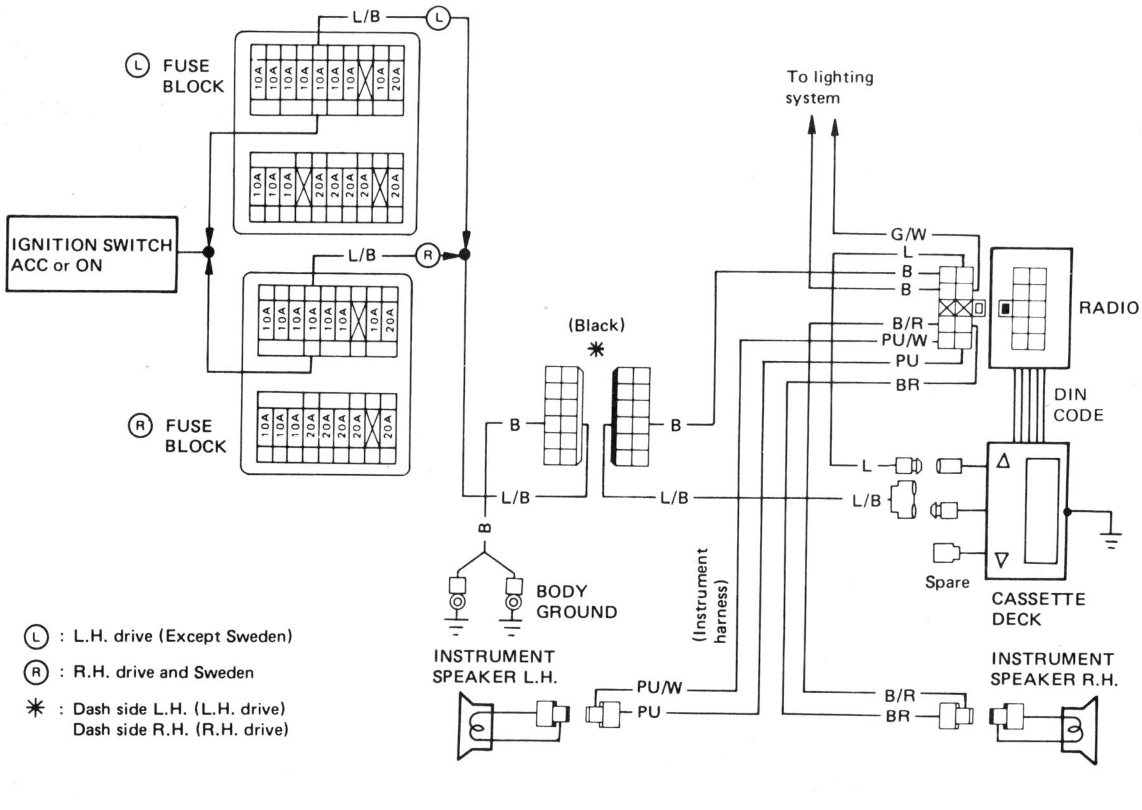

Two speaker audio circuit for UK non-Turbo models

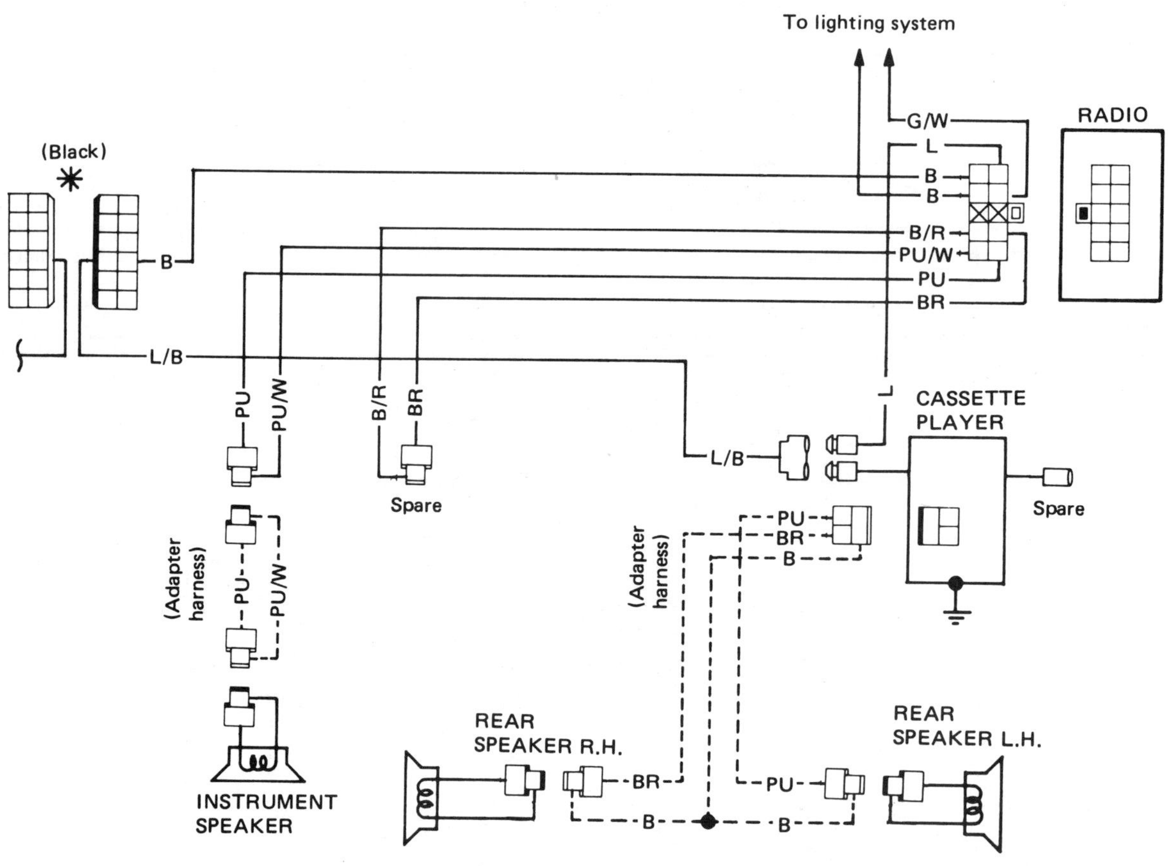

Three speaker audio circuit for UK non-Turbo models

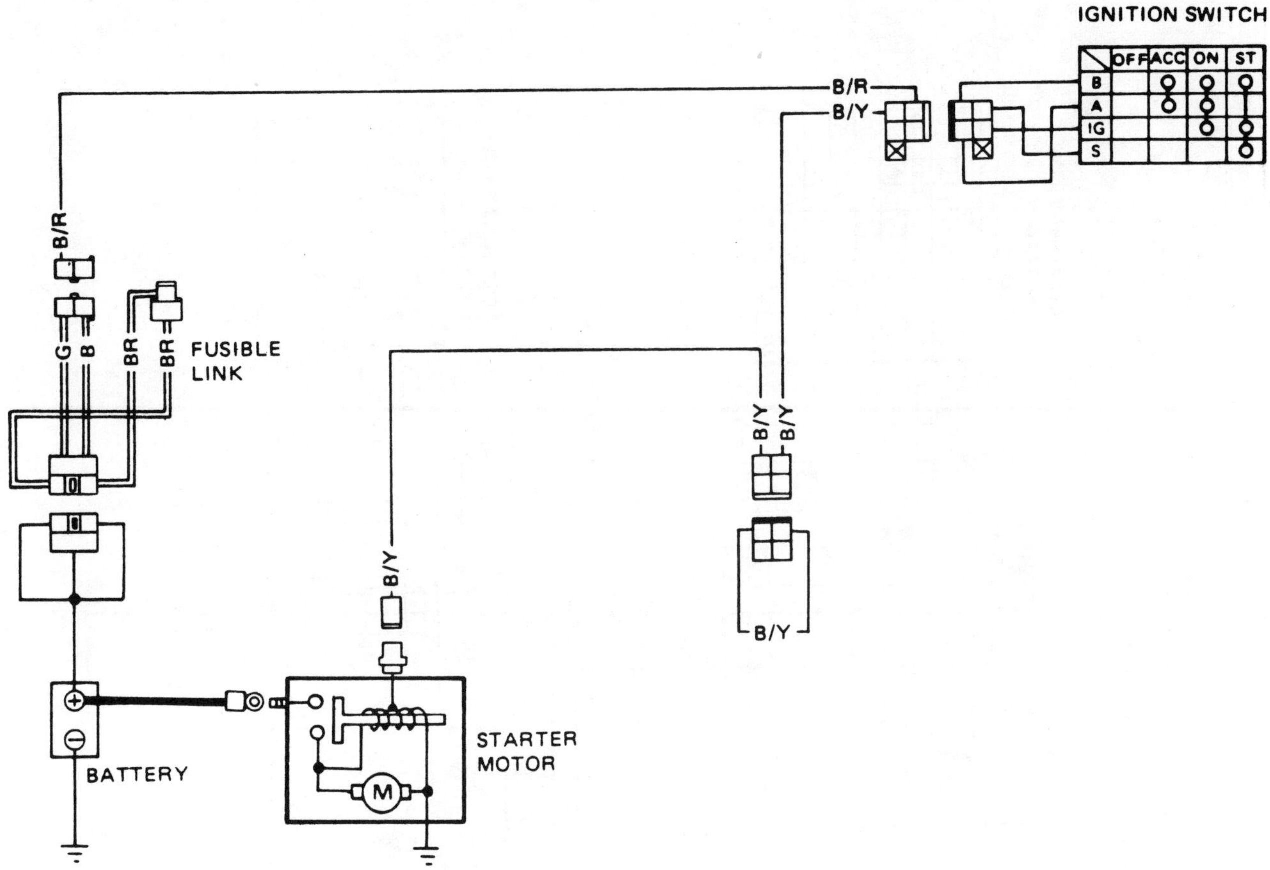

Starter motor circuit for UK Turbo models

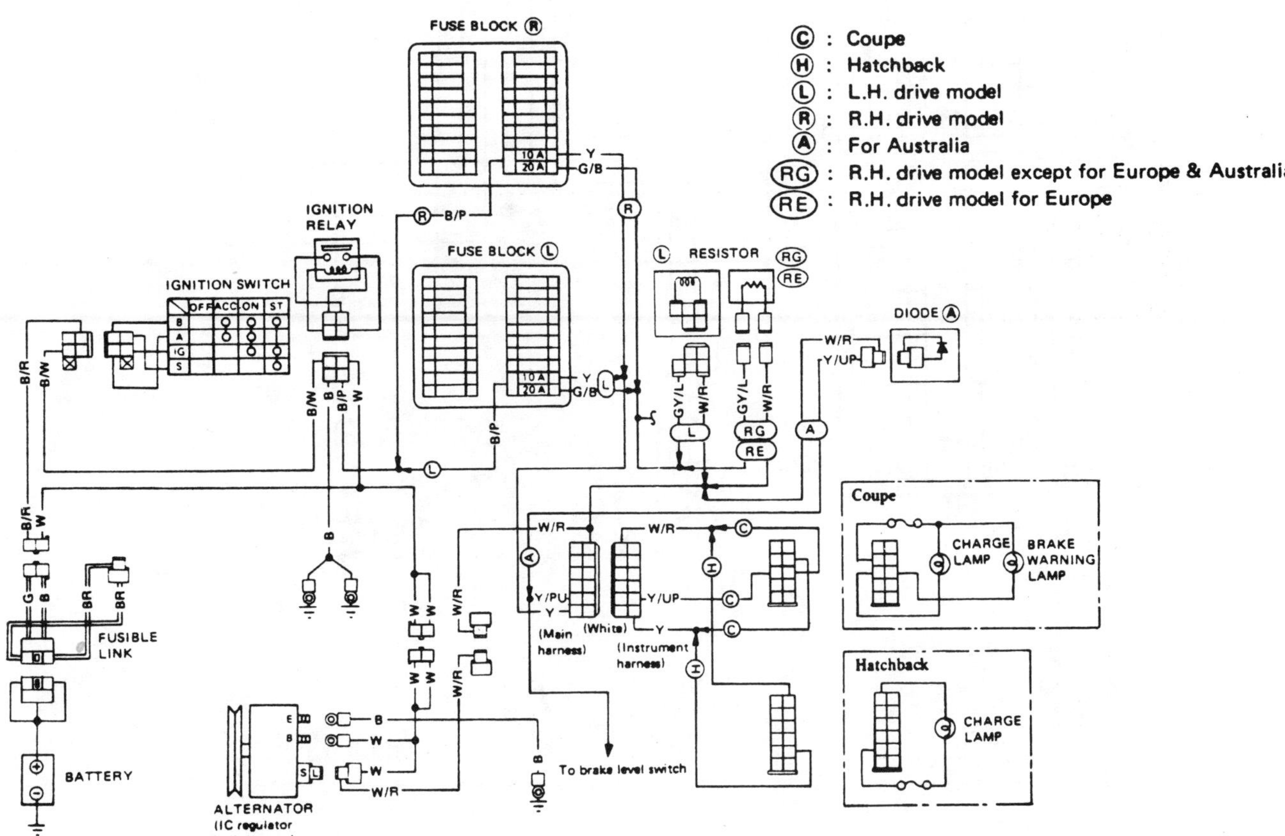

Charging circuit for UK Turbo models

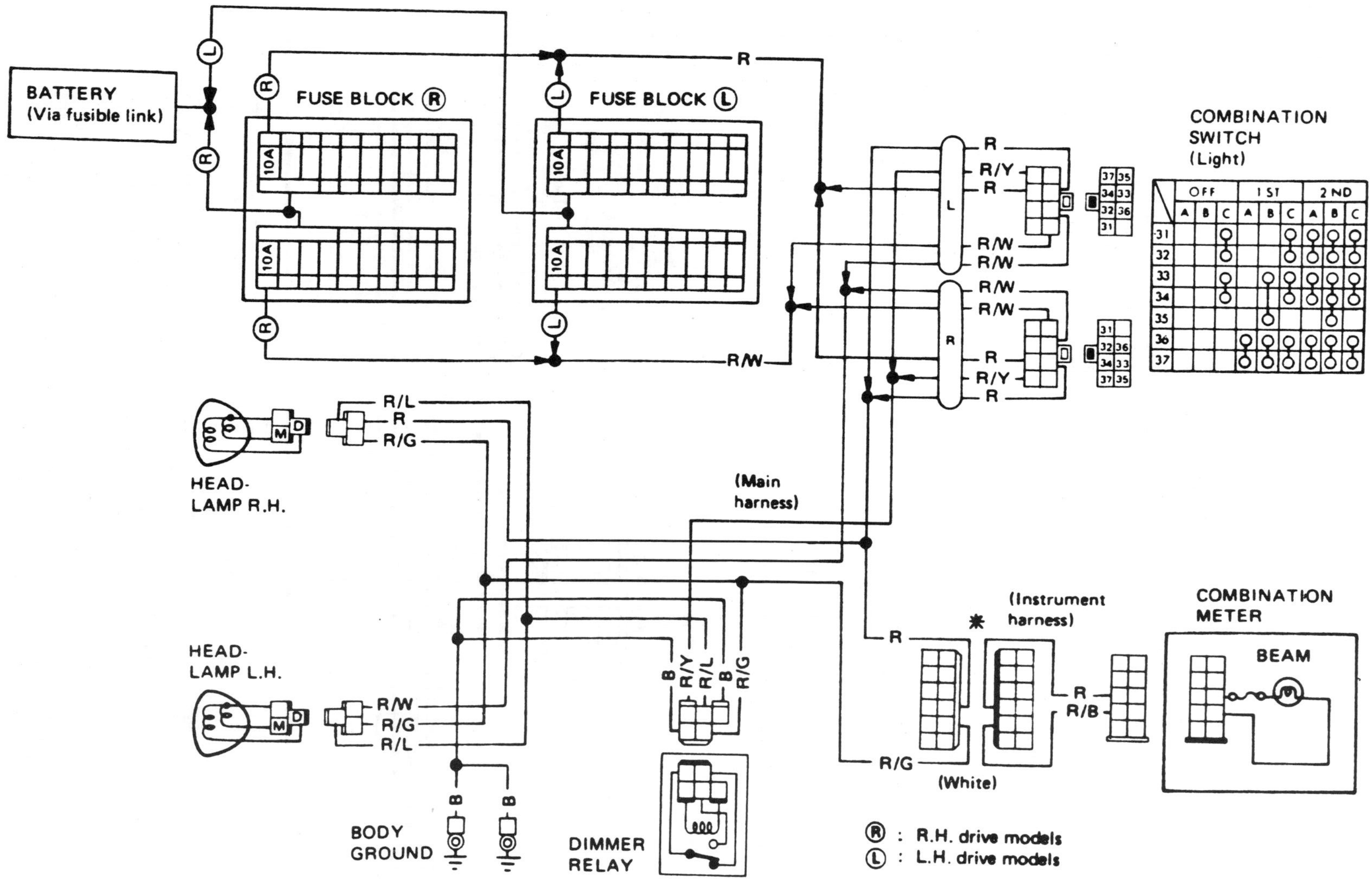

Headlamp circuit for UK Turbo saloon models

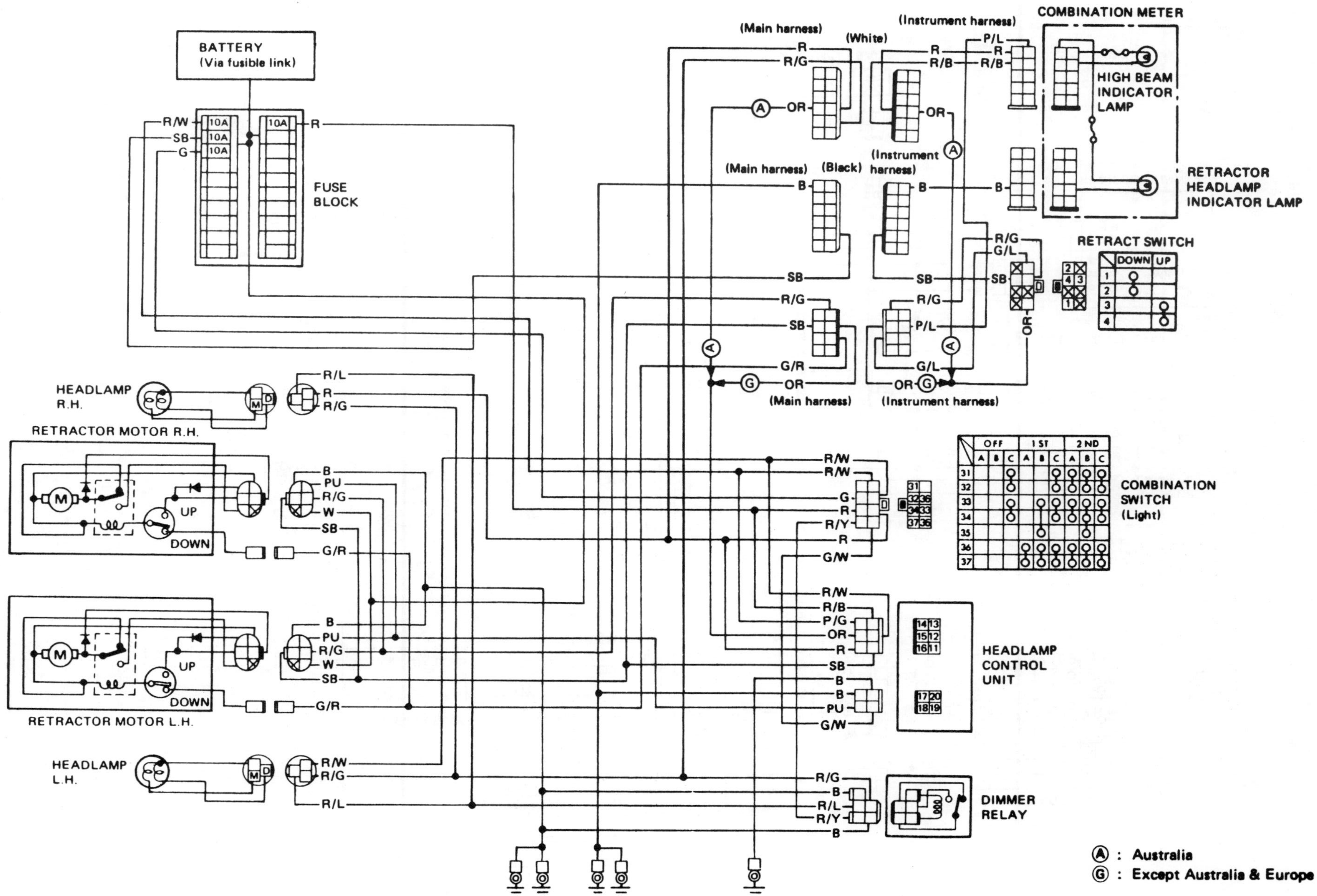

Headlamp circuit for UK Turbo coupe models

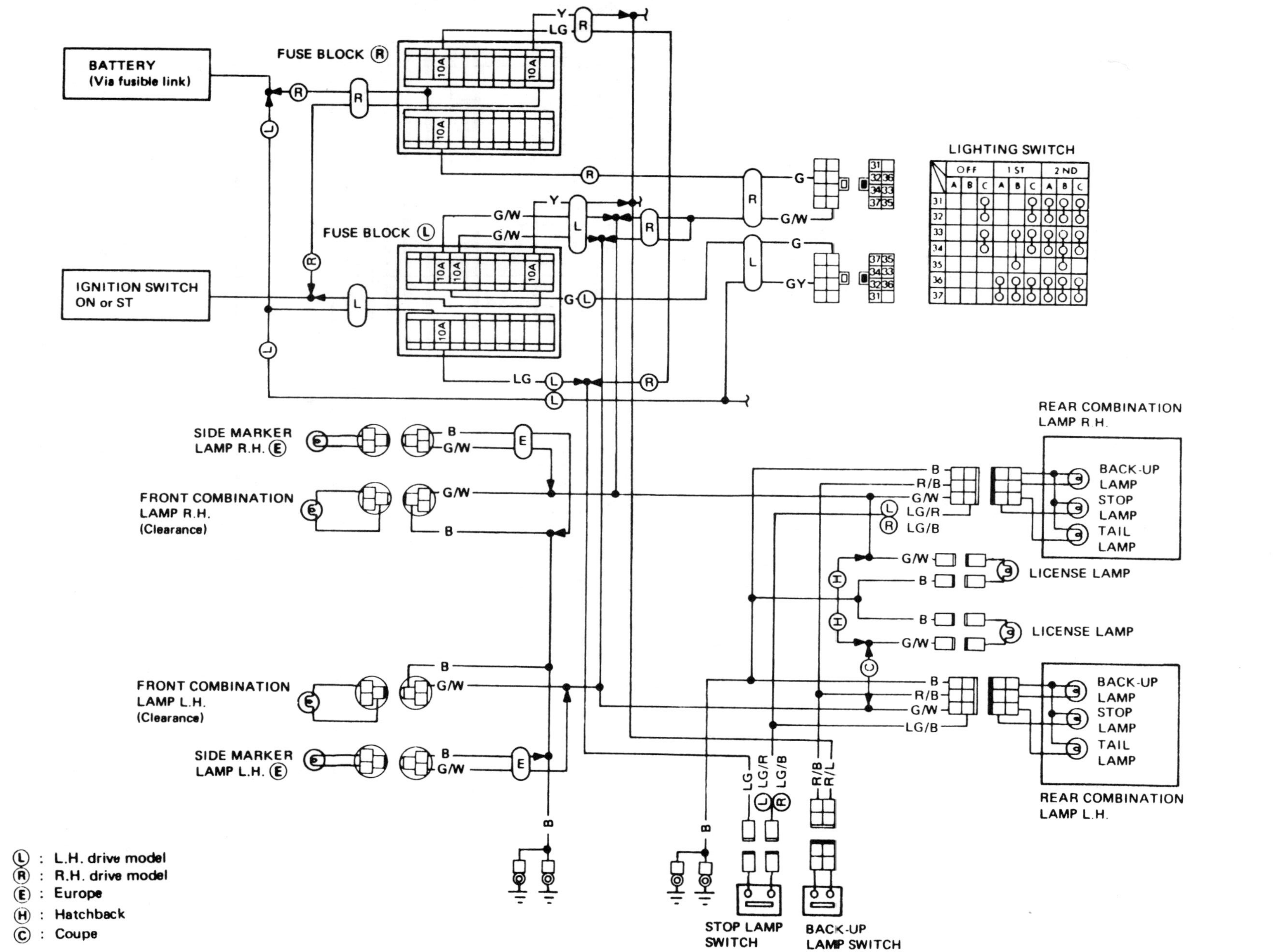

Side, side marker, number plate, tail, stop and reverse lamp circuit for UK Turbo models

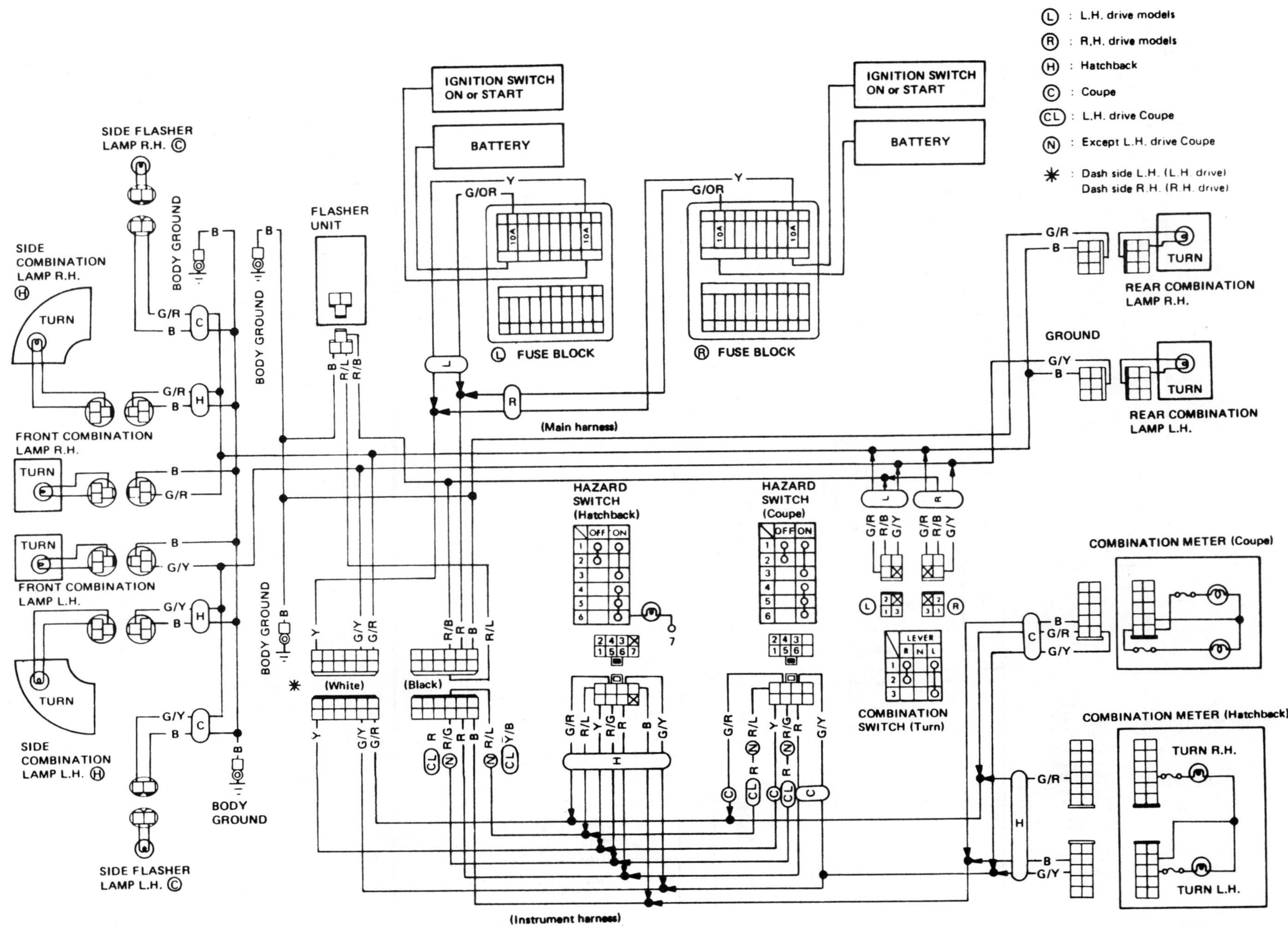

Direction indicator and hazard light circuit for UK Turbo models

Illumination lamp circuit for UK Turbo models

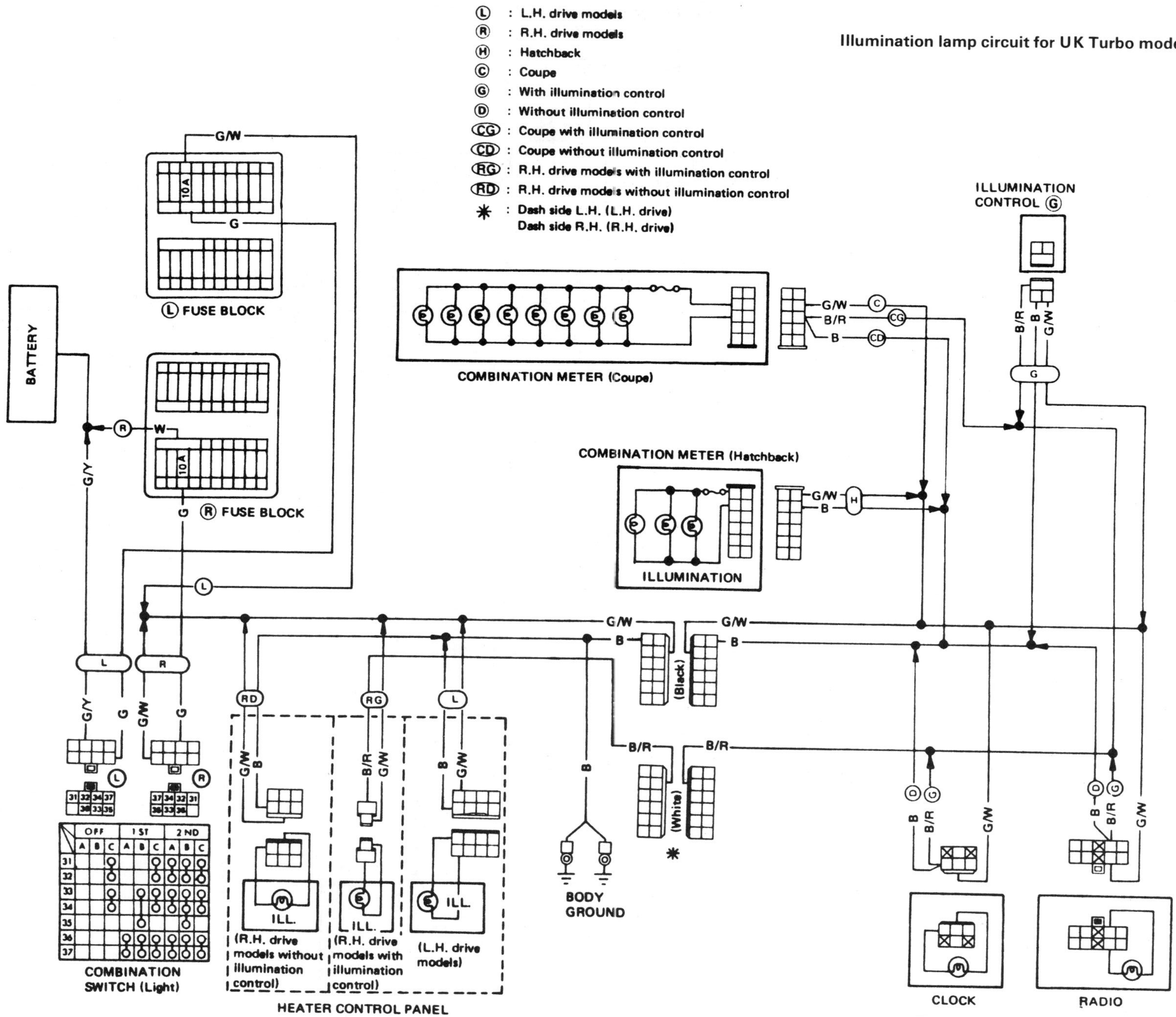

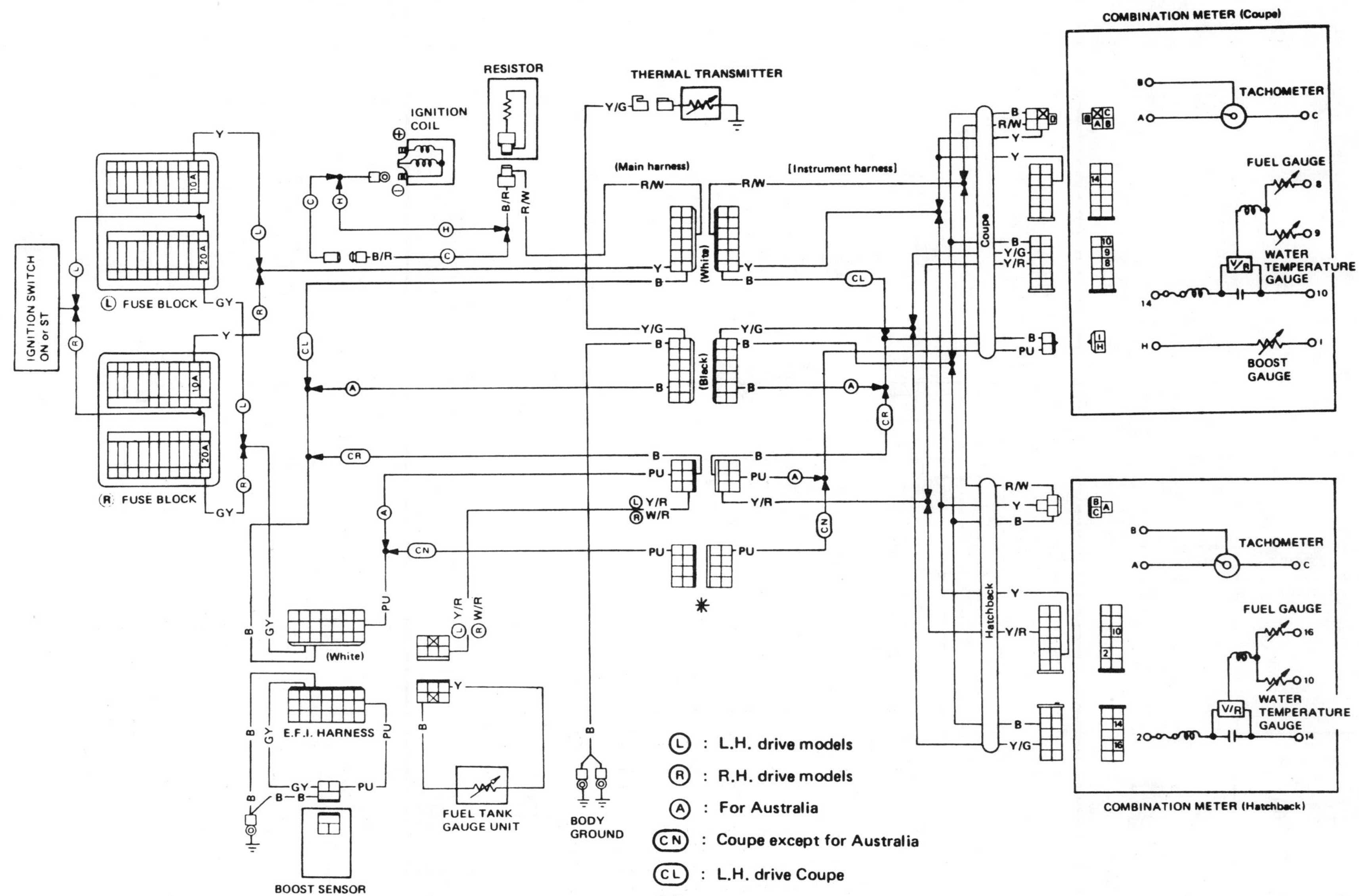

Ⓛ : L.H. drive models

Ⓡ : R.H. drive models

Ⓐ : For Australia

(CN) : Coupe except for Australia

(CL) : L.H. drive Coupe

(CR) : R.H. drive Coupe except for Australia

Ⓒ : Coupe

Ⓗ : Hatchback

✱ : Dash side L.H. (L.H. drive)
Dash side R.H. (R.H. drive)

Instrument panel and gauge circuit for UK Turbo models

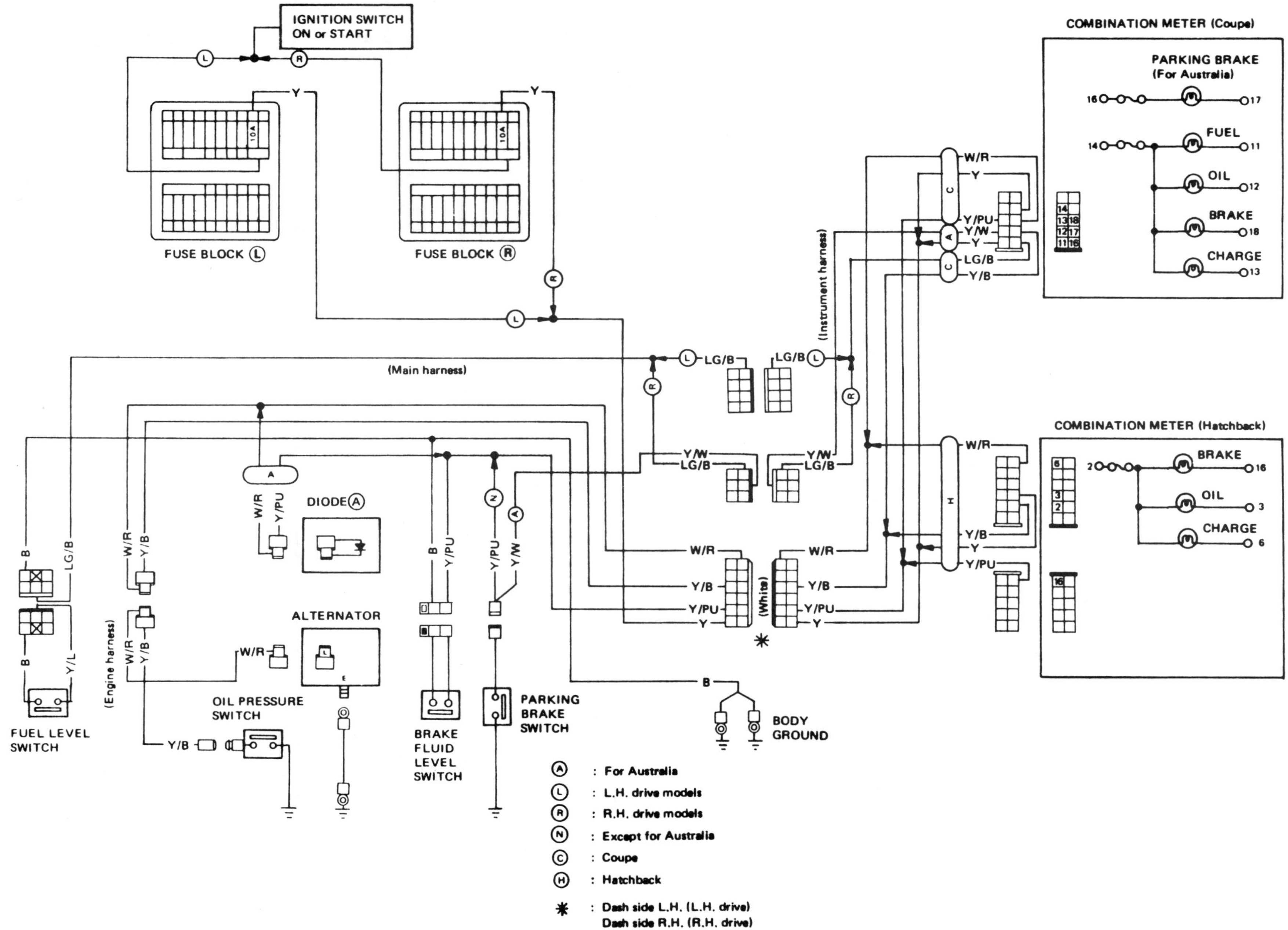

Warning lamp circuit for UK Turbo models

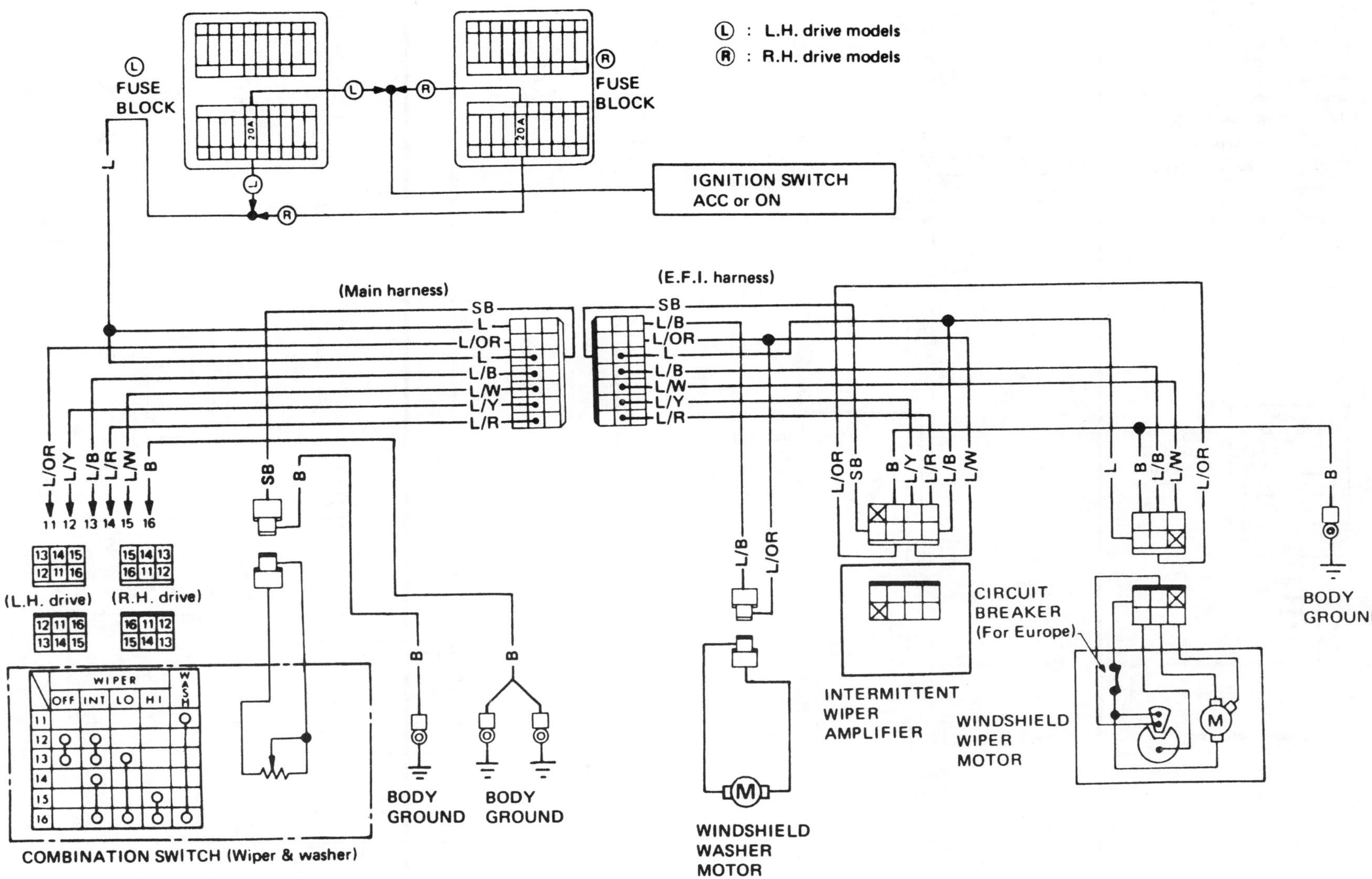

Windscreen wiper and washer circuit for UK Turbo models

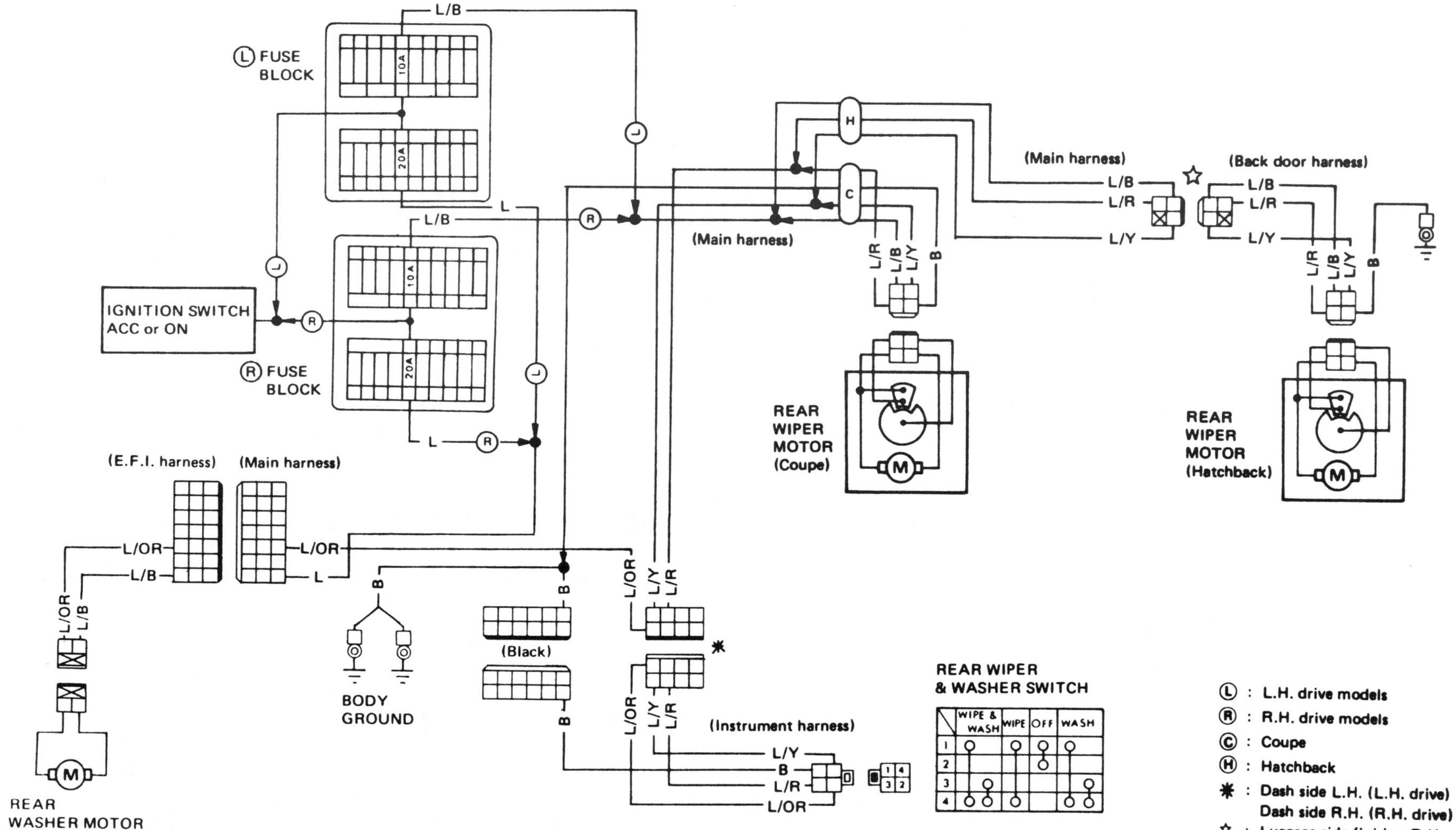

Tailgate wiper and washer circuit for UK Turbo models

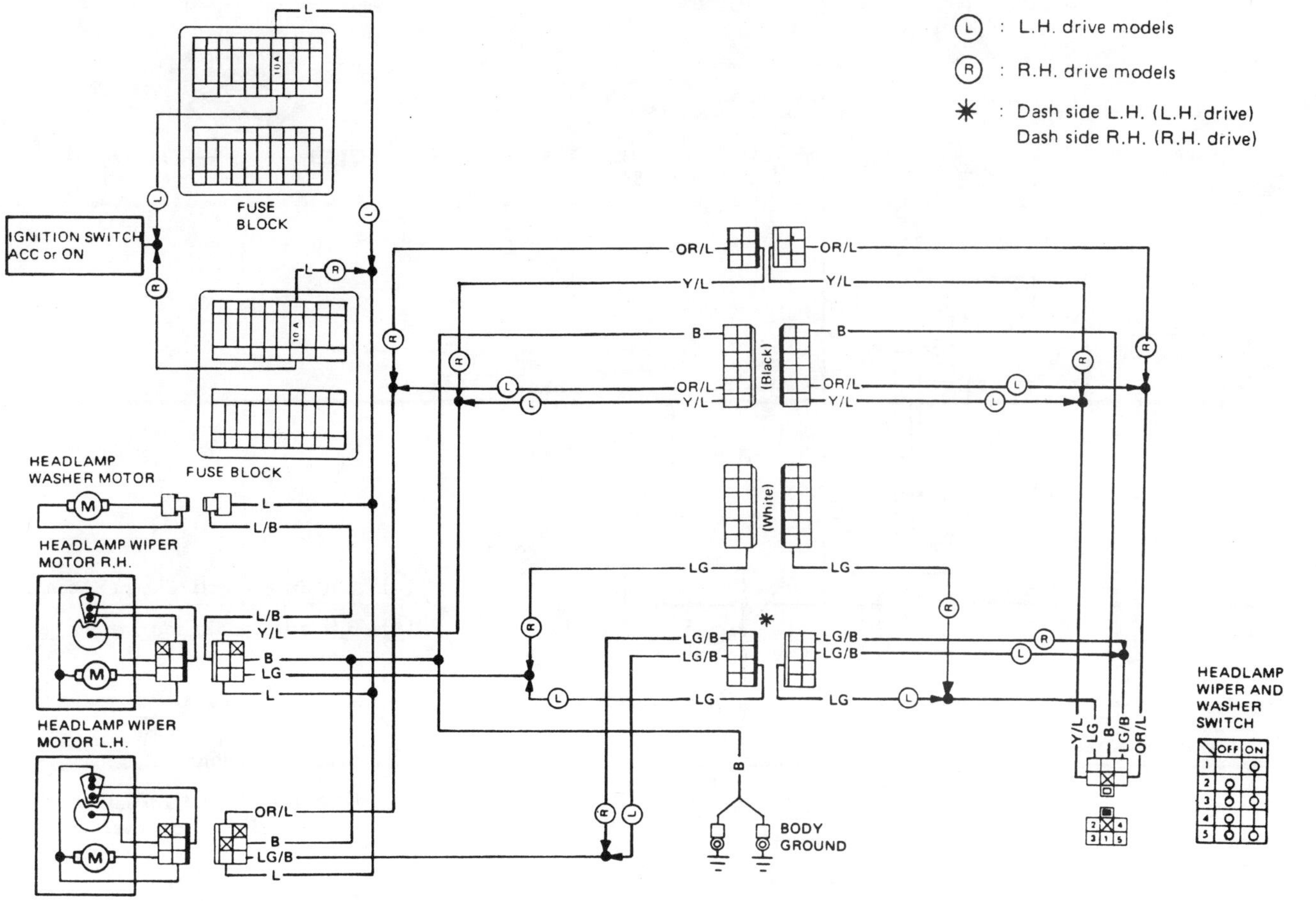

Headlamp wiper and washer circuit for UK Turbo models

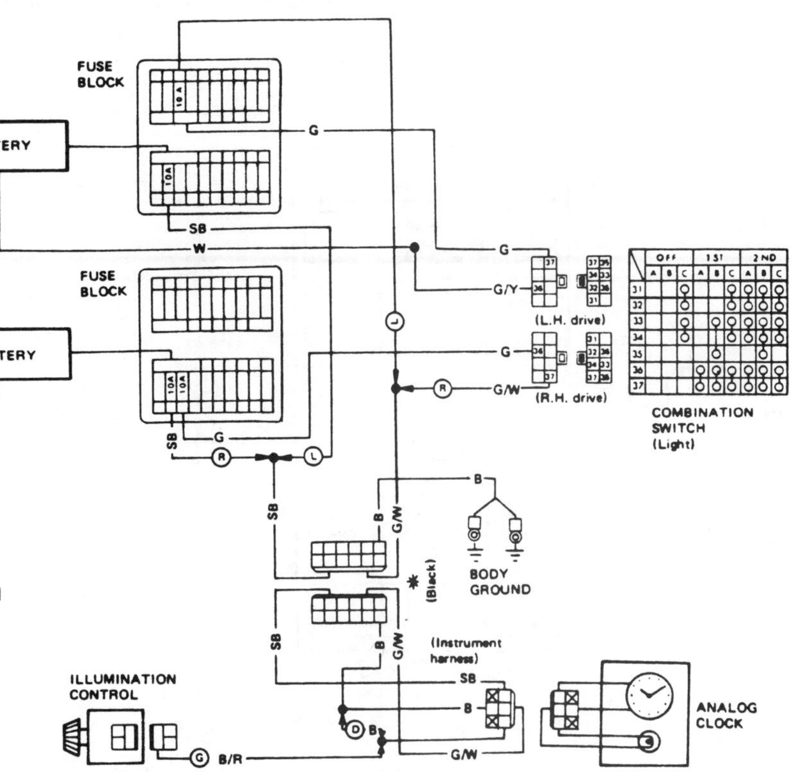

Ⓛ : L.H. drive models

Ⓡ : R.H. drive models

Ⓖ : With illumination control

Ⓓ : Without illumination control

✱ : Dash side L.H. (L.H. drive)
Dash side R.H. (R.H. drive)

Clock circuit for UK Turbo models

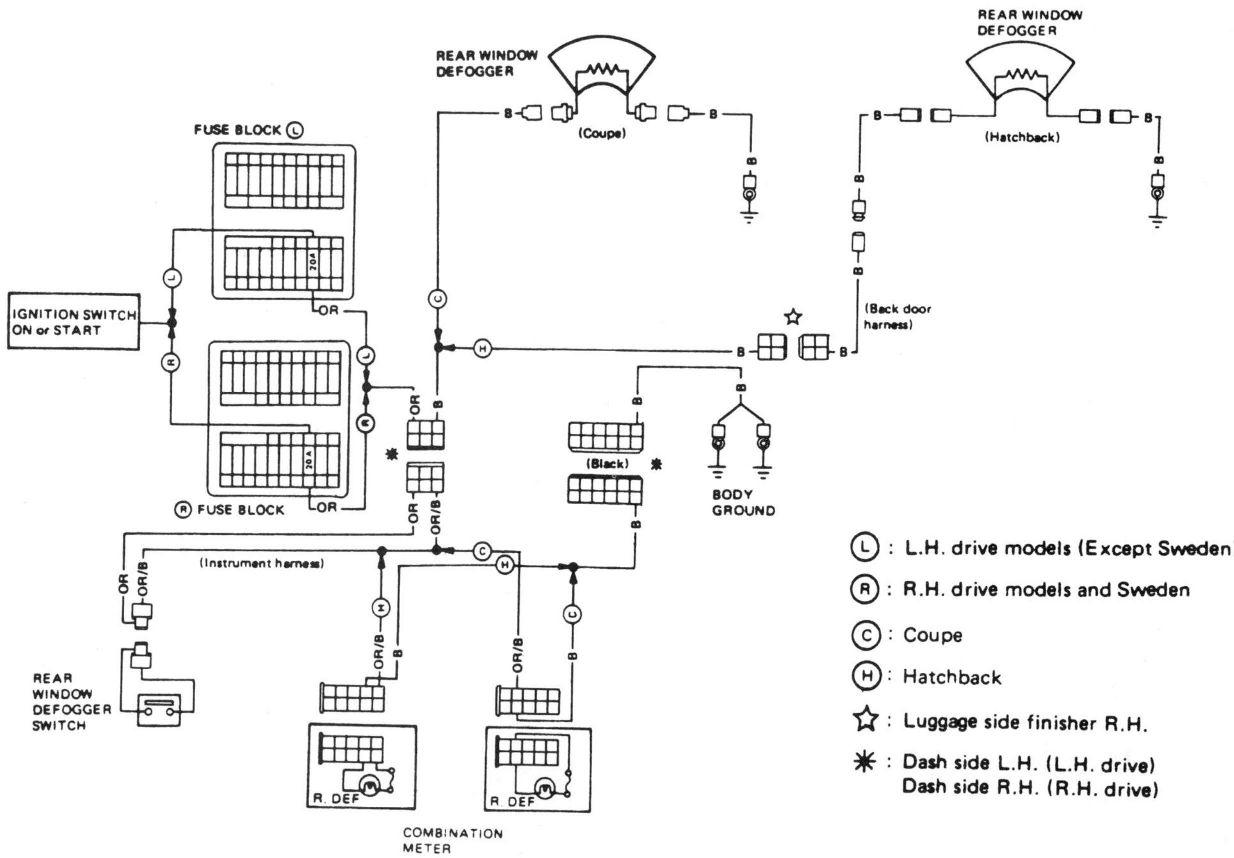

Heated rear window circuit for UK Turbo models

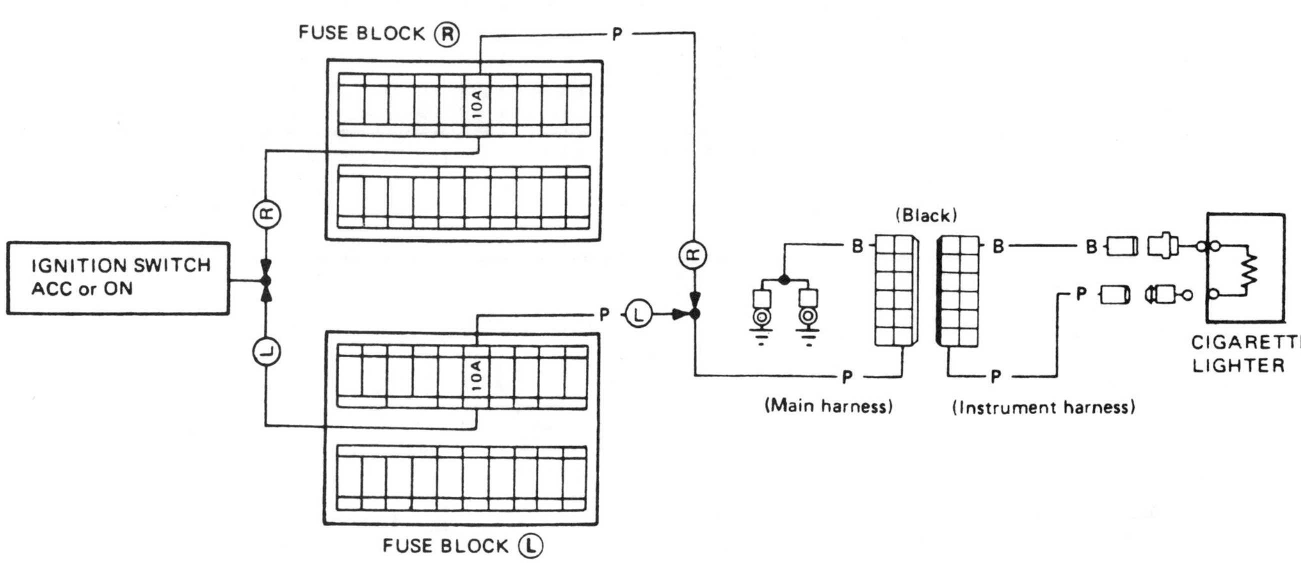

Cigarette lighter circuit for UK Turbo models

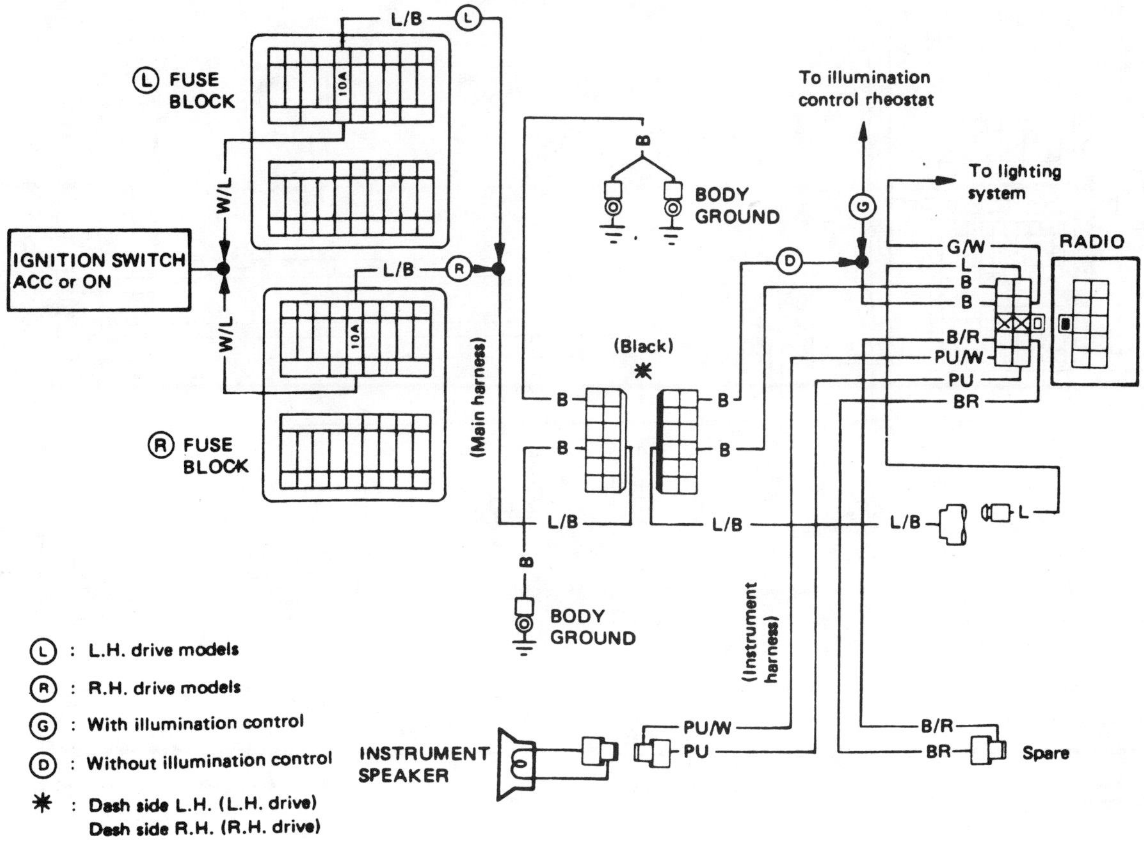

One speaker audio circuit for UK Turbo models

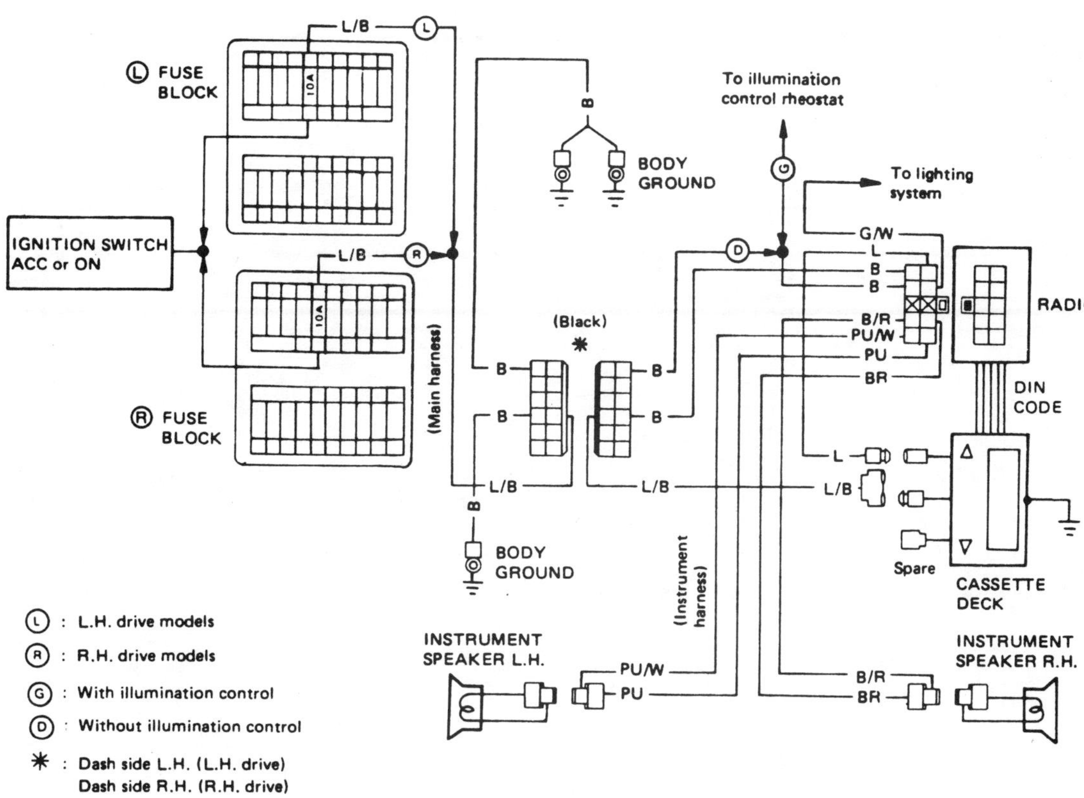

Two speaker audio circuit for UK Turbo models

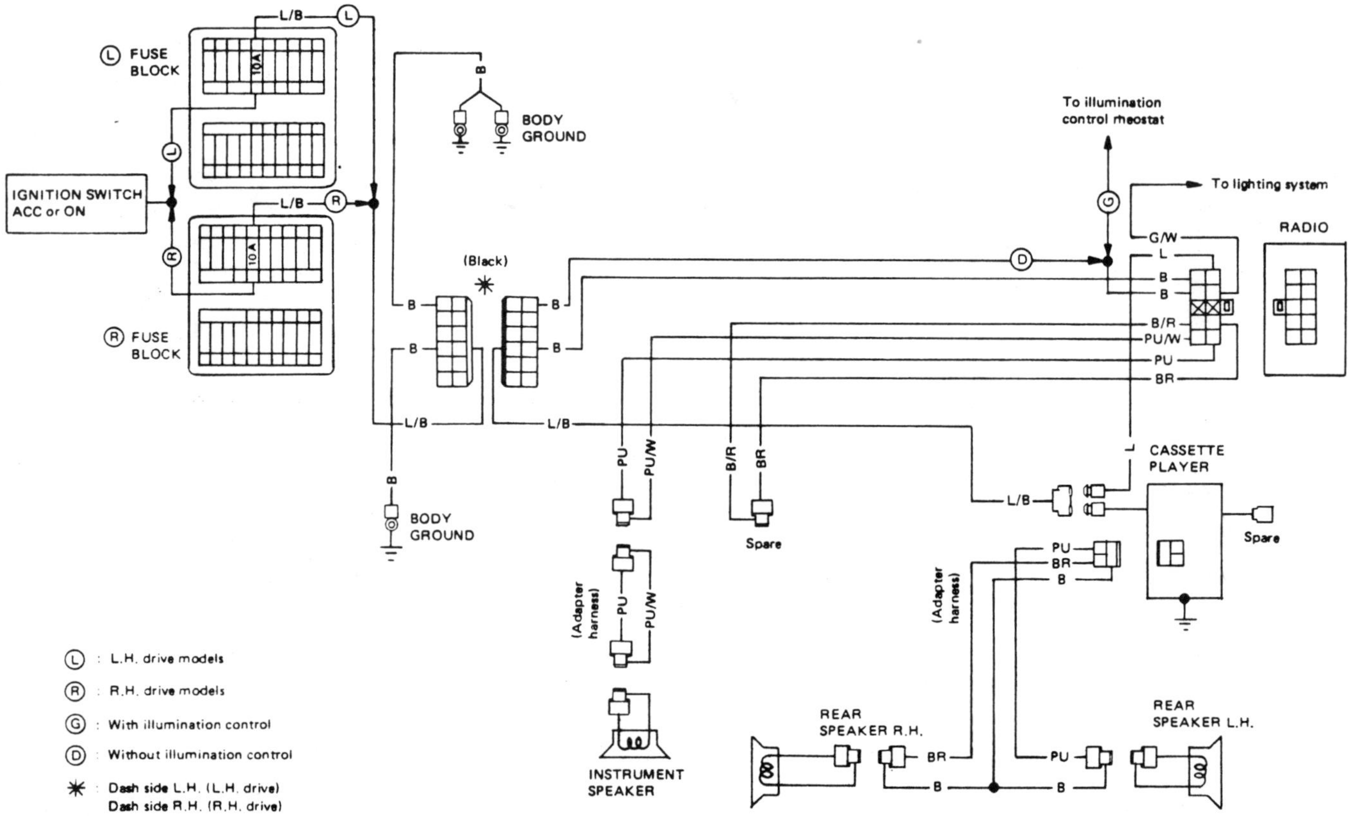

Three speaker audio circuit for UK Turbo models